Tao Cheng, Zhengwei Yang, Yoshio Inoue,
Yan Zhu, Weixing Cao (Eds.)

Recent Advances in
Remote Sensing for
Crop Growth Monitoring

MDPI

This book is a reprint of the Special Issue that appeared in the online, open access journal, *Remote Sensing* (ISSN 2072-4292) in 2015 (available at: http://www.mdpi.com/journal/remotesensing/special_issues/cropgrowth).

Guest Editors
Weixing Cao
Tao Cheng
Yan Zhu
National Engineering and Technology Center for
Information Agriculture (NETCIA),
Nanjing Agricultural University
China

Zhengwei Yang
USDA National Agricultural Statistics Service,
Research and Development Division
USA

Yoshio Inoue
National Institute for Agro-Environmental Sciences (NIAES)
Japan

Editorial Office
MDPI AG
St. Alban-Anlage 66
Basel, Switzerland

Publisher
Shu-Kun Lin

Managing Editor
Elvis Wang

1. Edition 2016

MDPI • Basel • Beijing • Wuhan • Barcelona

ISBN 978-3-03842-226-6 (Hbk)   ISBN 978-3-03842-227-3 (PDF)

Table of Contents

List of Contributors

Tsuyoshi Akiyama River Basin Research Center, Gifu University, 1-1 Yanagido, Gifu 501-1193, Japan.

Ramin Azar Institute for Electromagnetic Sensing of the Environment, National Research Council (IREA-CNR), via Bassini 15, Milan 20133, Italy.

Alfie P. Bacong Philippine Rice Research Institute (PhilRice), Muñoz, Nueva Ecija 3119, Philippines.

Massimo Barbieri Sarmap, Cascine di Barico 10, Purasca 6989, Switzerland.

Georg Bareth International Center for Agro-Informatics and Sustainable Development, College of Resources and Environmental Sciences, China Agricultural University, Beijing 100083, China; Institute of Geography, University of Cologne, 50923 Cologne, Germany.

Mirco Boschetti Institute for Electromagnetic Sensing of the Environment, Italian National Research Council, Via Bassini 15, Milan 20133, Italy.

Pietro Alessandro Brivio Institute for Electromagnetic Sensing of the Environment, National Research Council (IREA-CNR), via Bassini 15, Milan 20133, Italy.

Lorenzo Busetto Institute for Electromagnetic Sensing of the Environment, Italian National Research Council, Via Bassini 15, Milan 20133, Italy.

Dany Bylemans KU Leuven, Department of Biosystems, Division of Crop Biotechnics, Willem de Croylaan 34, BE-3001 Leuven, Belgium; Pcfruit research station, Fruittuinweg 1, BE-3800 Sint-Truiden, Belgium.

Weixing Cao National Engineering and Technology Center for Information Agriculture; Jiangsu Collaborative Innovation Center for Modern Crop Production; Jiangsu Key Laboratory for Information Agriculture, National Engineering and Technology Center for Information Agriculture, Nanjing Agricultural University, Nanjing 210095, China.

Tao Cheng Jiangsu Collaborative Innovation Center for Modern Crop Production; Jiangsu Key Laboratory for Information Agriculture; National Engineering and Technology Center for Information Agriculture, Nanjing Agricultural University, Nanjing 210095, China.

Pol Coppin KU Leuven, Department of Biosystems, Division of Crop Biotechnics, Willem de Croylaan 34, BE-3001 Leuven, Belgium.

Bo Dai School of Instrumentation Science and Opto-electronics Engineering, Beihang University, Beijing 100191, China.

Tom Deckers Pcfruit research station, Fruittuinweg 1, BE-3800 Sint-Truiden, Belgium.

Giacomo Fontanelli Institute for Electromagnetic Sensing of the Environment, National Research Council (IREA-CNR), via Bassini 15, Milan 20133, Italy.

Jian Gao Institute of Remote Sensing and Earth Sciences, Hangzhou Normal University, Hangzhou 311121, China.

Martin L. Gnyp International Center for Agro-Informatics and Sustainable Development, College of Resources and Environmental Sciences, China Agricultural University, Beijing 100083, China; Research Centre Hanninghof, Yara International, 48249 Duelmen, Germany.

Yiqing Guo School of Instrumentation Science and Opto-electronics Engineering, Beihang University, Beijing 100191, China.

Yongjiu Guo Jiangsu Key Laboratory for Information Agriculture, National Engineering and Technology Center for Information Agriculture, Nanjing Agricultural University, Nanjing 210095, China.

Pengyu Hao The State Key Laboratory of Remote Sensing Science, Institute of Remote Sensing and Digital Earth, Chinese Academy of Sciences, Beijing 100101, China.

Francesco Holecz Sarmap, Cascine di Barico 10, Purasca 6989, Switzerland.

Jianxi Huang College of Information & Electrical Engineering, China Agricultural University, No.17 Qinghua East Road, Haidian District, Beijing 100083, China; Key Laboratory of Agricultural Information Acquisition Technology, Ministry of Agriculture, Beijing 100083, China.

Jingfeng Huang Institute of Remote Sensing and Information Application, Zhejiang University, Hangzhou 310058, China; Institute of Agricultural Remote Sensing & Information Application, Zhejiang University, Hangzhou 310058, China.

Ran Huang College of Information & Electrical Engineering, China Agricultural University, No.17 Qinghua East Road, Haidian District, Beijing 100083, China; Key Laboratory of Agricultural Information Acquisition Technology, Ministry of Agriculture, Beijing 100083, China.

Shanyu Huang International Center for Agro-Informatics and Sustainable Development, College of Resources and Environmental Sciences, China Agricultural University, Beijing 100083, China; Institute of Geography, University of Cologne, 50923 Cologne, Germany.

Yanbo Huang Department of Agriculture-Agricultural Research Service, Crop Production Systems Research Unit, 141 Experiment Station Road, Stoneville, MS 38776, USA.

Yu Huang Jiangsu Key Laboratory for Information Agriculture; National Engineering and Technology Center for Information Agriculture, Nanjing Agricultural University, Nanjing 210095, China.

Yoshio Inoue National Institute for Agro-Environmental Sciences (NIAES), Tsukuba, Ibaraki 305-8604, Japan.

Mitsunori Ishihara National Institute for Agro-Environmental Sciences, 3-1-3 Kannondai, Tsukuba, Ibaraki 305-8604, Japan.

Pieter Janssens Soil Service of Belgium, Willem de Croylaan 48, BE-3001 Leuven, Belgium.

Kensuke Kawamura Graduate School for International Development and Cooperation, Hiroshima University, 1-5-1 Kagamiyama, Higashi-Hiroshima 739-8529, Japan.

Hyun-Ok Kim Earth Observation Research Team, Korea Aerospace Research Institute, 169-84 Gwahak-ro, Yuseong-Gu, Deajeon 305-806, Korea.

Alice Laborte International Rice Research Institute (IRRI), Los Baños, Laguna 4031, Philippines.

Victoria I.S. Lenz-Wiedemann International Center for Agro-Informatics and Sustainable Development, College of Resources and Environmental Sciences, China Agricultural University, Beijing 100083, China; Institute of Geography, University of Cologne, 50923 Cologne, Germany.

Xinxing Li Institute of Remote Sensing and Information Application, Zhejiang University, Hangzhou 310058, China.

Guang Liu Institute of Remote Sensing and Digital Earth, Chinese Academy of Sciences, No.9 Dengzhuang South Road, Haidian District, Beijing 100094, China.

Jia Liu Institute of Agricultural Resources and Regional Planning, Chinese Academy of Agricultural Sciences/Key Laboratory of Resources Remote Sensing and Digital Agriculture, Ministry of Agriculture, Beijing 100081, China.

Liangyun Liu Institute of Remote Sensing and Digital Earth, Chinese Academy of Sciences, No.9 Dengzhuang South Road, Haidian District, Beijing 100094, China.

Xiaobo Ma International Center for Agro-Informatics and Sustainable Development, College of Resources and Environmental Sciences, China Agricultural University, Beijing 100083, China.

Mary Rose O. Mabalay Philippine Rice Research Institute (PhilRice), Muñoz, Nueva Ecija 3119, Philippines.

Masayasu Maki Faculty of Engineering, Tohoku Institute of Technology, 35-1, YagiyamaKasumi-cho, Taihaku-ku, Sendai, Miyagi 982-8577, Japan.

Giorgos Mallinis Laboratory of Forest Remote Sensing, School of Agricultural and Forestry Sciences, Democritus University of Thrace, Orestiada 68200, Greece.

Giacinto Manfron Institute for Electromagnetic Sensing of the Environment, Italian National Research Council, Via Bassini 15, Milan 20133, Italy.

Shoji Matsuura National Agriculture and Food Research Organization Institute of Livestock and Grassland Science, 768 Senbonmatsu, Nasushiobara, Tochigi 329-2793, Japan.

Yuxin Miao International Center for Agro-Informatics and Sustainable Development, College of Resources and Environmental Sciences, China Agricultural University, Beijing 100083, China.

Yasunori Muramoto Gifu Prefectural Agricultural Technology Center, 729-1 Matamaru, Gifu 501-1152, Japan.

Andrew Nelson International Rice Research Institute (IRRI), Los Baños, Laguna 4031, Philippines.

Zheng Niu The State Key Laboratory of Remote Sensing Science, Institute of Remote Sensing and Digital Earth, Chinese Academy of Sciences, Beijing 100101, China.

Francesco Nutini Institute for Electromagnetic Sensing of the Environment, Italian National Research Council, Via Bassini 15, Milan 20133, Italy.

Keisuke Ono National Institute for Agro-Environmental Sciences, 3-1-3 Kannondai, Tsukuba, Ibaraki 305-8604, Japan.

Yehui Qin Jiangsu Key Laboratory for Information Agriculture, National Engineering and Technology Center for Information Agriculture, Nanjing Agricultural University, Nanjing 210095, China.

Eduardo Jimmy P. Quilang Philippine Rice Research Institute (PhilRice), Muñoz, Nueva Ecija 3119, Philippines.

Uwe Rascher International Center for Agro-Informatics and Sustainable Development, College of Resources and Environmental Sciences, China Agricultural University, Beijing 100083, China; Forschungszentrum Jülich, Institute of Bio-and Geosciences, IBG-2: Plant Sciences, D-52425 Jülich, Germany.

Jeny Raviz International Rice Research Institute (IRRI), Los Baños, Laguna 4031, Philippines.

Muhammad Shakir The State Key Laboratory of Remote Sensing Science, Institute of Remote Sensing and Digital Earth, Chinese Academy of Sciences, Beijing 100101, China.

Guiyan Shang Jiangsu Key Laboratory for Information Agriculture; National Engineering and Technology Center for Information Agriculture, Nanjing Agricultural University, Nanjing 210095, China.

Bao She Institute of Remote Sensing and Information Application, Zhejiang University, Hangzhou 310058, China.

Jingjing Shi Institute of Agricultural Remote Sensing & Information Application, Zhejiang University, Hangzhou 310058, China; School of Electronic and Information Engineering, Ningbo University of Technology, Ningbo 315016, China.

Mariko Shimizu Graduate School of Agriculture, Hokkaido University, Kita 9, Nishi 9, Kita-ku, Sapporo, Hokkaido 060-8589, Japan; Civil Engineering Research Institute for Cold Region, National Research and Development Agency Public Works Research Institute, 3-1-43 Hiragishi Ichijo, Toyohira-ku, Sapporo, Hokkaido 062-8602, Japan.

Sofia Siachalou Laboratory of Photogrammetry and Remote Sensing, School of Rural and Surveying Engineering, Aristotle University of Thessaloniki, Thessaloniki 54124, Greece.

Ben Somers KU Leuven, Department of Earth and Environmental Sciences, Division of Forest, Nature and Landscape Research, Celestijnenlaan 200E, BE-3001 Leuven, Belgium.

Xiaodong Song Institute of Remote Sensing and Information Application, Zhejiang University, Hangzhou 310058, China.

Daniela Stroppiana Institute for Electromagnetic Sensing of the Environment, National Research Council (IREA-CNR), via Bassini 15, Milan 20133, Italy.

Chuanxiang Tan International Center for Agro-Informatics and Sustainable Development, College of Resources and Environmental Sciences, China Agricultural University, Beijing 100083, China.

Shinya Tanaka Department of Forest Management, Forestry and Forest Products Research Institute, 1 Matsunosato, Tsukuba, Ibaraki 305-8687, Japan.

Yongchao Tian Jiangsu Collaborative Innovation Center for Modern Crop Production; National Engineering and Technology Center for Information Agriculture; Jiangsu Key Laboratory for Information Agriculture, National Engineering and Technology Center for Information Agriculture, Nanjing Agricultural University, Nanjing 210095, China.

Laurent Tits KU Leuven, Department of Biosystems, Division of Crop Biotechnics, Willem de Croylaan 34, BE-3001 Leuven, Belgium.

Maria Tsakiri-Strati Laboratory of Photogrammetry and Remote Sensing, School of Rural and Surveying Engineering, Aristotle University of Thessaloniki, Thessaloniki 54124, Greece.

Jonathan Van Beek KU Leuven, Department of Biosystems, Division of Crop Biotechnics, Willem de Croylaan 34, BE-3001 Leuven, Belgium.

Christiaan van der Tol Faculty of Geo-Information Science and Earth Observation (ITC), University of Twente, P.O. Box 217, Enschede 7500 AE, The Netherlands.

Wout Verhoef Faculty of Geo-Information Science and Earth Observation (ITC), University of Twente, P.O. Box 217, Enschede 7500 AE, The Netherlands,

Wim Verjans Pcfruit research station, Fruittuinweg 1, BE-3800 Sint-Truiden, Belgium.

Paolo Villa Institute for Electromagnetic Sensing of the Environment, National Research Council (IREA-CNR), via Bassini 15, Milan 20133, Italy.

Cuizhen Wang Department of Geography, University of South Carolina, 709 Bull St., Columbia, SC 29208, USA.

Jing Wang Institute of Remote Sensing and Information Application, Zhejiang University, Hangzhou 310058, China.

Li Wang The State Key Laboratory of Remote Sensing Science, Institute of Remote Sensing and Digital Earth, Chinese Academy of Sciences, Beijing 100101, China.

Limin Wang Institute of Agricultural Resources and Regional Planning, Chinese Academy of Agricultural Sciences/Key Laboratory of Resources Remote Sensing and Digital Agriculture, Ministry of Agriculture, Beijing 100081, China.

Chuanwen Wei Institute of Remote Sensing and Information Application, Zhejiang University, Hangzhou 310058, China.

Changshan Wu Department of Geography, University of Wisconsin-Milwaukee, Milwaukee, WI 53201, USA.

Zhengwei Yang USDA National Agricultural Statistics Service, Research and Development Division, 3251 Old Iee Highway, Room 305, Fairfax, VA 22030, USA.

Xia Yao Jiangsu Collaborative Innovation Center for Modern Crop Production; Jiangsu Key Laboratory for Information Agriculture; National Engineering and Technology Center for Information Agriculture, Nanjing Agricultural University, Nanjing 210095, China.

Jong-Min Yeom Earth Observation Research Team, Korea Aerospace Research Institute, 169-84 Gwahak-ro, Yuseong-Gu, Deajeon 305-806, Korea.

Kazuaki Yoshida Gifu Region Agriculture and Forestry Office, 5-14-53 YabutaMinami, Gifu 500-8384, Japan.

Weifeng Yu International Center for Agro-Informatics and Sustainable Development, College of Resources and Environmental Sciences, China Agricultural University, Beijing 100083, China.

Fei Yuan International Center for Agro-Informatics and Sustainable Development, College of Resources and Environmental Sciences, China Agricultural University, Beijing 100083, China; Department of Geography, Minnesota State University, Mankato, MN 56001, USA.

Yulin Zhan The State Key Laboratory of Remote Sensing Science, Institute of Remote Sensing and Digital Earth, Chinese Academy of Sciences, Beijing 100101, China.

Chao Zhang College of Information & Electrical Engineering, China Agricultural University, No.17 Qinghua East Road, Haidian District, Beijing 100083, China; Key Laboratory of Agricultural Information Acquisition Technology, Ministry of Agriculture, Beijing 100083, China.

Kangyu Zhang Institute of Remote Sensing and Information Application, Zhejiang University, Hangzhou 310058, China.

Ling Zhang Jiangsu Key Laboratory for Information Agriculture, National Engineering and Technology Center for Information Agriculture, Nanjing Agricultural University, Nanjing 210095, China.

Feng Zhao School of Instrumentation Science and Opto-electronics Engineering, Beihang University, Beijing 100191, China.

Guangming Zhao International Center for Agro-Informatics and Sustainable Development, College of Resources and Environmental Sciences, China Agricultural University, Beijing 100083, China.

Huijie Zhao School of Instrumentation Science and Opto-electronics Engineering, Beihang University, Beijing 100191, China.

Cheng Zhong Department of Geography, University of South Carolina, 709 Bull St., Columbia, SC 29208, USA.

Chen Zhou Jiangsu Key Laboratory for Information Agriculture;
National Engineering and Technology Center for Information Agriculture,
Nanjing Agricultural University, Nanjing 210095, China.

Dehai Zhu College of Information & Electrical Engineering, China Agricultural
University, No.17 Qinghua East Road, Haidian District, Beijing 100083, China;
Key Laboratory of Agricultural Information Acquisition Technology, Ministry of
Agriculture, Beijing 100083, China.

Yan Zhu Jiangsu Collaborative Innovation Center for Modern Crop Production;
National Engineering and Technology Center for Information Agriculture;
Jiangsu Key Laboratory for Information Agriculture, National Engineering and
Technology Center for Information Agriculture, Nanjing Agricultural University,
Nanjing 210095, China.

About the Guest Editors

Tao Cheng received his Ph.D. degree in Earth and Atmospheric Sciences from the University of Alberta, Edmonton, Canada in 2010. He is currently a Professor of agricultural remote sensing at Nanjing Agricultural University, Nanjing, China. From 2011 to 2013, he was a postdoctoral scholar in the Department of Land, Air and Water Resources at the University of California, Davis in the U.S. Dr. Cheng was appointed Jiangsu Distinguished Professor in 2014. He is a reviewer for *Remote Sensing of Environment, Precision Agriculture,* and many other international journals. He is a member of SPIE and IEEE Geoscience and Remote Sensing Society. His current research interests are in reflectance spectroscopy of vegetation, crop monitoring, unmanned aerial vehicle based remote sensing, and quantitative methods for vegetation characterization.

Yan Zhu received her PhD degree from the Nanjing Agricultural University in 2003. Currently, she is a full professor and the dean of College of Agriculture at NJAU. Prof. Zhu works mainly on information agriculture, specifically on crop modeling and crop monitoring. She has earned three National Second-Class Awards for Science and Technology Advancement and published about 250 papers, including over 70 papers in journals indexed by the Web of Science. Due to her outstanding achievements in the past few years, she was elected the "12 th Young Scientist Award of China" in 2011; "Leading Young Talents of China, Ministry of Science and Technology of China" in 2013; and "Youth Scholar of the Changjiang Scholars Program" in 2015.

Weixing Cao received his PhD degree in crop physiology from Oregon State University in the U.S. in 1989 and then worked as a postdoctoral fellow and a research scientist at the University of Wisconsin. He has been a professor at Nanjing Agricultural University in China since 1994. He is currently the Director of the National Engineering and Technology Center for Information Agriculture (NETCIA), Vice President of the Crop Science Society of China, and Vice Governor of Jiangsu Province. Prof. Cao's research interests are in the general fields of crop ecology and agro-informatics and he has accomplished outstanding achievements in the specific areas of crop growth modeling, growth monitoring, and precision management. Prof. Cao has earned three National Second-Class Awards for Science and Technology Advancement and the Achievement Award from the Crop Science Society of China. In addition, he was elected to the Distinguished Young Scholars by the National Science Foundation of China.

Zhengwei Yang is an information technology specialist with the Research and Development Division, National Agricultural Statistics Service, the United States Department of Agriculture. He received a Ph.D. in Electrical Engineering from Drexel University, Philadelphia in 1997. Dr. Yang was a recipient of the 2011 USDA Secretary's Honor Award for Excellence, and the 2010 NASS Administrator's Honor Award for Excellence. Dr. Yang's research interests include crop land cover, crop condition and growth monitoring and assessment, crop disaster monitoring and assessment, crop growth modeling and simulation, and geospatial application system design. Dr. Yang led the research and development of the web service, operational conterminous United States cropland crop cover geospatial information system—CropScape, and the operational crop vegetation condition monitoring geospatial information system—VegScape. He was one of the organizers of the First, Second and Third International Conferences on Agro-geoinformatics, and served as the chairman on the scientific committee.

Yoshio Inoue is a research scientist at the National Institute for Agro-Environmental Sciences, Japan. He received his Ph.D. degree in plant ecophysiology from the Kyoto University in 1988. His research fields include: (1) Remote sensing, modeling and geospatial analysis of agro-ecosystem dynamics; (2) Wide area monitoring and assessment of agro-ecosystems based on remote sensing and geospatial information systems; and (3) Methodological study of remote and non-destructive sensing of plant eco-physiological information. He has been a Professor at the Graduate School of Life and Environmental Sciences, University of Tsukuba, from 1995. He is contributing to both domestic and international research communities as editor-in-chief, editor and/or reviewer for more than 60 journals.

Preface to "Recent Advances in Remote Sensing for Crop Growth Monitoring"

Accurate and timely information on crop growth and conditions is critical for precision farming, crop management, crop yield estimation, crop disaster forecasting and mitigation, agricultural production planning, crop commodity trading, and food security decision support. Crop growth can be monitored with remotely sensed data acquired from various platforms including proximal devices, aircraft and satellites. While a large variety of studies focus on the crop growth parameters such as leaf nitrogen content and leaf area index, the community also shows huge interest in continuously monitoring crop spectral properties and large scale mapping of crop types and crop acreage. New analytical methods, instruments and applications for more accurate, reliable and efficient monitoring of crop conditions are continually reported in the literature.

Crop growth cycles are essential for condition monitoring and often vary with different crop types. This has generated a lot of research interest on the use of remotely sensed data to monitor all major growth stages and on the development of robust algorithms for generalization across stages. Numerous studies have used crop phenological information for crop monitoring, from the direct use of multi-temporal data to the addition of phenological metrics from time series data. To cope with the adverse effect of weather conditions, such as cloud cover and rainfall, on data acquisition, researchers have been working on integrating optical data with Synthetic Aperture Radar (SAR) data to avoid missing the observations at critical growth stages.

This Special Issue was initiated at the International Symposium on Crop Growth Monitoring (ISCGM) held in Nanjing, China from September 13–16, 2014. It covers a selection of work reporting on recent advances in crop status assessment and monitoring, crop type classification, and crop mapping based on remotely sensed data.

We are grateful for having the opportunity to lead this Special Issue. We would like to thank the journal editorial office and reviewers for conducting the review process and all the authors for submitting their work. In particular, the managing editor, Mr. Elvis Wang, provided tremendous assistance during the editing process. Our special thanks also go to the ISCGM attendees who contributed fruitful discussions on the topics covered in this Special Issue. The activities related to this Special Issue were supported by various funding agencies, including the National Natural Science Foundation of China (31201131, 31371534, 31371535, and 31470084), the Special Program for Agriculture Science and Technology of the Ministry of Agriculture in China (201303109), the Jiangsu Collaborative Innovation Center for Modern Crop Production, the Academic Program Development of Jiangsu Higher Education Institutions (PAPD) and the

Fundamental Research Funds for the Central Universities (KYZ201202-8 and KYRC201401).

Tao Cheng, Yan Zhu, Weixing Cao,
Zhengwei Yang and Yoshio Inoue
Guest Editors

Quantitative Estimation of Fluorescence Parameters for Crop Leaves with Bayesian Inversion

Feng Zhao, Yiqing Guo, Yanbo Huang, Wout Verhoef, Christiaan van der Tol, Bo Dai, Liangyun Liu, Huijie Zhao and Guang Liu

Abstract: In this study, backward and forward fluorescence radiance within the emission spectrum of 640–850 nm were measured for leaves of soybean, cotton, peanut and wheat using a hyperspectral spectroradiometer coupled with an integration sphere. Fluorescence parameters of crop leaves were retrieved from the leaf hyperspectral measurements by inverting the FluorMODleaf model, a leaf-level fluorescence model able to simulate chlorophyll fluorescence spectra for both sides of leaves. This model is based on the widely used and validated PROSPECT (leaf optical properties) model. Firstly, a sensitivity analysis of the FluorMODleaf model was performed to identify and quantify influential parameters to assist the strategy for the inversion. Implementation of the Extended Fourier Amplitude Sensitivity Test (EFAST) method showed that the leaf chlorophyll content and the fluorescence lifetimes of photosystem I (PSI) and photosystem II (PSII) were the most sensitive parameters among all eight inputs of the FluorMODleaf model. Based on results of sensitivity analysis, the FluorMODleaf model was inverted using the leaf fluorescence spectra measured from both sides of crop leaves. In order to achieve stable inversion results, the Bayesian inference theory was applied. The relative absorption cross section of PSI and PSII and the fluorescence lifetimes of PSI and PSII of the FluorMODleaf model were retrieved with the Bayesian inversion approach. Results showed that the coefficient of determination (R^2) and root mean square error (RMSE) between the fluorescence signal reconstructed from the inverted fluorescence parameters and measured in the experiment were 0.96 and 3.14×10^{-6} W·m^{-2}·sr^{-1}·nm^{-1}, respectively, for backward fluorescence, and 0.92 and 3.84×10^{-6} W·m^{-2}·sr^{-1}·nm^{-1} for forward fluorescence. Based on results, the inverted values of the fluorescence parameters were analyzed, and the potential of this method was investigated.

Reprinted from *Remote Sens.* Cite as: Zhao, F.; Guo, Y.; Huang, Y.; Verhoef, W.; van der Tol, C.; Dai, B.; Liu, L.; Zhao, H.; Liu, G. Quantitative Estimation of Fluorescence Parameters for Crop Leaves with Bayesian Inversion. *Remote Sens.* **2015**, *7*, 14179–14199.

1

1. Introduction

Chlorophyll fluorescence (ChlF) is considered a promising tool to effectively assess photosynthetic rates of green plants [1] and to monitor stress conditions of crops [2,3]. As a result, quantitative analysis of the ChlF signal using remote sensing techniques has been conducted extensively in recent years [1,4], along with development of leaf ChlF radiative transfer models that have improved understanding of the interactions of sunlight with plant leaves [5–7].

Leaf ChlF radiative transfer models can be used to simulate leaf backward (the emission direction opposite to the direction of the excitation light) and forward (the emission direction same as the direction of the excitation light) ChlF spectra as a function of the incident light, and the leaf biochemical and fluorescence parameters. The FluorMOD project began in 2002 with a goal of developing an integrated leaf-canopy fluorescence model [8]. As a subcomponent of the integrated model, FluorMODleaf [6,8] is a leaf-level fluorescence model based on the PROSPECT model [9,10] and can be used to calculate the radiative transfer of ChlF in plant leaves. Besides the FluorMODleaf model, other leaf ChlF models were also developed. For example, FLUSPECT [7] is another leaf ChlF radiative transfer model that is also based on the PROSPECT model and uses fluorescence quantum efficiencies of photosystem I (PSI) and photosystem II (PSII) as inputs. Computer-based Monte Carlo methods were also developed to simulate the leaf-level ChlF signal [5].

The FluorMODleaf model has a total of eight input parameters [6]. Besides five original parameters of the PROSPECT-5 model [9], $i.e.$, leaf structure parameter N, chlorophyll content C_{ab}, carotenoid content C_{ar}, water content C_w, and dry matter content C_m, three fluorescence parameters were newly introduced, $i.e.$, the relative absorption cross section of PSI and PSII, δ, and fluorescence lifetimes of PSI and PSII, τ_I and τ_{II}. Definitions, units, and descriptions of the eight input parameters of the FluorMODleaf model are illustrated in Table 1. Outputs of FluorMODleaf model are the forward and backward apparent spectral fluorescence yield (ASFY), besides leaf reflectance and transmittance. The FluorMODleaf model was evaluated using experimental datasets, and good agreement between the model-simulated and experimental data was shown [6]. However, the study on inversion of FluorMODleaf was not reported.

The relative absorption cross section of PSI and PSII (δ) and fluorescence lifetimes of PSI and PSII (τ_I and τ_{II}) are critical foliar parameters defining the fluorescence emission properties of plant leaves. However, these fluorescence parameters (1) are difficult to measure directly; (2) are species-dependent; and (3) vary greatly under different environmental conditions [6]. Therefore, quantitative retrieval of these fluorescence parameters from leaf hyperspectral fluorescence data by inverting a physically-based ChlF radiative transfer model would be a non-destructive and effective method to retrieve these parameters.

2

Table 1. The definitions, units, and descriptions of the eight input parameters of the FluorMODleaf model [6].

Parameter	Definition	Unit	Description
N	Leaf structure parameter	-	Number of compact layers specifying the average number of air/cell wall interfaces within the mesophyll.
C_{ab}	Chlorophyll a+b content	$\mu g \cdot cm^{-2}$	Mass of chlorophyll a+b per leaf area.
C_{ar}	Total carotenoid content	$\mu g \cdot cm^{-2}$	Mass of total carotenoid per leaf area.
C_w	Water content	$g \cdot cm^{-2}$	Mass of water per leaf area.
C_m	Dry matter content	$g \cdot cm^{-2}$	Mass of dry matter per leaf area.
δ	Relative absorption cross section ratio	-	The relative distribution of light between the two photosystems, which can be approximated by the product of the PSII/PSI antenna size ratio.
τ_I	Fluorescence lifetimes of photosystem I (PSI)	ns	Average time the chlorophyll molecule stays in its excited state before emitting a photon from isolated PSI complexes.
τ_{II}	Fluorescence lifetimes of photosystem II (PSII)	ns	Average time the chlorophyll molecule stays in its excited state before emitting a photon from isolated PSII complexes.

Compared with the reflected and transmitted signals of leaves, leaf ChlF is very weak. Therefore, in order to achieve stable inversion results, additional information and inversion strategy should be used to improve the accuracy of the inverted parameters. Bayesian inversion approach is a suitable alternative to impose *a priori* information on the inversion process and has shown potential for the inversion of remote sensing models [11]. By injecting reliable *a priori* information into the inversion process, a more stable solution for the unknown parameters can be achieved. As an effective way to alleviate ill-posed problems in the inversion process, the Bayesian inversion approach has been used in studies for the retrieval of terrestrial parameters from remote sensing data [12–14].

The objectives of this study were (1) to perform a sensitivity analysis of the FluorMODleaf model in order to identify and quantify influential parameters; (2) to retrieve the parameters of FluorMODleaf model using the experimental datasets. Firstly, a sensitivity analysis of the FluorMODleaf model was performed using the Extended Fourier Amplitude Sensitivity Test (EFAST) method. Based on sensitivity analysis results, the FluorMODleaf model was inverted using the experimental datasets acquired for four types of crop leaves. In order to achieve stable inversion results, Bayesian theory was introduced into the inversion process. The relative absorption cross section of PSI and PSII (δ) and fluorescence lifetimes of PSI and PSII (τ_I and τ_{II}) were then estimated with the Bayesian inversion approach of the FluorMODleaf model. Finally, the inversion results were validated and analyzed.

3

2. Materials and Methods

2.1. Experimental Datasets

Datasets at leaf level for four crop leaves were used in this study, and two field experiments were conducted. For each leaf, hyperspectral data and the biochemical content were obtained. Two leaves (for wheat) or three (for soybean, cotton and peanut) with similar color, height in the plant and physiological condition by visual inspection were chosen as a group for measurement of reflectance, transmittance, backward and forward fluorescence, and biochemical content. Then, data averages from these two or three leaves were used as a group for subsequent inversion.

The first experiment was conducted for winter wheat (*Triticum*) at the Beijing Academy of Agriculture and Forestry Sciences (39.942°N, 116.277°E) on 8 May 2014. Eight green leaves were measured with a hyperspectral spectroradiometer coupled with an integration sphere during 10:00–18:00 Beijing time.

The second experiment was conducted at the Huailai Remote Sensing Test Site (40.349°N, 115.785°E), Chinese Academy of Sciences, which is located at Huailai County, Hebei Province, China, during 15–19 September 2014. Three crops, soybean (*Glycine max*), cotton (*Gossypium*) and peanut (*Arachis hypogaea*), were targeted in the experiment. In the experiment, three leaves as a group were used for the experiment every hour from 9:00–18:00 for soybean, 9:00–15:00 for cotton, and 9:00–17:00 for peanut. Twenty-seven soybean leaves, 18 cotton leaves and 24 peanut leaves were measured in the experiment. Leaves from different heights and physiological conditions were measured in order to make the datasets more representative. Among them, a group of three senescent leaves with brown color for peanut was measured to compare with green leaves.

Similar to the measurement protocol of Zarco-Tejada *et al.* [15] and Zhang [16], the leaf hyperspectral data were measured using a LI-COR 1800-12 system integrating sphere apparatus (LI-COR Inc., Lincoln, NE, USA) coupled with an ASD FieldSpec Pro spectroradiometer (ASD Inc., Boulder, CO, USA) and removable filter, as shown in Figure 1. However, different from the protocol of using a long-pass filter by Zarco-Tejada *et al.* [15], a short-pass filter was used instead in front of the lamp in our experiment with irradiance longer than 640 nm being cut-off. Therefore, the reflected/transmitted signal should be filtered out in wavelengths longer than 640 nm, and the signal measured by the spectroradiometer within the wavelength range of 640–850 nm would be composed mainly of the emitted ChlF signal. The spectral data were measured *in situ* with leaves attaching to their stems. The hyperspectral measurements were conducted under guidance of the LI-COR integrating sphere manual [17]. The spectral resolution and spectral sampling interval of the spectroradiometer are 3 nm and 1 nm, respectively. An integration time of 1.09 s was used for all the measurements.

It is worth noting that the radiance of the lamp in the experiment was very low, compared with the solar radiance under natural conditions. Therefore, the fluorescence radiance measured in this experiment would be lower than that under natural conditions, because the magnitude of the fluorescence radiance is proportional to the magnitude of the excitation radiance [8]. A typical radiance distribution of the lamp with the short-pass filter is shown in Figure 2. Lamp's radiance passing through the filter is close to zero in the fluorescence emission wavelengths (640–850 nm), except for the initial parts of the cut-off wavelengths because of the instrument limit.

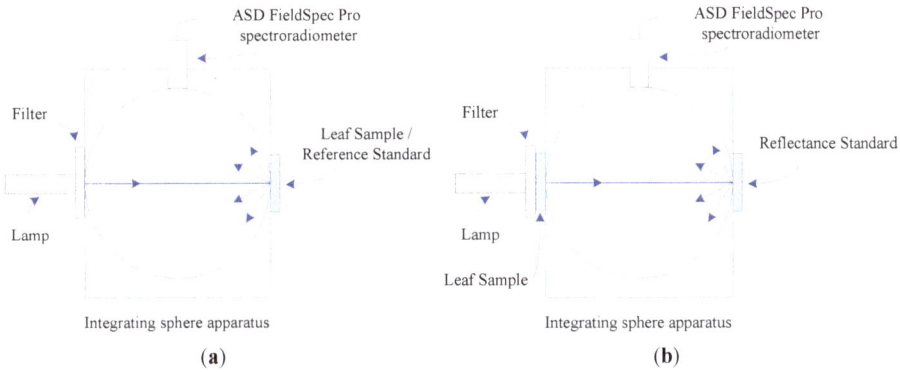

Figure 1. The experimental setup for leaf hyperspectral measurement. (**a**) Measurement configuration for leaf reflectance and backward fluorescence; (**b**) Measurement configuration for leaf transmittance and forward fluorescence.

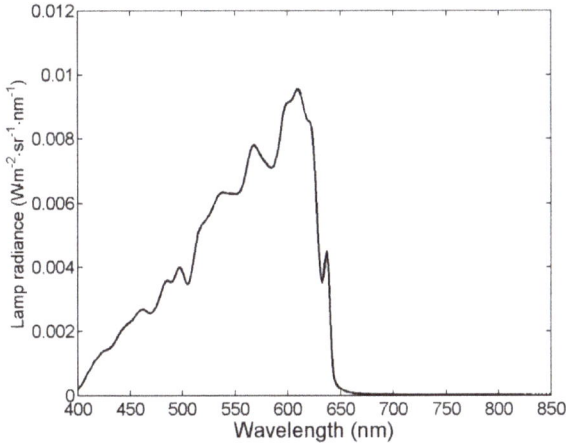

Figure 2. An example of radiance distribution of the lamp with the short-pass filter.

Three measurements of leaf reflectance and backward fluorescence were acquired by placing leaf sample as shown in Figure 1a: radiance of the leaf sample (Lb_{ls}^{on}), reference standard (Lb_{rs}^{on}), and dark current (Lb_{dc}^{on}). Then, another three measurements without the filter were acquired: radiance of the leaf sample (Lb_{ls}^{off}), reference standard (Lb_{rs}^{off}), and dark current (Lb_{dc}^{off}). The reflectance (R_{leaf}) and backward fluorescence radiance (F_b) of the leaf sample can be calculated as:

$$R_{leaf} = \frac{Lb_{ls}^{off} - Lb_{dc}^{off}}{Lb_{rs}^{off} - Lb_{dc}^{off}} \cdot R_{ref} \tag{1}$$

$$F_b = (Lb_{ls}^{on} - Lb_{dc}^{on}) - (Lb_{rs}^{on} - Lb_{dc}^{on}) \cdot R_{leaf} \tag{2}$$

where R_{ref} is the reflectance of the reference standard. The first part of the right side of the Equation (2) includes both mostly fluorescence emission by the leaf, and a small residual radiance reflected by the leaf, because transmittance of the filter is not exactly zero. The second part is added to correct the instrument limit.

To measure the transmittance and forward fluorescence, the leaf sample was moved to the front of the lamp, as shown in Figure 1b. Similarly, three measurements with the filter were acquired: radiance of the leaf sample (Lf_{ls}^{on}), reference standard (Lf_{rs}^{on}), and dark current (Lf_{dc}^{on}). Then, another three measurements without the filter were acquired: radiance of the leaf sample (Lf_{ls}^{off}), reference standard (Lf_{rs}^{off}), and dark current (Lf_{dc}^{off}). The transmittance (T_{leaf}) and forward fluorescence radiance (F_f) of the leaf sample can be calculated as:

$$T_{leaf} = \frac{Lf_{ls}^{off} - Lf_{dc}^{off}}{Lf_{rs}^{off} - Lf_{dc}^{off}} \cdot R_{ref} \tag{3}$$

$$F_f = (Lf_{ls}^{on} - Lf_{dc}^{on}) - (Lf_{rs}^{on} - Lf_{dc}^{on}) \cdot T_{leaf} \tag{4}$$

The output of the FluorMODleaf model is ASFY (in unit of nm^{-1}), which is defined as the ratio of the number of photons emitted by the leaf surface, per unit spectral bandwidth, to the number of incident photons [6,8], and not the fluorescence radiance measured in our experiment. Therefore, the output of the FluorMODleaf model was converted from ASFY into fluorescence radiance in order to be consistent with experimental data. The conversion was performed with the following formulae:

$$F_b(\lambda_{em}) = \int_{400}^{650} \frac{L_{lamp}^{on}(\lambda_{ex}) \cdot b_{mod}(\lambda_{ex}, \lambda_{em}) \cdot \lambda_{ex}}{\lambda_{em}} d\lambda_{ex} \tag{5}$$

$$F_f(\lambda_{em}) = \int_{400}^{650} \frac{L_{lamp}^{on}(\lambda_{ex}) \cdot f_{mod}(\lambda_{ex}, \lambda_{em}) \cdot \lambda_{ex}}{\lambda_{em}} d\lambda_{ex} \tag{6}$$

where λ_{ex} and λ_{em} represent the fluorescence excitation and emission wavelengths, respectively; b_{mod} and f_{mod} are the backward and forward ASFYs simulated by FluorMODleaf, respectively; L_{lamp}^{on} is the radiance of the lamp with the filter used in the integrating sphere apparatus; F_b and F_f are the backward and forward fluorescence radiance calculated from the output of the FluorMODleaf model, which are now directly comparable with the quantities calculated respectively by Equations (2) and (4) in the experiment.

After the spectral measurements, the leaves were immediately cut from the plants, placed into black plastic bags surrounded by ice lumps, and taken to the laboratory for biochemical analysis. Chlorophyll a + b content (C_{ab}, in unit of $\mu g/cm^2$), total carotenoid content (C_{ar}, in unit of $\mu g/cm^2$), water content (C_w, in unit of g/cm^2), and dry matter content (C_m, in unit of g/cm^2) were measured for each leaf in the laboratory. Six leaf disks of 15 mm diameter were punched from each leaf sample, chopped into small pieces, and then dropped into the vial with ethanol solution and covered with aluminum foil. After 48 h in the dark environment, the solution was used for measuring the chlorophyll content and carotenoid content using a Shimadzu UV160U Spectrophotometer (Shimadzu Corp., Kyoto, Japan), using the method described by Lichtenthaler and Buschmann [18]. In order to measure the water and dry matter contents, the remaining portions of the leaves were scanned to determine leaf area and weighed to measure their fresh weight. They were then oven-dried at 80 °C for 48 h, and reweighed to determine dry weight.

2.2. Sensitivity Analysis

Sensitivity analysis investigates the response of a model to variations of its input parameters by statistically calculating a limited, but representative number of simulations [19,20]. The analysis has been shown to be effective to help make strategy for the inversion of radiation transfer models [21]. Compared with the classic FAST (Fourier Amplitude Sensitivity Test) method for the sensitivity analysis of the models, which is only able to compute the first order sensitivity index, the Extended FAST (EFAST) method proposed by Saltelli *et al.* [20] allows the simultaneous computation of the first order and the total sensitivity indices for a given input parameter [21]. Therefore, in this study, the EFAST method was used for the sensitivity analysis of the FluorMODleaf model. The first order sensitivity index gives the independent effect of each parameter, while the total sensitivity index contains both independent effect of each parameter and the interaction effects with the others.

In the sensitivity analysis test, ranges of N, C_{ab}, C_{ar}, C_w, C_m, τ_I, τ_{II}, and δ were defined as 1–2.5, 0.4–76.8 $\mu g/cm^2$, 0–25.3 $\mu g/cm^2$, 0.0044–0.0340 g/cm^2, 0.0017–0.0331 g/cm^2, 0.034–0.1 ns, 0.3–2.0 ns, and 1.0–2.4, respectively, based on a previous study [6]. One thousand combinations of the parameters were randomly selected from their ranges as the inputs. Then, for each combination of the input

7

parameters, the spectra of the backward and forward fluorescence were simulated by the FluorMODleaf model. All simulated fluorescence spectra combined with the corresponding selected values of input parameters were used as input data for the sensitivity analysis. Detailed procedure and formulae can be found in our previous studies [21,22].

2.3. Inversion Procedure

The inversion procedure includes two steps. In the first step leaf structural and biochemical parameters were inverted. Then, they were fixed at their inverted values for the second step to retrieve the rest three fluorescence parameters. In each step, an efficient global optimization algorithm based on simulated annealing, which was constructed and used in our former study [21], was applied in the inversion procedure to minimize the merit functions described below.

In the first step, the leaf structure parameter N was firstly inverted from the measured data of leaf reflectance (ρ_{leaf}) and transmittance (τ_{leaf}) by minimizing the following merit function $Fn(N)$, which is defined in the near-infrared wavelengths (λ) of 750–1300 where N is the most sensitive parameter among the input parameters of the PROSPECT-5 model [22]:

$$Fn(N) = \sum_{\lambda \in [750,1300]} \left\{ \left[\rho_{leaf}(\lambda) - \rho_{simu}(N, C_{ab}, C_{ar}, C_w, C_m, \lambda) \right]^2 + \left[\tau_{leaf}(\lambda) - \tau_{simu}(N, C_{ab}, C_{ar}, C_w, C_m, \lambda) \right]^2 \right\} \quad (7)$$

where ρ_{simu} and τ_{simu} are the leaf reflectance and transmittance simulated by the FluorMODleaf model. During this step, the biochemical parameters, i.e., C_{ab}, C_{ar}, C_w, and C_m, were all maintained at their measured values.

Then, the other four parameters of the PROSPECT-5 model, including C_{ab}, C_{ar}, C_w, and C_m, were inverted by minimizing the following merit function $Fp(C_{ab}, C_{ar}, C_w, C_m)$, with the leaf structure parameter N being maintained at its inverted value obtained in the first step:

$$Fp(C_{ab}, C_{ar}, C_w, C_m)$$
$$= \sum_{\lambda \in [400,2500]} \left\{ \left[\rho_{leaf}(\lambda) - \rho_{simu}(N, C_{ab}, C_{ar}, C_w, C_m, \lambda) \right]^2 + \left[\tau_{leaf}(\lambda) - \tau_{simu}(N, C_{ab}, C_{ar}, C_w, C_m, \lambda) \right]^2 \right\} \quad (8)$$

The merit function is defined on the spectral region of the PROSPECT-5 model (i.e., 400–2500 nm).

In the second step of the inversion procedure, the fluorescence parameters τ_I, τ_{II}, and δ were retrieved from the measured leaf fluorescence spectra by minimizing the following merit function $Ff(\tau_I, \tau_{II}, \delta)$, while the other parameters were all maintained at their inverted values obtained in the first step. The merit function $Ff(\tau_I,\tau_{II},\delta,N,C_{ab},C_{ar})$ was constructed with the Bayesian inversion theory [11,23]:

$$Ff(\tau_I,\tau_{II},\delta) = \tfrac{1}{2}(Fb_{simu} - Fb_{meas})^T C_{nb}^{-1}(Fb_{simu} - Fb_{meas})$$
$$+ \tfrac{1}{2}(Ff_{simu} - Ff_{meas})^T C_{nf}^{-1}(Ff_{simu} - Ff_{meas}) + \tfrac{1}{2}(x - x_{priori})^T C_x^{-1}(x - x_{priori}) \tag{9}$$

where Fb_{meas} and Ff_{meas} are the backward and forward fluorescence measured in the experiment, respectively; Fb_{simu} and Ff_{simu} are the forward and backward fluorescence calculated by the output of FluorMODleaf model, as shown in Equations (5) and (6) respectively; C_{nb} and C_{nf} are the inaccuracy of model simulations and the noise covariance matrices for the measurements of the backward and forward fluorescence, respectively; x contains the unknown variables; x_{priori} is the a priori guess of the unknown variables; and C_x is the covariance matrix of the a priori variables. The expressions of these vectors and matrices are:

$$Fb_{meas} = [\ Fb_{meas}(\lambda_1)\quad Fb_{meas}(\lambda_2)\quad \cdots\quad Fb_{meas}(\lambda_{211})\ ]^T$$
$$Ff_{meas} = [\ Ff_{meas}(\lambda_1)\quad Ff_{meas}(\lambda_2)\quad \cdots\quad Ff_{meas}(\lambda_{211})\ ]^T$$
$$Fb_{simu} = [\ Fb_{simu}(\tau_I,\tau_{II},\delta,\lambda_1)\quad Fb_{simu}(\tau_I,\tau_{II},\delta,\lambda_2)\quad \cdots\quad Fb_{simu}(\tau_I,\tau_{II},\delta,\lambda_{211})\ ]^T$$
$$Ff_{simu} = [\ Ff_{simu}(\tau_I,\tau_{II},\delta,\lambda_1)\quad Ff_{simu}(\tau_I,\tau_{II},\delta,\lambda_2)\quad \cdots\quad Ff_{simu}(\tau_I,\tau_{II},\delta,\lambda_{211})\ ]^T$$
$$C_{nb} = \mathrm{diag}[\ \sigma_b(\lambda_1)^2\quad \sigma_b(\lambda_2)^2\quad \cdots\quad \sigma_b(\lambda_{211})^2\ ]$$
$$C_{nf} = \mathrm{diag}[\ \sigma_f(\lambda_1)^2\quad \sigma_f(\lambda_2)^2\quad \cdots\quad \sigma_f(\lambda_{211})^2\ ]$$
$$x = [\ \tau_I\quad \tau_{II}\quad \delta\ ]^T$$
$$x_{priori} = [\ \tau_I^{priori}\quad \tau_{II}^{priori}\quad \delta^{priori}\ ]^T$$
$$C_x = \mathrm{diag}[\sigma(\tau_I)^2\sigma(\tau_{II})^2\sigma(\delta)^2]$$

where $\lambda_1, \lambda_2, \dots, \lambda_{211}$ represent the wavelengths of 640 nm, 641 nm, $\dots$, 850 nm, respectively. The τ_I, τ_{II}, and δ are the variables during the inversion process. The τ_I^{priori}, τ_{II}^{priori}, and δ^{priori} are the a priori guesses of τ_I, τ_{II}, and δ, respectively. The $\sigma(\tau_I)^2$, $\sigma(\tau_{II})^2$, and $\sigma(\delta)^2$ are the variances of the a priori guesses of τ_I, τ_{II}, and δ, respectively. The σ_b and σ_f represent the measurement noise of backward and forward fluorescence and uncertainty of model accuracy. The covariance matrices of observation and model uncertainty (C_{nb} and C_{nf}) and of the a priori variables (C_x) determine the respective weights from the measurements and a priori knowledge to the cost function. However, their determinations are difficult and somewhat subjective. Detailed discussion on this can be found in [14]. Here, the leaf fluorescence measurements are considered high quality, especially for the spectral range of 670–800 nm. Therefore, higher weights for these leaf measurements are given than those for a priori knowledge.

The first and second terms of the merit function Ff (τ_I,τ_{II},δ) in Equation (9) aim to search for values for the unknown fluorescence parameters (τ_I, τ_{II}, and δ) that best match the simulated backward and forward fluorescence to their correspondingly measured ones, respectively. The third term of the merit function is to inject *a priori* knowledge to the merit function. The *a priori* guesses of τ_I, τ_{II}, and δ (*i.e.*, τ_I^{priori}, τ_{II}^{priori}, and δ^{priori}) were selected as the standard values of τ_I, τ_{II}, and δ given by Pedrós *et al.* [6]. Variances of the *a priori* guesses of τ_I, τ_{II}, and δ (*i.e.*, $\sigma\,(\tau_I)^2$, $\sigma\,(\tau_{II})^2$, and $\sigma\,(\delta)^2$) were estimated by assuming these parameters were uniformly distributed within the variation ranges with the reference given in [6]. The *a priori* knowledge assigned in this study for the unknown parameters is shown in Table 2.

Table 2. *A priori* knowledge for the relative absorption cross section of photosystem I (PSI) and photosystem II (PSII) (δ), the fluorescence lifetimes of PSI and PSII (τ_I and τ_{II}) with the reference given in [6] for the Bayesian inversion of the FluorMODleaf model. The *a priori* knowledge is provided as the *a priori* guesses and the variances of these *a priori* guesses.

Parameter	τ_I	τ_{II}	δ
A priori guess	0.035	0.5	1
Variances of the *a priori* guess	0.0833	0.3333	0.48

The results of the Bayesian inversion procedure contain both the posterior estimates of the unknown parameters of τ_I, τ_{II}, and δ, which are obtained by minimizing the merit function as defined in Equation (9), and the covariance matrix of the posterior estimates, which contains the posterior variances of the inverted values of the unknown parameters. The covariance matrix of the posterior estimates is calculated as:

$$C_{post} = [h_b(x^*)^T C_{nb}^{-1} h_b(x^*) + h_f(x^*)^T C_{nf}^{-1} h_f(x^*) + C_x^{-1}]^{-1} \qquad (10)$$

where x^* is a vector that contains the posterior estimates of the unknown parameters of τ_I, τ_{II}, and δ; $h(x^*)$ is the Jacobian matrix for the FluorMODleaf model at the point of x^* and expressed as:

$$h_b(x^*) = \begin{bmatrix} \left.\frac{\partial Fb_{simu}(\lambda_1)}{\partial \tau_I}\right|_{\tau_I=\tau_I^*} & \left.\frac{\partial Fb_{simu}(\lambda_1)}{\partial \tau_{II}}\right|_{\tau_{II}=\tau_{II}^*} & \left.\frac{\partial Fb_{simu}(\lambda_1)}{\partial \delta}\right|_{\delta=\delta^*} \\ \left.\frac{\partial Fb_{simu}(\lambda_2)}{\partial \tau_I}\right|_{\tau_I=\tau_I^*} & \left.\frac{\partial Fb_{simu}(\lambda_2)}{\partial \tau_{II}}\right|_{\tau_{II}=\tau_{II}^*} & \left.\frac{\partial Fb_{simu}(\lambda_2)}{\partial \delta}\right|_{\delta=\delta^*} \\ \vdots & \vdots & \vdots \\ \left.\frac{\partial Fb_{simu}(\lambda_{211})}{\partial \tau_I}\right|_{\tau_I=\tau_I^*} & \left.\frac{\partial Fb_{simu}(\lambda_{211})}{\partial \tau_{II}}\right|_{\tau_{II}=\tau_{II}^*} & \left.\frac{\partial Fb_{simu}(\lambda_{211})}{\partial \delta}\right|_{\delta=\delta^*} \end{bmatrix}$$

$$
h_f(x^*) =
\begin{bmatrix}
\left.\dfrac{\partial Ff_{simu}(\lambda_1)}{\partial \tau_I}\right|_{\tau_I=\tau_I^*} & \left.\dfrac{\partial Ff_{simu}(\lambda_1)}{\partial \tau_{II}}\right|_{\tau_{II}=\tau_{II}^*} & \left.\dfrac{\partial Ff_{simu}(\lambda_1)}{\partial \delta}\right|_{\delta=\delta*} \\[2mm]
\left.\dfrac{\partial Ff_{simu}(\lambda_2)}{\partial \tau_I}\right|_{\tau_I=\tau_I^*} & \left.\dfrac{\partial Ff_{simu}(\lambda_2)}{\partial \tau_{II}}\right|_{\tau_{II}=\tau_{II}^*} & \left.\dfrac{\partial Ff_{simu}(\lambda_2)}{\partial \delta}\right|_{\delta=\delta*} \\[2mm]
\vdots & \vdots & \vdots \\[2mm]
\left.\dfrac{\partial Ff_{simu}(\lambda_{211})}{\partial \tau_I}\right|_{\tau_I=\tau_I^*} & \left.\dfrac{\partial Ff_{simu}(\lambda_{211})}{\partial \tau_{II}}\right|_{\tau_{II}=\tau_{II}^*} & \left.\dfrac{\partial Ff_{simu}(\lambda_{211})}{\partial \delta}\right|_{\delta=\delta*}
\end{bmatrix}
$$

The posterior standard deviations of the inverted parameters are contained in the main diagonal of C_{post}:

$$
C_{post} =
\begin{bmatrix}
v_{\tau_I}^2 & v_{12}^2 & v_{13}^2 \\
v_{21}^2 & v_{\tau_{II}}^2 & v_{23}^2 \\
v_{31}^2 & v_{32}^2 & v_{\delta}^2
\end{bmatrix}
\tag{11}
$$

where $v_{\tau I}$, $v_{\tau II}$, and v_δ are the posterior standard deviations of τ_I^*, τ_{II}^*, and δ, respectively; and the other elements in C_{post} are the covariance values between each two inverted parameters.

The flow diagram of the inversion procedure is illustrated in Figure 3.

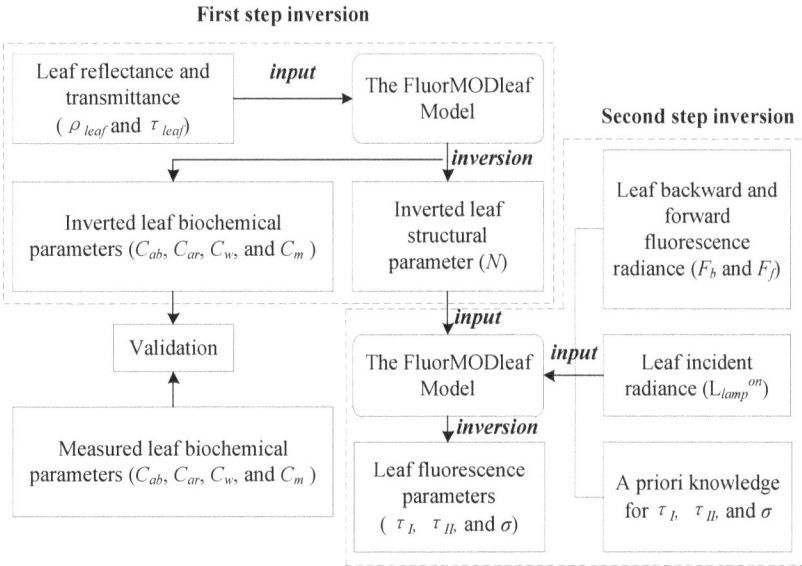

Figure 3. Flow diagram of the retrieval of the fluorescence parameters of plant leaves.

3. Results and Discussion

3.1. Distributions of the Fluorescence Spectra

In Figure 4a, examples of the measured leaf radiance spectra of the sample with and without the filter are shown for the measurement of backward fluorescence; Figure 4b illustrates measurement of forward fluorescence. Corresponding measured leaf fluorescence radiance is also shown in the inset using a finer scale.

In Figure 5, the curves show the mean fluorescence spectra measured in the experiments for crop leaves, and the corresponding shaded areas represent standard deviations of the measured spectra. For the backward fluorescence spectra (Figure 5a,c), two peaks can be observed, with the left one being located approximately at 690 nm and the right one at 740 nm; the right peak higher than the left peak. For the forward fluorescence spectra (Figure 5b,d), the left peak is weak, and almost unnoticeable for wheat leaves (Figure 5b). Highest contrasts between the left and right peaks for both backward and forward fluorescence are observed for soybean leaves (Figure 5c,d). Cotton (Figure 5a,b) and peanut (Figure 5c,d) leaves show relatively lower magnitude of fluorescence, especially for the former. For the peanut leaves (Figure 5c,d), higher variance for both backward and forward fluorescence spectra can be observed. This higher variance was probably caused by the inclusion of the spectra of senescent leaves, whose left peaks for both backward and forward fluorescence are higher than the right peaks (not shown herein).

Generally, shapes of the fluorescence spectra and positions of left peak (occurs in the range of 686–691 nm) and right peak (in the range of 739–743 nm) measured in this study are consistent with the spectra measured by a specifically designed equipment (FluoWat) to measure leaf fluorescence reported in other studies [24–26]. However, intensity of the lamp with the filter used in this study is much weaker than that of FluoWat. Thus, values for fluorescence radiance measured here are lower and not directly comparable with those by FluoWat. It can be observed that fluorescence radiance is higher for backward measurements compared with forward measurements for all four crop leaves because absorption and scattering effect are stronger for the forward measurements [27].

It can also be seen that the fluorescence radiance at right peak is generally higher than the one at left peak. This phenomenon is probably caused by the fact that most leaves chosen in the experiment are green and healthy ones, whose fluorescence emission around left peak subjects to strong re-absorption due to the overlap with red region of chlorophyll absorption. This is especially evident for the forward fluorescence spectra with relatively weaker left peaks, since emitted fluorescence travels from the adaxial to the abaxial leaf side and experiences stronger re-absorption. However for the senescent peanut leaves with low chlorophyll contents, as noted above, the left peaks are higher than the right peaks of both backward and forward fluorescence spectra.

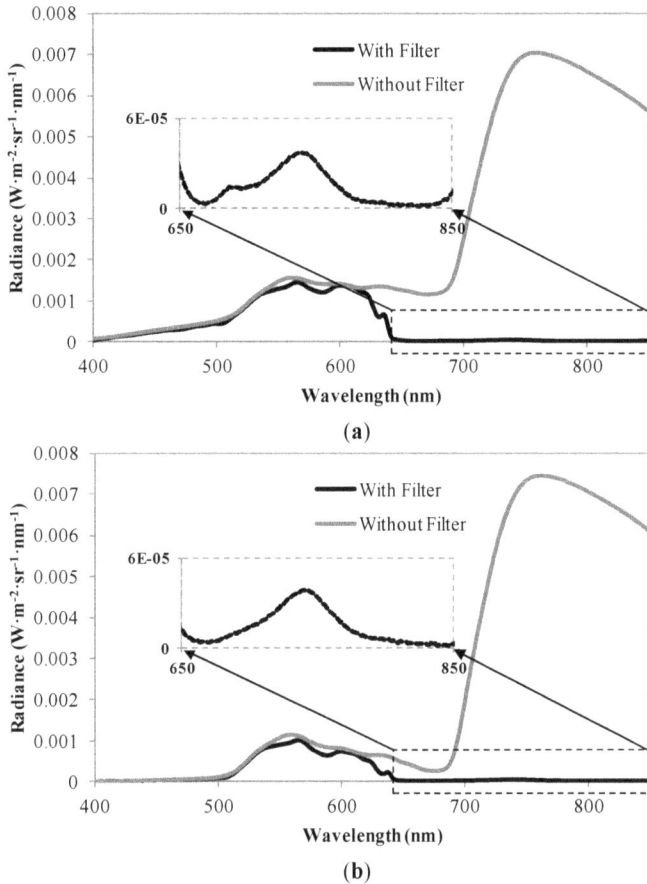

Figure 4. Examples of the radiance spectra of the leaf sample with and without the filter when measuring (**a**) backward fluorescence and (**b**) forward fluorescence. Insets: distributions of measured leaf backward and forward fluorescence radiance with the same unit but in finer scale.

The differences of peak distributions may also be caused by actual engagement of two photosystems. The left peak originates mainly from PS II, while the right peak originates from both PS I and PS II. Since factors from physiological drivers to environmental drivers can trigger dynamic regulation of the two photosystems [8], magnitudes of the two peaks will be changing accordingly. This reason may explain why the distributions of backward and forward fluorescence between the peaks for wheat leaves are slightly different from those by other three types of crop leaves.

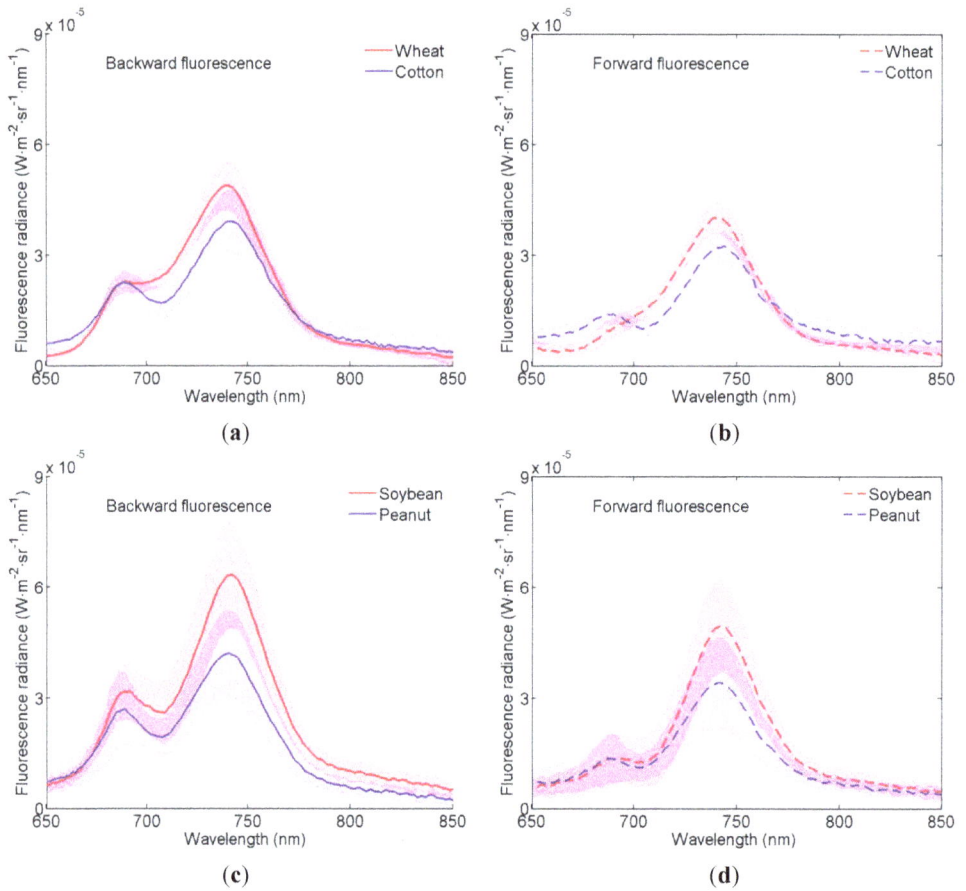

Figure 5. The mean fluorescence spectra measured in the experiments. (**a**) The backward fluorescence spectra for wheat and cotton leaves; (**b**) The forward fluorescence spectra for wheat and cotton leaves; (**c**) The backward fluorescence spectra for soybean and peanut leaves; (**d**) The forward fluorescence spectra for soybean and peanut leaves. The shaded portions represent standard deviation of the measured spectra.

Removal of light with the cut-off filter of 640 nm is biasing the performance of photosynthetic apparatus towards the PS II center, which may additionally affect the peak distributions. Therefore, the noticeable differences in magnitude and subtle distributions of fluorescence for different crops may result from differences in leaf structure of species, pigment contents, and crop physiological conditions.

3.2. Results of Sensitivity Analysis for the FluorMODleaf Model

Results of sensitivity analysis for the FluorMODleaf model are illustrated in Figure 6. Figure 6a shows the first order sensitivity indices of the input parameters of the FluorMODleaf model to the backward fluorescence. The total sensitivity indices of the input parameters are similar to the first order sensitivity indices, and, therefore, are not shown here. It shows that τ_I, τ_{II}, and C_{ab} are the most sensitive parameters among all eight input parameters. The τ_I is more sensitive in the near-infrared region where the PSI contributes the major fluorescence emission, while the τ_{II} is more sensitive in the red region where the PSII is the main photosystem that emits fluorescence. The C_{ab} is a sensitive parameter within the spectral range of 640–850 nm, because it not only has an absorption effect for the emitted ChlF, but it also determines the excitation efficiency of leaves. The C_{ar} is also a relatively sensitive parameter because it partially transfers the absorbed energy to chlorophylls for ChlF emission [6].

For the first order sensitivity indices of the forward fluorescence (Figure 6b), τ_I, τ_{II}, and C_{ab} are still the most influential parameters. It can also be seen that the model becomes relatively sensitive to leaf structural parameter N in the red region compared with its sensitivity of the backward fluorescence. It is because the absorption effect of the leaf biochemical contents (mainly the C_{ab} and C_{ar}) can be indirectly affected by the leaf thickness through the photon's path length, and this effect is more obvious for the forward fluorescence than for the backward fluorescence.

It was also found that the model is relatively insensitive to parameters C_w and C_m with sensitivity indices lower than 0.05 in the wavelength region of 640–850 nm for both the forward and backward fluorescence. This is because the absorption effects of C_w and C_m are relatively insignificant within the ChlF emission region of 640–850 nm.

From the results of sensitivity analysis for FluoMODleaf model, it can be observed that all three fluorescence parameters are relatively influential, although the extents are different for different spectral bands. Thus, it is feasible to invert these parameters from the leaf fluorescence measurements. Three other parameters, C_{ab}, C_{ar}, and N are also sensitive to the leaf fluorescence. However, as inversion studies by using leaf reflectance and transmittance show [9,21], these parameters and other two insensitive parameters, C_w and C_m, for FluorMODleaf, can be successfully inverted by the PROSPECT model. Thus, two stages of inversion were employed: in the first stage, five parameters, N, C_{ab}, C_{ar}, C_w and C_m, were inverted by leaf reflectance and transmittance, and they were fixed at these inverted values; in the second stage, only three fluorescence parameters are changed to optimize the cost function.

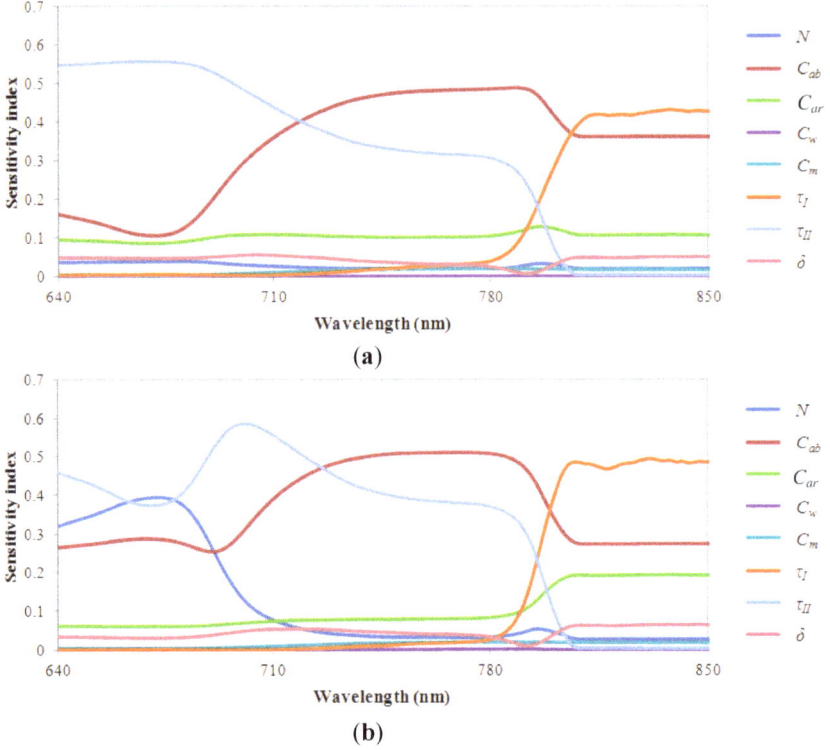

Figure 6. The sensitivity analysis results of the FluorMODleaf model. (a) The first order sensitivity indices of the input parameters to the backward fluorescence; (b) The first order sensitivity indices of the input parameters to the forward fluorescence.

3.3. Retrieval Results of the Leaf Biochemical Contents

Figure 7 shows the results of the first step inversion for chlorophyll content (C_{ab}), carotenoid content (C_{ar}), water content (C_w), and dry matter content (C_m) for four crops' leaves. The dashed 1:1 line and the equation of regression line are also presented in the figures. It can be observed that retrieved values agree well with their corresponding measured values for C_{ab} (Figure 7a), C_{ar} (Figure 7b) and C_m (Figure 7d). For C_w (Figure 7c), measured values are generally lower than the inverted ones, which is probably caused by the water loss during the later weighting process in the laboratory before oven-drying. It can be found that biochemical contents of peanut leaves cover relatively larger ranges, notably for a low value of C_{ab} around 15 µg/cm^2, which corresponds to the senescent leaves. The coefficient of determination (R^2) and root mean square error (RMSE) between the retrieved and measured values are 0.90 and 3.38 µg/cm^2, 0.83 and 0.93 µg/cm^2, 0.60 and 0.00379 g/cm^2, and 0.61 and 0.00326 g/cm^2, for C_{ab}, C_{ar}, C_w, and C_m, respectively.

16

This generally good agreement between retrieved and measured leaf biochemical contents, especially for the two sensitive parameters for fluorescence, C_{ab} and C_{ar}, assists the second step inversion for the fluorescence parameters.

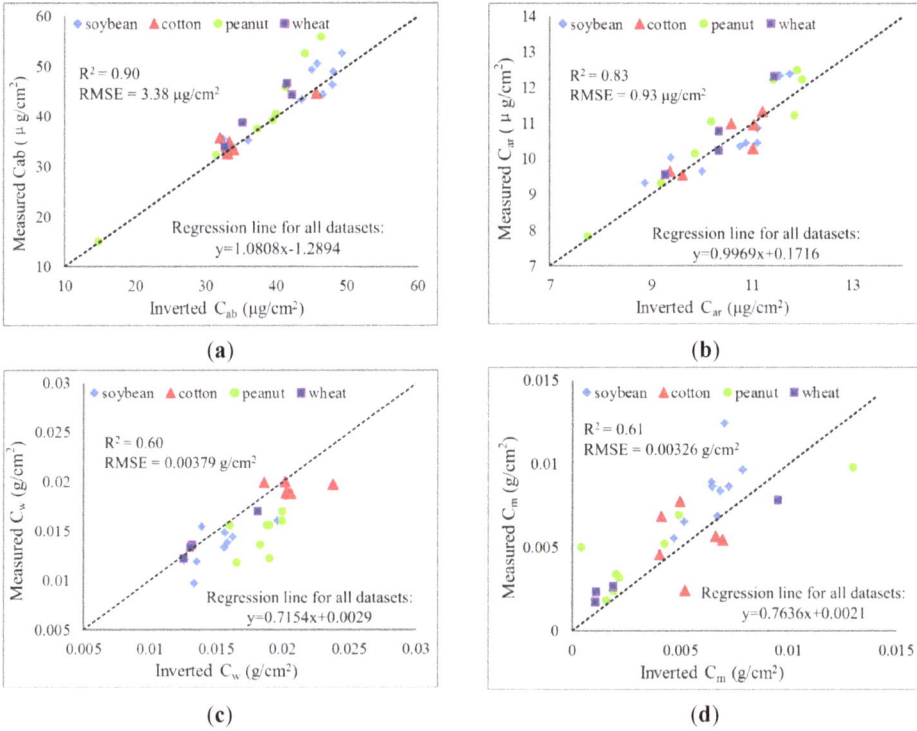

Figure 7. Scatter diagram of inverted and measured values of the chlorophyll content (**a**); carotenoid content (**b**); water content (**c**); and dry matter content (**d**) for four crop leaves. The coefficient of determination (R^2) and root mean square error (RMSE) between the retrieved and measured values are also provided.

3.4. Inversion Results of the Fluorescence Parameters

The fluorescence parameters were retrieved from the leaf spectral measurements by the Bayesian inversion approach. Figure 8 shows the retrieved fluorescence lifetimes of PSI and PSII (τ_I and τ_{II}), the relative absorption cross section of PSI and PSII (δ), and their standard deviations by inverting the FluorMODleaf model for soybean, cotton, peanut and wheat leaves. It can be observed that τ_I is more stable for all four crop types, predominantly in the range of 0.05–0.15 ns. This relatively weak variation is consistent with the assumption that PSI fluorescence does not change with photochemistry, though may change with species [6]. However, for τ_{II}, larger variations within and between species are observed. τ_{II} for soybean is much larger

17

than the other three crops. By comparing the distributions of fluorescence spectra of four crops (Figure 5), we can see that values in the red parts (around the left peak) of the leaf fluorescence spectra for soybean are more distinct and higher than those for other three crops. Since fluorescence emission in this spectral part mainly originates from PSII, higher values of τ_{II}, corresponding to higher contribution from PSII, are obtained. In the FluorMODleaf model, the relative absorption cross section ratio δ affects the fractions of contributions by PSI and PSII to the total fluorescence, with lower value corresponding to larger contributions from PSI, and higher one to larger contributions from PSII. For our measurements, most leaves show a higher right peak than the left peak, except for some leaves with low chlorophyll contents. Thus, for soybean with the more distinct contrast of fluorescence spectra and wheat with a bit less extent, generally low δ values were obtained. For cotton and peanut leaves with relatively weak contrast between the left and right peaks, δ values are generally higher. For the senescent peanut leaves, inverted δ reaches 1.71. The inverted values of δ here are generally lower than the values suggested by Pedrós et al. [6]. Besides, the aforementioned features of measured fluorescence data, this difference may also be caused by the different experimental setup and light source used in our experiment. Another output of the Bayesian inversion with the inverted parameters is their corresponding posterior standard deviations. These posterior standard deviations are always lower than the standard deviations of the *a priori* guess, which shows the reduction of uncertainty of model parameters during the inversion.

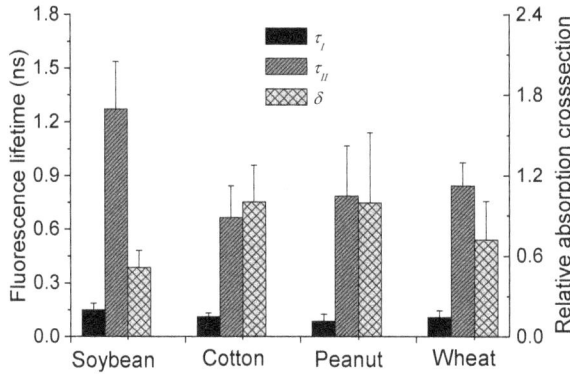

Figure 8. The fluorescence lifetimes of PSI and PSII (τ_I and τ_{II}), the relative absorption cross section of PSI and PSII (δ), and the standard deviations by inverting the FluorMODleaf model for four crops' leaves.

Although different leaves of the crops at different times in a day were sampled in the measurement, the results show that the fluorescence parameters are species-dependent and sensitive to biochemical contents and environmental factors. Because of the complexity of the relationship between fluorescence emission and the plant physiology, it is difficult at this stage to quantitatively interpret physiological meaning of these inverted fluorescence parameters. Further studies with simultaneous measurement of photosynthetic functions and fluorescence emission spectra are needed to better understand these parameters.

These fluorescence parameters are difficult to measure directly, and consequently it is difficult to evaluate the inversion results through measurements. As an alternative, the fluorescence spectra reconstructed from the inverted fluorescence parameters and measured in the experiment, both with a step of 1 nm, were then compared. The comparison results are shown in Figure 9a,b for leaf backward and forward fluorescence, respectively. R^2 and RMSE are 0.96 and 3.14×10^{-6} W$\cdot$m$^{-2}\cdot$sr$^{-1}\cdot$nm^{-1}, respectively, for backward fluorescence, and 0.92 and 3.84×10^{-6} W$\cdot$m$^{-2}\cdot$sr$^{-1}\cdot$nm^{-1} for forward fluorescence, which indicates a high accuracy of the inversion results. The reconstructed and measured fluorescence radiances at two peaks (690 and 740 nm) are also presented in the insets, which do not show systematic deviations between them.

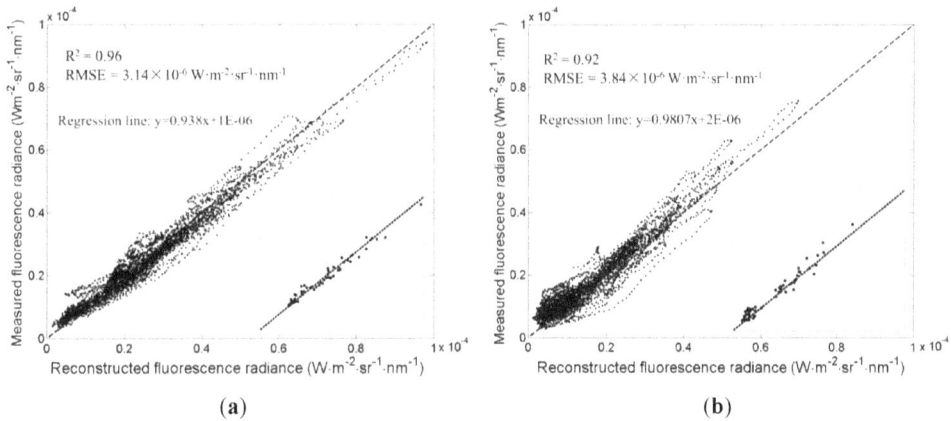

Figure 9. Comparison between the leaf fluorescence radiance spectra reconstructed from the inverted fluorescence parameters and the leaf fluorescence radiance spectra measured in the experiment for (**a**) backward and (**b**) forward fluorescence radiance. Insets: Comparison between reconstructed and measured fluorescence radiances at 690 and 740 nm with the same unit.

3.5. Potential and Limitations of Applying Model Inversion for the Retrieval of Leaf Fluorescence Parameters

The ChlF signal can provide critical information about the growth status of plants, and therefore it has been used as an effective tool to monitor plant stress induced by air pollution [25,26], water deficit [28,29], herbicide treatment [2], and salt and drought [30]. Quantitative estimation of the fluorescence parameters for crop leaves would be of high importance in assessing the photosynthetic rates of green plants and monitoring the stress conditions of crops. In this study, the leaf-level FluorMODleaf model was inverted using the leaf fluorescence spectra measured in the experiments. Results indicate that, even though the ChlF signal is relatively weak, the fluorescence parameters can be reliably inverted by introducing two stages inversion and adopting the Bayesian-based inversion strategy. However, this conclusion comes from an indirect way: inverted fluorescence parameters are generally in the reasonable ranges, there are no high and systematic deviations between measured fluorescence and re-constructed fluorescence, and the posterior standard deviations are always lower than the standard deviations of the *a priori* guess. More experiments can be designed and conducted to further evaluate the inversion strategy and better investigate the potential of the inverted fluorescence parameters in crop stress detections and growth status monitoring. Moreover, for practical applications of remote sensing technique, canopy-level ChlF model can be simulated in order to interpret the canopy fluorescence signal from the airborne and space-borne observations. With the fast development of the vegetative canopy models based on the radiative transfer theory [8,31–33] and the computer simulation methods [34], coupling the leaf-level ChlF model (e.g., FluorMODleaf) with a canopy-level ChlF models can become a promising tool for the growth status monitoring of crops in precision agriculture.

Indeed, the incident radiance between 640–700 nm can also excite fluorescence. However, the processes to emit fluorescence and reflect (and transmit) the incident radiation occur simultaneously, thus making the separation of the fluorescence from the total radiation very challenging. In order to ensure that the entire leaf fluorescence spectra of 640–850 nm could be obtained, the short-pass filter with the cut-off wavelength of 640 nm was used in the experiment, which blocked the lamp radiance between 640–700 nm and consequently the reflected and transmitted radiance from the lamp. This experimental setup provides an effective and efficient method to non-destructively obtain the leaf ChlF spectra. The intensity of the lamp used in this study is weak enough to avoid the influence to the photosynthetic process and induction of variable fluorescence. However, the removal of excitation radiation from 640–700 may induce potential bias in the measured ChlF spectra, which needs further investigation. In the future studies, filters with different cut-off wavelengths can be used to measure leaf ChlF spectra to compare the inversion results.

4. Conclusions

Leaf ChlF is closely related to the photosynthetic conditions of green plants. In this study, a sensitivity analysis of the FluorMODleaf model was performed using the EFAST method. Based on the sensitivity analysis results, the FluorMODleaf model was inverted using the experimental datasets. Bayesian theory was introduced to the inversion process aiming to achieve a stable inversion results. Results showed that R^2 and RMSE between the fluorescence simulated from the inverted fluorescence parameters and measured in the experiment were 0.96 and 3.14×10^{-6} W·m^{-2}·sr^{-1}·nm^{-1}, respectively, for backward fluorescence, and 0.92 and 3.84×10^{-6} W·m^{-2}·sr^{-1}·nm^{-1} for forward fluorescence. Based on results, it can be concluded that the Bayesian inversion approach can be used to retrieve the fluorescence parameters of plant leaves by inverting the FluorMODleaf model. The retrieved fluorescence parameters have the potential for agricultural applications.

Acknowledgments: This work is supported by the Chinese Natural Science Foundation under Project 41371325. Thanks go to Xu Dai and Peng Zhang for their assistance during the experiment. The authors are grateful to Yves Goulas, Roberto Pedrós, and Fabrice Daumard, for providing the codes of the FluorMODleaf model and helpful comments. Feng Zhao would like to express his appreciation for the assistance given by E.L. Butt-Castro (Tina) and J. de Koning (Anke) during his visit at Faculty of Geo-Information Science and Earth Observation (ITC), University of Twente. The authors thank Steven J. Thomson for polishing the manuscript. The authors also thank the reviewers for thoroughly reading the paper and providing useful suggestions.

Author Contributions: Feng Zhao conceived the research, proposed the research method, conducted the data analysis, prepared the manuscript and made the revision. Yiqing Guo contributed to the data analysis and the manuscript preparation. Yanbo Huang contributed to the research method and the manuscript revision. Wout Verhoef and Christiaan van der Tol provided suggestions for the research method and manuscript revision. Bo Dai contributed to the data analysis. Liangyun Liu contributed to the design of field experiment, and provided suggestions for the manuscript revision. Huijie Zhao provided suggestions for the research and manuscript revision. Guang Liu contributed to the design of field experiment and the manuscript revision.

References

1. Meroni, M.; Rossini, M.; Guanter, L.; Alonso, L.; Rascher, U.; Colombo, U.; Moreno, J. Remote sensing of solar-induced chlorophyll fluorescence: Review of methods and applications. *Remote Sens. Environ.* **2009**, *113*, 2037–2051.
2. Huang, Y.; Thomson, S.J.; Molin, W.T.; Reddy, K.N.; Yao, H. Early detection of soybean plant injury from glyphosate by measuring chlorophyll reflectance and fluorescence. *J. Agric. Sci.* **2012**, *4*, 117–124.
3. Zhao, F.; Guo, Y.; Huang, Y.; Reddy, K.N.; Zhao, Y.; Molin, W.T. Detection of the onset of glyphosate-induced soybean plant injury through chlorophyll fluorescence signal extraction and measurement. *J. Appl. Remote Sens.* **2015**, *9*.
4. Guanter, L.; Zhang, Y.; Jung, M.; Joiner, J.; Voigta, M.; Berry, J.A.; Frankenberg, C.; Huete, A.R.; Zarco-Tejada, P.; Lee, J.-E.; *et al.* Global and time-resolved monitoring of crop photosynthesis with chlorophyll fluorescence. *Proc. Natl. Acad. Sci. USA* **2014**, *111*, E1327–E1333.

5. Sušila, P.; Nauš, J. A Monte Carlo study of the chlorophyll fluorescence emission and its effect on the leaf spectral reflectance and transmittance under various conditions. *Photochem. Photobiol. Sci.* **2007**, *6*, 894–902.

6. Pedrós, R.; Goulas, Y.; Jacquemoud, S.; Louis, J.; Moya, I. FluorMODleaf: A new leaf fluorescence emission model based on the PROSPECT model. *Remote Sens. Environ.* **2010**, *114*, 155–167.

7. Verhoef, W. Modeling vegetation fluorescence observations. In Proceedings of the EARSel 7th SIG-Imaging Spectroscopy Workshop, Edinburgh, UK, 11–13 April 2011.

8. Miller, J.R.; Berger, M.; Goulas, Y.; Jacquemoud, S.; Louis, J.; Mohammed, G.; Moise, N.; Moreno, J.; Moya, I.; Pedrós, R.; *et al. Development of A Vegetation Fluorescence Canopy Model*; ESTEC Contract No. 1635/02/NL/FF; ESA Scientific and Technical Publications Branch, ESTEC: Paris, French, 2015.

9. Feret, J.-B.; François, C.; Asner, G.P.; Gitelson, A.A.; Martin, R.E.; Bidel, L.P.R.; Ustin, S.L.; Le Maire, G.; Jacquemoud, S. PROSPECT-4 and 5: Advances in the leaf optical properties model separating photosynthetic pigments. *Remote Sens. Environ.* **2008**, *112*, 3030–3043.

10. Jacquemoud, S.; Baret, F. PROSPECT: A model of leaf optical properties spectra. *Remote Sens. Environ.* **1990**, *34*, 75–91.

11. Li, X.; Gao, F.; Wang, J.; Strahler, A. A priori knowledge accumulation and its application to linear BRDF model inversion. *J. Geophys. Res. D: Atmos.* **2001**, *106*, 11925–11935.

12. Laurent, V.C.E.; Schaepman, M.E.; Verhoef, W.; Weyermann, J.; Chávez, R.O. Bayesian object-based estimation of LAI and chlorophyll from a simulated Sentinel-2 top-of-atmosphere radiance image. *Remote Sens. Environ.* **2014**, *140*, 318–329.

13. Wang, Y.; Li, X.; Nashed, Z.; Zhao, F.; Yang, H.; Guan, Y.; Zhang, H. Regularized kernel-based BRDF model inversion method for ill-posed land surface parameter retrieval. *Remote Sens. Environ.* **2007**, *111*, 36–50.

14. Laurent, V.C.E.; Verhoef, W.; Damm, A.; Schaepman, M.E.; Clevers, J.G.P.W. A Bayesian object-based approach for estimating vegetation biophysical and biochemical variables from at-sensor APEX data. *Remote Sens. Environ.* **2013**, *139*, 6–17.

15. Zarco-Tejada, P.J.; Miller, J.R.; Mohammed, G.H.; Noland, T.L. Chlorophyll fluorescence effects on vegetation apparent reflectance: I. Leaf-Level measurements and model simulation. *Remote Sens. Environ.* **2000**, *74*, 582–595.

16. Zhang, Y. Studies on Passive Sensing of Plant Chlorophyll Florescence and Application of Stress Detection. Ph.D. Thesis, Zhejiang University, Hangzhou, China, May 2006. pp. 24–27.

17. LI-COR Inc. *LI-COR LI-1800-12 Integrating Sphere Instruction Manual*; LI-COR Inc.: Lincoln, NE, USA, 1983.

18. Lichtenthaler, H.K.; Buschmann, C. Extraction of photosynthetic tissues: Chlorophylls and carotenoids. In *Current Protocols in Food Analytical Chemistry*; Wrolstad, R.E., Acree, T.E., Decker, E.A., Penner, M.H., Reid, D.S., Schwarts, S.J., Eds.; John Wiley and Sons: New York, NY, USA, 2001; p. F4.2.1-6.

19. Saltelli, A.; Ratto, M.; Andres, T.; Campolongo, F.; Cariboni, J.; Gatelli, D.; Saisana, M.; Tarantola, S. *Global Sensitivity Analysis, the Primer*; John Wiley & Sons Ltd.: West Sussex, UK, 2008.

20. Saltelli, A.; Tarantola, S.; Chan, K. A quantitative, model independent method for global sensitivity analysis of model output. *Technometrics* **1999**, *41*, 39–56.

21. Zhao, F.; Guo, Y.; Huang, Y.; Reddy, K.N.; Lee, M.A.; Fletcher, R.S.; Thomson, S.J. Early detection of crop injury from herbicide glyphosate by leaf biochemical parameter inversion. *Int. J. Appl. Earth Observ. Geoinf.* **2014**, *31*, 78–85.

22. Zhao, F.; Huang, Y.; Guo, Y.; Reddy, K.N.; Lee, M.A.; Fletcher, R.S.; Thomson, S.J.; Zhao, H. Early detection of crop injury from glyphosate on soybean and cotton using plant leaf hyperspectral data. *Remote Sens.* **2014**, *6*, 1538–1563.

23. Liu, Q. Study on Component Temperature Inversion Algorithm and the Scale Structure for Remote Sensing Pixel. Ph.D. Thesis, Institute of Remote Sensing Applications, Chinese Academy of Sciences, Beijing, China, May 2002.

24. Van Wittenberghe, S.; Alonso, L.; Verrelst, J.; Moreno, J.; Samson, R. Bidirectional sun-induced chlorophyll fluorescence emission is influenced by leaf structure and light scattering properties—A bottom-up approach. *Remote Sens. Environ.* **2015**, *158*, 169–179.

25. Van Wittenberghe, S.; Alonso, L.; Verrelst, J.; Verrelst, I.; Delegido, J.; Veroustraete, F.; Veroustraete, R.; Moreno, J.; Samson, R. Upward and downward solar-induced chlorophyll fluorescence yield indices of four tree species as indicators of traffic pollution in Valencia. *Environ. Poll.* **2013**, *173*, 29–37.

26. Van Wittenberghe, S.; Alonso, L.; Verrelst, J.; Hermans, I.; Valcke, R.; Veroustraete, F.; Moreno, J.; Samson, R. A field study on solar-induced chlorophyll fluorescence and pigment parameters along a vertical canopy gradient of four tree species in an urban environment. *Sci. Total Environ.* **2014**, *466–467*, 185–194.

27. Louis, J.; Cerovic, Z.G.; Moya, I. Quantitative study of fluorescence excitation and emission spectra of bean leaves. *J. Photochem. Photobiol. B* **2006**, *85*, 65–71.

28. Zarco-Tejada, P.J.; Berni, J.A.J.; Suárez, L.; Sepulcre-Cantó, G.; Morales, F.; Miller, J.R. Imaging chlorophyll fluorescence with an airborne narrow-band multispectral camera for vegetation stress detection. *Remote Sens. Environ.* **2009**, *113*, 1262–1275.

29. Zarco-Tejada, P.J.; González-Dugo, V.; Berni, J.A.J. Fluorescence, temperature and narrow-band indices acquired from a UAV platform for water stress detection using a micro-hyperspectral imager and a thermal camera. *Remote Sens. Environ.* **2012**, *117*, 322–337.

30. Naumann, J.C.; Young, D.R.; Anderson, J.E. Linking leaf chlorophyll fluorescence properties to physiological responses for detection of salt and drought stress in coastal plant species. *Physiol. Plant.* **2007**, *131*, 422–433.

31. Verhoef, W. Light scattering by leaves with application to canopy reflectance modelling: The SAIL model. *Remote Sens. Environ.* **1984**, *16*, 125–178.

32. Van der Tol, C.; Verhoef, W.; Timmermans, J.; Verhoef, A.; Su, X. An integrated model of soil-canopy spectral radiances, photosynthesis, fluorescence, temperature and energy balance. *Biogeosciences* **2009**, *6*, 3109–3129.

33. Zhao, F.; Gu, X.; Verhoef, W.; Wang, Q.; Yu, T.; Liu, Q.; Huang, H.; Qin, W.; Chen, L.; Zhao, H. A spectral directional reflectance model of row crops. *Remote Sens. Environ.* **2010**, *114*, 265–285.
34. Zhao, F.; Li, Y.; Dai, X.; Verhoef, W.; Guo, Y.; Shang, H.; Gu, X.; Huang, Y.; Yu, T.; Huang, J. Simulated impact of sensor field of view and distance on field measurements of bidirectional reflectance factors for row crops. *Remote Sens. Environ.* **2015**, *156*, 129–142.

Evaluation of Six Algorithms to Monitor Wheat Leaf Nitrogen Concentration

Xia Yao, Yu Huang, Guiyan Shang, Chen Zhou, Tao Cheng, Yongchao Tian, Weixing Cao and Yan Zhu

Abstract: The rapid and non-destructive monitoring of the canopy leaf nitrogen concentration (LNC) in crops is important for precise nitrogen (N) management. Nowadays, there is an urgent need to identify next-generation bio-physical variable retrieval algorithms that can be incorporated into an operational processing chain for hyperspectral satellite missions. We assessed six retrieval algorithms for estimating LNC from canopy reflectance of winter wheat in eight field experiments. These experiments represented variations in the N application rates, planting densities, ecological sites and cultivars and yielded a total of 821 samples from various places in Jiangsu, China over nine consecutive years. Based on the reflectance spectra and their first derivatives, six methods using different numbers of wavelengths were applied to construct predictive models for estimating wheat LNC, including continuum removal (CR), vegetation indices (VIs), stepwise multiple linear regression (SMLR), partial least squares regression (PLSR), artificial neural networks (ANNs), and support vector machines (SVMs). To assess the performance of these six methods, we provided a systematic evaluation of the estimation accuracies using the six metrics that were the coefficients of determination for the calibration (R^2_C) and validation (R^2_V) sets, the root mean square errors of prediction ($RMSE_P$) for the calibration and validation sets, the ratio of prediction to deviation (RPD), the computational efficiency (CE) and the complexity level (CL). The following results were obtained: (1) For the VIs method, $SAVI(R_{1200}, R_{705})$ produced a more accurate estimation of the LNC than other indices, with R^2_C, R^2_V, $RMSE_P$, RPD and CE values of 0.844, 0.795, 0.384, 2.005 and 0.10 min, respectively; (2) For the SMLR, PLSR, ANNs and SVMs methods, the SVMs using the first derivative canopy spectra (SVM-FDS) offered the best accuracy in terms of R^2_C, R^2_V, $RMSE_P$, RPD, and CE, at 0.96, 0.78, 0.37, 2.02, and 21.17, respectively; (3) The PLSR-FDS, ANN-OS and SVM-FDS methods yield similar accuracies if the CE and CL are not considered, however, ANNs and SVMs performed better on calibration set than the validation set which indicate that we should take more caution with the two methods for over-fitting. Except PLS method, the performance for most methods did not enhance when the spectrum were operated by the first derivative. Moreover, the evaluation of the robustness demonstrates that SVM method may be better suited than the other methods to cope with potential confounding factors for most varieties, ecological site and growth stage; (4) The prediction accuracy was found to be higher when more wavelengths were used, though at the cost of a lower CE. The findings are of interest to the remote sensing community for the development of improved inversion schemes for hyperspectral applications concerning other types of vegetation. The examples provided in this paper may also serve to illustrate the advantages and shortcomings of empirical hyperspectral models for mapping important vegetation biophysical properties of other crops.

Reprinted from *Remote Sens.* Cite as: Yao, X.; Huang, Y.; Shang, G.; Zhou, C.; Cheng, T.; Tian, Y.; Cao, W.; Zhu, Y. Evaluation of Six Algorithms to Monitor Wheat Leaf Nitrogen Concentration. *Remote Sens.* **2015**, *7*, 14939–14966.

1. Introduction

In cereal crops, nitrogen (N) is the most important element for maintaining growth status and enhancing grain yield [1]. Therefore, the real-time, nondestructive and accurate monitoring of the nitrogen (N) concentration in crops has become a key technique for timely diagnosis of problems, precise fertilization and productivity estimation [2–10]. Remote sensing has been widely applied in recent decades to determine the biophysical and chemical parameters of crops [2,11,12]. Many forthcoming hyperspectral satellite missions will be dedicated to land and crop monitoring. Hence, there is an urgent need to identify next-generation bio-geophysical variable retrieval algorithms that can be incorporated into an operational processing chain.

Considerable progress has been made using multispectral and hyperspectral data acquired from ground and aerial platforms to estimate the N concentration of crops [8,13–19]. Existing reports indicate that in most previous work, the core wavelengths have first been determined and then used to construct a sensitive spectral index, as in the case of the continuum removal (CR) and the vegetation index (VI) method. The CR method can be used to effectively isolate individual absorption features of interest and estimate the chemical concentration in dried leaves [20–22]. However, one must determine the spectral range each time when the CR operation is performed, which results in unstable performance in monitoring of the chemical concentration of crops [23]. In addition to the CR method, various vegetation indices, such as the Normalized Difference Vegetation Index (NDVI), the Ratio Vegetation Index (RVI), the Soil-Adjusted Vegetation Index (SAVI), Modified Normalized Difference (mND), and the Photochemical Reflectance Index (PRI), have been widely used to characterize chemical concentration of plants because these indices have simple forms and are easy to calculate [10,12,24–26]. However, most researchers use only a limited number of wavelengths in specific spectral regions to calculate these indices and have not exploited the full spectrum information in hyperspectral data. In addition, many of these vegetation indices are strongly influenced by the soil background, resulting in soil-dependent VI-biophysical relationships. Linear regression models are typically analyzed based on individual input variables of the characteristic wavelength or vegetation index. Therefore, several researchers have suggested that multivariable input parameters should be considered when constructing such linear regressions.

Presently, the commercial instruments that are used to monitor crop N concentrations, such as ASD [27] and hyperspectral imager, are not suitable for future use on family farms or for individual users because of their high cost and relatively complex operational procedures. A number of other portable devices, such as the SPAD (650 and 940 nm) [28], can only work on a single leaf each time and therefore cannot be applied to large populations of plants. The LNC models that are currently developed with specific wavelengths on portable devices, such as the GreenSeeker (656 and 770 nm) and the Crop Circle (450,550,650,670,730, and 760 nm) [29–31], may not be accurately transferrable among ecological sites and crop varieties. For the development of instruments with lower manufacturing cost and higher accuracy, it is unclear how many input variables should be used and which type of regression algorithms offers the best stability and computational efficiency.

A comprehensive multivariable linear regression could be performed to establish N predictive models for modern crop production. Several studies have addressed various multivariate models, such as stepwise multiple linear regression (SMLR) and partial least squares regression (PLSR) [5,16]. The SMLR is likely to suffer from multicollinearity when applied to canopy hyperspectral data [32,33]. Grossman *et al.* [33] have found that the best wavelengths selected with SMLR might not be related to the absorption characteristics of the compounds of interest and do not produce consistent results between datasets. Hence, care should be taken when using SMLR to select wavelengths and estimate N concentration. Alternatively, the PLSR approach has been adopted to reduce the large number of measured collinear spectral variables to a few non-correlated latent variables (LVs), thereby avoiding the potential overfitting problems that are typically associated with SMLR [16,33].

A number of spectrometric studies have been undertaken concerning the estimation of the N content of plants using CR, vegetation indices (VIs), SMLR and PLSR [8,10–12,16,33,34]. These approaches use an inconsistent number of wavelengths to estimate the N concentrations or estimate the chlorophyll status. Apart from these linear regression methods, some recent studies have investigated non-linear regression methods from the machine learning field such as artificial neural networks (ANNs) and support vector machines (SVMs) [34,35].

To date, the performance, advantages and disadvantages of leaf nitrogen concentration (LNC) estimation for wheat crops using ANN and SVM algorithms remain unclear. Currently, the ANN method is widely used in remote sensing to predict vegetation parameters and crop yields [6,34,35]. However, it inevitably suffers from the overfitting problem. Fortunately, some researchers reported the

SVM method resolves the problem of overfitting encountered when analyzing high-dimensional data [36] and has been used to soil moisture [37], hourly typhoon rainfall [38], long-lead stream flows [39], leaf area index, and leaf chlorophyll density [40,41]. These studies have shown that the SVM approach is preferable to the ANN approach for these applications because of its greater generalizability. In addition to the conventional application, ANN and SVM methods should be assessed in a comparative way in terms of their performance and potential for the estimation of wheat LNC.

Currently, the first derivative is often used to decompose a mixed spectrum and reduce the noise in the hyperspectral region [41,42]. Mauser and Bach [43] have concluded that derivative spectral indices are very sensitive to LAI. Yoder and Pettigrew-Crosby [4] have found that first-order derivative spectra are the best predictors of the N and chlorophyll contents of big-leaf maples grown under different fertilization treatments. Johnson and Billow [44] have examined Douglas fir needles grown using various fertilization treatments and also found the first-order derivatives of the fresh leaf spectra to be strongly correlated with the total N concentration. Many studies have demonstrated the potential of derivative spectra for estimating chemical concentrations of non-crop vegetation types. However, few studies have examined the performance of first-order derivative spectra with respect to the LNC of fresh wheat crop leaves.

To the best of our knowledge, no studies in the literature have provided an evaluation of all these methods and their predictive equations for wheat LNC using a large number of samples accumulated over nine consecutive years of field trial experiments with a total of 821 wide representatively samples. Moreover, previous evaluations have focused on the prediction accuracies and have not reported results on computational efficiency and complex level, which may be a serious problem when using hyperspectral imaging data. To address these research gaps, this study presents the results of a comparative assessment of six retrieval methods applied to *in situ* measurements acquired over eight years for seven varieties, four eco-sites, and 821 samples. The main objectives were (1) to evaluate the ANNs and performance of various linear (CR, VIs, SMLR and PLSR) and nonlinear (ANN$_S$ and SVM$_S$) regression methods based on the original and first derivative spectra for LNC estimation; and (2) to determine which method, input variable and model could estimate the LNC in winter wheat with higher accuracy, better robustness, less time, and less complexity.

2. Materials and Methods

2.1. Design of Field Experiments

Eight field experiments were conducted over eight growing seasons, with four located in Nanjing (32°03′N, 118°42′E), two in Rugao (32°15′N, 120°38′E), one in Hai'an (32°32′N, 120°28′E) and one in Yancheng (33°29′N, 120°28′E) in Jiangsu Province of eastern China. The experimental variables included different N fertilization rates and different cultivars of winter wheat. Each experiment consisted of a randomized complete block design with three replications. For all treatments, sufficient $Ca(H_2PO_4)_2$ and KCl were applied (150 kg· ha^{-1}) prior to seeding. Crop management followed local standard practices for wheat production. Additional details regarding the experimental design are provided in Table 1.

Table 1. Details of the eight field experiments.

Experi-ment(Exp.)	Year	Ecological Site	Wheat Cultivar	N Application Rates (kg· ha^{-1})	Sampling Dates	Number of Samples	Data Function
Exp. 1	04–05	Nanjing	Ningmai 9, Yangmai 12, Yumai 34	0, 75, 150, 225	19 March, 13/26 April, 3/6/12/24 May, 1 July	102	Validation
Exp. 2	05–06	Nanjing	Ningmai 9, Yumai 34 Yangmai 12	0, 75, 150, 225	19 March, 13/26 April, 3/6/12/24 May, 1 July	110	Calibration
Exp. 3	06–07	Yancheng	Yanmai 4110	0, 75, 150, 225 300	23 April, 17 May	103	Calibration
Exp. 4	07–08	Nanjing	Ningmai 9	90, 180, 270	8/23 April, 17 May	88	Validation
Exp. 5	08–09	Rugao	Yangmai 13	225, 275, 325	6/22 April, 6 May	120	Calibration
Exp. 6	09–10	Hai'an	Ningmai 13	0, 75, 150, 225	6/22 April, 6 May	122	Calibration
Exp. 7	10–11	Nanjing	Yangmai 18	150, 300	2/14/26 April, 5/17 May	93	Validation
Exp. 8	12–13	Rugao	Yangmai 18, Shengxuan 6	0, 100, 300	14/26 April, 3 May	83	Validation

2.2. Measurements of Hyperspectral Reflectance

All canopy spectral measurements were performed using an ASD FieldSpec Pro FR2500 spectrometer (Analytical Spectral Devices, Boulder, CO, USA) [27]. This spectrometer was fitted with 25° field-of-view fiber optics operating in the 350–2500 nm spectral range with a sampling interval of 1.4 nm and spectral resolution of 3 nm between 350 and 1050 nm, and of 2 nm and 10 nm, respectively, between 1050 and 2500 nm. The spectrometer was equipped with three separate holographic diffraction gratings and three different detectors: VNIR (350–1000 nm), SWIR1 (1001–1800 nm), and SWIR2 (1801–2500 nm). Because the SWIR2 detector was influenced by water vapor in the field tests, the spectral response in the visible and near-infrared bands (350–1800 nm) was used to monitor the wheat LNC in this study. The measurements were conducted 1 m above the wheat canopy with a view diameter of 0.44 m under clear sky conditions between 10:00 a.m. and 2:00 p.m. (Beijing time). Measurements of vegetation irradiance were performed at five sample sites in each plot. Each sample consisted of an average of three scans at an optimized integration time. The resulting spectral file contained the continuous spectral reflectance data collected in 1 nm steps in the band region of 350–2500 nm. Panel irradiance measurements (two scans each) were performed before and after each vegetation measurement. The smoothing procedure of Savitzky and Golay [31], which uses a five-point moving window, was applied to preprocess the spectrum. After smoothing, the first derivative was calculated to eliminate background effects and reduce noise.

2.3. Determination of Leaf N Concentration

After each measurement of the canopy spectral reflectance, wheat plants from a 0.25 m^2 area (two 0.5 m rows) were collected from each plot to determine their LNC values (%). For each sample, all green leaves were separated from the stems, oven-dried at 70 °C to constant weight, and then weighed. The dried leaf samples were ground, passed through a 1 mm screen, and stored in plastic bags for subsequent chemical analysis. The total N concentration in the leaf tissues was determined using the micro-Kjeldahl method.

2.4. Data Analysis

In this study, six different algorithms (CR, SI, SMLR, PLSR, ANN, and SVM) were comparatively analyzed using MATLAB (2010b).

2.4.1. Continuum Removal (CR)

The CR method was first applied to isolate individual absorption features of interest [21]. Based on the N-absorption characteristics, a local starting point (550 nm) and ending point (750 nm) were selected for CR analysis in this study. The selected region is primarily influenced by chlorophyll absorption, represented by an exponential function [23] that is used for the retrieval of biochemical and biophysical parameters [15,22,23]. Three CR parameters were used: (1) the band depth (BD); (2) the band depth ratio (BDR) and (3) the normalized band depth index (NBDI) [23]. These three CR parameters were calculated using the methods of Curran [27] and Mutanga [23,45].

2.4.2. Vegetation Indices (VIs)

Three types of vegetation indices, including the normalized difference vegetation index (NDVI, $(R_{\lambda 1} - R_{\lambda 2})/(R_{\lambda 1} + R_{\lambda 2})$), ratio vegetation index (RVI, $(R_{\lambda 1}/R_{\lambda 2})$), and soil-adjusted vegetation index (SAVI, $[1.5 * (R_{\lambda 1} - R_{\lambda 2})/(R_{\lambda 1} + R_{\lambda 2} - 0.5)]$), were calculated using the presented equations for all possible two-band combinations in the full spectral range. $R_{\lambda 1}$ and $R_{\lambda 2}$ represent those spectral reflectances drawn from the full spectral range.

2.4.3. Stepwise Multiple Linear Regression (SMLR)

SMLR was first proposed by Chatterjee and Price [46]. Using SMLR to filter the independent variables and construct regression models is a good approach to the current problem. With y as the independent variable and x as the dependent variable, the result is a linear relationship between the independent and dependent variables. Then, the multiple linear regression models take the following form:

$$y = b_0 + b_1 x_1 + b_2 x_2 + \ldots + b_k x_k + \varepsilon \tag{1}$$

where b_0 is a constant term, ε is a regression coefficient, and $b_1, b_2, \ldots, b_k$ are bands.

2.4.4. Partial Least-Squares Regression (PLSR)

The PLSR approach is a new type of multivariate statistical analysis algorithm that primarily considers a single dependent variable among the multiple variables of the regression model. In addition, PLSR is more effective under conditions in which the number of samples is fewer than the number of variables. Although the PLSR method is similar to principal component regression (PCR), PLSR actually involves decomposing both the spectra and the response variables simultaneously [47]. In this study, the spectral data were mean-centered before analysis, and the number of latent variables (LVs) was determined following the guidelines prescribed by Esbensen [48]. The optimal number of LVs was determined based on the relationship between the percentage variance captured by the model and the number of latent variables. With an increasing number of LVs, the percentage variance captured gradually changed, and the value indicated the optimal number of LVs. The basic PLSR methodology has been described in previous studies [46,49]. The objective of PLSR is to construct a linear model as follows:

$$Y = X\beta + \varepsilon \tag{2}$$

where Y is a mean-centered vector of a dependent variable, X is a mean-centered matrix of the independent variables, β is a matrix of regression coefficients, and ε is a matrix of residuals.

2.4.5. Artificial Neural Networks (ANNs)

Multi-layer perceptron networks constitute one of the most widely used types of neural networks in the remote sensing community [50]. A typical ANN is composed of various layers (an input layer, an output layer, and several hidden layers), and each layer contains a number of interconnected nodes and activation functions [7]. In this study, the optimum number of hidden layer nodes (HLNs) was determined based on the minimum value of RMSE$_P$, and gradient descent with momentum was used to train the network using 5000 iterations.

2.4.6. Support Vector Machines (SVMs)

The SVM technique is a universal theory of machine learning originally developed by Vapnik and Cortes for pattern recognition and classification [51,52]. SVM regression models can map low-dimensional nonlinear input to high-dimensional linear output with good results. The SVM approach has many unique advantages in pattern recognition for small samples as well as nonlinear and high-dimensional cases. The kernel function is particularly important for SVM analysis. In this study, the sigmoid tanh kernel was used for SVM analysis, with the equation shown below (Equation (3)) [36]. The SVM parameters were selected based on the mean square error (MSE). The parameters with the lowest MSE in the SVM regression were considered the best.

$$K(x,y) = tanh(k(x,z) + v), k > 0, v < 0 \qquad (3)$$

where k is a scalar and v is a displacement parameter.

2.4.7. Calibration and Validation

Six algorithms (CR, SI, SMLR, PLSR, ANN, and SVM) using different numbers of wavelengths were applied to construct models for monitoring the wheat LNC. The data from Exp. 2, 3, 5, and 6 were used as the calibration set because they contained a wider range of representative data, including a higher number of samples of different cultivars, more ecological sites and more growth stages. Exp. 1, 4, 7, and 8 were used as the validation set (Table 2). The fitness was evaluated from a 1:1 plot of the predicted and observed data.

Table 2. The statistical parameters of the calibration and validation sets for the wheat leaf nitrogen content (LNC).

Dataset	Number of Samples	Names of Cultivars	Ecological Sites	Minimum (%)	Maximum (%)	Mean (%)	SD	CV
Calibration (Exp. 2, 3, 5, 6)	456	Ningmai 9, Yumai 34, Yangmai 12, Yanmai 4110, Yangmai 13, Ningmai 13	Nanjing, Yancheng, Rugao, Hai'an	0.45	4.52	2.66	0.98	0.37
Validation (Exp. 1, 4, 7, 8)	366	Ningmai 9, Yangmai 12, Yumai 34,Yangmai 18, Shengxuan 6	Nanjing, Rugao	0.98	4.29	2.92	0.77	0.26
All data	822	All of the above		0.45	4.52	2.78	0.87	0.32

The performances of all models were evaluated based on several statistical parameters, including the calibration R^2 (R^2_C), the root mean square error of calibration (RMSE$_C$; see Equation (4)), validation R^2 (R^2_V), and the root mean square error of prediction (RMSE$_P$). All calculations were performed using custom-written MATLAB (2010b) scripts. Higher values of R^2_C, R^2_V, and PDP and lower values of RMSE$_C$ and RMSE$_P$ indicated higher precision and accuracy of the model. The running time was calculated using MATLAB 10b, and the level of operating complexity was determined based on the algorithm used to construct the model and the number of wavelengths.

$$RMSE_c = \sqrt{\sum_{i=1}^{n} (Y_{est,i} - Y_{mea,i})^2 / n} \qquad (4)$$

where $Y_{est,i}$ is the estimated LNC$_i$, $Y_{mea,i}$ is the measured LNC$_i$, and n is the number of samples. $RMSE_P$ was also calculated using Equation (4).

The ratio of prediction to deviation (RPD) was calculated as follows:

$$RPD = \frac{SD}{RMSE_P} \qquad (5)$$

where SD is the standard deviation. A value of $RPD > 2.0$ indicates a stable and accurate predictive model, an RPD value between 1.4 and 2.0 indicates a fair model that could be improved by more accurate prediction techniques, and a value of $RPD < 1.4$ indicates poor predictive capacity [53].

3. Results

3.1. Changes in the Canopy Spectral Reflectance and Its Relationship with the LNC for Wheat

The Yumai 34 cultivar at the various N rates used in Experiment 3 is used as an example of the analysis of the spectral variations in Figure 1A. The results show that the reflectance decreases in the visible region with increasing N concentration because of the increased absorption of the pigments and increases in the near-infrared region because of the effects of moisture and leaf structure. Further analysis of the relationships between the LNC and the reflectance determined from the original and first derivative canopy spectra was also performed (Figure 1B). A negative correlation was found in the visible region (350–710 nm) for the original spectra, whereas a positive correlation was observed in the near-infrared range (710–1410 nm), which was regarded as a higher reflectance platform ($R^2 > 0.78$, between 760 and 1100 nm). The first derivative canopy spectrum exhibited a strong correlation throughout a wavelength range that was similar to that of the original canopy spectrum but contained more prominent peaks.

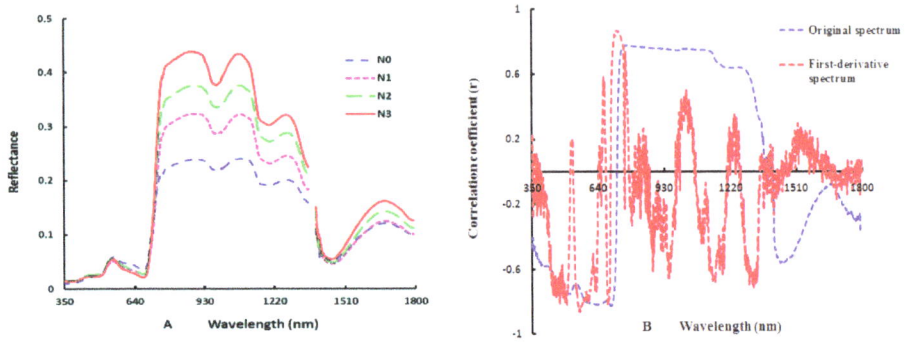

Figure 1. (**A**) Canopy spectral reflectance under four N rates at booting for Yumai 34 in Experiment 3; (**B**) Correlation of the LNC with the original and first derivative spectra.

3.2. Models for Estimating the LNC Based on Six Algorithms Using Different Numbers of Wavelengths

3.2.1. CR with One Wavelength

Figure 2A displays the original canopy spectrum, continuum line, and CR spectrum of Yumai 34 in the booting stage at an N rate of 150 kg/ha in Experiment 3. Figure 2B shows the correlation coefficients between the BD, BDR, and NBDI and the canopy LNC of the wheat. We found that the correlation coefficient between the BD and the canopy LNC exhibited a less distinct variation and that the correlation coefficients between the BDR and NBDI and the canopy LNC exhibited their lowest values between 550 and 750 nm. Figure 2B indicates that BD_{709}, BDR_{713}, and $NBDI_{727}$ showed the highest correlations.

Table 3 shows the values of the BDR, BD, and NBDI along with those of R^2_C, $RMSE_C$, R^2_V, $RMSE_P$, and RDP for the LNC model. The three CR parameters indicate a good slope value for the 1:1 line and also require little running time. Among the three indices, BD_{709} was the most effective parameter because it yielded not only the highest precision on the calibration set but also had the highest accuracy on the validation set. $NBDI_{727}$ was the least effective parameter because of its poor stability. Figure 3 shows a scatter diagram of the LNC values from the model obtained using BD_{709} from the original canopy spectra.

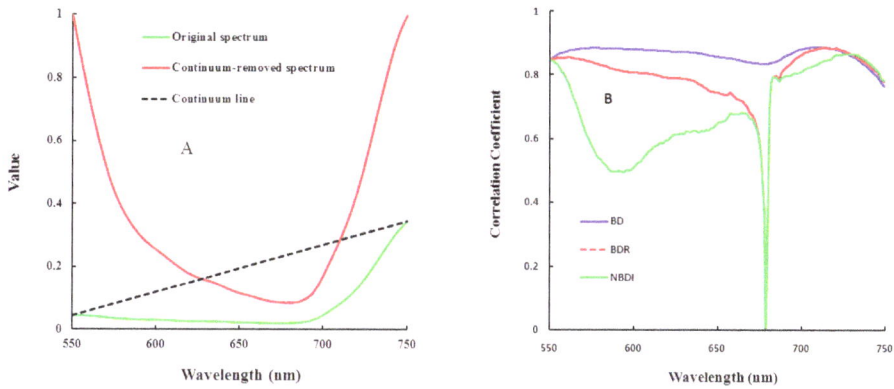

Figure 2. (**A**) The original spectrum, continuum line, and continuum-removed spectrum of Yumai 34 at the booting stage at an N rate of 150 kg/ha in Experiment 3. (**B**) Correlation coefficients between the band depth (BD), the band depth ratio (BDR), and the normalized band depth index (NBDI) and the LNC in the range of 550–750 nm.

Table 3. The best-performing LNC models based on the continuum removal (CR) parameters for the calibration and validation sets.

Band Range	Input Parameter	Calibration			Validation				
		Equation	R^2_C	$RMSE_C$	R^2_V	$RMSE_P$	RPD	CE (min)	CL
550–750	BD_{709}	$y = 0.823 \times e^{2.056x}$	0.78	0.46	0.78	0.42	1.84	0.07 min	Low
	BDR_{713}	$y = 0.536 \times e^{2.588x}$	0.78	0.48	0.74	0.45	1.72	0.08 min	Low
	$NBDI_{727}$	$y = 9.147 \times e^{2.51x}$	0.76	0.54	0.71	0.49	1.56	0.07 min	Low

Notes: BD: band depth; BDR, band depth ratio; NBDI: normalized band depth index.

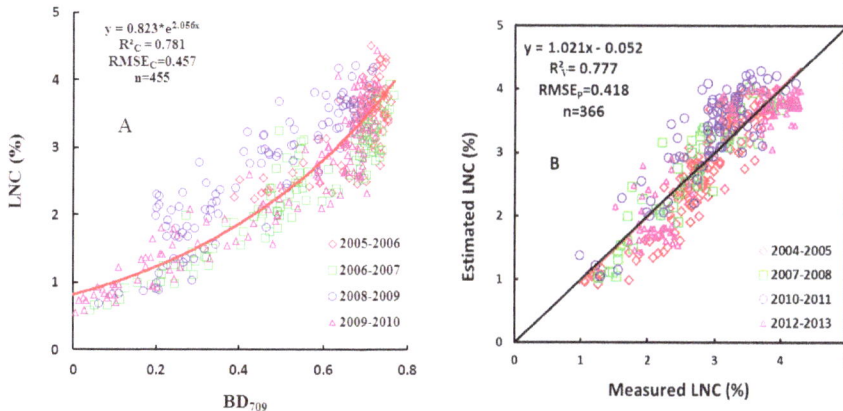

Figure 3. Calibration (**A**) and validation (**B**) of the model based on BD_{709} from the original canopy spectra.

3.2.2. VI with Two Wavelengths

Figure 4 shows the coefficients of determination (R^2) of the linear regressions between the LNC and the NDVI, RVI, and SAVI constructed from arbitrary two-band combinations based on the original and first derivative canopy spectra. The maximum R^2_C values for the NDVI, RVI, and SAVI based on the original canopy spectra were 0.830, 0.828, and 0.844, respectively, and those based on the first derivative canopy spectra were 0.858, 0.864, and 0.851, respectively. For the original spectra, the strongest correlation ($R^2 > 0.75$) between the arbitrary two-band combinations and the wheat LNC was found in the visible and near-infrared ranges. For the first derivative canopy spectra, the best band combination ($R^2 > 0.75$) was in the visible range. In the contour maps of the coefficient of determination ($R^2 > 0.5$), more regions were identified based on the original canopy spectra than were identified based on the first derivative canopy spectra.

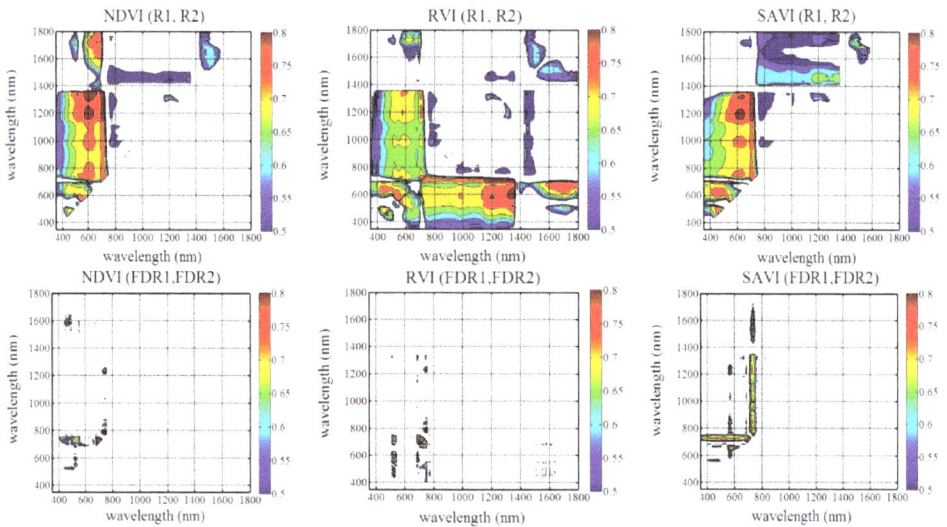

Figure 4. Contour maps of the coefficients of determination ($R^2 > 0.5$) between the normalized difference vegetation index (NDVI), ratio vegetation index (RVI), and soil-adjusted vegetation index (SAVI) and the canopy LNC based on the original and first derivative canopy spectra.

Based on the statistical parameters of R^2_C, $RMSE_C$, R^2_V, $RMSE_P$, and RDP for the calibration and validation sets and the spectrum principle, we selected the optimal wavelength and spectral index (Table 4). Table 4 shows that three types of VI$_S$ yielded better precision for the first derivative canopy spectra than for the original canopy spectra in the calibration set. The optimal wavelengths selected based on the NDVI, RVI, and SAVI were very similar. For the original spectra, the performance of SAVI(R_{1200}, R_{705}) was significantly better than that of NDVI(R_{1340}, R_{700}), which was very similar to that of RVI(R_{700}, R_{1335}). For the first derivative canopy spectra, the optimal wavelength combinations were observed between 695 and 700 nm in the visible range. According to a comprehensive evaluation of the calibration and validation performance, the SAVI obtained using the original canopy spectra performed best and exhibited good stability. In particular, the adjustable index L(L = 0.5) for the SAVI yielded superior results for the reduction of soil noise. Figure 5 shows a scatter diagram of SAVI(R_{1200}, R_{705}) and the validation performance on the original canopy spectra.

Table 4. The best-performing LNC models based on the vegetation indices (VIs) for the calibration and validation sets.

VI	λ1 (nm)	λ2 (nm)	Calibration			Validation				
			Equation	R^2_C	$RMSE_C$	R^2_V	$RMSE_P$	RPD	CE (min)	CL
NDVI	1340	700	y = 5.58x − 0.02	0.83	0.39	0.76	0.41	1.86	0.11	Low
RVI	700	1335	y = −5.63x + 4.69	0.83	0.39	0.76	0.40	1.95	0.10	Low
SAVI	1200	705	y = 8.72x + 0.10	0.84	0.38	0.80	0.38	2.01	0.10	Low
NDVI *	710	690	y = 3.59x + 1.38	0.86	0.36	0.72	0.51	1.53	0.11	Low
RVI *	700	695	y = 3.19x − 1.38	0.86	0.36	0.68	0.57	1.35	0.11	Low
SAVI *	710	695	y = 283.2x + 1.8	0.85	0.37	0.76	0.40	1.92	0.10	Low

Note: * vegetation indices calculated with the first derivatives of reflectance spectra.

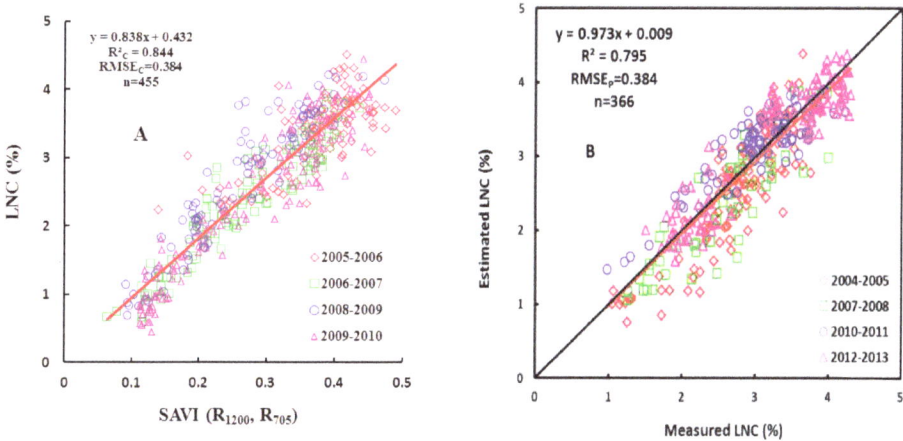

Figure 5. The relationship between SAVI(R_{1200}, R_{705}) and the wheat LNC (**A**); and the 1:1 relationship between the measured LNC and those estimated values based on SAVI(R_{1200}, R_{705}) (**B**).

3.2.3. SMLR with Multiple Wavelengths

The wavelengths selected by SMLR were 384, 492, 695, 1339, and 1369 nm for the original canopy spectra and 508, 681, 722, 960, and 1264 nm for the first derivative canopy spectra. The R^2_C values for the SMLR-OS and SMLR-FDS models were 0.869 and 0.855, respectively. The values of the statistical parameters for calibration and validation (R^2_C, $RMSE_C$, R^2_V, $RMSE_P$, and RDP) and the wavelengths selected are summarized in Table 5. The results show that SMLR based on the original canopy spectra offered a higher accuracy in the monitoring of the wheat LNC ($R^2_C = 0.869$, $RMSE_C = 0.353$, $R^2_V = 0.778$, $RMSE_P = 0.390$, RDP = 1.974); however, no significant difference was observed between the SMLR-OS and SMLR-FDS models. These two models could be expressed as follows:

SMLR-OS model:
$$y = 1.941 - 92.315 * b_{384} + 122.732 * b_{492} - 55.338 * b_{695} + 8.591 * b_{1339} - 0.321 * b_{1369} \quad (6)$$

SMLR-FDS model:
$$y = 2.189 - 980.699 * b_{FD508} - 1034.799 * b_{FD681} + 303.223 * b_{FD722} +$$
$$195.538 * b_{FD960} + 451.419 * b_{FD1264} \quad (7)$$

where b and b_{FD} represent the reflectance of the original and first derivative wavelength spectra, e.g., b_{695} is the reflectance at 695 nm and b_{FD722} is the first derivative reflectance at 722 nm.

Table 5. The best-performing LNC models based on stepwise multiple linear regressions (SMLR) for the calibration and validation sets.

Model	Selected Wavelengths (nm)	Calibration		Validation				
		R^2_c	$RMSE_C$	R^2_v	$RMSE_P$	RDP	CE (min)	CL
SMLR-OS	695, 1339, 492, 384, 1369	0.87	0.35	0.78	0.39	1.97	32.15	Middle
SMLR-FDS	722, 681, 1264, 508, 960	0.86	0.37	0.76	0.39	1.95	33.16	Middle

Notes: OS: original canopy spectra; FDS: first derivative canopy spectra.

3.2.4. PLSR with All Wavelengths

Figure 6 shows the changes in the percentage variance captured with an increasing number of latent variables (LVs) using the original and first derivative canopy spectra. When the number of latent variables (LVs) was greater than five or seven for the original or first derivative spectra, respectively, the percentage variance captured by the model decreased only minimally. Therefore, we selected five and seven latent variables (LVs) for the PLSR analyses based on the original and first derivative canopy spectra, respectively.

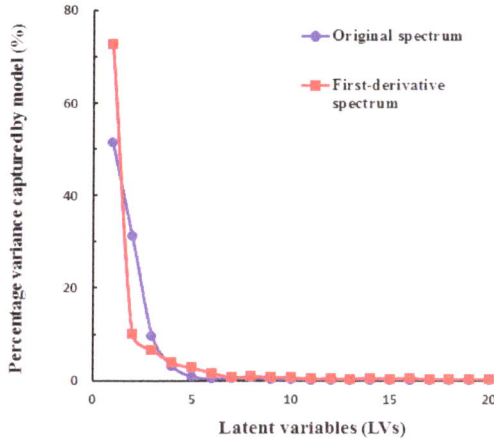

Figure 6. Changes in the variance explained by the latent variables (LVs) based on the original and first derivative canopy spectra.

The results of the PLSR analyses are shown in Table 6. With all wavelengths used as input variables, the PLSR analysis based on the first derivative canopy spectra (FDS) demonstrated a higher estimation accuracy for the canopy LNC than did the analysis based on the original canopy spectra for both the calibration and validation sets, with statistical parameters of $R^2_C = 0.908$, $RMSE_C = 0.298$, $R^2_V = 0.815$, $RMSE_P = 0.385$, and RDP = 2.000. Figure 7 shows the results of predicting the LNC for winter wheat based on the calibration and validation sets using the PLSR-FDS model.

Table 6. The best-performing LNC models based on partial least-squares regression (PLSR) for the calibration and validation sets.

Model	Input Variables	Calibration			Validation				
		LVs	R^2_c	$RMSE_C$	R^2_v	$RMSE_P$	RDP	CE (min)	CL
PLSR-OS	all wavelengths	5	0.85	0.37	0.81	0.35	2.22	6.10	High
PLSR-FDS	all wavelengths	7	0.91	0.30	0.815	0.39	2.00	5.50	High

Notes: OS: original canopy spectra; FDS: first derivative canopy spectra.

3.2.5. ANN with All Wavelengths

Figure 8 shows the changes in $RMSE_P$ as a function of the number of hidden layer neurons (HLNs). The results indicate that the value of $RMSE_P$ is lowest when the number of hidden neurons is equal to twelve. Therefore, we selected 12 as the optimal number of HLNs for the ANN analyses based on the original and first derivative canopy spectra.

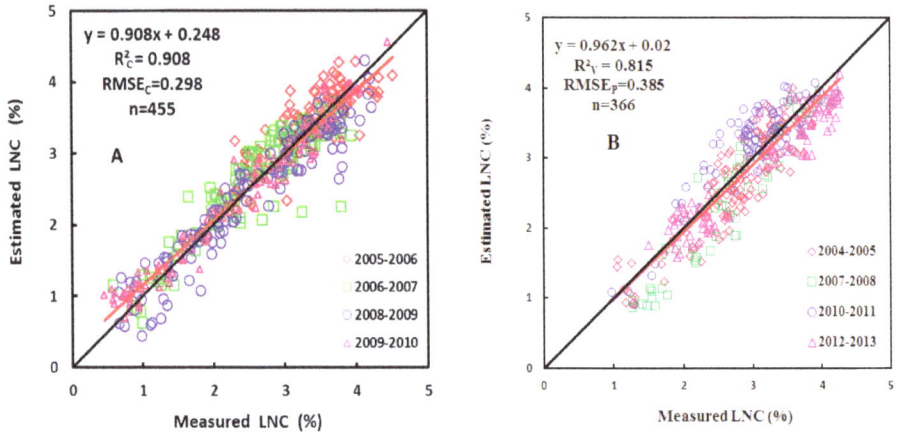

Figure 7. The 1:1 relationship between the measured LNC and those estimated values using the PLSR analysis on the first derivative canopy spectra (PLSR-FDS) model for the calibration (**A**) and validation (**B**) sets.

Table 7 shows the results of the ANN-based LNC models for both the calibration and validation sets. According to Table 7, when all wavelengths were used as input variables for the ANN analysis, the ANN-FDS model offered a higher estimation accuracy for LNC monitoring than did the ANN-OS model for the calibration set. However, for the validation set, the ANN-OS model exhibited higher estimation accuracy than the ANN-FDS model and the slope value for the ANN-OS model was closer to 1 than that for the ANN-FDS model. Overall, the model based on all wavelengths in the first derivative canopy spectra yielded the higher estimation accuracy for the calibration set ($R^2_C = 0.987$, $RMSE_C = 0.111$), but for the validation set, it exhibited the lower estimation accuracy ($R^2_V = 0.734$, $RMSE_P = 0.512$, $RDP = 1.504$). The difference in performance between the calibration and validation sets indicates that the ANN method appears to suffer from overfitting when many input variables are used.

Table 7. The best-performing artificial neural networks (ANNs) -based LNC models for the calibration and validation sets.

Inputs	Optimal Numbers of Neurons			Calibration		Validation				
	Input	Hidden	Output	R^2_C	$RMSE_C$	R^2_V	$RMSE_P$	RDP	CE (min)	CL
ANN-OS	1451	12	1	0.95	0.22	0.76	0.45	1.72	71.50	High
ANN-FDS	1451	12	1	0.99	0.18	0.73	0.51	1.50	67.20	High
ANN-PCA-OS	9	4	1	0.94	0.25	0.796	0.35	1.44	15.60	High
ANN-PCA-FDS	11	5	1	0.95	0.22	0.72	0.41	1.48	14.80	High

Note: OS: original canopy spectra; FDS: first derivative canopy spectra; ANN-PCA-OS indicates that we used PCA to select the primary factor and then used the PCA-derived factor to execute the ANN model.

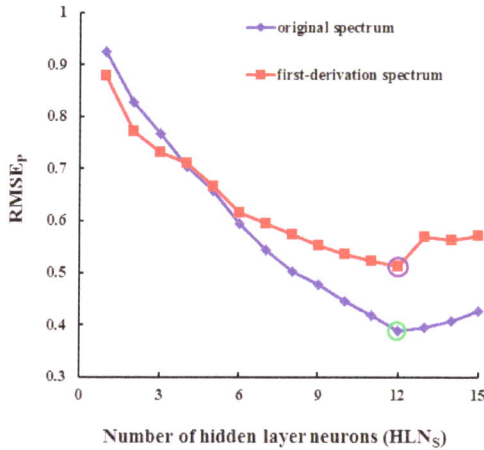

Figure 8. Changes in RMSE$_P$ as a function of the number of hidden layer neurons (HLNs) for the original and first derivative canopy spectra.

3.2.6. SVM with All Wavelengths

Table 8 summarizes the performance of the SVM-based LNC models with different input variables. The results show that the SVM-based models using all wavelengths in the first derivative spectra demonstrated better performance on the calibration set; however, the SVM-OS model offered slightly better performance on the validation set, with slightly higher R^2V and RDP values and a shorter running time. Figure 9 shows the 1:1 relationship between the measured LNC and those estimated using the SVM-FDS model for the calibration and validation sets.

Table 8. The best-performing support vector machines (SVMs)-based LNC models for the calibration and validation sets.

Model	Input Variables	Calibration		Validation				
		R^2C	RMSE$_C$	R^2V	RMSE$_V$	RDP	CE (min)	CL
SVM-OS	All wavelengths	0.96	0.21	0.80	0.38	2.05	20.34	High
SVM-FDS	All wavelengths	0.96	0.19	0.78	0.37	2.02	21.17	High
ANN-PCA-OS	9PCA	0.94	0.20	0.67	0.47	1.64	5.64	High
ANN-PCA-FDS	11PCA	0.92	0.27	0.55	0.57	1.36	5.76	High

Notes: OS, original canopy spectra; FDS, first derivative canopy spectra.

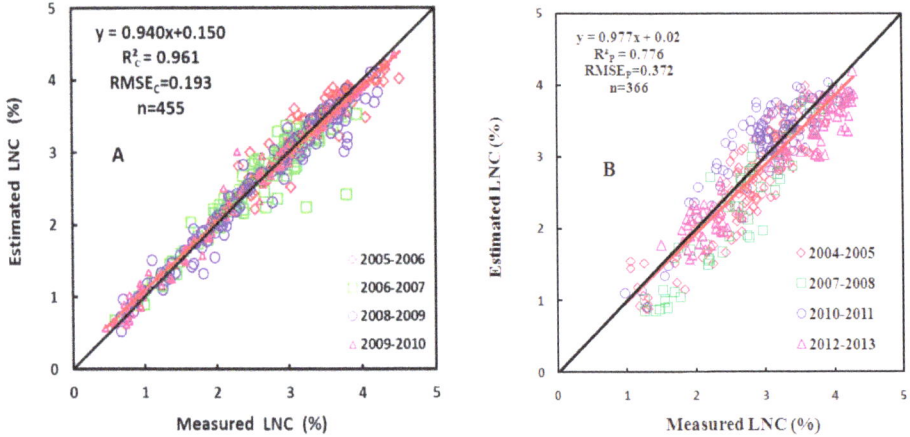

Figure 9. The 1:1 relationship between the measured LNC and those estimated using the SVM-FDS model for the calibration (**A**) and validation (**B**) sets.

3.3. Evaluation and Comparison of the Robustness of the Six Algorithms

We compared the robustness of the six algorithms based on the statistical parameters R^2_C, $RMSE_C$, R^2_V, $RMSE_P$, CE and CL (Table 9). The results show that with an increasing number of wavelengths, the value of R^2_C increased from 0.78 for BD_{709} to 0.96 for SVM-FDS. However, the value of R^2_V did not exhibit a similar increase. The CR algorithms used only one wavelength and demonstrated the poorest performance on both the calibration and validation sets, although they also required less running time and had lower complexity. The $SAVI(R_{1200}, R_{705})$ method required only two wavelengths and offered better performance on both the calibration and validation sets ($R^2_C = 0.844$, $RMSE_C = 0.384$, $R^2_V = 0.795$, $RMSE_P = 0.384$, $RDP = 2.005$, and running time = 0.10 min). The SMLR-OS method used five wavelengths, whereas the PLSR-FDS, ANN-OS and SVM-FDS methods used all available wavelengths. Although PLSR-FDS demonstrated the best R^2_V performance on the validation set, with a value of 0.82, the errors in calibration and validation sets were higher than those for SVM-FDS. Therefore, the SVM-based method yielded a higher prediction accuracy than the other methods on the calibration set ($R^2_C = 0.961$, $RMSE_C = 0.193$, $R^2_V = 0.776$, $RMSE_P = 0.382$, $RDP = 2.016$, and running time = 21.17 min). In addition, we found that with an increasing number of wavelengths, the running time increased; the BD_{709} method exhibited the shortest running time (0.07 min), whereas the ANN-OS method required the longest running time (71.50 min), and the operational complexity also correspondingly increased. With regard to the slope of the 1:1 line, the SMLR-OS method yielded the smallest slope value, whereas the BD_{709} method produced the greatest slope value. The SVM-FDS and $SAVI(R_{1200}, R_{705})$ methods offered higher accuracy. However, SVM-FDS incurred a higher cost, as reflected in its use of multiple wavelengths, its higher complexity level, and its longer running time.

Table 9. The robustness evaluation of the wheat LNC models based on the six considered algorithms.

Method	Wavelengths (nm)	Calibration			Validation			
		R^2_C	$RMSE_C$	R^2_V	$RMSE_P$	RDP	CE (min)	CL
BD_{709}	709	0.78	0.46	0.78	0.42	1.84	0.07	Low
$SAVI(R_{1200}, R_{705})$	1200, 705	0.84	0.38	0.80	0.38	2.01	0.10	Low
SMLR-OS	695, 1339,492,384,1369	0.87	0.35	0.78	0.39	1.97	33.16	Middle
PLSR-FDS	All wavelengths	0.91	0.30	0.82	0.39	2.00	5.50	High
ANN-OS	All wavelengths	0.95	0.22	0.76	0.45	1.72	71.50	High
SVM-FDS	All wavelengths	0.96	0.19	0.78	0.37	2.02	21.17	High

We further categorized the samples using three grouping variables (variety, ecological site, and growth stage) to compare the robustness of the optimal LNC model algorithms (Table 10). The results show that the prediction accuracy was always improved with an increasing number of wavelengths for each of the three grouping variables. However, the CE and CL also substantially increased. The results also show that the six algorithms were suitable and robust for the Ningmai 9 and Shengxuan 6 varieties, with maximum R^2_V values of 0.86 and 0.88, respectively, and that the SVM-FDS algorithm offered the best overall performance, with a mean R^2_V value of 0.79 for all five varieties. However, a suitable algorithm could not be found for the Yangmai 12 and Yumai 34 varieties, for which the R^2_V values ranged from 0.65 to 0.80. These results demonstrate that the Ningmai 9 variety represents a generally adaptable variety and that the SVM-FDS method may be better suited than the other methods to cope with potential confounding factors for most varieties. Of the two ecological-site-based groups, Rugao yielded better results than did Nanjing for all six algorithms, with R^2_V values ranging from 0.80 to 0.90. The robustness of the PLSR-FDS and SVM-FDS methods was particularly strong; these methods were suitable for both ecological sites, with R^2_V values of 0.85 and 0.84, respectively. For the two growth-stage-based groups, the six algorithms all yielded better and more stable results for the stage of heading and anthesis, with R^2_V values ranging from 0.78 to 0.86. The statistical parameters indicated poorer performance in the stage of jointing and booting.

Table 10. Robustness of the LNC models based on the six algorithms when the samples are categorized using three grouping variables (variety, ecological site, and growth stage).

Grouping Variable	Algorithm	Sub-Group	Validation				R^2_V–RMSEP
			R^2_V	$RMSE_P$	CE (min)	CL	
Variety	BD_{709}	Ningmai 9	0.84	0.37	0.07	Low	0.47
		Yangmai 12	0.73	0.38	0.07	Low	0.36
		Yumai 34	0.80	0.28	0.07	Low	0.52
		Yangmai 18	0.71	0.48	0.07	Low	0.23
		Shengxuan 6	0.76	0.43	0.07	Low	0.33
	$SAVI(R_{1200}, R_{705})$	Ningmai 9	0.80	0.43	0.10	Low	0.37
		Yangmai 12	0.75	0.41	0.10	Low	0.34
		Yumai 34	0.75	0.36	0.10	Low	0.40
		Yangmai 18	0.86	0.31	0.10	Low	0.55
		Shengxuan 6	0.77	0.36	0.10	Low	0.41
	SMLR-OS	Ningmai 9	0.83	0.47	16.23	Middle	0.36
		Yangmai 12	0.67	0.40	15.21	Middle	0.27
		Yumai 34	0.67	0.35	17.32	Middle	0.32
		Yangmai 18	0.81	0.36	16.46	Middle	0.45
		Shengxuan 6	0.77	0.38	15.35	Middle	0.39
	PLSR-FDS	Ningmai 9	0.86	0.38	5.21	High	0.48
		Yangmai 12	0.76	0.34	5.74	High	0.42
		Yumai 34	0.67	0.32	5.32	High	0.35
		Yangmai 18	0.81	0.36	5.21	High	0.45
		Shengxuan 6	0.88	0.35	5.56	High	0.53
	ANN-OS	Ningmai 9	0.85	0.44	65.23	High	0.41
		Yangmai 12	0.78	0.44	64.32	High	0.34
		Yumai 34	0.65	0.36	65.45	High	0.30
		Yangmai 18	0.76	0.42	63.23	High	0.34
		Shengxuan 6	0.77	0.55	62.89	High	0.22
	SVM-FDS	Ningmai 9	0.84	0.40	19.21	High	0.44
		Yangmai 12	0.79	0.36	18.32	High	0.43
		Yumai 34	0.67	0.42	18.21	High	0.25
		Yangmai 18	0.79	0.36	19.72	High	0.43
		Shengxuan 6	0.87	0.35	19.32	High	0.52
Ecological site	BD_{709}	Nanjing	0.76	0.42	0.08	Low	0.34
		Rugao	0.80	0.41	0.08	Low	0.39
	$SAVI(R_{1200}, R_{705})$	Nanjing	0.78	0.40	0.11	Low	0.38
		Rugao	0.85	0.34	0.11	Low	0.51
	SMLR-OS	Nanjing	0.77	0.42	28.22	Middle	0.35
		Rugao	0.86	0.37	12.32	Middle	0.49
	PLSR-FDS	Nanjing	0.80	0.38	5.86	High	0.42
		Rugao	0.90	0.31	5.30	High	0.59
	ANN-OS	Nanjing	0.76	0.45	68.21	High	0.31
		Rugao	0.83	0.45	62.20	High	0.39
	SVM-FDS	Nanjing	0.77	0.41	19.98	High	0.36
		Rugao	0.90	0.33	18.32	High	0.57
Growth stage	BD_{709}	Jointing, Booting	0.73	0.40	0.08	Low	0.33
		Heading, Anthesis	0.78	0.42	0.08	Low	0.36
	$SAVI(R_{1200}, R_{705})$	Jointing, Booting	0.75	0.37	0.11	Low	0.38
		Heading, Anthesis	0.80	0.40	0.11	Low	0.40
	SMLR-OS	Jointing, Booting	0.71	0.39	17.82	Middle	0.32
		Heading, Anthesis	0.83	0.44	17.86	Middle	0.39
	PLSR-FDS	Jointing, Booting	0.73	0.37	5.78	High	0.36
		Heading, Anthesis	0.85	0.35	5.83	High	0.50
	ANN-OS	Jointing, Booting	0.66	0.46	68.68	High	0.20
		Heading, Anthesis	0.86	0.43	67.98	High	0.43
	SVM-FDS	Jointing, Booting	0.75	0.36	18.79	High	0.39
		Heading, Anthesis	0.85	0.49	18.89	High	0.36

3.4. Performance Comparison of the Best Models Identified in the Present Study with Previous Models

To determine whether the estimation models established in the present study based on the SAVI and SVM approaches were comparable to previously reported LNC models for wheat, all of the observed calibration and validation data considered in the present study were used to compare the performance of these models with those proposed in previous reports (Table 11, [54–57]). The results showed that the SAVI and SVM models not only exhibited better performance on the calibration set, with R^2_C values of 0.844 and 0.961, respectively, but also offered higher prediction accuracy on the validation set, with R^2_V values of 0.795 and 0.776, respectively. In addition, the $RMSE_C$, $RMSE_P$, and RPD values also demonstrated that the model based on the SASI exhibited higher stability and reliability. Therefore, the SAVI calculation is a potentially useful algorithm for monitoring wheat canopy LNC that offers almost identically high levels of prediction accuracy, stability, and complexity while requiring fewer wavelengths and less running time.

Table 11. Comparison of the SAVI(R_{1200}, R_{705}) and SVM approaches with previous models for LNC estimation.

Method	Equation	Calibration		Validation			Source
		R^2_C	$RMSE_C$	R^2_V	$RMSE_P$	RPD	
OSAVI	$1.16*(R_{810} - R_{680})/(R_{810} - R_{680} + 0.16)$	0.74	0.49	0.66	0.73	1.06	Rondeaux *et al.* (1996) [54]
ND$_{705}$	$(R_{750} - R_{705})/(R_{750} + R_{705})$	0.79	0.44	0.73	0.42	1.83	Gitelson *et al.* (1994) [55]
	$(R_{924} - R_{703} + 2*R_{423})/(R_{924} - R_{703} - 2*R_{423})$	0.79	0.42	0.72	0.45	1.71	Wang *et al.* (2012) [56]
mND$_{705}$	$(R_{750} - R_{705})/(R_{750} + R_{705} - 2*R_{445})$	0.80	0.43	0.74	0.41	1.90	Sims *et al.* (2002) [57]
SAVI	$1.5*(R_{1200} - R_{705})/(R_{1200} + R_{705} - 0.5)$	0.84	0.38	0.80	0.38	2.01	This paper
SVM	-	0.96	0.19	0.78	0.37	2.02	This paper

4. Discussion

4.1. Wavelength Selection for the Six Algorithms

According to previous reports, the most informative feature bands may differ in different crop types and experimental conditions. Therefore, the selection and exploration of new key-wavebands is an important task in the field of the remote sensing of vegetation and has been performed for a number of different cases [16]. Further investigations are needed to identify consistent feature bands with wider applicability for the estimation of the N concentration in crops. In the present study, all possible two-wavelength combinations of hyperspectral indices throughout the entire spectral range of 350–1800 nm were considered in matrix form. Based on the R^2 and RMSE values and the absorption principle, we found that the wavelengths selected by the CR and VI methods were 690/695, 709/710, 700/705, 713/727, 1200, and 1335/1340 nm, which are predominantly located in the red-edge and near-infrared regions, as noted in many previous studies [16,17,26,58,59]. The selected wavelengths differed for the CR and VIs methods, perhaps because CR can be used to determine the absorbing positions of chlorophyll or carotenoids, whereas for the VI-based approach, the much more sensitive wavelength of N can be used because of the different calculation formulas for the two spectral indices. These wavelengths are suitable for estimating the canopy LNC because they are less sensitive to soil background and atmospheric effects and are strongly absorbed by plant chlorophyll and carotenoids for photosynthetic production and thus can be regarded as representative spectral wavelengths [60]. The corresponding spectral indices (BD_{709}, BDR_{713}, $NBDI_{727}$, NDVI (R_{1340}, R_{700}), SAVI(R_{1200}, R_{705}), and NDVI (FD_{1340}, FD_{700}) were constructed for wheat LNC estimation, and these indices demonstrated good performance. Thus, these key wavelengths and indices should be regarded as new alternatives to the previously reported indicator wavelengths used to monitor the LNC of crop plants.

For the PLSR, ANNs, and SVMs algorithms, we used all wavelengths in the original and first derivative canopy spectra as input variables to select the best bands and input variable to construct the multivariate linear model for canopy LNC monitoring. When using the SMLR method, we chose five wavelengths from the original and first derivative canopy spectra to predict the wheat LNC. The selected wavelengths were 384, 492, 695, 1339, and 508 nm and 681, 722, 960, 1264 and 1369 nm, respectively. The wavelengths of 492 and 508 nm lie in the visible range and are often strongly absorbed by plant chlorophyll and carotenoids in green plants [42]. The wavelengths of 681, 695, and 722 nm lie in the red range and are sensitive indicators of the LNC and chlorophyll [16,17,26,58,59]. The wavelengths of 960, 1264, 1339, and 1369 nm are located in the shortwave infrared range and are indicators of proteins [27]. Atzberger [60] has reported that a close relationship exists between the N and chlorophyll concentrations as well as between the N and protein concentrations. Therefore, many researchers have used these relationships to monitor the LNC in crops based on crop canopy spectra [61].

4.2. The Reliability and Practicability of the Six Algorithms

The result indicated that the VIs are superior to the CR parameters for canopy LNC monitoring because of their good precision, high stability, shorter running time and lower level of operational complexity, similar to previous results [48]. Because the noise had little chance to cancel out when only two bands were used for modeling. Indeed, with this better index (SAVI(R_{1200}, R_{705})) one band is still located on the near-infrared (1200 nm), however, the second band is located at 720 nm, and thus in the red-edge where the chlorophyll absorption is strongly reduced compared to the red wavelength. This increases the sensitivity of the index and explains the relatively good results obtained in this study. Another advantage for the VIs is that was used as a baseline approach. The advantage of the VI_S method is that it is easily implemented in stand (image) processing software. However, using the VIs with only part of the available spectral information (*i.e.*, two bands) resulted in a strong loss of predictive power, and the classical NDVI easily saturated explaining the poor performance of this widely used indicator.

For the SMLR, PLSR, ANNs, and SVMs methods, when all wavelengths were used as input variables, these models showed higher precisions than that of the SAVI(R_{1200}, R_{705}) model, which requires only two wavelengths, for LNC estimation on the calibration set in the following order: SVM-FDS > ANN-FDS > PLSR-FDS > SMLR-OS. However, their stability in terms of validation performance was generally not as good, with an overall ranking of PLSR-FDS > SAVI(R_{1200}, R_{705}) > SVM-FDS = SMLR-OS > ANN-FDS, which may have resulted from overfitting in the multiple regression methods. The advantage for the PLSR, ANNs, and SVMs was not easily saturated which explain the good performance, and demonstrates the potential of chemometric techniques for mapping some important biophysical variable. However, many software packages don't yet include routines for calibrating and applying those models.

Among the six algorithms, as the number of wavelengths increased, the value of R^2_C also increased on the calibration set. However, the value of R^2_V on the validation set did not increase, indicating that the stability of all algorithms was not good, which is consistent with the results of a study by Qi [62]. In addition, the running times exhibited an increasing trend with an increasing number of wavelengths, with the BD method requiring the shortest and the ANN-based model requiring the longest running time; a corresponding increase in operational complexity was also observed. These results indicate that for the design of future portable spectrometer instruments with low cost and high accuracy for LNC monitoring, the SAVI approach may be the best choice. However, for the development of a software program executed by a computer, the SVM-based algorithm is a better selection.

4.3. The Applicability of the Six Algorithms to Different Groups of Samples

It is well known that statistical models developed for specific applications sometimes lack transferability to other sites with different vegetation or to other types of image or acquisition conditions [60,61]. Additional disadvantages of statistical models include the facts that they require a set of *in situ* data and that their robustness depends on the properties of these datasets (*i.e.*, the number, quality and representativeness of the available reference samples), especially when extrapolated to other varieties, ecological sites, and growth stages [10,12,15–17,20,25]. However, statistical models offer certain advantages that promote their widespread use. For example, several of the cited statistical models are easy to apply. In addition, suitable software is often readily available [62–64]. This study was conducted on field experimental data acquired over nine consecutive years that included seven varieties of wheat, four eco-sites, and 455 samples in the calibration set and 366 samples in the validation set, corresponding to different N levels and growth stages. Through a systematic analysis, we compared the performance of the six algorithms. The selected samples were highly representative, and the findings may be applicable to other sites or similar crops, including the other crops.

The results presented here also indicate that the PLSR-FDS method and especially the SVM-FDS method may be better suited than the other methods to cope with potential confounding factors for most varieties. The SMLR-OS and ANN-OS methods exhibited the worst performance, as indicated by the fact that they yielded the lowest R^2-$RMSE_P$ values and mid-to-high computational efficiency. In the future, the newly developed algorithms should be adapted to the Yangmai 12 and Yumai 34 varieties, which mostly showed the worst performance for all of the algorithms. Regarding to the differences among the six models at two ecological sites, the Rugao location yielded better performance than did the Nanjing location. This may have occurred because the data collected at the Nanjing sites contained more noise produced by clouds than that from the Rugao sites. However, this result should be confirmed in the future. The robustness of the PLSR-FDS method was sufficiently strong that it displayed good performance for both ecological sites, which is consistent with findings of previous studies conducted at various ecological sites [16]. Previous researchers have reported that LNC models tend to yield varying results at different growth stages, with better performance in the later growth stage [65]. In this paper, the six algorithms also exhibited better and more stable results in the later stage of growth than in the early stage. This may have occurred because of the noise generated by the soil background exposed by the open canopy during the early growth stage [14]. The relatively good LNC correlations that we observed suggest that the SVM and SAVI methods could be applied across different varieties, ecological sites and growth stages without extensive calibration.

5. Conclusions

In this study, we demonstrated the performance, advantages, shortcomings, and robustness of six statistical modeling approaches for wheat canopy LNC. The PLSR-FDS, ANN-OS and SVM-FDS methods yield similar accuracies with SVM-FDS as the best if the CE and CL are not considered, however, ANNs and SVMs performed better on calibration set than the validation set which indicate that we should take more caution with the two methods for over-fitting. Except PLS method, the performance for most methods did not enhance when the spectrum were operated by the first derivative. The prediction accuracy was found to be higher when more wavelengths were used, though at the cost of a lower CE. Moreover, the evaluation of the robustness demonstrates that SVMs method may be better suited than the other methods to cope with potential confounding factors for most varieties, ecological site and growth stage. However, when the estimation accuracy, the CE, the number of wavelengths, and the CL of each model are systematically considered for the design of hardware devices, the SAVI(R_{1200}, R_{705}) model is found to be the best option for estimating the LNC in wheat. Although it might generally be preferable to make use of the full spectral resolution, our study demonstrated that even with two spectral bands, it is possible to (locally) obtain very good results. Hence, it remains to be proven that the full wavelength spectrum contains substantially more information than do narrow-band vegetation indices.

The current study focused on the six most widely used algorithms for the considered task. The results of this study are of interest to the remote sensing community for the development of improved inversion schemes for hyperspectral applications concerning other types of vegetation using empirical models, such as mapping important vegetation biophysical properties of other crops. The examples provided in this paper may also serve as illustrations of the advantages and disadvantages of empirical models. Although statistical models have been developed and successfully applied across various growth stages, varieties and eco-sites, the use of these methods is not always possible. Those methods in this paper established for vegetation variable retrieval, which are frequently applied in terrestrial bio-physical products, proving a high potential of hyperspectral measurement in the future. Because our study was performed using a specific dataset, our findings necessarily have certain limitations in applicability. In order to develop accurate, robust and fast model with high reliability, practicability and applicability, the next step should be to confirm these findings for a broader range of species and environments. A simulation experiment based on synthetic spectra generated by physically-based radiative transfer model will be conducted. Physical accuracy estimates are mandatory and should be provided using comprehensive validation datasets collected on more various sites and varieties. Except parametric regression and non-parametric regression, the hybrid methods combine generic capability of physically-based methods with flexible and computationally efficient methods should be tested. What is more, the impact of feature selection and randomly generated noise should be considered to study the stability of the developed statistical models to unfavorable measuring conditions with different sites and varieties in the future. Additionally, the theoretical uncertainties of the biophysical parameter products should be analysis in the study. The associated uncertainty estimates also provide information on the success of transporting a locally trained model to other sites and/or observation conditions, which are not intended to replace true accuracy estimates, but instead provide complementary information.

Acknowledgments: This work was supported by the National Natural Science Foundation of China (31201131, 31201130, 31371534, 31470084), the Special Program for Agriculture Science and Technology of the Ministry of Agriculture in China (201303109), the Natural Science Foundation of Jiangsu Province (BK20141371), the Jiangsu Collaborative Innovation Center for Modern Crop Production (JCICMCP), the Academic Program Development of Jiangsu Higher

Education Institutions (PAPD), the Fundamental Research Funds for the Central Universities (KYRC201401, KYZ201202-8), and the Experiment and Technology Talent Funding in the Platform of Science and Technology (STTFPST-Njau). We acknowledge the support provided by the National Engineering and Technology Center for Information Agriculture in Nanjing. We are very grateful to Yingxue Li, Wei Feng, Yan Huang, Haijian Ren and Gang Han for their assistance with data collection. Finally, we acknowledge the anonymous reviewers who provided useful comments regarding this manuscript.

Author Contributions: Xia Yao and Yu Huang analyzed the data and wrote the manuscript. Tao Cheng, Guiyan Shang, Chen Zhou, Yongchao Tian, Weixing Cao and Yan Zhu offered comments on and proofread the manuscript. Xia Yao provided the data and the means of data acquisition.

Conflicts of Interest: The authors declare no conflicts of interest.

References

1. Clevers, J.G.P.W.; Gitelson, A.A. Remote estimation of crop and grass chlorophyll and nitrogen content using red-edge bands on Sentinel-2 and -3. *Int. J. Appl. Earth Obs. Geoinform.* **2013**, *23*, 344–351.

2. Thomas, J.R.; Oerther, G.F. Estimating nitrogen content of sweet pepper leaves by reflectance measurements. *Agron. J.* **1972**, *64*, 11–13.

3. Jensen, A.; Lorenzen, B.; Østergaard, H.S.; Hvelplund, E.K. Radiometric estimation of biomass and nitrogen content of barley grown at different nitrogen levels. *Int. J. Remote Sens.* **1990**, *11*, 1809–1820.

4. Yoder, B.J.; Pettigrew-Crosby, R.E. Predicting nitrogen and chlorophyll content and concentrations from reflectance spectra (400–2500 nm) at leaf and canopy scales. *Remote Sens. Environ.* **1995**, *53*, 199–211.

5. Kokaly, R.; Clark, R.N. Spectroscopic determination of leaf biochemistry using band-depth analysis of absorption features and stepwise multiple linear regression. *Remote Sens. Environ.* **1999**, *67*, 267–287.

6. Pinter, P.J.; Hatfield, J.L.; Schepers, J.S.; Barnes, E.M.; Moran, M.S.; Daughtry, C.S.T.; Craig, S.T.; Upchurch, D.R. Remote sensing for crop management. *Photogramm. Eng. Remote Sens.* **2003**, *69*, 647–664.

7. Dobermann, A.; Blackmore, S.; Cook, S.E.; Adamchuk, V.I. Precision Farming: Challenges and Future Directions. In Proceedings of the 4th International Crop Science Congress, Brisbane, Australia, 26 September–1 October 2004.

8. Huang, Z.; Turner, B.J.; Dury, S.J.; Wallis, I.R.; Foley, W.J. Estimating foliage nitrogen concentration from HYMAP data using continuum removal analysis. *Remote Sens. Environ.* **2004**, *93*, 18–29.

9. Ladha, J.K.; Pathak, H.; Krupnik, T.J.; Six, J.; van Kessel, C. Efficiency of fertilizer nitrogen in cereal production: Retrospects and prospects. *Adv. Agron.* **2005**, *87*, 85–156.

10. Zhu, Y.; Li, Y.X.; Zhou, D.Q.; Tian, Y.C.; Yao, X.; Cao, W.X. Quantitative relationship between leaf nitrogen concentration and canopy reflectance spectra in rice and wheat. *Acta Ecol. Sin.* **2006**, *26*, 3463–3469.

11. Haboudane, D.; Tremblay, N.; Miller, J.R.; Vigneault, P. Remote estimation of crop chlorophyll content using spectral indices derived from hyperspectral data. *IEEE Trans. Geosci. Remote Sens.* **2008**, *46*, 423–437.

12. Tian, Y.C.; Yao, X.; Yang, J.; Cao, W.X.; Hannaway, D.B.; Zhu, Y. Assessing newly developed and published vegetation indices for estimating rice leaf nitrogen concentration with ground- and space-based hyperspectral reflectance. *Field Crops Res.* **2011**, *120*, 299–310.

13. Ecarnot, M.; Compan, F.; Roumet, P. Assessing leaf nitrogen content and leaf mass per unit area of wheat in the field throughout plant cycle with a portable spectrometer. *Field Crops Res.* **2013**, *140*, 44–50.

14. Yao, X.; Ren, H.; Cao, Z.; Tian, Y.; Cao, W.; Zhu, Y.; Cheng, T. Detecting leaf nitrogen content in wheat with canopy hyperspectrum under different soil backgrounds. *Int. Appl. Earth Obs. Geoinform.* **2014**, *32*, 114–124.

15. Curran, P.J.; Dungan, J.L.; Peterson, D.L. Estimating the foliar biochemical concentration of leaves with reflectance spectrometry. *Remote Sens. Environ.* **2001**, *76*, 349–359.

16. Hansen, P.M.; Schjoerring, J.K. Reflectance measurement of canopy biomass and nitrogen status in wheat crops using normalized difference vegetation indices and partial least squares regression. *Remote Sens. Environ.* **2003**, *86*, 542–553.

17. Li, Y.X.; Zhu, Y.; Tian, Y.C.; Yao, X.; Qin, X.D.; Cao, W.X. Quantitative relationship between leaf nitrogen concentration and canopy reflectance spectra in wheat. *Acta Agron. Sin.* **2006**, *32*, 358–362.

18. Darvishzadeh, R.; Skidmore, A.; Atzberger, C. Estimation of vegetation LAI from hyperspectral reflectance data: Effects of soil type and plant architecture. *Int. J. Appl. Earth Obs. Geoinform.* **2008**, *10*, 358–373.

19. Darvishzadeh, R.; Skidmore, A.; Schlerf, M.; Atzberger, C.; Corsi, F.; Cho, M. LAI and chlorophyll estimation for a heterogeneous grassland using hyperspectral measurements. *ISPRS J. Photogramm.* **2008**, *63*, 409–426.

20. Vigneau, N.; Ecarnot, M.; Rabatel, G.; Roumet, P.; Roumet, P. Potential of field hyperspectral imaging as a non-destructive method to assess leaf nitrogen content in wheat. *Field Crops Res.* **2011**, *122*, 25–31.

21. Clark, R.N.; Roush, T.L. Reflectance spectroscopy: Quantitative analysis techniques for remote sensing applications. *J. Geophys. Res.* **1984**, *89*, 6329–6340.

22. Kokaly, R.F. Investigating a physical basis for spectroscopic estimates of leaf nitrogen concentration. *Remote Sens. Environ.* **2001**, *75*, 153–161.

23. Mutanga, O.; Skidmore, A.K.; Prins, H.H.T. Predicting *in situ* pasture quality in the Kruger National Park, South Africa, using continuum-removed absorption features. *Remote Sens. Environ.* **2004**, *89*, 393–408.

24. Gitelson, A.; Merzlyak, M.N. Spectral reflectance changes associated with autumn senescence of *Aesculus hippocastanum* L. and *Acer platanoides* L. leaves. Spectral features and relation to chlorophyll estimation. *J. Plant Physiol.* **1994**, *143*, 286–292.

25. Chen, P.F.; Wang, J.S.; Peng, P. Remote detection of wheat grain protein content using nitrogen nutrition index. *Soc. Agric. Eng.* **2011**, *9*, 75–80.

26. Yao, X.; Zhu, Y.; Tian, Y.C.; Feng, W.; Cao, W.X. Research of optimum hyperspectral vegetation indices on monitoring the nitrogen content in wheat leaves. *Sci. Agric. Sin.* **2009**, *42*, 2716–2725.

27. Hatchell, D. *(ASD) Technical Guide*; Analytical Spectral Devices. Inc.: Boulder, CO, USA, 1999.

28. Lin, F.F.; Qiu, L.F.; Deng, J.S.; Shi, Y.Y.; Chen, L.S.; Wang, K. Investigation of SPAD meter-based indices for estimating rice nitrogen status. *Comput. Electron. Agric.* **2010**, *71*, S60–S65.

29. Cao, Q.; Miao, Y.; Wang, H.; Huang, S.; Cheng, S.; Khosla, R.; Jiang, R. Non-destructive estimation of rice plant nitrogen status with Crop Circle multispectral active canopy sensor. *Field Crops Res.* **2013**, *154*, 133–144.

30. Cao, Q.; Miao, Y.; Feng, G.; Gao, X.; Li, F.; Liu, B.; Yue, S.; Cheng, S.; Ustin, S.L.; Khosla, R.; *et al.* Active canopy sensing of winter wheat nitrogen status: An evaluation of two sensor systems. *Comput. Electron. Agric.* **2015**, *112*, 54–67.

31. Savitzky, A.; Golay, M.J.E. Smoothing and differentiation of data by simplified least squares procedures. *Anal. Chem.* **1964**, *36*, 1627–1632.

32. Bolster, K.L.; Martin, M.E.; Aber, J.D. Determination of carbon fraction and nitrogen concentration in tree foliage by near infrared reflectances: A comparison of statistical methods. *Can. J. For. Res.* **1996**, *26*, 590–600.

33. Grossman, Y.L.; Ustin, S.L.; Jacquemoud, S.; Sanderson, E.W.; Schmuck, G.; Verdebout, J. Critique of stepwise multiple linear regression for the extraction of leaf biochemistry information from leaf reflectance data. *Remote Sens. Environ.* **1996**, *56*, 182–193.

34. Yu, K.; Li, F.; Gnyp, M.L.; Miao, Y.; Bareth, G.; Chen, X. Remotely detecting canopy nitrogen concentration and uptake of paddy rice in the Northeast China Plain. *ISPRS J. Photogramm.* **2013**, *78*, 102–115.

35. Farifteh, J.; van der Meer, F.; Atzberger, C.; Carranza, E.J.M. Quantitative analysis of salt-affected soil reflectance spectra: A comparison of two adaptive methods (PLSR and ANN). *Remote Sens. Environ.* **2007**, *110*, 59–78.

36. Brown, M.P.; Grundy, W.N.; Lin, D.; Cristianini, N.; Sugnet, C.W.; Furey, T.S.; Ares, M.; Haussler, D. Knowledge-based analysis of microarray gene expression data by using support vector machines. *Proc. Natl. Acad. Sci. USA* **2000**, *97*, 262–267.

37. Gill, M.K.; Asefa, T.; Kemblowski, M.W.; McKee, M. Soil moisture prediction using support vector machines. *J. Am. Water Resour. Assoc.* **2006**, *42*, 1033–1046.

38. Lin, G.; Chen, G.; Wu, M.; Chou, Y. Effective forecasting of hourly typhoon rainfall using support vector machines. *Water Resour. Res.* **2009**, *45*, W08440.

39. Kalra, A.; Ahmad, S. Using oceanic-atmospheric oscillations for long lead time streamflow forecasting. *Water Resour. Res.* **2009**, *45*, W03413.

40. Yang, X.; Huang, J.; Wu, Y.; Wang, J.; Wang, P.; Wang, X.; Huete, A.R. Estimating biophysical parameters of rice with remote sensing data using support vector machines. *Sci. China Life Sci.* **2011**, *54*, 272–281.

41. Wang, Y.; Huang, J.F.; Wang, F.M.; Liu, Z.Y. Predicting nitrogen concentrations from hyperspectral reflectance at leaf and canopy for rape. *Guang Pu* **2008**, *28*, 273–277. (In Chinese)
42. Filella, I.; Peñuelas, J. The red edge position and shape as indicators of plant chlorophyll content, biomass and hydric status. *Int. J. Remote Sens.* **1994**, *15*, 1459–1470.
43. Mauser, W.; Bach, H. Imaging spectroscopy in hydrology and agriculture-determination of model parameters. In *Imaging Spectrometry—A Tool for Environmental Observations*; Hill, J., Megier, J., Eds.; Kluwer Academic Publishing: Dordrecht, The Netherlands, 1995; pp. 261–283.
44. Johnson, L.F.; Billow, C.R. Spectrometry estimation of total nitrogen concentration in Douglas-fir foliage. *Int. J. Remote Sens.* **1996**, *17*, 489–500.
45. Mutanga, O.; Skidmore, A.K. Narrow band vegetation indices solve the saturation problem in biomass estimation. *Int. J. Remote Sens.* **2004**, *25*, 1–16.
46. Chatterjee, S.; Price, B. *Regression Analysis by Example*; Wiley: New York, NY, USA, 1977.
47. Geladi, P.; Kowalski, B.R. Partial least-squares regression: A tutorial. *Anal. Chim. Acta* **1986**, *185*, 1–17.
48. Esbensen, K.H.; Guyot, D.; Westad, F.; Houmøller, L.P. *Multivariate Data Analysis—In Practice*, 5th ed.; CAMO Software: Oslo, Norway, 2010; pp. 241–293.
49. Ehsani, M.R.; Upadhyaya, S.K.; Slaughter, D.; Shafii, S.; Pelletier, M. A NIR technique for rapid determination of soil mineral nitrogen. *Precis. Agric.* **1999**, *1*, 219–236.
50. Atkinson, P.M.; Tatnall, A.R.L. Introduction neural networks in remote sensing. *Int. J. Remote Sens.* **1997**, *18*, 699–709.
51. Vapnik, V.N. *Statistical Learning Theory*; John Wiley & Sons Inc.: New York, NY, USA, 1998.
52. Cortes, C.; Vapnik, V. Support-vector networks. *Mach. Learn.* **1995**, *20*, 273–297.
53. Chang, C.; Laird, D.A.; Mausbach, M.J.; Hurburgh, C.R. Near infrared reflectance spectroscopy: Principal components regression analysis of soil properties. *Soil Sci. Soc. Am. J.* **2001**, *65*, 480–490.
54. Rondeaux, G.; Steven, M.; Baret, F. Optimization of soil-adjusted vegetation indices. *Remote Sens. Environ.* **1996**, *55*, 95–107.
55. Gitelson, A.A.; Merzlyak, M.N. Quantitative estimation of chlorophyll a using reflectance spectra: Experiments with autumn chestnut and maple leaves. *J Photochem Photobiol(B).* **1994**, *22*, 247–252.
56. Wang, W.; Yao, X.; Yao, X.F.; Tian, Y.C.; Liu, X.J.; Ni, J.; Cao, W.X.; Zhu, Y. Estimating leaf nitrogen concentration with three-band vegetation indices in rice and wheat. *Field Crops Res.* **2012**, *129*, 90–98.
57. Sims, D.A.; Gamon, J.A. Relationships between leaf pigment content and spectral reflectance across a wide range of species, leaf structures and developmental stages. *Remote Sens. Environ.* **2002**, *81*, 331–354.
58. Lamb, D.W.; Steyn-Ross, M.; Schaare, P.; Hanna, M.M.; Silvester, W.; Steyn-Ross, A. Estimating leaf nitrogen concentration in ryegrass (*Lolium spp.*) pasture using the chlorophyll red-edge: Theoretical modelling and experimental observations. *Int. J. Remote Sens.* **2002**, *23*, 3619–3648.

59. Jongschaap, R.E.E.; Booij, R. Spectral measurements at different spatial scales in potato: Relating leaf, plant and canopy nitrogen status. *Int. J. Appl. Earth Obs. Geoinform.* **2004**, *5*, 205–218.

60. Atzberger, C.; Guérif, M.; Baret, F.; Werner, W. Comparative analysis of three chemometric techniques for the spectroradiometric assessment of canopy chlorophyll content in winter wheat. *Comput. Electron. Agric.* **2010**, *73*, 165–173.

61. Pan, W.C.; Li, S.K.; Wang, K.R.; Xiao, H.; Chen, B.; Wang, F.Y.; Su, Y.; Chen, J.L.; Lai, J.C.; Huang, F.D. Monitoring soil nitrogen and plant nitrogen based on hyperspectral of cotton canopy. *Cotton Sci.* **2010**, *22*, 70–76.

62. Qi, X.M.; Zhang, L.D.; Du, X.L.; Song, Z.J.; Zhang, Y.; XU, S.Y. Quantitative analysis using NIR by building PLSR-BP model. *Spectrosc. Spectr. Anal.* **2003**, *23*, 870–872.

63. Vuolo, F.; Neugebauer, N.; Bolognesi, S.; Atzberger, C.; D'Urso, G. Estimation of leaf area index using DEIMOS-1 data: Application and transferability of a semi-empirical relationship between two agricultural areas. *Remote Sens.* **2013**, *5*, 1274–1291.

64. Rivera, J.; Verrelst, J.; Delegido, J.; Veroustraete, F.; Moreno, J. On the semi-automatic retrieval of biophysical parameters based on spectral index optimization. *Remote Sens.* **2014**, *6*, 4927–4951.

65. Tian, Y.; Gu, K.; Chu, X.; Yao, X.; Cao, W.; Zhu, Y. Comparison of different hyperspectral vegetation indices for canopy leaf nitrogen concentration estimation in rice. *Plant Soil* **2014**, *376*, 193–209.

Spectral Index for Quantifying Leaf Area Index of Winter Wheat by Field Hyperspectral Measurements: A Case Study in Gifu Prefecture, Central Japan

Shinya Tanaka, Kensuke Kawamura, Masayasu Maki, Yasunori Muramoto, Kazuaki Yoshida and Tsuyoshi Akiyama

Abstract: Timely and nondestructive monitoring of leaf area index (LAI) using remote sensing techniques is crucial for precise and efficient management of crops. In this paper, a new spectral index (SI) for estimating LAI of winter wheat (*Triticum aestivum* L.) is proposed on the basis of field hyperspectral measurements. A simple index based on the empirical relationships between LAIs and SIs of all available two-waveband combinations from hyperspectral data is developed by considering the difference between reflectance values at 760 and 739 nm ($DSI_{R760-R739} = R_{760} - R_{739}$). Among published and newly developed SIs, $DSI_{R760-R739}$ exhibited a significant and strong linear relationship with LAI and showed outstanding performance in LAI assessments. The permissible bandwidths for broad-band $DSI_{R760-R739}$ investigated using simulated reflectance were 5 nm for both 760 and 739 nm center wavelengths. The results indicate that the linear regression model based on the narrow-band and broad-band $DSI_{R760-R739}$ is a simple but accurate method for timely and nondestructive monitoring of LAI.

Reprinted from *Remote Sens.* Cite as: Tanaka, S.; Kawamura, K.; Maki, M.; Muramoto, Y.; Yoshida, K.; Akiyama, T. Spectral Index for Quantifying Leaf Area Index of Winter Wheat by Field Hyperspectral Measurements: A Case Study in Gifu Prefecture, Central Japan. *Remote Sens.* **2015**, *7*, 5329–5346.

1. Introduction

In remote sensing, among all canopy variables, leaf area index (LAI) is identified as a key biophysical parameter for crop growth diagnosis and pre-harvest grain yield prediction [1,2], as well as having a key role in terrestrial ecosystem processes [3]. The green LAI, defined as the one-sided green leaf area per unit horizontal ground area [4], is directly related to the growth status of crops [5] and largely influences the spectral reflectance of vegetation canopies. Thus, for site-specific crop management, it is crucial to be able to estimate LAI in a timely and nondestructive manner using remote sensing, since the site-specific crop management requires both high-quality crop production and the minimization of adverse environmental effects via better fertilizer management [2,6–8].

To date, empirical regression models based on the spectral indices (SIs) have been widely used for estimating LAI in crop fields because SI is a simpler, more convenient, and lesser restrictive approach than multivariate statistical techniques or radiative transfer model inversions [6,9–11]. For example, the normalized difference vegetation index (NDVI) [12] has been widely used in the assessment of above-ground biomass (AGB) and LAI [2]. The modified soil-adjusted vegetation index (MSAVI), optimized soil-adjusted vegetation index (OSAVI), and enhanced vegetation index (EVI) were developed to minimize the effects of varying background soil reflectance and atmospheric influences in measuring vegetation signal [13–16]. Gitelson [17] proposed the wide dynamic range vegetation index (WDRVI) to accurately assess crop biomass and LAI under conditions of moderate to high AGB. Recently, Viña et al. [10] found that chlorophyll indices (the red-edge chlorophyll index [$CI_{red-edge}$] and the green chlorophyll index [CI_{green}]), devised for chlorophyll assessment at the leaf scale [18], are more accurate for LAI assessments of maize and soybean crops than for the above-mentioned SIs. These SI-based studies have successfully predicted LAI at various spatial scales using commercially available digital cameras, field spectroradiometers, or airborne and satellite-borne sensors [8,11,19–24].

Winter wheat (*Triticum aestivum* L.) is one of the most important crops in Japan and has been planted in more than 212,600 ha [25]. At the canopy scale, previous research has reported that the LAI for wheat can be accurately estimated by field spectral measurements [20,26,27]. However, SI-based empirical regression models have often been growth-stage-specific or year-specific [9,21,28,29]. Therefore, quantitative assessments of LAI remain uncertain. For example, Haboudane et al. found useful SIs (e.g., the modified triangular vegetation index, MTVI2) for LAI predictions of wheat, corn, and soybean based on simulated data using radiative transfer models [20]. However, the response of MTVI2 of airborne hyperspectral data with respect to LAI were different at maturity and senescence growth stages compared with that at early and mid-growth stage for wheat because of the dominance of the heads of the wheat plants and the increase in yellow and dry leaves [20,30]. These results suggest that the use of datasets representing actual canopy characteristics is important for development of new SIs.

Previous research demonstrated the suitability of hyperspectral remote sensing for monitoring crop growth [31–33]. Particularly for LAI predictions, the saturation problem of NDVI under moderate to high LAI conditions has been extensively investigated by developing new SIs using hyperspectral data [34,35]. The advantage of hyperspectral data is that it can be used for exploring useful SIs via various waveband combinations. However, hyperspectral data are costly to collect. Therefore, investigating the impact of new SI bandwidths on predictive accuracy is important—from an economic standpoint—for designing sensors [36]. In addition, the large numbers of hyperspectral bands are redundant; selection of important bands for crop monitoring is key to maximizing the efficiency of spectral data use [33].

The aim of this study was to identify simple and accurate SIs for LAI assessment of winter wheat. We explored new SIs based on the empirical relationships between the LAIs and SIs of all available two-waveband combinations. This exploration was conducted using field datasets collected at several growth stages of winter wheat. Then, the predictive ability and sensitivity of the newly developed and existing SIs were evaluated. Finally, the impacts of new SI bandwidths were investigated using simulated reflectance.

2. Materials

2.1. Experimental Site

Field experiments were conducted during two growing seasons in 2006 and 2007 at two experimental dried paddy fields of typical size in the Gifu Prefectural Agricultural Technology Center (GPATC) in southern Gifu prefecture, Japan (35°26.6'N, 136°42'E) (Figure 1). The mean annual temperature and annual precipitation at GPATC in 2004 were 17.7 °C and 1903 mm, respectively [37].

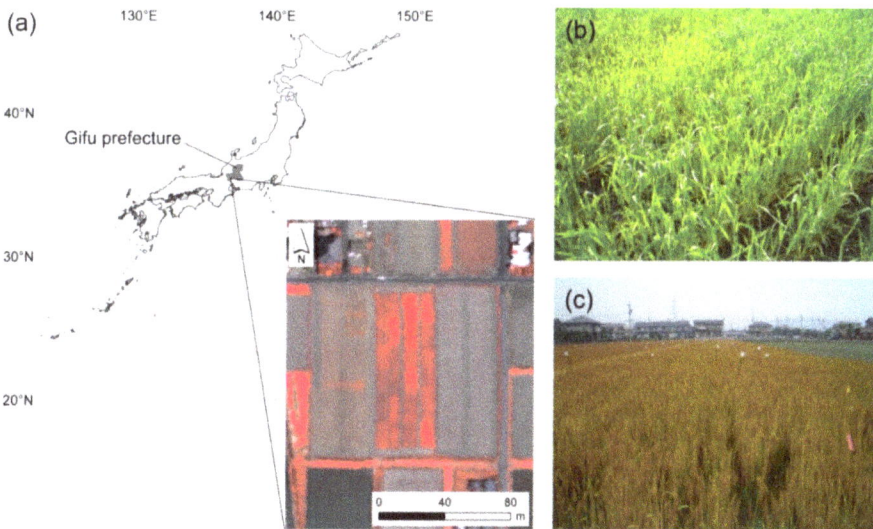

Figure 1. Location and photographs of the experimental site. (a) Location of the experimental site; (b) Photograph of the winter wheat on 25 April 2006 (just before heading); and (c) Photograph of the winter wheat on 2 June 2006 (10 days before harvest).

Two major wheat cultivars (cv. "Norin 61" and "Iwainodaichi") were sown at a 30 cm ridge width at an 80 kg·ha^{-1} seeding rate (standard cultivation practice). We then investigated the "Norin 61" in 2006 and the "Norin 61" and "Iwainodachi" in 2007. The sowing and heading dates are summarized in Table 1. Wheat was harvested between late May and early June in both the 2006 and 2007 seasons. In this region, a rotational cropping system for rice, wheat, and soybean has been widely adopted, and accordingly, wheat was seeded in the experimental fields after the paddy rice had been harvested. Field measurements were performed in different dried paddy fields at GPATC in 2006 and 2007.

Table 1. Sowing and heading dates of winter wheat in the experimental field.

Sowing Date/Heading Date	Norin 61 (2006)	Norin 61 (2007)	Iwainodaichi (2007)
Sowing date	9 November 2005	7 November 2006	17 October 2006 27 October 2006 6 November 2006 16 November 2006
Heading date	29 April 2006	9 April 2007	2 April 2007

Source: [38,39].

2.2. Ground-Based Radiometric Measurements

Canopy reflectance measurements were performed for 10:00–15:00 LST (GMT+9) under clear-sky conditions during the mid (stem extension growth stage) to late (1–2 weeks before harvest) growing stages (Table 2). The canopy spectra were measured using two portable hyperspectral spectroradiometers (ASD FieldSpec Handheld [FSHH] or FieldSpec 3 [FS3]; Analytical Spectral Devices, Boulder, CO, USA). The spectral range was 325–1075 nm for the FSHH and 350–2500 nm for the FS3. For all measurements, the sensor heads were positioned to look vertically downward, centered over the wheat hill, and were kept at a constant 1.3 m above the ground with a commercially available tripod. The radiometers had a 25° field of view, for a viewing area of 58 cm in diameter at the canopy level.

Using the FSHH, we recorded the upwelling radiance of the wheat canopies, as well as that of the white Spectralon reflectance standard (Labsphere, Inc., North Sutton, NH, USA) at ~15–30 min intervals to determine the canopy reflectance. For the FS3 on 10 April 2007, the upwelling radiance of the white Spectralon reflectance standard was used to calibrate the instrument at 15–30 min intervals, and then the reflectance values of the wheat canopy were recorded. Finally, the spectral data stored in a personal computer were resampled at 1 nm intervals and exported as text files using computer software (RS2 for Windows; Analytical Spectral Devices, Boulder, CO, USA).

Table 2. Overview of the field spectral measurements.

Date	n	Measured Cultivar	Growth Stage	Spectroradiometer
4 Apr. 2006	15	Norin 61	Stem extension	FSHH
7 Apr. 2006	5	Norin 61	Stem extension	FSHH
17 Apr. 2006	15	Norin 61	Stem extension	FSHH
24 Apr. 2006	15	Norin 61	Stem extension	FSHH
21 May 2006	14	Norin 61	Maturing	FSHH
24 May 2006	6	Norin 61	Maturing	FSHH
10 Apr. 2007	9	Norin 61	Heading	FS3
17 Apr. 2007	9	Iwainodaichi	Anthesis	FSHH
26 Apr. 2007	6	Norin 61 and Iwainodaichi	Grain filling	FSHH

2.3. Determination of Field LAI Value

Agronomic survey was carried out on either the same day or the day following the hyperspectral measurements. Above-ground plant samples were obtained by cutting plants at the soil surface level in 50 cm lengths for one hill at each sampling point where the ground-spectral measurements had been made. All plant samples were transported to the laboratory immediately after sampling, where they were then divided into green leaves, yellow leaves, stems, and panicles. The surface area of all the green leaves was determined using a leaf area meter (LI-3100; Li-Cor Inc., Lincoln, NE, USA). The LAI values for a unit ground area were determined by multiplying with a conversion factor (6.67 for 50 cm length samples) in consideration of the ridge width of wheat.

3. Methods

3.1. Contour-Map Approach for Exploring New Useful Spectral Indices

Previous studies used contour maps of the coefficient of determination (R^2) obtained by a linear regression analysis between agronomic variables and all possible two-waveband combinations of reflectance values to explore useful SIs [6,26,37,40–43]. This procedure is inadequate when the relationship between the LAI and the SI is nonlinear, and it cannot run with nonlinear fitting because of the requirement for initial parameter values [26]. Conversely, this contour-map approach has the advantage of providing an efficient selection of the optimal combination and width for use in existing sensors and for designing future sensors [6,41], and its results can easily be compared with those from other studies. Therefore, we used this approach with formulae that take the difference (Difference Spectral Index; DSI), ratio (Ratio Spectral Index; RSI), and normalized difference (Normalized Difference Spectral Index; NDSI) of the reflectance values to generate new useful SIs for predicting the

LAI. In this analysis, spectral ranges of <400 nm and >1000 nm were omitted due to noise. The DSI, RSI, and NDSI are defined as follows:

$$\text{DSI}_{R_i-R_j} = R_i - R_j \tag{1}$$

$$\text{RSI}_{R_i-R_j} = R_i/R_i \tag{2}$$

$$\text{NDSI}_{R_i-R_j} = (R_i - R_j)/(R_i + R_j) \tag{3}$$

where R_i and R_j are the reflectance values at i and j nm. In this study, both the R^2 value (highest-R^2 criteria) and the root mean square error (RMSE) (minimum-RMSE criteria) from the leave-one-out cross-validation (LOOCV) procedure were used to explore useful SIs. The RMSE was calculated by the following formula:

$$\text{RMSE} = \sqrt{\frac{1}{n} \sum_{i=1}^{n} (y_i - y_i)^2} \tag{4}$$

where y_i and $\hat{y}_i$ are the observed and predicted value of sample data i, respectively, and n is the number of sample data. To compare the predictive ability of these three new SIs, we selected nine other major and potentially useful SIs for LAI prediction: NDVI, EVI, OSAVI, WDRVI, CI$_\text{red-edge}$, CI$_\text{green}$, MSAVI, MTVI1 and MTVI2 (Table 3).

Table 3. Existing spectral indices (SIs) used for comparison with the DSI, RSI, and NDSI.

Spectral Index	Formulation	Reference
NDVI	$(R_{800} - R_{670}) / (R_{800} + R_{670})$	[12,20]
EVI	$2.5\,[(R_{800} - R_{670}) / (R_{800} + 6R_{670} - 7.5R_{445})]$	[14]
OSAVI	$(R_{800} - R_{670}) / (R_{800} + R_{670} + 0.16)$	[16]
WDRVI ($\alpha = 0.1$)	$(\alpha R_{800} - R_{670}) / (\alpha R_{800} + R_{670})$	[17]
CI$_\text{red-edge}$	$R_{800}/R_{710} - 1$	[10,18,44]
CI$_\text{green}$	$R_{800}/R_{550} - 1$	[10,18]
MSAVI	$0.5\left[2R_{800} + 1 - \sqrt{(2R_{800} + 1)^2 - 8\,(R_{800} - R_{670})}\right]$	[15,20]
MTVI1	$1.2\,[1.2\,(R_{800} - R_{550}) - 2.5\,(R_{670} - R_{550})]$	[20]
MTVI2	$\dfrac{1.5[1.2(R_{800}-R_{550})-2.5(R_{670}-R_{550})]}{\sqrt{(2R_{800}+1)^2 - \left(6R_{800}-5\sqrt{R_{670}}\right) - 0.5}}$	[20]

3.2. Model Construction and Validation

For each SI, linear and non-linear regression models between SI and LAI were constructed [10], and the two models were compared with the Akaike Information Criterion (AIC; [45]) to select the better model.

Then, a bootstrap procedure similar to that in previous studies [46,47] was performed to evaluate the predictive ability of the SIs. First, the data were divided into a calibration dataset (66.7%) and a validation dataset (33.3%) by stratified random sampling (four stratums; $0 \leqslant LAI < 1$, $1 \leqslant LAI < 2$, $2 \leqslant LAI < 3$, and $LAI \geqslant 3$) because there was less data in the high LAI range. Next, the linear or nonlinear model for each SI was fitted to the calibration dataset, and a set of best-fitted values was determined. Finally, the validation dataset was bootstrapped 1000 times, and for each repetition, the inverted linear or nonlinear model with the best-fitted parameters for each SI was used to predict LAI for the validation subsamples. In this study, same calibration dataset and bootstrapped samples in all SIs were used. To assess the predictive accuracy, the RMSE was used.

For our sensitivity analysis, the noise equivalent (NE) ΔLAI [10,48] was used to represent the sensitivity of the SI in detecting changes in LAI:

$$NELAI = RMSE(SI\ vs.\ LAI)/[d(SI)/d(LAI)] \tag{5}$$

where $d(SI)/d(LAI)$ is the first derivative of the SI with respect to LAI, and the RMSE(SI *vs.* LAI) is the RMSE of the SI *versus* LAI relationship. The NEΔLAI has the advantage of allowing a direct comparison of different SIs [49].

All data handling and statistical analyses were performed using the R software (version 2.15.0) [50] and the nonlinear fitting was made using the "nls" function in R.

3.3. Determination of Bandwidths for Broad-Band SI

To investigate the performance of new SI under different bandwidths (full width at half maximum; FWHM), simulated reflectance (R_{sim}) was used. In accordance with a previous study [36], the R_{sim} was simulated by Equation (6) with the Gaussian response function (Equation (7)):

$$R_{sim} = \frac{\sum_{\lambda_s}^{\lambda_e} R_\lambda\, f(\lambda,\, \sigma)}{\sum_{\lambda_s}^{\lambda_e} f(\lambda,\, \sigma)} \tag{6}$$

$$f(\lambda,\, \sigma) = \exp\left(-\frac{(\lambda - \lambda_c)^2}{2\sigma^2}\right) \tag{7}$$

where λ is wavelength in the range of spectral response for simulated bandwidth, λ_c is the central wavelength, and $\sigma = \frac{FWHM}{2\sqrt{2\ln2}}$. Different bandwidths in the range of 1–61 nm were simulated to investigate changes in predictive accuracy using the RMSE values calculated by the bootstrap procedure described in the previous section. The bandwidths for which the RMSE was not greater than 5% of the smallest RMSE value (optimal$_{1.05}$ criterion; e.g., [51]) were adopted as the permissible bandwidths for economical sensor design [36].

4. Results

4.1. Agronomic Data

Summary statistics of the field LAI values are shown in Table 4. The ranges of LAI in the entire dataset were 0.3–5.5. These ranges were sufficiently broad to evaluate the predictive ability of the SIs. Conversely, the ranges of LAI in each dataset, *i.e.*, dataset of the cv. "Norin 61" in 2006 and 2007 and the cv. "Iwainodaichi" in 2007, were insufficient for analysis by year or cultivar. In addition, the 2007 data was collected mainly at the specific growth stage (Table 2). For these reasons, all the data were pooled and then used for statistical analysis.

Table 4. Summary statistics of the field LAI values.

Variable	Statistic	Entire Dataset	Norin 61 (2006)	Norin 61 (2007)	Iwainodaichi (2007)
LAI	Average	1.7	1.4	2.7	2.3
	Max	5.5	3.0	5.5	3.9
	Min	0.3	0.3	1.4	0.7
	Range	5.2	2.7	4.1	3.1
n		94	70	12	12

4.2. Contour Maps of R^2 Value

Figure 2 shows contour maps of the R^2 values from the linear regression analysis between LAI and all possible two-waveband combinations of DSI, RSI, and NDSI, respectively. Similar results were also obtained in the RMSE values (data not shown). In DSI (Figure 2a), a higher R^2 (>0.75) and smaller RMSE (<0.45) areas were found at the combination of red-edge wavelengths (720–750 nm) and red-edge to NIR wavelengths (740–840 nm). The maximum R^2 value (0.860; $p < 0.001$) and minimum RMSE value (0.345) were obtained by the difference of the reflectance values at 760 nm and 739 nm, *i.e.*, $DSI_{R760-R739}$. In RSI (Figure 2b), major $R^2 > 0.75$ areas with RMSE < 0.45 included a combination of those around 500 nm and 760 or 990 nm wavelengths, 680 nm with red-edge to NIR wavelengths (735–930 nm, 960–1000 nm), and 730–760 nm wavelengths. The maximum R^2 value (0.785; $p < 0.001$) and minimum RMSE value (0.428) were obtained by $RSI_{R760-R730}$. In NDSI (Figure 2c), a major $R^2 > 0.75$ area with RMSE < 0.45 was found in region with a combination of around 760 nm and 730 nm wavelengths. The maximum R^2 (0.788; $p < 0.001$) and minimum RMSE (0.425) values were obtained by $NDSI_{R760-R730}$.

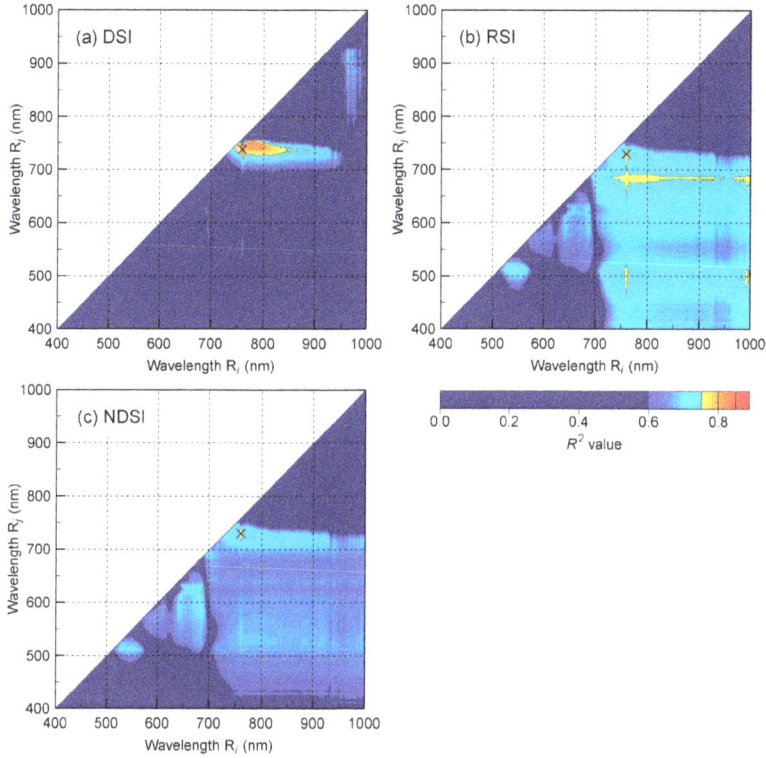

Figure 2. Contour maps of the coefficient of determination (R^2) between LAI and (**a**) DSI; (**b**) RSI; and (**c**) NDSI. The crosses (x) indicate the points with the highest R^2 values.

Among the three contour maps, RSI showed $R^2 > 0.70$ in a larger number of combinations than the other SIs for estimating LAI. However, we obtained no $R^2 > 0.80$ with RSI. Overall, the $DSI_{R760-R739}$ yielded the best R^2 values with respect to LAI ($R^2 = 0.860$). Moreover, the best waveband-combination for DSI, RSI, and NDSI that were determined by the minimum-RMSE criteria were coincident with the results of the highest-R^2 criteria. These results indicate that these three new SIs are useful for LAI prediction; therefore, were further used in this study.

4.3. LAI Prediction and Validation

Figure 3 shows the relationships between LAIs and SIs. NDVI, EVI, OSAVI, MSAVI and MTVI1 each exhibited an asymptotic relationship with LAI. In contrast, the other SIs (*i.e.*, WDRVI, $CI_{red-edge}$, CI_{green}, MTVI2, $DSI_{R760-R739}$, $RSI_{R760-R730}$, and $NDSI_{R760-R730}$) showed a more linear relationship with LAI. In particular,

DSI$_{R760-R739}$ had the most linear relationship with LAI. Based on the AIC, nonlinear model was selected for all SIs except for DSI$_{R760-R739}$.

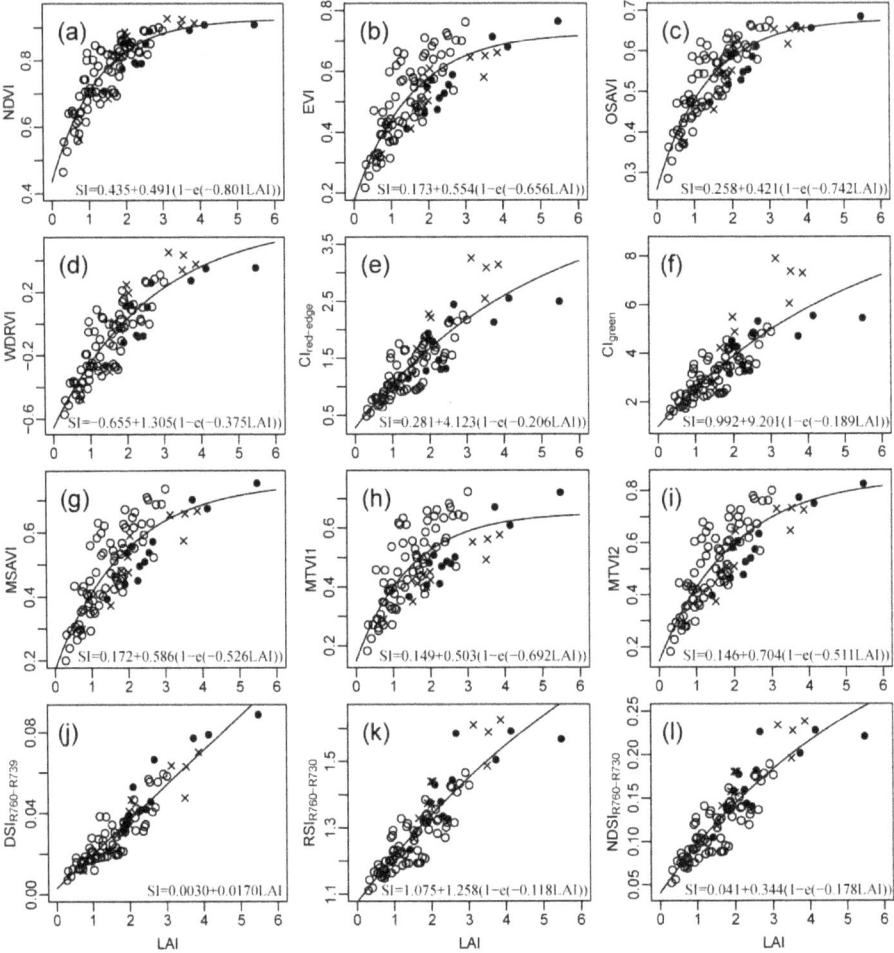

Figure 3. Relationships between LAI and **(a)** NDVI, **(b)** EVI, **(c)** OSAVI, **(d)** WDRVI, **(e)** CI$_{red-edge}$, **(f)** CI$_{green}$, **(g)** MSAVI, **(h)** MTV1, **(i)** MTV2, **(j)** DSI$_{R760-R739}$, **(k)** RSI$_{R760-R730}$, and **(l)** NDSI$_{R760-R730}$. Open circles, filled circles, and crosses indicate data from the cv. "Norin 61" in 2006, cv. "Norin 61" in 2007, and cv. "Iwainodaichi" for 2007, respectively. Solid lines indicate the best-fitted lines.

Subsequently, we compared the predictive ability of SIs for assessing LAI by RMSEs obtained via the modified bootstrap procedure. Table 5 shows the point-estimated mean values of RMSE and the 95% confidence intervals (95% CI). Lower RMSE values (RMSE $\leqslant 0.457$) were obtained from the three newly explored

SIs ($DSI_{R760-R739}$, $RSI_{R760-R730}$, and $NDSI_{R760-R730}$) than those from previously known SIs (RMSE > 0.457). The linear predictive model based on $DSI_{R760-R739}$ showed the best performance (RMSE = 0.372; 0.280–0.487, 95% CI).

Table 5. Best-fitted parameters, point-estimated mean values of RMSE, and 95% confidence intervals (CIs) of RMSE for narrow-band and broad-band SIs calculated by a modified bootstrap procedure.

Spectral index	Model [a]	Best Fitted Parameter			RMSE	95% CI
		Y_0	a	b		
NDVI	nonlinear	0.431	0.499	0.811	0.466	0.357–0.546
EVI	nonlinear	0.183	0.589	0.574	0.656	0.535–0.847
OSAVI	nonlinear	0.263	0.434	0.691	0.492	0.404–0.617
WDRVI	nonlinear	−0.674	1.300	0.409	0.487	0.378–0.566
$CI_{red-edge}$	nonlinear	0.242	3.770	0.252	0.516	0.404–0.605
CI_{green}	nonlinear	0.933	8.493	0.225	0.572	0.442–0.692
MSAVI	nonlinear	0.174	0.633	0.475	0.582	0.469–0.753
MTVI1	nonlinear	0.165	0.545	0.555	0.824	0.622–1.046
MTVI2	nonlinear	0.144	0.753	0.474	0.541	0.434–0.687
$DSI_{R760-R739}$	linear	0.003	0.017	NA	0.372	0.280–0.487
$RSI_{R760-R730}$	nonlinear	1.071	0.994	0.165	0.457	0.371–0.551
$NDSI_{R760-R730}$	nonlinear	0.039	0.300	0.224	0.455	0.368–0.553
Broad-band $DSI_{R760-R739}$	linear	0.006	0.017	NA	0.390	0.302–0.477

[a] Inverted regression model, $LAI = \dfrac{\ln\left(1/(1-\frac{SI-Y_0}{a})\right)}{b}$, was used for nonlinear models, while the model, $LAI = \frac{SI-Y_0}{a}$, was used for the linear model in the LAI prediction.

On the basis of the sensitivity analysis, we found large differences in sensitivity when LAI value exceeded 3.0 but only minor differences among SIs when LAI values were below 3.0 (Figure 4). Overall, WDRVI, $CI_{red-edge}$, CI_{green}, $DSI_{R760-R739}$, $RSI_{R760-R730}$, and $NDSI_{R760-R730}$ showed higher sensitivities with respect to LAI at moderate to high LAI values (3.0–5.5). $DSI_{R760-R739}$, in particular, revealed the highest sensitivity when LAI exceeded 1.5, indicating that it is effective in predicting high LAI values.

On the basis of its high predictive ability, high sensitivity, and high degree of linearity, we consider $DSI_{R760-R739}$ to be the most useful SI for estimating LAI in our dataset.

4.4. Impact of Bandwidths on Predictive Accuracy

The impact of bandwidths on predictive accuracy was investigated using $DSI_{R760-R739}$ with simulated reflectance R_{sim}. A best predictive accuracy was found to be the narrow-band $DSI_{R760-R739}$ (*i.e.*, bandwidths are 1 nm for 760 and 739 nm center wavelengths). The predictive accuracy decreased with increases in bandwidths, as shown in Figure 5. The downward trends of predictive accuracy were different in 760 and 739 nm center wavelengths. When the bandwidth at the 760 nm center wavelength was as narrow as 9 nm, the impact of increases in bandwidths was relatively small for the 739 nm center wavelength. Although increases in bandwidths decreased predictive accuracy, the broad-band $DSI_{R760-R739}$ (e.g., 15 nm for both wavelengths) had higher predictive accuracy (*i.e.*, RMSE < 0.455) than the existing SIs listed in Table 5. The permissible bandwidths determined by the optimal$_{1.05}$ criterion were found to be 5 nm for 760 and 739 nm center wavelengths (Table 5).

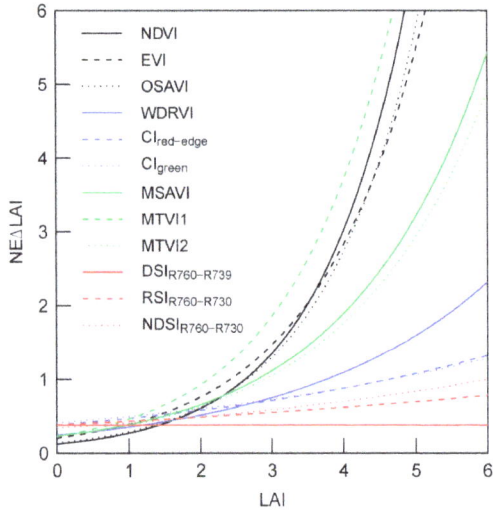

Figure 4. Sensitivity (NEΔLAI) of the SIs tested in the study.

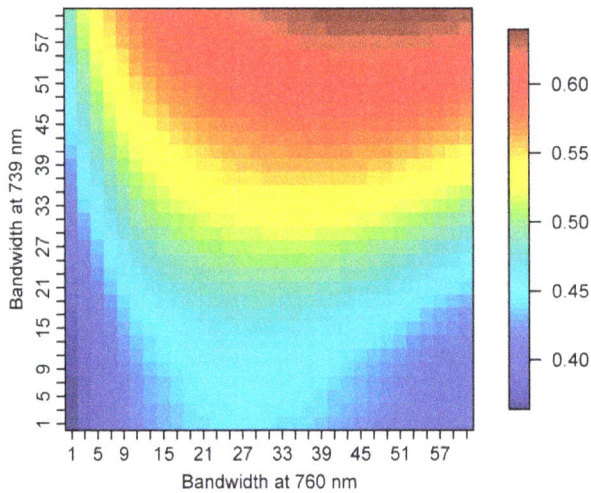

Figure 5. Contour map of RMSE value estimated from predictive model under different bandwidths (1–61 nm) based on the $DSI_{R760-R739}$.

66

5. Discussion

Using the contour map approach, three SIs ($DSI_{R760-R739}$, $RSI_{R760-R730}$, and $NDSI_{R760-R730}$) were found to be closely related to LAI (Figure 2). In particular, $DSI_{R760-R739}$ was most significantly related to LAI ($R^2 = 0.860$; RMSE = 0.345). As shown in Figure 2, the wavelengths between 730 and 760 nm showed the most important spectral signatures for LAI assessment. The red-edge-NIR wavelengths are recognized as the most important spectral signatures for the assessment of LAI and canopy chlorophyll content [52–55]. In general, because of the multiple scattering of light in canopies, the reflectance values of the NIR wavelengths increase as LAI increases [56,57]. First-derivative processing is well known to be effective for enhancing the spectral signature from a target by removing noise. Difference processing of two close wavelengths yields results similar to those of the first-derivative processing. Therefore, we considered the spectral signature of the NIR wavelengths, which are sensitive to changes in LAI, to be enhanced by difference processing. As the canopy reflectance values of red-edge wavelengths are closely related to canopy chlorophyll content [55], and canopy chlorophyll content is strongly related to LAI [58,59], previous studies successfully predicted LAI by using chlorophyll-related SIs such as $CI_{red-edge}$ [10,53]. According to such results, another possible reason for the success of $DSI_{R760-R739}$ for LAI prediction is the high sensitivity of reflectance values at red-edge wavelengths (739 nm in this study) to the canopy chlorophyll content.

The analyses revealed that NDVI, EVI, and OSAVI (normalized difference indexing) are in a nonlinear relationship with LAI (Figure 3a–c) and are less sensitive to changes in LAI at high LAI values (Figure 4). The main reason for the low sensitivity of these SIs could be their normalized-difference processing; this processing of the ρ_{NIR} and ρ_{red} values makes the SIs insensitive to variations in the ρ_{NIR} when $\rho_{NIR} \gg \rho_{red}$ [17]. SIs that take normalized difference processing tend to be insensitive to changes in LAI at high LAI values. In contrast, the WDRVI, a linear transformation of the normalized difference of ρ_{NIR} and ρ_{red}, is more sensitive to changes in LAI at moderate to high LAI values (Figure 4). Thus, introducing a weight coefficient such as the WDRVI's "α" is a simple and efficient approach to enhance sensitivity to LAI under moderate to high LAI conditions. Although its sensitivity at high LAI values was slightly inferior to that of $DSI_{R760-R739}$, the WDRVI is useful for LAI assessment.

We also confirmed the applicability of chlorophyll-related SIs (e.g., $CI_{red-edge}$ and CI_{green}) at the leaf scale for LAI assessment (Figure 3e,f and Figure 4). These results are consistent with a previous study [10,53]. When a reflectance value at 760 nm is used as the ρ_{NIR} of $CI_{red-edge}$, the $CI_{red-edge}$ becomes an index similar to $RSI_{R760-R730}$, which indicates higher predictive accuracy and sensitivity. Therefore, $CI_{red-edge}$, computed by the red-edge and the shortest part of the NIR wave region, is more useful for LAI assessment than SIs that use the longer part of the NIR wave region. The $CI_{red-edge}$ has been successfully applied to LAI assessment of maize, soybeans [10], and wheat crops [53]. Thus, we believe that the $CI_{red-edge}$ is useful for LAI assessment regardless of crop type. Nevertheless, note that a high spectral resolution was required to observe optimal wavelengths because the optimal bandwidth of this SI was observed to be narrow, as shown in the RSI contour map (Figure 2b).

Among the SIs tested in this study, $DSI_{R760-R739}$ was the most linearly and closely related SI to LAI and demonstrated the best performance for the ground-based dataset. The advantage demonstrated by $DSI_{R760-R739}$ was its higher sensitivity in detecting changes in LAI at moderate to high LAI values (3.0–5.5). The ground-based dataset was collected at multiple growth stages during the two-year experiment (Table 3). In spite of the clear differences in ground and canopy conditions, $DSI_{R760-R739}$ could accurately predict the LAI.

In monitoring the crop nitrogen status of rice and wheat, Wang *et al.* [36] reported that the relatively wide bandwidths (36, 15, and 21 nm for 924, 703, and 423 nm wavelengths, respectively) in their three broad-band vegetation index are ideal for sensor design. However, our study found rapid decreases in predictive accuracy, especially for the 760 nm center wavelength. The permissible bandwidths for $DSI_{R760-R739}$ determined by the optimal$_{1.05}$ criterion were found to be 5 nm at both the 760 nm and 739 nm center wavelengths. The result of this study is consistent with previous studies showing that narrow-band SIs from hyperspectral remote sensing are suitable for monitoring crop growth (e.g., [33,60]). However, even though the bandwidths of $DSI_{R760-R739}$ are wider (e.g., 15 nm for both wavelengths), the predictive accuracy of the model was still higher than existing SIs (Figure 5). Therefore, like the narrow-band $DSI_{R760-R739}$, which is suitable for LAI assessment, the broad-band $DSI_{R760-R739}$ may also prove to be useful for LAI assessment.

Le Maire *et al.* [61] conducted a systematic study aimed at finding efficient hyperspectral indices for the estimation of forest sun leaf chlorophyll content, sun leaf mass per area, LAI, and leaf canopy biomass using radiative transfer models, canopy reflectance data, and Hyperion images. They found a reliable index that uses difference of reflectance value at 1725 and 970 nm (*i.e.*, $R_{1725} - R_{970}$). In the spectral range that we tested, and in the form of normalized difference, they reported that the combination of NIR and blue reflectance is the most useful spectral region for LAI estimation. However, the results of the present study suggested that the combination of NIR and blue reflectance are not useful for LAI assessment of winter wheat. A possible reason for this significant difference is the difference in ranges in LAI values because le Maire *et al.* estimated LAI in ranges >3.0 in their study. Although this study was site-specific and did not include wavelengths longer than 1000 nm, the ranges of LAI in the datasets were 0.3–5.5, sufficiently broad to evaluate the predictive ability of SIs (Table 4). The $DSI_{R760-R739}$ developed for the wide ranges of LAI values in this study would therefore be reliable for LAI assessment of wheat crops.

Similar to our study, some previous studies have reported that the first-derivative reflectance of red-edge wavelengths or SIs using red-edge wavelengths and wavelengths at 730–760 nm are closely related to the LAI of paddy rice [40], wheat [26,53], and pasture biomass [35]. Another study suggested that the first-derivative reflectance at 740 nm has high sensitivity to the difference in LAI of paddy rice at the panicle-formation stage [6]. The results of further studies investigating the spectral response of $DSI_{R760-R739}$ for LAI assessment of multiple crops or vegetation types would be interesting and significant to confirm the robustness of the $DSI_{R760-R739}$ predictive model because of the differing spectral responses among crops [32,42,43]. For example, an analysis comparing wheat and paddy rice, which are the two major gramineous crops that have relatively similar canopy structures, would be very interesting and required to determine the suitability of $DSI_{R760-R739}$ for remotely assessing LAI. In addition, since all the ground datasets in our study were pooled and then used for statistical analysis, this study could not clarify the applicability of $DSI_{R760-R739}$ for determining differences in wheat varieties and fertilization. In the future, we will examine other fields with different wheat varieties and fertilizer management techniques by using a larger dataset with a wider range of LAI values.

6. Conclusions

To identify simple and accurate SIs for LAI assessment of winter wheat, this study evaluated the predictive ability and sensitivity of several SIs with respect to LAI assessment. Hyperspectral and ground data collected at the middle and late growing stages were used to identify useful SIs in the present study.

During the study, three new SIs ($DSI_{R760-R739}$, $RSI_{R760-R730}$, and $NDSI_{R760-R730}$) were developed based on the empirical relationships between LAIs and SIs of all available two-waveband combinations from hyperspectral data. Of the 12 SIs that were tested, $DSI_{R760-R739}$ was the most linearly and closely related to LAI and the most sensitive to changes in LAI at moderate to high LAI conditions. The permissible bandwidths for broad-band $DSI_{R760-R739}$ were identified as 5 nm at both center wavelengths. The narrow-band and broad-band $DSI_{R760-R739}$ could be used for LAI assessment with portable spectroradiometers, thus providing useful information for farmers to conduct improved site-specific crop management and sustainable agricultural decisions.

Since all the ground datasets in our study were pooled and then used for statistical analysis, this study could not clarify the applicability of $DSI_{R760-R739}$ for determining differences in wheat varieties and fertilization. In the future, we will examine other fields with different wheat varieties and fertilizer management techniques by using a larger dataset with a wider range of LAI values. In addition, we will also examine the response of the $DSI_{R760-R739}$ for LAI assessments for multiple crops or vegetation types to confirm the robustness of the predictive model based on $DSI_{R760-R739}$.

Acknowledgments: All data used in this study were collected when the first author was a graduate student at the River Basin Research Center, Gifu University. We are grateful to the members of this center for their encouragement. Comments from G. Takao, M. Takahashi, T. Nishizono, and T. Takahashi (FFPRI), and from anonymous reviewers served to greatly improve the early drafts of this manuscript. This study was partially supported by a grant from the Ministry of Education, Culture, Sports, Science and Technology, Japan ("Satellite Ecology", the 21st Century COE Program at Gifu University), and the Forestry and Forest Products Research Institute (FFPRI).

Author Contributions: Shinya Tanaka designed this study, collected field data, and performed all statistical analyses, with the co-authors providing mentorship throughout the study. Shinya Tanaka and Kensuke Kawamura jointly wrote the manuscript. Masayasu Maki collected the reflectance data on 10 April 2007 and helped interpretation of results. Yasunori Muramoto and Kazuaki Yoshida managed wheat growth. Tsuyoshi Akiyama conceived the present study. All authors have read and approved the final manuscript.

Conflicts of Interest: The authors declare no conflict of interest.

References

1. Dente, L.; Satalino, G.; Mattia, F.; Rinaldi, M. Assimilation of leaf area index derived from ASAR and MERIS data into CERES-Wheat model to map wheat yield. *Remote Sens. Environ.* **2008**, *112*, 1395–1407.
2. Pinter, P.J.; Hatfield, J.L.; Schepers, J.S.; Barnes, E.M.; Moran, M.S.; Daughtry, C.S.; Upchurch, D.R. Remote sensing for crop management. *Photogramm. Eng. Remote Sens.* **2003**, *69*, 647–664.
3. Baret, F.; Buis, S. Estimating canopy characteristics from remote sensing observations: Review of methods and associated problems. In *Advances in Land Remote Sensing*; Liang, S., Ed.; Springer: Dordrecht, The Netherlands, 2008; pp. 173–201.
4. Campbell, G.S.; Norman, J.M. The description and measurement of plant canopy structure. In *Plant Canopies: Their Growth, Form and Function*; Cambridge University Press: Cambridge, UK, 1989; pp. 1–10.
5. Casa, R.; Varella, H.; Buis, S.; Guérif, M.; de Solan, B.; Baret, F. Forcing a wheat crop model with LAI data to access agronomic variables: Evaluation of the impact of model and LAI uncertainties and comparison with an empirical approach. *Eur. J. Agron.* **2012**, *37*, 1–10.
6. Inoue, Y.; Sakaiya, E.; Zhu, Y.; Takahashi, W. Diagnostic mapping of canopy nitrogen content in rice based on hyperspectral measurements. *Remote Sens. Environ.* **2012**, *126*, 210–221.
7. Moran, M.S.; Inoue, Y.; Barnes, E.M. Opportunities and limitations for image-based remote sensing in precision crop management. *Remote Sens. Environ.* **1997**, *61*, 319–346.

8. Shibayama, M.; Sakamoto, T.; Takada, E.; Inoue, A.; Morita, K.; Takahashi, W.; Kimura, A. Estimating paddy rice leaf area index with fixed point continuous observation of near infrared reflectance using a calibrated digital camera. *Plant Prod. Sci.* **2011**, *14*, 30–46.

9. Maki, M.; Homma, K. Empirical regression models for estimating multiyear leaf area index of rice from several vegetation indices at the field scale. *Remote Sens.* **2014**, *6*, 4764–4779.

10. Viña, A.; Gitelson, A.A.; Nguy-Robertson, A.L.; Peng, Y. Comparison of different vegetation indices for the remote assessment of green leaf area index of crops. *Remote Sens. Environ.* **2011**, *115*, 3468–3478.

11. Wu, J.; Wang, D.; Bauer, M.E. Assessing broadband vegetation indices and QuickBird data in estimating leaf area index of corn and potato canopies. *Field Crops Res.* **2007**, *102*, 33–42.

12. Rouse, J.W.; Haas, R.H.; Schell, J.A.; Deering, D.W. Monitoring vegetation systems in the Great Plains with ERTS. *Third ERTS Symp. NASA SP-351 I* **1973**, *351*, 309–317.

13. Huete, A.R. A soil-adjusted vegetation index (SAVI). *Remote Sens. Environ.* **1988**, *25*, 295–309.

14. Huete, A.; Didan, K.; Miura, T.; Rodriguez, E.P.; Gao, X.; Ferreira, L.G. Overview of the radiometric and biophysical performance of the MODIS vegetation indices. *Remote Sens. Environ.* **2002**, *83*, 195–213.

15. Qi, J.; Chehbouni, A.; Huete, A.R.; Kerr, Y.H.; Sorooshian, S. A modified soil adjusted vegetation index. *Remote Sens. Environ.* **1994**, *48*, 119–126.

16. Rondeaux, G.; Steven, M.; Baret, F. Optimization of soil-adjusted vegetation indices. *Remote Sens. Environ.* **1996**, *55*, 95–107.

17. Gitelson, A.A. Wide Dynamic Range Vegetation Index for remote quantification of biophysical characteristics of vegetation. *J. Plant Physiol.* **2004**, *161*, 165–173.

18. Gitelson, A.A.; Gritz, Y.; Merzlyak, M.N. Relationships between leaf chlorophyll content and spectral reflectance and algorithms for non-destructive chlorophyll assessment in higher plant leaves. *J. Plant Physiol.* **2003**, *160*, 271–282.

19. Chen, J.M.; Pavlic, G.; Brown, L.; Cihlar, J.; Leblanc, S.G.; White, H.P.; Hall, R.J.; Peddle, D.R.; King, D.J.; Trofymow, J.A.; *et al.* Derivation and validation of Canada-wide coarse-resolution leaf area index maps using high-resolution satellite imagery and ground measurements. *Remote Sens. Environ.* **2002**, *80*, 165–184.

20. Haboudane, D.; Miller, J.R.; Pattey, E.; Zarco-Tejada, P.J.; Strachan, I.B. Hyperspectral vegetation indices and novel algorithms for predicting green LAI of crop canopies: Modeling and validation in the context of precision agriculture. *Remote Sens. Environ.* **2004**, *90*, 337–352.

21. Potithep, S.; Nagai, S.; Nasahara, K.N.; Muraoka, H.; Suzuki, R. Two separate periods of the LAI–VIs relationships using *in situ* measurements in a deciduous broadleaf forest. *Agric. For. Meteorol.* **2013**, *169*, 148–155.

22. Sakamoto, T.; Gitelson, A.A.; Nguy-Robertson, A.L.; Arkebauer, T.J.; Wardlow, B.D.; Suyker, A.E.; Verma, S.B.; Shibayama, M. An alternative method using digital cameras for continuous monitoring of crop status. *Agric. For. Meteorol.* **2012**, *154–155*, 113–126.

23. Shibayama, M.; Akiyama, T. Seasonal visible, near-infrared and mid-infrared spectra of rice canopies in relation to LAI and above-ground dry phytomass. *Remote Sens. Environ.* **1989**, *27*, 119–127.

24. Welles, J.M.; Cohen, S. Canopy structure measurement by gap fraction analysis using commercial instrumentation. *J. Exp. Bot.* **1996**, *47*, 1335–1342.

25. Ministry of Agriculture. Forestry and Fisheries of Japan (MAFF) Statistics of Agriculture, Forestry and Fisheries. Available online: http://www.maff.go.jp/j/tokei/kouhyou/sakumotu/sakkyou_kome/pdf/syukaku_mugi_14.pdf (accessed on 14 November 2014).

26. Hansen, P.M.; Schjoerring, J.K. Reflectance measurement of canopy biomass and nitrogen status in wheat crops using normalized difference vegetation indices and partial least squares regression. *Remote Sens. Environ.* **2003**, *86*, 542–553.

27. Jin, X.; Diao, W.; Xiao, C.; Wang, F.; Chen, B.; Wang, K.; Li, S. Estimation of wheat agronomic parameters using new spectral indices. *PLoS ONE* **2013**, *8*, e72736.

28. Inoue, Y.; Iwasaki, K. Spectral estimation of radiation absorptance and leaf area index in corn canopies as affected by canopy architecture and growth stage. *Jpn. J. Crop Sci.* **1991**, *60*, 578–580.

29. Li, F.; Miao, Y.; Hennig, S.D.; Gnyp, M.L.; Chen, X.; Jia, L.; Bareth, G. Evaluating hyperspectral vegetation indices for estimating nitrogen concentration of winter wheat at different growth stages. *Precis. Agric.* **2010**, *11*, 335–357.

30. Smith, A.M.; Bourgeois, G.; Teillet, P.M.; Freemantle, J.; Nadeau, C. A comparison of NDVI and MTVI2 for estimating LAI using CHRIS imagery: A case study in wheat. *Can. J. Remote Sens.* **2008**, *34*, 539–548.

31. Thenkabail, P.S.; Lyon, J.G.; Huete, A. *Hyperspectral Remote Sensing of Vegetation*; CRC Press: Boca Raton, FL, USA, 2012.

32. Mariotto, I.; Thenkabail, P.S.; Huete, A.; Slonecker, E.T.; Platonov, A. Hyperspectral *versus* multispectral crop-productivity modeling and type discrimination for the HyspIRI mission. *Remote Sens. Environ.* **2013**, *139*, 291–305.

33. Thenkabail, P.S.; Gumma, M.K.; Teluguntla, P.; Mohammed, I.A. Hyperspectral remote sensing of vegetation and agricultural crops. *Photogramm. Eng. Remote Sens.* **2014**, *80*, 697–709.

34. Delegido, J.; Verrelst, J.; Meza, C.M.; Rivera, J.P.; Alonso, L.; Moreno, J. A red-edge spectral index for remote sensing estimation of green LAI over agroecosystems. *Eur. J. Agron.* **2013**, *46*, 42–52.

35. Mutanga, O.; Skidmore, A.K. Narrow band vegetation indices overcome the saturation problem in biomass estimation. *Int. J. Remote Sens.* **2004**, *25*, 3999–4014.

36. Wang, W.; Yao, X.; Yao, X.; Tian, Y.; Liu, X.; Ni, J.; Cao, W.; Zhu, Y. Estimating leaf nitrogen concentration with three-band vegetation indices in rice and wheat. *Field Crops Res.* **2012**, *129*, 90–98.

37. Kanemoto, M.; Tanaka, S.; Kawamura, K.; Matsufuru, H.; Yoshida, K.; Akiyama, T. Wavelength selection for estimating biomass, LAI, and leaf nitrogen concentration in winter wheat of Gifu prefecture using *in situ* hyperspectral data. *J. Jpn. Agric. Syst. Soc.* **2008**, *24*, 43–56.

38. Muramoto, Y.; Yoshida, K. A study of production technique for high-quality wheat grains using controlled-release coated urea fertilizer. *Bull. Gifu Prefect. Agric. Technol. Cent.* **2010**, *10*, 1–9.

39. Tanaka, S.; Goto, S.; Maki, M.; Akiyama, T.; Muramoto, Y.; Yoshida, K. Estimation of leaf chlorophyll concentration in winter wheat before maturing stage by a newly developed vegetation index-rbNDVI. *J. Jpn. Agric. Syst. Soc.* **2007**, *23*, 297–303.

40. Evri, M.; Akiyama, T.; Kawamura, K. Optimal visible and near-infrared waveband used in hyperspectral indices to predict crop variables of rice. *J. Jpn. Agric. Syst. Soc.* **2008**, *24*, 19–29.

41. Inoue, Y.; Miah, G.; Sakaiya, E.; Nakano, K.; Kawamura, K. NDSI map and IPLS using hyperspectral data for assessment of plant and ecosystem variables. *J. Remote Sens. Soc. Jpn.* **2008**, *28*, 317–330.

42. Thenkabail, P.S.; Smith, R.B.; De Pauw, E. Hyperspectral vegetation indices and their relationships with agricultural crop characteristics. *Remote Sens. Environ.* **2000**, *71*, 158–182.

43. Marshall, M.; Thenkabail, P. Biomass modeling of four leading world crops using hyperspectral narrowbands in support pf HyspIRI mission. *Photogramm. Eng. Remote Sens.* **2014**, *80*, 757–772.

44. Ciganda, V.; Gitelson, A.; Schepers, J. Non-destructive determination of maize leaf and canopy chlorophyll content. *J. Plant Physiol.* **2009**, *166*, 157–167.

45. Akaike, H. A new look at the statistical model identification. *IEEE Trans. Autom. Control* **1974**, *19*, 716–723.

46. Kawamura, K.; Watanabe, N.; Sakanoue, S.; Lee, H.-J.; Inoue, Y.; Odagawa, S. Testing genetic algorithm as a tool to select relevant wavebands from field hyperspectral data for estimating pasture mass and quality in a mixed sown pasture using partial least squares regression. *Grassl. Sci.* **2010**, *56*, 205–216.

47. Mutanga, O.; Skidmore, A.K.; Prins, H.H. T. Predicting *in situ* pasture quality in the Kruger National Park, South Africa, using continuum-removed absorption features. *Remote Sens. Environ.* **2004**, *89*, 393–408.

48. Gitelson, A.A. Remote estimation of crop fractional vegetation cover: The use of noise equivalent as an indicator of performance of vegetation indices. *Int. J. Remote Sens.* **2013**, *34*, 6054–6066.

49. Viña, A.; Gitelson, A.A. New developments in the remote estimation of the fraction of absorbed photosynthetically active radiation in crops. *Geophys. Res. Lett.* **2005**, *32*, L17403.

50. R Development Core Team. *R: A Language and Environment for Statistical Computing*; R Foundation for Statistical Computing: Vienna, Austria, 2012.

51. McRoberts, R.E.; Nelson, M.D.; Wendt, D.G. Stratified estimation of forest area using satellite imagery, inventory data, and the *k*-Nearest Neighbors technique. *Remote Sens. Environ.* **2002**, *82*, 457–468.

52. Herrmann, I.; Pimstein, A.; Karnieli, A.; Cohen, Y.; Alchanatis, V.; Bonfil, D.J. LAI assessment of wheat and potato crops by VENμS and Sentinel-2 bands. *Remote Sens. Environ.* **2011**, *115*, 2141–2151.

53. Li, X.; Zhang, Y.; Bao, Y.; Luo, J.; Jin, X.; Xu, X.; Song, X.; Yang, G. Exploring the best hyperspectral features for LAI estimation using partial least squares regression. *Remote Sens.* **2014**, *6*, 6221–6241.

54. Schlemmer, M.; Gitelson, A.; Schepers, J.; Ferguson, R.; Peng, Y.; Shanahan, J.; Rundquist, D. Remote estimation of nitrogen and chlorophyll contents in maize at leaf and canopy levels. *Int. J. Appl. Earth Obs. Geoinf.* **2013**, *25*, 47–54.

55. Filella, I.; Serrano, L.; Serra, J.; Peñuelas, J. Evaluating wheat nitrogen status with canopy reflectance indices and discriminant analysis. *Crop Sci.* **1995**, *35*, 1400–1405.

56. Filella, I.; Penuelas, J. The red edge position and shape as indicators of plant chlorophyll content, biomass and hydric status. *Int. J. Remote Sens.* **1994**, *15*, 1459–1470.

57. Daughtry, C.S. T.; Walthall, C.L. Spectral discrimination of *Cannabis sativa L.* leaves and canopies. *Remote Sens. Environ.* **1998**, *64*, 192–201.

58. Peng, Y.; Gitelson, A.A.; Keydan, G.; Rundquist, D.C.; Moses, W. Remote estimation of gross primary production in maize and support for a new paradigm based on total crop chlorophyll content. *Remote Sens. Environ.* **2011**, *115*, 978–989.

59. Gitelson, A.A.; Peng, Y.; Arkebauer, T.J.; Schepers, J. Relationships between gross primary production, green LAI, and canopy chlorophyll content in maize: Implications for remote sensing of primary production. *Remote Sens. Environ.* **2014**, *144*, 65–72.

60. Thenkabail, P.S.; Smith, R.B.; de Pauw, E. Evaluation of narrowband and broadband vegetation indices for determining optimal hyperspectral wavebands for agricultural crop characterization. *Photogramm. Eng. Remote Sens.* **2002**, *68*, 607–621.

61. Le Maire, G.; François, C.; Soudani, K.; Berveiller, D.; Pontailler, J.-Y.; Bréda, N.; Genet, H.; Davi, H.; Dufrêne, E. Calibration and validation of hyperspectral indices for the estimation of broadleaved forest leaf chlorophyll content, leaf mass per area, leaf area index and leaf canopy biomass. *Remote Sens. Environ.* **2008**, *112*, 3846–3864.

Satellite Remote Sensing-Based In-Season Diagnosis of Rice Nitrogen Status in Northeast China

Shanyu Huang, Yuxin Miao, Guangming Zhao, Fei Yuan, Chuanxiang Tan, Weifeng Yu, Martin L. Gnyp, Victoria I.S. Lenz-Wiedemann, Uwe Rascher and Georg Bareth

Abstract: Rice farming in Northeast China is crucially important for China's food security and sustainable development. A key challenge is how to optimize nitrogen (N) management to ensure high yield production while improving N use efficiency and protecting the environment. Handheld chlorophyll meter (CM) and active crop canopy sensors have been used to improve rice N management in this region. However, these technologies are still time consuming for large-scale applications. Satellite remote sensing provides a promising technology for large-scale crop growth monitoring and precision management. The objective of this study was to evaluate the potential of using FORMOSAT-2 satellite images to diagnose rice N status for guiding topdressing N application at the stem elongation stage in Northeast China. Five farmers' fields (three in 2011 and two in 2012) were selected from the Qixing Farm in Heilongjiang Province of Northeast China. FORMOSAT-2 satellite images were collected in late June. Simultaneously, 92 field samples were collected and six agronomic variables, including aboveground biomass, leaf area index (LAI), plant N concentration (PNC), plant N uptake (PNU), CM readings and N nutrition index (NNI) defined as the ratio of actual PNC and critical PNC, were determined. Based on the FORMOSAT-2 imagery, a total of 50 vegetation indices (VIs) were computed and correlated with the field-based agronomic variables. Results indicated that 45% of NNI variability could be explained using Ratio Vegetation Index 3 (RVI3) directly across years. A more practical and promising approach was proposed by using satellite remote sensing to estimate aboveground biomass and PNU at the panicle initiation stage and then using these two variables to estimate NNI indirectly (R^2 = 0.52 across years). Further, the difference between the estimated PNU and the critical PNU can be used to guide the topdressing N application rate adjustments.

Reprinted from *Remote Sens.* Cite as: Huang, S.; Miao, Y.; Zhao, G.; Yuan, F.; Ma, X.; Tan, C.; Yu, W.; Gnyp, M.L.; Lenz-Wiedemann, V.I.S.; Rascher, U.; Bareth, G. Satellite Remote Sensing-Based In-Season Diagnosis of Rice Nitrogen Status in Northeast China. *Remote Sens.* **2015**, *7*, 10646–10667.

1. Introduction

Rice (*Oryza sativa* L.) is one of the most important crops in the world, and more than two-thirds of China's population relies on rice as the staple food [1]. Nitrogen (N) is an important element in chlorophyll constitution. Its supply rate affects biomass production and yield to a large extent. Farmers tend to apply high rates of N fertilizer in order to get a high yield. In the past 50 years, Chinese cereal production increased by 3.2 times, mainly due to an increased input of synthetic fertilizers, especially N fertilizer [2]. The agronomic efficiency of N fertilizer for rice is only 11.7 kg·kg^{-1} in China, much lower than those in developed countries (20–25 kg·kg^{-1}) [3,4]. The over-application of N fertilizer increases the risks of environmental pollution due to N loss into the surface water bodies, groundwater or atmosphere, resulting in water eutrophication, increased nitrate content in the groundwater and greenhouse gas emissions [5]. Precision N management strategies are developed to improve fertilizer N use efficiency by matching the fertilizer N input to crop N demand in proper time and space [6]. This requires the development of technologies for real-time and site-specific diagnosis of crop N status in the field for guiding the topdressing N applications [7].

Plant N concentration (PNC) and uptake (PNU) have been commonly used as crop N status indicators. To improve crop N status diagnosis, the concept of critical N concentration (N_c) has been proposed as the minimum PNC necessary to achieve maximum aboveground biomass production [8,9]. N_c decreases with increasing biomass. Their relationship can be described using a negative power function, called the critical N dilution curve [10]. Thus, the N_c at any given biomass value can be calculated by this dilution curve. The actual PNC (N_a) can then be compared to N_c, and their ratio is termed the N nutrition index (NNI). NNI is a better indicator for diagnosing crop N status than PNC or PNU [10]. If N_a is greater than N_c (NNI > 1), this indicates an over-supply of N, while the opposite is true if N_a is smaller than N_c (NNI < 1) [10]. An NNI value of one indicates an optimal N supply. The calculation of NNI requires destructive sampling and chemical analysis to determine biomass and plant N concentration, which is time and cost consuming and, thus, impractical for in-season site-specific N management across large areas. Therefore, there is an increasing interest in using proximal and remote sensing technologies to non-destructively estimate the crop NNI [10–13]. Several researchers have successfully used chlorophyll meter (CM) data to estimate the NNI of wheat (*Triticum aestivum* L.) [14–17] and maize (*Zea mays* L.) [18]. However, CM data are point measurements at the leaf level and unsuitable for precision N management across large areas [19].

Crop canopy sensors are more efficient and promising than leaf sensors for monitoring crop N status across large fields [7,13]. Mistele and Schmidhalter [20] used a passive hyperspectral canopy sensor to estimate NNI. They found that the red edge inflection point (REIP) could explain 95% of winter wheat NNI variability. A passive hyperspectral canopy sensor was also applied to estimate maize NNI by Chen *et al.* [21]. They reported that a model based on principal component analysis and a back propagation artificial neural network approach performed the best by explaining 81% of NNI variability. However, passive canopy sensors are constrained by the time and cloud cover of the acquisition day. Such hyperspectral sensors are also very expensive; therefore, they may be more suitable for research than for on-farm applications.

Active optical crop canopy sensors, unlike passive sensors, have modulated light emitting diodes that irradiate a plant canopy and measure a portion of the reflected radiation, without relying on ambient sunlight [22]. They are not influenced by environmental light conditions and do not need frequent calibrations. The GreenSeeker active canopy sensor (Trimble Navigation Limited, Sunnyvale, CA, USA) has a red (R) and near-infrared (NIR) band and provides two vegetation indices (VIs), the Normalized Difference Vegetation Index (NDVI) and the Ratio Vegetation Index (RVI). It was found that GreenSeeker NDVI and RVI explained 47% and 44% of winter wheat NNI variability, respectively, across site years and growth stages [7]. The Crop Circle ACS 470 sensor (Holland Scientific, Inc., Lincoln, NE, USA) is a configurable active crop canopy sensor with three wavebands. It was found that two VIs calculated with the Crop Circle wavebands, the Green Re-normalized Difference Vegetation Index (GRDVI) and the Modified Green Soil Adjusted Vegetation Index (MGSAVI), were effective for estimating winter wheat NNI across site years and growth stages ($R^2 = 0.77 - 0.78$) [7]. For rice, the GreenSeeker sensor explained 25%–34% and 30%–31% of NNI variability at the stem elongation and heading stage, respectively [13]. Using the Crop Circle ACS 470 sensor, four red edge-based indices, including the Red Edge Soil Adjusted Vegetation Index (RESAVI), the Modified RESAVI (MRESAVI), the Red Edge Difference Vegetation Index (REDVI) and the Red Edge Re-normalized Difference Vegetation Index (RERDVI), performed equally well for estimating rice NNI across growth stages ($R^2 = 0.76$) [12]. Active crop sensors have been mounted on fertilizer applicators, and on-the-go sensing and variable rate N applications have been realized for maize and wheat, but not for rice, considering the challenges for fertilizer application machines to enter paddy fields flooded with water.

Aerial and satellite remote sensing is a promising technology to monitor crop N status for large production fields [23]. Aerial hyperspectral remote sensing and CM data were combined to diagnose maize N status using the N Sufficiency Index (NSI) approach [19]. Cilia *et al.* [24] applied aerial hyperspectral sensing to estimate maize NNI indirectly. They calculated the Modified Chlorophyll Absorption Ratio Index/Modified Triangular Vegetation Index 2 (MCARI/MTVI2) and MTVI2 to estimate maize PNC ($R^2 = 0.59$) and biomass ($R^2 = 0.80$), respectively. Then, they combined the predicted PNC and biomass maps to generate an NNI map, which agreed well with the NNI obtained by destructive sampling and analysis ($R^2 = 0.70$). The improvements in spatial and temporal resolutions of satellite remote sensing make it possible to monitor crop N status at key crop growth stages. Wu *et al.* [25] compared QuickBird data with CM readings and petiole nitrate concentration. They found that the QuickBird-VIs differed significantly for different N input treatments at the late growing season. Yang *et al.* [26] found that the NDVI derived from FORMOSAT-2 satellite imagery was highly correlated to the NDVI calculated from a ground canopy reflectance sensor ($R^2 = 0.79$). Darvishzadeh *et al.* [27] used the inversion of the PROSAIL model with a lookup table approach and multispectral satellite image data of ALOS AVNIR-2. The method explained 65% of rice plant chlorophyll content variability with a low root mean square error (RMSE) of $0.45 \, g \cdot m^{-2}$.

So far, little has been reported on rice NNI estimation using satellite remote sensing. Therefore, the objective of this study was to evaluate the potential of using FORMOSAT-2 satellite remote sensing to estimate rice NNI at a key growth stage for guiding panicle N fertilizer application in Northeast China.

2. Materials and Methods

2.1. Study Site

The study site is located at the Qixing Farm in the Sanjiang Plain, Heilongjiang Province, Northeast China. The Sanjiang Plain used to be a wild natural wetland formed by the alluvia of three river systems—Heilong River, Songhua River and Wusuli River. During the past 50 years, the natural wetland was reclaimed for arable land, especially paddy rice fields. Due to the small population density in this region, each farmer's household has about a 20–30-ha cultivation area, making it the leading large-scale farming region in China. The main soil type is Albic soil, classified as Mollic Planosols in the FAO-UNESCO system, and typical Argialbolls in the Soil Taxonomy [28]. This area has a typical cool-temperature sub-humid continental monsoon climate. During the growing season (April–October), the average rainfall is about 400 mm, which accounts for approximately 70% of yearly precipitation. The mean annual temperature is about $2 \, °C$ [29]. The annual sunshine duration is 2300–2600 h, and the whole year frost-free period ranges from 120–140 days [30].

2.2. Field Information

This study was conducted to diagnose rice N status at a key growth stage to guide panicle fertilizer application based on satellite images. For cold region rice, the crucial period for panicle fertilizer topdressing is during the stem elongation stage. Considering the time it takes for satellite image acquisition and processing, the best diagnosis stage is at panicle initiation, which is about 7–10 days before the stem elongation stage [12,31,32]. Three farmers' fields in 2011 and two in 2012 were selected for this study. The cultivars and transplanting densities varied (Table 1). The seedlings were prepared in greenhouses and then transplanted at the 3.1–3.5 leaf stage into the fields.

The regional optimal N rate recommended by the local extension service was around 100 kg·ha^{-1}. Field 1 (F1) was managed by an experienced farmer. The best rice management practice of the region, supported by the Jiansanjiang Experiment Station of the China Agricultural University, was applied for this field. Other fields were managed by individual farmers following their own practices.

Table 1. Detailed information about the farmers' fields selected for this study, Heilongjiang Province, China, 2011–2012.

Field	Year	Number of Samples	Area (ha)	N Rate (kg·ha^{-1})	Variety	Number of Leaves	Transplanting Date	Plant Density (hills·m^{-2})
F1	2011	33	29.6	97.9	Kendao 6	12	17 May 2011	27
F2	2011	4	13.1	105.9	Longjing 26	11	20 May 2011	30
F3	2011	4	31.0	101.0	Kendao 6	12	12 May 2011	27
F4	2012	14	10.7	120.2	Longjing 31	11	16 May 2012	28
F5	2012	37	21.6	98.3	Longjing 31	11	20 May 2012	30

2.3. Remote Sensing Images and Preprocessing

For this study, we selected the FORMOSAT-2 satellite, which belongs to the National Space Organization of Taiwan (NSPO). It runs on a Sun-synchronous orbit with an orbit altitude of 891 km and collects images at the same local hour with a constant observation angle for the same site [33]. The multispectral image of FORMOSAT-2 covers four spectral band regions with a ground resolution of 8 m: blue (B) (450–520 nm), green (G) (520–600 nm), red (R) (630–690 nm) and NIR (760–900 nm) [34]. One image scene covers an area of 24 km × 24 km. The panchromatic image with 2-m ground resolution is collected simultaneously. The daily revisit interval makes FORMOSAT-2 one of the most suitable satellites for precision agriculture applications. Images were obtained on 25 June 2011 and 26 June 2012. These two images were almost cloud-free, especially in the study area.

The images were geometrically corrected and radiometrically calibrated using ENVI 4.8 (ENVI, Boulder, CO, USA). The radiometric calibration was performed using the satellite calibration parameters in the following formula for each band:

$$L = DN/a + L_0 \tag{1}$$

where L stands for radiance; DN is the abbreviation of digital number; a is the absolute calibration coefficients, which is also called gain; and L_0 stands for the offset. After the linear transformation, the DN values were converted to radiance values in units of $W \cdot m^{-2} \cdot sr^{-1} \cdot \mu m^{-1}$. For geometric correction, high precision ground control points were used. The rectification accuracy was less than 0.5 pixels (<4 m), which was acceptable for this research.

2.4. Field Data Collection and Analysis

A total of 41 and 51 ground samples were collected in 2011 and 2012, respectively. The samples were collected from sites representing different crop growth conditions (N deficient, optimum and surplus conditions), based on visual observations. The sampling dates were 25 June 2011, the same acquisition date as the satellite image, and 28 June 2012, two days after the FORMOSAT-2 image collection. At each sampling site, a hand-held differential Trimble Ag332 GPS was used for geo-referencing. Ground truth data included rice cultivar, plant density, tiller numbers and relative chlorophyll concentration measured with the SPAD-502 instrument (Soil-Plant Analysis Development Section, Minolta, Osaka, Japan). Twenty rice plants were selected at each sampling site for CM measurements in the middle part of the top second leaf for each individual plant. At each sampling site, the aboveground biomass was collected destructively by clipping three hills (each hill consisting of 4–6 rice plants). These samples were taken to the laboratory and rinsed with water. The roots were removed, and the samples were separated into leaves and stems. The Leaf Area Index (LAI) was determined by the dry weight method as described by Bei *et al.* [35]. All parts of the samples were put into the oven for deactivation of enzymes at 105 °C for half an hour and then dried at 80 °C until constant weight. After being weighted, the sub-samples were ground to particles smaller than 1 mm and analyzed for N concentration using the Kjeldahl method [36,37].

For the NNI, the N_c was calculated by the following equations developed for rice in this region according to Justes *et al.* [38], based on data from N rate experiments conducted in this region from 2008–2013:

$$N_c = 2.77W^{-0.34} \tag{2}$$

where N_c is the critical N concentration (%) in the aboveground biomass and W is the shoot dry weight expressed in $t \cdot ha^{-1}$. For aboveground biomass larger than $1 \, t \cdot ha^{-1}$, the N_c was calculated by the above equation, otherwise the N_c was set to 2.77%.

2.5. Data Analysis

Many spectral VIs have been developed to estimate plant biophysical variables, such as chlorophyll concentration or content, LAI and biomass. However, many of them use narrow bands based on the research results of proximal hyperspectral sensing. In this study, the potential of using broad band satellite remote sensing images for estimating rice N status indicators was evaluated using the broad bands of FORMOSAT-2 satellite images. A total of 50 VIs were evaluated (Table 2, [39–60]). The software ENVI and ArcGIS 9 (ESRI, Redlands, CA, USA) were used to extract the pixel values from the FORMOSAT-2 satellite images and to calculate the VIs for corresponding sampling sites.

The regression analysis considered the 50 VIs and each of the 6 field-measured agronomic variables separately. The correlation and regression analyses were performed using SPSS V.20.0 (SPSS, Chicago, IL, USA). The RMSE and relative error (RE) were also calculated to evaluate model performances.

Table 2. Vegetation indices evaluated in this study for estimating rice N status indicators, Heilongjiang Province, China, 2011–2012.

Vegetation Index	Formula	Ref.
Two-band vegetation indices		
Ratio Vegetation Index 1 (RVI1)	NIR/B	[39]
Ratio Vegetation Index 2 (RVI2)	NIR/G	[40]
Ratio Vegetation Index 3 (RVI3)	NIR/R	[39]
Difference Index1 (DVI1)	NIR − B	[39]
Difference Index2 (DVI2)	NIR − G	[39]
Difference Index3 (DVI3)	NIR − R	[39]
Normalized Difference Vegetation Index 1 (NDVI1)	(NIR − R)/(NIR + R)	[40]
Normalized Difference Vegetation Index 2 (NDVI2)	(NIR − G)/(NIR + G)	[41]
Normalized Difference Vegetation Index 3 (NDVI3)	(NIR − B)/(NIR + B)	[40]
Renormalized Difference Vegetation Index 1 (RDVI1)	(NIR − B)/SQRT(NIR + B)	[42]
Renormalized Difference Vegetation Index 2 (RDVI2)	(NIR − G)/SQRT(NIR + G)	[42]
Renormalized Difference Vegetation Index 3 (RDVI3)	(NIR − R)/SQRT(NIR + R)	[42]
Chlorophyll Index (CI)	NIR/G − 1	[43]
Wide Dynamic Range Vegetation Index 1 (WDRVI1)	$(0.12\,NIR - R)/(0.12{\cdot}NIR + R)$	[44]
Wide Dynamic Range Vegetation Index 2 (WDRVI2)	$(0.12\,NIR - G)/(0.12{\cdot}NIR + G)$	[44]
Wide Dynamic Range Vegetation Index 3 (WDRVI3)	$(0.12\,NIR - B)/(0.12{\cdot}NIR + B)$	[44]
Soil Adjusted Vegetation Index (SAVI)	1.5(NIR − R)/(NIR + R + 0.5)	[45]
Green Soil Adjusted Vegetation Index (GSAVI)	1.5(NIR − G)/(NIR + G + 0.5)	[45]
Blue Soil Adjusted Vegetation Index (BSAVI)	1.5(NIR − B)/(NIR + B + 0.5)	[45]
Modified Simple Ratio (MSR)	(NIR/R − 1)/SQRT(NIR/R + 1)	[46]
Optimal Soil Adjusted Vegetation Index (OSAVI)	(1 + 0.16)[(NIR − R)/(NIR + R + 0.16)]	[47]
Green Optimal Soil Adjusted Vegetation Index (GOSAVI)	(1 + 0.16)[(NIR − G)/(NIR + G + 0.16)]	[47]
Blue Optimal Soil Adjusted Vegetation Index (BOSAVI)	(1 + 0.16)[(NIR − B)/(NIR + B + 0.16)]	[47]
Modified Soil Adjusted Vegetation Index (MSAVI)	$0.5\{2{\cdot}NIR + 1 - SQRT[(2{\cdot}NIR + 1)^2 - 8(NIR - R)]\}$	[48]
Modified Green Soil Adjusted Vegetation Index (MGSAVI1)	$0.5\{2{\cdot}NIR + 1 - SQRT[(2{\cdot}NIR + 1)^2 - 8(NIR - G)]\}$	[48]
Modified Blue Soil Adjusted Vegetation Index (MBSAVI)	$0.5\{2{\cdot}NIR + 1 - SQRT[(2{\cdot}NIR + 1)^2 - 8(NIR - B)]\}$	[48]
Three-band vegetation indices		
Simple Ratio Vegetation Index (SR)	R/G × NIR	[49]
Modified Normalized Difference Vegetation Index 1 (mNDVI1)	$(NIR - R + 2{\cdot}G)/(NIR + R - 2{\cdot}G)$	[50]
Modified Normalized Difference Vegetation Index 2 (mNDVI2)	$(NIR - R + 2{\cdot}B)/(NIR + R - 2{\cdot}B)$	[50]

Table 2. *Cont.*

Vegetation Index	Formula	Ref.
New Modified Simple Ratio (mSR)	$(NIR - B)/(R - B)$	[51]
Visible Atmospherically-Resistant Index (VARI)	$(G - R)/(G + R - B)$	[52]
Structure Insensitive Pigment Index (SIPI)	$(NIR - B)/(NIR - R)$	[53]
Structure Insensitive Pigment Index 1 (SIPI1)	$(NIR - B)/(NIR - G)$	[53]
Normalized Different Index (NDI)	$(NIR - R)/(NIR - G)$	[49]
Plant Senescence Reflectance Index (PSRI)	$(R - B)/NIR$	[51]
Plant Senescence Reflectance Index 1 (PSRI1)	$(R - G)/NIR$	[51]
Modified Chlorophyll Absorption in Reflectance Index (MCARI)	$[(NIR - R) - 0.2(R - G)] \times (NIR/R)$	[54]
Modified Chlorophyll Absorption in Reflectance Index 1 (MCARI1)	$1.2[2.5(NIR - R) - 1.3(NIR - G)]$	[55]
Modified Chlorophyll Absorption in Reflectance Index 2 (MCARI2)	$1.2[2.5(NIR - R) - 1.3(R - G)]/SQRT[(2 \cdot NIR + 1)^2 - (6 \cdot NIR - 5 \cdot SQRT(R) - 0.5]$	[55]
Triangular Vegetation Index (TVI)	$0.5[120(NIR - G) - 200(R - G)]$	[57]
Modified Triangular Vegetation Index 1 (MTVI1)	$1.2[1.2(NIR - G) - 2.5(R - G)]$	[55]
Modified Triangular Vegetation Index 2 (MTVI2)	$1.5[1.2(NIR - G) - 2.5(R - G)]/SQRT[(2 \cdot NIR + 1)^2 - (6 \cdot NIR - 5 \cdot SQRT(R) - 0.5]$	[55]
Modified Triangular Vegetation Index 3 (MTVI3)	$1.5[1.2(NIR - B) - 2.5(R - B)]/SQRT[(2\ NIR + 1)^2 - (6\ NIR - 5\ SQRT(R) - 0.5]$	[55]
Enhanced Vegetation Index (EVI)	$2.5(NIR - R)/(1 + NIR + 6\ R - 7.5\ B)$	[58]
Transformed Chlorophyll Absorption in Reflectance Index (TCARI)	$3[(NIR - R) - 0.2(NIR - G)(NIR/R)]$	[56]
Triangular Chlorophyll Index (TCI)	$1.2(NIR - G) - 5(R - G)(NIR/R)\char`\^0.5$	[59]
TCARI/OSAVI	TCARI/OSAVI	[56]
MCARI/MTVI2	MCARI/MTVI2	[60]
TCARI/MSAVI	TCARI/MSAVI	[56]
TCI/OSAVI	TCI/OSAVI	[59]

2.6. The Estimation of NNI

The rice NNI can be estimated directly and indirectly. The direct method is to use the selected VI to estimate NNI directly based on the established relationships. The indirect method is to first use the selected VIs to estimate rice biomass and PNU. With the critical N dilution curve developed for rice in this region, the N_c can be derived for each biomass value. The estimated biomass and N_c can then be used together to calculate critical PNU (biomass $\times$ N_c). The NNI can then be estimated using PNU and critical PNU, because PNU/critical PNU equal (biomass $\times$ N_a)/(biomass $\times$ N_c), which can be further simplified to N_a/N_c. Considering practical applications, we classified the rice N status into three categories based on NNI values: deficient N status (NNI < 0.95), optimal N status (NNI = 0.95–1.05) and surplus N status (NNI > 1.05).

The indirect method was used in this study to create NNI maps of selected fields at the pixel-level. For irrigation purpose, each rice field was divided into many smaller plots, which were also used as management units for fertilizer application. Therefore, the pixel-level NNI values were averaged for each small plot to create plot-level NNI maps using ArcGIS 9.

3. Results

3.1. Variability of Rice N Status Indicators

The variability of rice biomass, LAI and PNU (CV = 23%–28%) was consistently larger than that of PNC, SPAD values and NNI (CV = 4%–14%) (Table 3). In addition, larger variability of PNC and NNI was found in 2012 (CV = 11% and 14%, respectively) than in 2011 (CV = 5%). Likewise, the values of biomass, LAI and PNU were significantly higher in 2012 than in 2011. The NNI ranged from 0.89–1.17 in 2011, with an average of 1.01. This indicated that in general, the N status of these fields was optimal. In 2012, the NNI ranged from 0.83–1.50, with an average of 1.15, revealing a surplus N status (Table 3).

An examination of each individual field indicated that the average PNC and SPAD values were the highest in Filed 1 (F1), the biomass value was the lowest, while the average NNI was optimal. In contrast, F4 had the lowest PNC, but the highest average NNI and biomass, indicating a surplus N status (Table 4). These results indicated the importance of using NNI for N status diagnosis, rather than PNC.

Table 3. Descriptive statistics of rice N status indicators for 2011 (41 field samples) and 2012 (51 field samples), Heilongjiang Province, China.

	Mean	Minimum	Maximum	SD	CV (%)
2011					
Biomass (t·ha^{-1})	0.87	0.50	1.55	0.22	25
Leaf Area Index	0.84	0.52	1.51	0.20	23
Plant N concentration (%)	2.76	2.45	3.06	0.14	5
SPAD value	42.30	37.03	44.08	1.80	4
Plant N uptake (kg·ha^{-1})	23.86	12.97	43.25	5.80	24
Nitrogen Nutrition Index	1.01	0.89	1.17	0.05	5
2012					
Biomass (t·ha^{-1})	2.91	1.45	4.68	0.79	27
Leaf Area Index	3.34	1.77	5.66	0.86	26
Plant N concentration (%)	2.24	1.75	2.77	0.25	11
SPAD Value	40.60	37.07	43.40	1.68	4
Plant N uptake (kg·ha^{-1})	65.00	30.11	114.9	17.93	28
Nitrogen Nutrition Index	1.15	0.83	1.50	0.16	14

Table 4. Descriptive statistics of rice N status indicators for different fields, Heilongjiang Province, China, 2011–2012. NNI stands for N Nutrition Index.

Field	Biomass (t·ha^{-1})	Plant N Concentration (%)	SPAD Value	NNI
F1	0.81 ± 0.16	2.77 ± 0.14	43.07 ± 0.62	1.00 ± 0.05
F2	1.27 ± 0.25	2.63 ± 0.14	37.89 ± 0.89	1.03 ± 0.10
F3	0.97 ± 0.17	2.62 ± 0.11	39.83 ± 0.65	1.00 ± 0.04
F4	3.89 ± 0.41	2.12 ± 0.28	40.90 ± 1.08	1.21 ± 0.16
F5	2.53 ± 0.53	2.29 ± 0.23	40.49 ± 1.85	1.13 ± 0.16

3.2. Vegetation Index Analysis

The performance of the VIs differed with N status indicators. The top 10 VIs for estimating different N status indicators in each year are listed in Table 5.

Table 5. The top 10 coefficients of determination (R^2) for the relationships between vegetation indices based on the FORMOSAT-2 satellite images and rice N status indicators in Heilongjiang Province, China, 2011–2012. Only significant R^2 values are listed.

Index	2011	2012	2011 + 2012	Index	2011	2012	2011 + 2012
	Aboveground Biomass (t·ha^{-1})				LAI		
MCARI	0.67 **	0.62 **	0.90 **	MCARI	0.67 **	0.58 **	0.90 **
DVI3	0.65 **	0.63 **	0.90 **	DVI2	0.67 **	0.58 **	0.91 **
TVI	0.64 **	0.64 **	0.90 **	RVI3	0.65 **	0.60 **	0.90 **
RVI3	0.64 **	0.63 **	0.90 **	DVI3	0.65 **	0.60 **	0.91 **
MTVI1	0.63 **	0.64 **	0.90 **	RDVI2	0.65 **	0.58 **	0.90 **
MCARI1	0.63 **	0.64 **	0.90 **	WDRVI1	0.65 **	0.60 **	0.90 **
TCARI	0.63 **	0.64 **	0.89 **	MSR	0.65 **	0.60 **	0.90 **
WDRVI1	0.63 **	0.64 **	0.89 **	RDVI3	0.64 **	0.60 **	0.90 **
MSR	0.63 **	0.64 **	0.90 **	SAVI	0.63 **	0.61 **	0.88 **
SAVI	0.61 **	0.64 **	0.87 **	NDVI1	0.63 **	0.61 **	0.88 **
	Plant N Concentration (%)				SPAD Values		
DVI4			0.55 **	TCI	0.27 **	0.17 **	0.13 **
RDVI4			0.53 **	PSRI	0.19 **		0.10 **
NDVI4			0.49 **	MTVI2	0.18 **	0.22 **	0.16 **
RDVI2			0.49 **	TCARI	0.16 **	0.22 **	0.14 **
RVI4			0.49 **	MCARI2	0.15 *	0.23 **	0.15 **
MGSAVI			0.48 **	WDRVI1	0.14 *	0.20 **	0.12 **
NDVI2			0.48 **	MTVI3	0.10 *	0.25 **	0.13 **
GOSAVI			0.48 **	TCARI/OSAVI		0.14 **	
WDRVI2			0.47 **	EVI		0.14 **	
mNDVI1			0.30 **	DVI	0.13*		0.19
	Plant N Uptake (kg·ha^{-1})				NNI		
RVI3	0.66 **	0.61 **	0.87 **	RDVI1	0.18 **	0.32 **	0.41 **
TVI	0.66 **	0.61 **	0.87 **	DVI2	0.17 **	0.33 **	0.43 **
WDRVI1	0.66 **	0.62 **	0.87 **	RVI2	0.17 **	0.33 **	0.44 **
RDVI3	0.66 **	0.62 **	0.87 **	WDRVI2	0.16 **	0.34 **	0.43 **
TCARI	0.65 **	0.63 **	0.86 **	DVI3	0.16 **	0.34 **	0.43 **
MSR	0.65 **	0.62 **	0.87 **	RDVI2	0.16 **	0.34 **	0.42 **
MCARI1	0.65 **	0.62 **	0.87 **	RVI3	0.16 **	0.34 **	0.45 **
MTVI1	0.65 **	0.62 **	0.87 **	WDRVI1	0.15 *	0.35 **	0.44 **
SAVI	0.64 **	0.62 **	0.85 **	RDVI3	0.15 *	0.35 **	0.43 **
OSAVI	0.64 **	0.62 **	0.85 **	TVI	0.15 *	0.34 **	0.44 **

** Correlation is significant at the 0.01 level; * Correlation is significant at the 0.05 level.

For aboveground biomass, the top 10 VIs performed similarly in 2011 (R^2 = 0.63–0.67) and 2012 (R^2 = 0.63–0.64). This was also true for PNU for both years. For LAI, the top 10 VIs performed slightly better in 2011 (R^2 = 0.63–0.67) than in 2012 (R^2 = 0.58–0.60). Four VIs that are based on the combinations of NIR and red bands, including Ratio Vegetation Index 3 (RVI3), Wide Dynamic Range Vegetation Index 1 (WDRVI1), Soil Adjusted Vegetation Index (SAVI) and Modified Simple Ratio (MSR), were consistently among the top 10 indices for biomass, PNU and LAI. The MCARI index, based on the combination of NIR, red and green bands, had the highest correlation with aboveground biomass (R^2 = 0.67) and LAI (R^2 = 0.67) in 2011. Four VIs, which included MCARI1, Triangular Vegetation Index (TVI), Modified TVI1 (MTVI1) and Transformed Chlorophyll Absorption in Reflectance Index (TCARI), were also among the top 10 indices for both aboveground biomass and PNU.

Lower correlations were found between the VIs and NNIs, with R^2 of 0.15–0.18 in 2011 and 0.33–0.35 in 2012 for the 10 best models. None of the VIs was significantly correlated with PNC in a specific year, although 30–55% of the PNC variability was explained across the two years (Table 5). The relationships between VIs and SPAD values were also weak, with R^2 being 0.10–0.27 and 0.14–0.23 in 2011 and 2012, respectively.

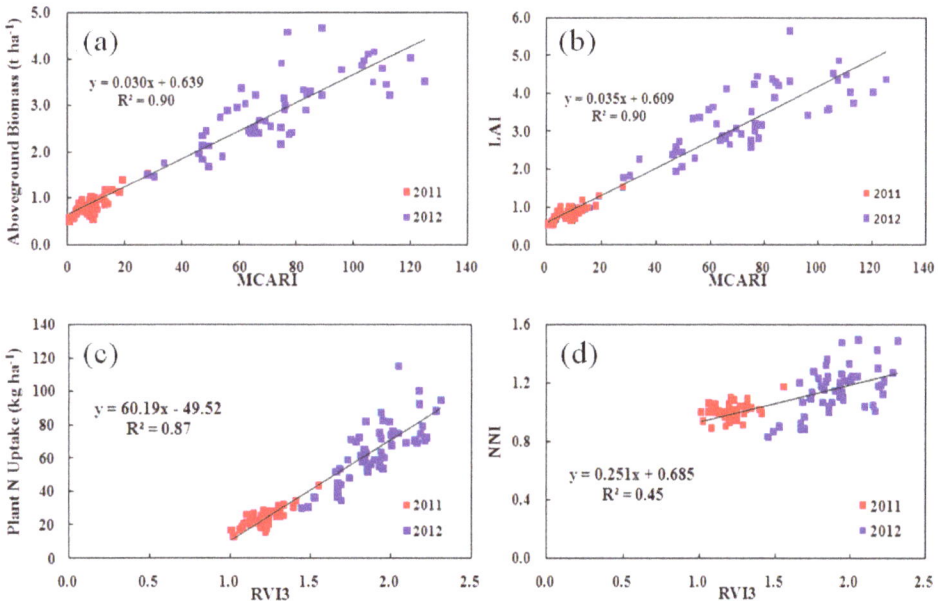

Figure 1. Selected VI regression *vs.* rice aboveground biomass (**a**) LAI; (**b**) plant N uptake (PNU); (**c**) and NNI; (**d**) Heilongjiang Province, China, 2011–2012.

Figure 1 shows selected VI models with the best performance in estimating rice aboveground biomass, LAI, PNU and NNI across years. The values for 2011 samples were all smaller than those of 2012. Most samples in 2011 had NNI values close to optimum, and the variability was very small, with CV being only 5%. As a result, a cluster was formed at the lower end of Figure 1d. This may explain why the relationships between VIs and NNI were quite weak in 2011 (Table 5).

3.3. Nitrogen Status Diagnosis

According to the above results, an indirect NNI estimation method was used in this study. The NNI values estimated this way were moderately correlated with measured NNI across 2011 and 2012 (R^2 = 0.52, RMSE = 0.10 and RE = 9.14%) (Figure 2). By comparing the regression line to the 1:1 line in Figure 2, a systematic bias can be identified in the regression model. In particular, when the observed NNI was less than 1.08, the model overestimated the NNI, while the opposite was true when the NNI was greater than 1.08.

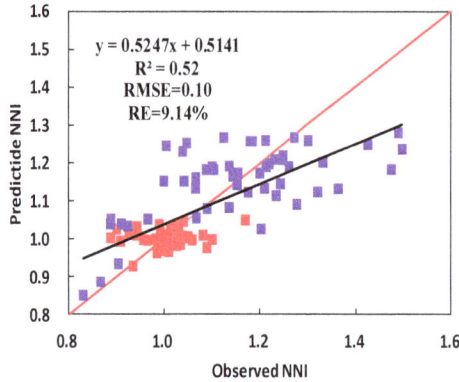

Figure 2. Relationship between observed and predicted NNI using MCARI-estimated biomass and RVI3-estimated plant N uptake in 2011 and 2012, Heilongjiang Province, China. The red line is the 1:1 line.

The NNI maps created using the indirect method for two farmers' fields are shown in Figure 3 as an example. Figure 3a,b shows the NNI maps at the pixel level and the plot level, respectively. The first (Figure 3, left) is a well-managed field, with 92% of the field being in the optimal N status category. In contrast, the second field (Figure 3, right) had only 35% in the optimal N category and about 51% in the deficient N category.

A more quantitative and preferable approach is to produce a PNU difference map (ΔPNU) by subtracting the critical PNU map from the predicted PNU map. This ΔPNU map can not only tell us if the N status is deficient, optimal or surplus, but also the amount of deficiency or surplus. This further can be used to produce a prescription map for topdressing N application rates (NR) at the stem elongation stage. Specifically, the prescription map will be the planned topdressing panicle NR map based on regional best management practice minus the ΔPNU map. Figure 4 displays a ΔPNU map of the second field shown in Figure 3. About 12% of the field had an N surplus of over 5 kg·ha^{-1}, while 20% of the field had an N deficiency of over 5 kg·ha^{-1}.

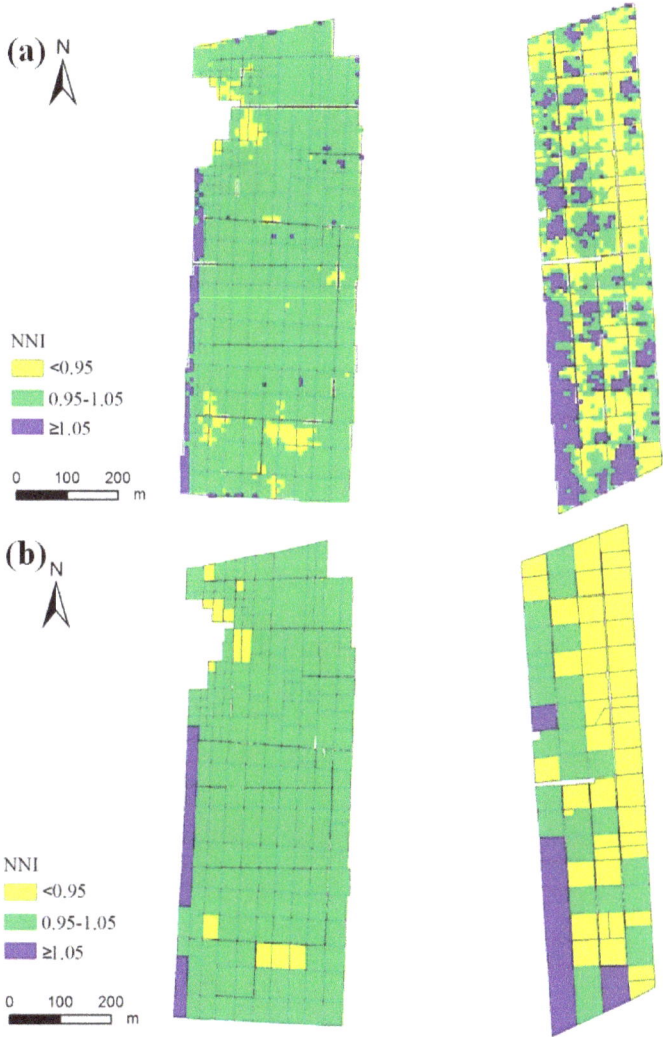

Figure 3. Examples of predicted rice nitrogen nutrition index (NNI) maps of two fields at the pixel level (**a**) and the plot-level (**b**), Heilongjiang Province, China.

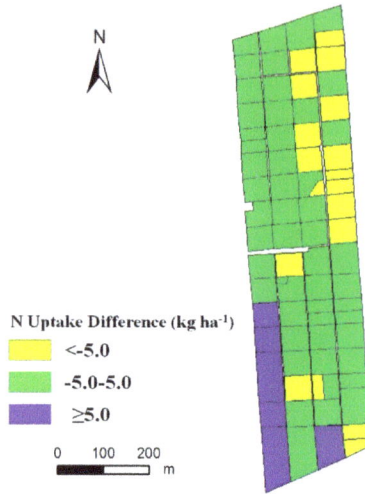

Figure 4. Example of a plant N uptake difference map of a farmer's field, Heilongjiang Province, China.

4. Discussion

4.1. Direct Estimation of NNI

Using satellite remote sensing to estimate rice plant NNI for diagnosing rice N status and guiding in-season site-specific N management across large areas is an attractive idea. How well can we estimate NNI directly using FORMOSAT-2 satellite data? The results of this study indicated that all of the top 10 VIs were significantly correlated with NNI, explaining 18% and 35% of the NNI variability in 2011 and 2012, respectively. Across years, 45% of NNI variability was explained with RVI3. This result is slightly better than what Yao et al. [13] found using the handheld GreenSeeker NDVI and RVI, which explained 25% and 34% of rice NNI variability at the stem elongation stage, respectively. It was found that the top 10 VIs obtained with the three-band Crop Circle ACS 470 sensor explained 61%–69% of rice NNI variability across the panicle initiation and stem elongation stages [12]. However, our study only used data from the panicle initiation stage in 2011, which was expected to be more influenced by the water background than the stem elongation stage. In general, it is not satisfactory to use satellite images to directly estimate rice plant NNI at this stage. At later stages when the rice plants reach canopy closure, this approach may work better. However, it may then be too late for guiding in-season N application.

4.2. Indirect Estimation of NNI

An alternative approach is to use remote sensing to estimate key parameters and indirectly estimate NNI. Cilia *et al.* [24] used aerial hyperspectral remote sensing to estimate maize N concentration and biomass and then estimated NNI indirectly. Our study indicated that biomass could be reliably estimated using satellite remote sensing at the panicle initiation and stem elongation stages, with over 60% of its variability being explained by the top 10 VIs in both 2011 and 2012. We selected MCARI for further analysis. This index was initially developed for estimating leaf chlorophyll variation, but it was also significantly related to LAI [54,55]. In this study, the MCARI index was highly correlated with rice aboveground biomass and LAI (R^2 = 0.58–0.67). The results agree with those of Cao *et al.* [12], who also identified a modified MCARI as the best index for estimating rice biomass (R^2 = 0.79) and plant N uptake (R^2 = 0.83) across growth stages. The top 10 Crop Circle VIs in their study explained 50%–54% of rice biomass variability across the panicle initiation and stem elongation stages. Our results were comparable to the results (R^2 = 0.68–0.69) of Gnyp *et al.* [61] that were obtained with optimized narrow band RVI and NDVI for estimating rice biomass at the stem elongation stage. However, estimating rice PNC before canopy closure is a great challenge. We did not find any significant correlation between VIs and rice PNC in this study. This was also stated by Yao *et al.* [13]. They found that the GreenSeeker NDVI and RVI were not significantly correlated with rice PNC at the stem elongation stage. Cao *et al.* [12] found that the three-band Crop Circle ACS 470 sensor at best explained 33% of rice PNC across the panicle initiation and stem elongation stages using the Red Edge Green Difference Vegetation Index (REGDVI). Even with hyperspectral remote sensing, Yu *et al.* [62] only explained 39% of rice PNC variability across the tillering and heading stages using the Optimized Simple Ratio or Normalized Difference Index. Before canopy closure, soil and water backgrounds in paddy rice fields can influence plant reflectance [63]. In addition, plant biomass dominates canopy reflectance before the heading stage, making the estimation of chlorophyll and N concentration at early growth stages difficult [20]. Therefore, the approach adopted by Cilia *et al.* [24] did not work for rice monitoring at the panicle initiation and stem elongation stages in our study.

A practical approach is to use satellite remote sensing to estimate rice biomass and PNU. From the estimated biomass and the critical N dilution curve, the critical PNU can be determined, and NNI will be calculated using the estimated PNU and the critical PNU. The results of this study supported this idea. Over 60% of rice PNU variability was explained by RVI3 in both years. This was even better than the result obtained with the GreenSeeker sensor for estimating rice PNU at the stem elongation stage (R^2 = 0.40–0.41) by Yao *et al.* [13] and similar to the results (R^2 = 0.63–0.65) obtained with the Crop Circle ACS 470 sensor for estimating rice PNU across the panicle initiation and stem elongation stages by Cao *et al.* [12]. The estimated NNI obtained this way explained 52% of the measured NNI variability across 2011 and 2012, which was slightly better than the direct estimation of NNI using VIs obtained from satellite images ($R^2 = 0.45$).

4.3. Applications for Rice N Status Diagnosis and Topdressing N Recommendation

After the NNI map is generated, it is necessary to define the NNI thresholds for N status diagnosis. The current thresholds (NNI < 1: deficient; NNI = 1: optimal; NNI > 1: surplus) may need to be further refined for practical applications. For example, the NNI values of 0.99 and 1.01 are very close to each other and are all quite optimal, but they will be classified as deficient and surplus N status, respectively, based on current thresholds. Cilia *et al.* [24] proposed to classify NNI into five classes (NNI $\leqslant$ 0.7, 0.7 < NNI $\leqslant$ 0.9, 0.9 < NNI $\leqslant$ 1.1, 1.1 < NNI $\leqslant$ 1.3, NNI > 1.3) and regarded NNI $\leqslant$ 0.9 as N deficient, 0.9 < NNI $\leqslant$ 1.1 as N optimal and NNI > 1.1 as N surplus. Based on the rice N management situations in the study region, we proposed the following thresholds for rice: NNI $\leqslant$ 0.95 as N deficient, 0.95 < NNI $\leqslant$ 1.05 as N optimal and NNI > 1.05 as N surplus. These threshold values can be used to delineate a field into three regions with different N nutritional status. The diagnosis results shown in Figure 3 indicated that the first field (Figure 3 left) was well managed, with the majority of the field having an optimal N status, while about 51% of the second field (Figure 3, right) was deficient in N. These agreed quite well with the two farmers' management practices. However, these threshold values are empirical, and more studies are needed to further test and refine these thresholds by relating NNI to relative grain yield.

The NNI-based rice N status map can be used to guide in-season topdressing N application. For the optimal N zone, 30 kg$\cdot$N$\cdot$ha^{-1} was recommended based on the regional best N management practice. For the deficient N zone, 35 or 40 kg$\cdot$ha^{-1} can be recommended, and for the surplus N zone 25 or 20 kg$\cdot$ha^{-1}. This approach is commonly used in site-specific N management of rice based on CM diagnosis developed by the International Rice Research Institute [64]. It is empirical, but very practical for on-farm applications in small-scale farming areas of Asia. A more quantitative approach is to produce a PNU difference map using the estimated PNU map minus the critical PNU map. The recommended N topdressing application rate can be determined using the regional optimum topdressing N application rate minus the PNU difference. This approach is different from the variable rate N application strategy proposed by Cilia *et al.* [24]. They first computed the average PNU from the optimal NNI pixels and then used this average value together with the estimated PNU to calculate the difference, and for N deficient pixels, the deficient amounts were used as variable N application rates. For pixels with optimal and surplus N, no N fertilizers were recommended. In our approach, we did not analyze the pixel scale, because in rice farming, the field is divided into many small plots for irrigation purpose. These plots also serve as management units. We applied plot-average NNI values to diagnose the rice N status of each plot. Our precision N management strategy takes the regional optimal N rate as the initial total N rate, with 40% and 30% being applied as basal and tillering N fertilizers, respectively. For topdressing N application at the stem elongation stage, 30% of the initial total N rate should be applied if the N status is optimal. Otherwise, the topdressing N rates can be adjusted based on deficient or surplus N amounts. Even if the N status is optimal at the stem elongation stage, it only indicates the N status at that stage, which is more than two months prior to harvest, and a certain amount of N fertilizers should still be recommended to meet the N requirements from stem elongation to harvest.

4.4. Challenges and Future Research Needs

The proposed approach discussed above requires the satellite imagery to be collected in a narrow time window, preferably one week before topdressing N application at the stem elongation stage for rice in the study region. If the image is collected too early, the diagnosis result may not match the true rice N status at the stem elongation stage. In addition, rice plants will be too small, and the water background will strongly influence the plant reflectance. If the image is collected too close to the stem elongation stage, it may be too late to use the diagnosis result for guiding the topdressing N application. Therefore, a satellite with a high temporal resolution is required. The daily revisit time of the FORMOSAT-2 satellite makes it ideal for this purpose. Its 8-m spatial resolution may be too coarse for small-scale farming in other parts of China, such as in the North China Plain [65], but is good enough for large-scale farming in the Sanjiang Plain of Northeast China.

It should be noted that there are 7–10 days between the panicle initiation and stem elongation stages, and the rice plants are fast developing, so the rice biomass and plant N uptake determined at the panicle initiation stage are smaller than the values at the stem elongation stage. Studies are needed to determine the influence of this difference on the recommended topdressing N application rates.

Year to year weather variability poses a challenge to use satellite remote sensing for in-season rice N status diagnosis and guiding topdressing application. The satellite imageries were collected at similar times in both years. However, the temperature in 2012 was higher than 2011. The accumulated temperature from transplanting date to the sampling date of 2012 was about 100 °C higher than that in 2011. As a result, rice plants grew faster in 2012 and already reached the stem elongation stage when the image was collected on 26 June 2012. This was reflected by the larger biomass, LAI and plant N uptake values in 2012 than 2011 (Table 3). Another factor to consider is that there are many cloudy and rainy days during the growing season in many parts of the major rice planting regions, which can prevent us from getting the needed satellite images within the narrow time window [13] in some years. Such uncertainty in year to year weather variability makes it very difficult to collect the satellite images at the right time for guiding in-season N management.

To overcome this limitation, multi-temporal and dual-polarimetric TerraSAR-X satellite data were evaluated for monitoring rice crop growth, and very promising results were obtained for rice biomass estimation [66]. Low-altitude remote sensing based on unmanned aerial vehicles (UAVs) may also be an alternative way for diagnosing in-season rice N status and guiding variable rate N management [67–69]. Due to the quick turn-around time, UAV-based remote sensing images can be collected 1–2 days before the topdressing N application, and the diagnosis result will be more representative. Nevertheless, due to the much smaller coverage and bigger data volume of UAV images, they are still not very practical for regional studies over large areas.

The FORMOSAT-2 satellite images only have four commonly-used wavebands (B, G, R and NIR). Previous research indicated that red edge-based vegetation indices performed better for estimating crop N status NNI than traditional red light-based indices [7,12,70]. According to Li *et al.* [70], the red edge-based Canopy Chlorophyll Content Index (CCCI) was reported to have the best performance among all of the indices evaluated for estimating summer maize N concentration and uptake at V6, V7 and V10–V12 stages, based on the simulation of Crop Circle ACS 470 active sensor, RapidEye and WorldView 2 satellite images. It is necessary to evaluate the potential improvements in estimating rice NNI using RapidEye and WorldView 2 satellite images. Hyperspectral sensing has the potential to further improve the estimation of crop NNI, as demonstrated in winter wheat [20] and summer maize [21], and more studies are needed to explore the potential of hyperspectral sensing for monitoring crop NNI.

In summary, the proposed satellite remote sensing approach can achieve comparable performance as ground-based active canopy sensors for estimating rice N status and is applicable to other rice planting regions. It is more efficient for large area applications, but is more influenced by weather conditions, while active canopy sensors are independent of environmental light conditions. It requires special training to process satellite remote sensing data, while active canopy sensors are easy to use, but are not suitable for large area applications. The UAV-based approach, coupled with red edge-based indices and hyperspectral remote sensing, has the potential to overcome the disadvantages of the ground active sensing and satellite remote sensing approaches. Therefore, it deserves further studies.

5. Conclusions

This study evaluated the potential of using FORMOSAT-2 satellite images to estimate rice NNI at the panicle initiation stage for guiding topdressing N application at the stem elongation stage in Northeast China. Across years, 45% of NNI variability could be explained using the RVI3 index directly. On the other hand, the indirect approach using FORMOSAT-2 images to estimate the aboveground biomass, PNU and, consequently, NNI achieved slightly better results (R^2 = 0.52 across years). Moreover, the calculated difference between the estimated PNU and the critical PNU based on the indirect method can be used to guide the topdressing N application rate adjustments, which demonstrated that FORMOSAT-2 images have the potential to estimate rice N status for guiding panicle N fertilizer applications in Northeast China. However, more studies are needed to further evaluate and improve the proposed method of in-season rice N status diagnosis and precision N management strategy under different on-farm conditions using different types of satellite data. The potential of UAV-based remote sensing, coupled with red edge-based indices and hyperspectral sensors, for improving rice NNI monitoring also needs to be studied in future research.

Acknowledgments: This study was financially supported by the National Basic Research Program (2015CB150405), the Natural Science Foundation of China (31071859), the Innovative Group Grant of the Natural Science Foundation of China (31421092), the European Aeronautic Defense and Space Company (EADS), the CHN-2152, 14-0039 SINOGRAIN project and the Sino-UK Newton AgriTech Remote Sensing for Sustainable Intensification project. We thank the staff of the Jiansanjiang Bureau of Agricultural Land Reclamation, Qixing Farm, and the Jiansanjiang Institute of Agricultural Sciences for their support. We also would like to thank all of the farmers for their cooperation in this research, and Prasad Thenkabail, Tao Cheng and the anonymous reviewers for their suggestions to improve the manuscript.

Author Contributions: Yuxin Miao and Georg Bareth conceived and guided the study. Shanyu Huang, Guangming Zhao, Xiaobo Ma, Chuanxiang Tan, Weifeng Yu and Martin L. Gnyp conducted the field experiments. Shanyu Huang performed the image processing and data analysis, Yuxin Miao and Shanyu Huang wrote the paper, Georg Bareth, Fei Yuan, Victoria I.S. Lenz-Wiedemann, Martin L. Gnyp and Uwe Rascher provided suggestions for the study, reviewed and edited the manuscript. All authors read and approved the manuscript.

Conflicts of Interest: The authors declare no conflict of interest.

References

1. Dawe, D. The contribution of rice research to poverty alleviation. *Stud. Plant Sci.* **2000**, *7*, 3–12.
2. Zhang, F.; Cui, Z.; Fan, M.; Zhang, W.; Chen, X.; Jiang, R. Integrated soil-crop system management: Reducing environmental risk while increasing crop productivity and improving nutrient use efficiency in China. *J. Environ. Qual.* **2011**, *40*, 1051–1057.
3. Zhang, F.; Wang, J.; Zhang, W.; Cui, Z.; Ma, W.; Chen, X.; Jiang, R. Nutrient use efficiencies of major cereal crops in China and measures for improvement. *Acta Pedolog. Sin.* **2008**, *5*, 915–924.
4. Jin, J. Changes in the efficiency of fertiliser use in China. *J. Sci. Food Agric.* **2012**, *92*, 1006–1009.

5. Ju, X.; Xing, G.; Chen, X.; Zhang, S.; Zhang, L.; Liu, X.; Cui, Z.; Yin, B.; Christie, P.; Zhu, Z.; *et al.* Reducing environmental risk by improving N management in intensive Chinese agricultural systems. *Proc. Natl. Acad. Sci. USA.* **2009**, *106*, 3041–3046.

6. Doberman, A.; Witt, C.; Dawe, D.; Abdulrachman, S.; Gines, H.C.; Nagarajan, R.; Satawathananont, S.; Son, T.T.; Tan, P.S.; Wang, G.H.; *et al.* Site-specific nutrient management for intensive rice cropping systems in Asia. *Field Crops Res.* **2002**, *74*, 37–66.

7. Cao, Q.; Miao, Y.; Feng, G.; Gao, X.; Li, F.; Liu, B.; Yue, S.; Cheng, S.; Ustin, S.; Khosla, R. Active canopy sensing of winter wheat nitrogen status: An evaluation of two sensor systems. *Comput. Electron. Agric.* **2015**, *112*, 54–67.

8. Greenwood, D.J.; Neeteson, J.J.; Draycott, A. Quantitative relationships for the dependence of growth rate of arable crops on their nitrogen content, dry weight and aerial environment. *Plant Soil* **1986**, *91*, 281–301.

9. Greenwood, D.J.; Gastal, F.; Lemaire, G.; Draycott, A.; Millard, P.; Neeteson, J.J. Growth rate and % N of field grown crops: Theory and experiments. *Ann. Bot.* **1991**, *67*, 181–190.

10. Lemaire, G.; Jeuffroy, M.H.; Gastal, F. Diagnosis tool for plant and crop N status in vegetative stage: Theory and practices for crop N management. *Eur. J. Agron.* **2008**, *28*, 614–624.

11. Houlès, V.; Guerif, M.; Mary, B. Elaboration of a nitrogen nutrition indicator for winter wheat based on leaf area index and chlorophyll content for making nitrogen recommendations. *Eur. J. Agron.* **2007**, *27*, 1–11.

12. Cao, Q.; Miao, Y.; Wang, H.; Huang, S.; Cheng, S.; Khosla, R.; Jiang, R. Non-destructive estimation of rice plant nitrogen status with crop circle multispectral active canopy sensor. *Field Crops Res.* **2013**, *154*, 133–144.

13. Yao, Y.; Miao, Y.; Cao, Q.; Wang, H.; Gnyp, M.L.; Bareth, G.; Khosla, R.; Yang, W.; Liu, F.; Liu, C. In-season estimation of rice nitrogen status with an active crop canopy sensor. *IEEE J. Sel. Topics Appl. Earth Observ. Remote Sens.* **2014**, *7*, 4403–4413.

14. Debaeke, P.; Rouet, P.; Justes, E. Relationship between the normalized SPAD index and the nitrogen nutrition index: Application to durum wheat. *J. Plant Nutr.* **2006**, *29*, 75–92.

15. Prost, L.; Jeuffroy, M.H. Replacing the nitrogen nutrition index by the chlorophyll meter to assess wheat N status. *Agron. Sustain. Dev.* **2007**, *27*, 321–330.

16. Ziadi, N.; Bélanger, G.; Claessens, A.; Lefebvre, L.; Tremblay, N.; Cambouris, A.N.; Nolin, M.C.; Parent, L.É. Plant-based diagnostic tools for evaluating wheat nitrogen status. *Crop Sci.* **2010**, *50*, 2580–2590.

17. Cao, Q.; Cui, Z.; Chen, X.; Khosla, R.; Dao, T.H.; Miao, Y. Quantifying spatial variability of indigeneous nitrogen supply for precision nitrogen management in small sacle farming. *Precis. Agric.* **2012**, *13*, 45–61.

18. Ziadi, N.; Brassard, M.; Bélanger, G.; Claessens, A.; Tremblay, N.; Cambouris, A.N.; Nolin, M.C.; Parent, L.-É. Chlorophyll measurements and nitrogen nutrition index for the evaluation of corn nitrogen status. *Agron. J.* **2008**, *100*, 1264–1273.

19. Miao, Y.; Mulla, D.J.; Randall, G.W.; Vetsch, J.A.; Vintila, R. Combining chlorophyll meter readings and high spatial resolution remote sensing images for in-season site-specific nitrogen management of corn. *Precis. Agric.* **2009**, *10*, 45–62.

20. Mistele, B.; Schmidhalter, U. Estimating the nitrogen nutrition index using spectral canopy reflectance measurements. *Eur. J. Agron.* **2008**, *29*, 184–190.

21. Chen, P.; Wang, J.; Huang, W.; Tremblay, N.; Ouyang, Z.; Zhang, Q. Critical nitrogen curve and remote detection of nitrogen nutrition index for corn in the Northwestern Plain of Shandong Province, China. *IEEE J. Sel. Topics Appl. Earth Observ. Remote Sens.* **2013**, *6*, 682–689.

22. Holland, K.H.; Lamb, D.W.; Schepers, J.S. Radiometry of proximal active optical sensors (AOS) for agricultural sensing. *IEEE J. Sel. Topics Appl. Earth Observ. Remote Sens.* **2012**, *5*, 1793–1802.

23. Mulla, D.J. Twenty five years of remote sensing in precision agriculture: Key advances and remaining knowledge gaps. *Biosyst. Eng.* **2013**, *114*, 358–371.

24. Cilia, C.; Panigada, C.; Rossini, M.; Meroni, M.; Busetto, L.; Amaducci, S.; Boschetti, M.; Picchi, V.; Colombo, R. Nitrogen status assessment for variable rate fertilization in maize through hyperspectral imagery. *Remote Sens.* **2014**, *6*, 6549–6565.

25. Wu, J.; Wang, D.; Rosen, C.J.; Bauer, M.E. Comparison of petiole nitrate concentrations, SPAD chlorophyll readings, and QuickBird satellite imagery in detecting nitrogen status of potato canopies. *Field Crops Res.* **2007**, *101*, 96–103.

26. Yang, C.M.; Liu, C.C.; Wang, Y.W. Using FORMOSAT-2 satellite data to estimate leaf area index of rice crop. *J. Photogram. Remote Sens.* **2008**, *13*, 253–260.

27. Darvishzadeh, R.; Matkan, A.A.; Ahangar, A.D. Inversion of a radiative transfer model for estimation of rice canopy chlorophyll content using a lookup-table approach. *IEEE J. Sel. Topics Appl. Earth Observ. Remote Sens.* **2012**, *5*, 1222–1230.

28. Xing, B.; Dudas, M.J.; Zhang, Z.; Qi, X. Pedogenetic characteristics of albic soils in the three river plain, Heilongjiang Province. *Acta Pedolog. Sin.* **1994**, *31*, 95–104.

29. Wang, Y.; Yang, Y. Effects of agriculture reclamation on hydrologic characteristics in the Sanjiang Plain. *Chin. Geogr. Sci.* **2001**, *11*, 163–167.

30. Yan, M.; Deng, W.; Chen, P. Climate change in the Sanjiang Plain disturbed by large-scale reclamation. *J. Geogr. Sci.* **2002**, *12*, 405–412.

31. Yao, Y.; Miao, Y.; Huang, S.; Gao, L.; Zhao, G.; Jiang, R.; Chen, X.; Zhang, F.; Yu, K.; *et al.* Active canopy sensor-based precision N management strategy for rice. *Agron. Sustain. Dev.* **2012**, *32*, 925–933.

32. Zhao, G.; Miao, Y.; Wang, H.; Su, M.; Fan, M.; Zhang, F.; Jiang, R.; Zhang, Z.; Liu, C.; Liu, P.; *et al.* A preliminary precision rice management system for increasing both grain yield and nitrogen use efficiency. *Field Crops Res.* **2013**, *154*, 23–30.

33. Chern, J.S.; Wu, A.M.; Lin, S.F. Lesson learned from FORMOSAT-2 mission operations. *Acta Astronaut.* **2006**, *59*, 344–350.

34. Liu, C.C. Processing of FORMOSAT-2 daily revisit imagery for site surveillance. *IEEE Trans. Geosci. Remote Sens.* **2006**, *44*, 3206–3214.

35. Bei, J.H.; Wang, K.R.; Chu, Z.D.; Chen, B.; Li, S.K. Comparitive study on the measurement methods of the leaf area. *J. Shihezi Uni.* **2005**, *23*, 216–218.

36. Lv, W.; Ge, Y.; Wu, J.; Chang, J. Study on the method for the determination of nitric nitrogen, ammoniacal nitrogen and total nitrogen in plant. *Spectrosc. Spect. Anal.* **2004**, *24*, 204–206.

37. Li, N. The comparison on various methods for determing different proteins. *J. Shanxi Agric. Univ.* **2006**, *26*, 132–134.

38. Justes, E.; Mary, B.; Meynard, J.M.; Machet, J.M.; Thelier-Huche, L. Determination of a critical nitrogen dilution curve for winter wheat crops. *Ann. Bot.* **1994**, *74*, 397–407.

39. Tucker, C.J. Red and photographic infrared linear combinations for monitoring vegetation. *Remote Sens. Environ.* **1979**, *8*, 127–150.

40. Buschmann, C.; Nagel, E. In vivo spectroscopy and internal optics of leaves as basis for remote sensing of vegetation. *Int. J. Remote Sens.* **1993**, *14*, 711–722.

41. Gitelson, A.A.; Kaufman, Y.J.; Merzlyak, M.N. Use of a green channel in remote sensing of global vegetation from EOS-MODIS. *Remote Sens. Environ.* **1996**, *58*, 289–298.

42. Roujean, J.L.; Breon, F.M. Estimating PAR absorbed by vegetation from bidirectional reflectance measurements. *Remote Sens. Environ.* **1995**, *51*, 375–384.

43. Gitelson, A.A.; Gritz, Y.; Merzlyak, M.N. Relationships between leaf chlorophyll content and spectral reflectance and algorithms for non-destructive chlorophyll assessment in higher plant leaves. *J. Plant Physiol.* **2003**, *160*, 271–282.

44. Gitelson, A.A. Wide dynamic range vegetation index for remote quantification of biophysical characteristics of vegetation. *J. Plant Physiol.* **2004**, *161*, 165–173.

45. Huete, A.R. A soil-adjusted vegetation index (SAVI). *Remote Sens. Environ.* **1988**, *25*, 295–309.

46. Chen, J.M. Evaluation of vegetation indices and a modified simple ratio for boreal applications. *Can. J. Remote Sens.* **1996**, *22*, 229–242.

47. Rondeaux, G.; Steven, M.; Baret, F. Optimization of soil-adjusted vegetation indices. *Remote Sens. Environ.* **1996**, *55*, 95–107.

48. Qi, J.; Chehbouni, A.; Huete, A.R.; Kerr, Y.H.; Sorooshian, S. A modified soil adjusted vegetation index. *Remote Sens. Environ.* **1994**, *48*, 119–126.

49. Datt, B. Visible/near infrared reflectance and chlorophyll content in *Eucalyptus* leaves. *Int. J. Remote Sens.* **1999**, *20*, 2741–2759.

50. Wang, W.; Yao, X.; Yao, X.F.; Tian, Y.C.; Liu, X.J.; Ni, J.; Cao, W.X.; Zhu, Y. Estimating leaf nitrogen concentration with three-band vegetation indices in rice and wheat. *Field Crops Res.* **2012**, *129*, 90–98.

51. Sims, D.A.; Gamon, J.A. Relationships between leaf pigment content and spectral reflectance across a wide range of species, leaf structures and developmental stage. *Remote Sens. Environ.* **2002**, *81*, 337–354.

52. Gitelson, A.A.; Kaufman, Y.J.; Stark, R.; Rundquist, D. Novel algorithms for remote estimation of vegetation fraction. *Remote Sens. Environ.* **2002**, *80*, 76–87.

53. Peñuelas, J.; Baret, F.; Filella, I. Semi-empirical indices to assess carotenoids/chlorophyll a ratio from leaf spectral reflectance. *Photosynthetica.* **1995**, *32*, 221–230.

54. Daughtry, C.S.T.; Walthall, C.L.; Kim, M.S.; Colstoun, E.B.; McMurtrey, J.E., III. Estimating corn leaf chlorophyll concentration from leaf and canopy reflectance. *Remote Sens. Environ.* **2000**, *74*, 229–239.

55. Haboudane, D.; Miller, J.R.; Pattey, E.; Zarco-Tejada, P.J.; Strachan, I.B. Hyperspectral vegetation indices and novel algorithms for predicting green LAI of crop canopies: Modeling and validation in the context of precision agriculture. *Remote Sens. Environ.* **2004**, *90*, 337–352.

56. Haboudane, D.; Miller, J.R.; Tremblay, N.; Zarco-Tejada, P.J.; Dextraze, L. Integrated narrow-band vegetation indices for prediction of crop chlorophyll content for application to precision agriculture. *Remote Sens. Environ.* **2002**, *81*, 416–426.

57. Broge, N.H.; Leblanc, E. Comparing prediction power and stability of broadband and hyperspectral vegetation indices for estimation of green leaf area index and canopy chlorophyll density. *Remote Sens. Environ.* **2000**, *76*, 156–172.

58. Huete, A.; Didan, K.; Miura, T.; Rodriguez, E.P.; Gao, X.; Ferreira, L.G. Overview of the radiometric and biophysical performance of the MODIS vegetation indices. *Remote Sens. Environ.* **2002**, *83*, 195–213.

59. Haboudane, D.; Tremblay, N.; Miller, J.R.; Vigneault, P. Remote estimation of crop chlorophyll content using spectral indices derived from hyperspectral data. *IEEE Trans. Geosci. Remote. Sens.* **2008**, *46*, 423–437.

60. Eitel, J.U.H.; Long, D.S.; Gessler, P.E.; Smith, A.M.S. Using in-situ measurements to evaluate the new RapidEye™ satellite series for prediction of wheat nitrogen status. *Int. J. Remote Sens.* **2007**, *28*, 4183–4190.

61. Gnyp, M.L.; Miao, Y.; Yuan, F.; Ustin, S.L.; Yu, K.; Yao, Y.; Huang, S.; Bareth, G. Hyperspectral canopy sensing of paddy rice aboveground biomass at different growth stages. *Field Crops Res.* **2014**, *155*, 42–55.

62. Yu, K.; Li, F.; Gnyp, M.L.; Miao, Y.; Bareth, G.; Chen, X. Remotely detecting canopy nitrogen concentration and uptake of paddy rice in the Northeast China Plain. *ISPRS J. Photogramm. Remote Sens.* **2013**, *78*, 102–115.

63. Van Niel, T.G.; McVicar, T.R. Current and potential uses of optical remote sensing in rice-based irrigation systems: A review. *Aust. J. Agric. Res.* **2004**, *55*, 155–185.

64. Peng, S.; Buresh, R.J.; Huang, J.; Zhong, X.; Zou, Y.; Yang, J.; Wang, G.; Liu, Y.; Hu, R.; Tang, Q.; Cui, K.; Zhang, F.; Dobermann, A. Improving nitrogen fertilization in rice by site-specific N management. A review. *Agron. Sustain. Dev.* **2010**, *30*, 649–656.

65. Shen, J.; Cui, Z.; Miao, Y.; Mi, G.; Zhang, H.; Fan, M.; Zhang, C.; Jiang, R.; Zhang, W.; Li, H.; *et al.* Transforming agriculture in China: From solely high yield to both high yield and high resource use efficiency. *Global Food Sec.* **2013**, *2*, 1–8.

66. Koppe, W.; Gnyp, M.L.; Hütt, C.; Yao, Y.; Miao, Y.; Chen, X.; Bareth, B. Rice monitoring with multi-temporal and dual-polarimetric TerraSAR-X data. *Int. J. Appl. Earth Observ. Geoinf.* **2013**, *21*, 568–576.

67. Zhang, C.; Kovacs, J.M. The application of small unmanned aerial systems for precision agricuture: A review. *Precis. Agric.* **2012**, *13*, 693–712.

68. Huang, Y.; Thomson, S.J.; Hoffmann, W.C.; Lan, Y.; Fritz, B.K. Development and prospect of unmanned aerial vehicle technologies for agricultural production management. *Int. J. Agric. Biol. Eng.* **2013**, *6*, 1–10.

69. Uto, K.; Seki, H.; Saito, G.; Kosugi, Y. Characterization of rice paddies by a UAV-mounted miniature hyperspectral sensor system. *IEEE J. Sel. Topics Appl. Earth Observ. Remote Sens.* **2013**, *6*, 851–860.

70. Li, F.; Miao, Y.; Feng, G.; Yuan, F.; Yue, S.; Gao, X.; Liu, Y.; Liu, B.; Ustin, S.L.; Chen, X. Improving estimation of summer maize nitrogen status with red edge-based spectral vegetation indices. *Field Crops Res.* **2014**, *157*, 111–123.

Temporal Dependency of Yield and Quality Estimation through Spectral Vegetation Indices in Pear Orchards

Jonathan Van Beek, Laurent Tits, Ben Somers, Tom Deckers, Wim Verjans, Dany Bylemans, Pieter Janssens and Pol Coppin

Abstract: Yield and quality estimations provide vital information to fruit growers, yet require accurate monitoring throughout the growing season. To this end, the temporal dependency of fruit yield and quality estimations through spectral vegetation indices was investigated in irrigated and rainfed pear orchards. Both orchards were monitored throughout three consecutive growing seasons, including spectral measurements (*i.e.*, hyperspectral canopy reflectance measurements) as well as yield determination (*i.e.*, total yield and number of fruits per tree) and quality assessment (*i.e.*, fruit firmness, total soluble solids and fruit color). The results illustrated a clear association between spectral vegetation indices and both fruit yield and fruit quality ($|r| > 0.75$; $p < 0.001$). However, the correlations between vegetation indices and production variables varied throughout the growing season, depending on the phenological stage of fruit development. In the irrigated orchard, index values showed a strong association with production variables near time of harvest ($|r| > 0.6$; $p < 0.001$), while in the rainfed orchard, index values acquired during vegetative growth periods presented stronger correlations with fruit parameters ($|r| > 0.6$; $p < 0.001$). The improved planning of remote sensing missions during (rainfed orchards) and after (irrigated orchards) vegetative growth periods could enable growers to more accurately predict production outcomes and improve the production process.

Reprinted from *Remote Sens.* Cite as: Van Beek, J.; Tits, L.; Somers, B.; Deckers, T.; Verjans, W.; Bylemans, D.; Janssens, P.; Coppin, P. Temporal Dependency of Yield and Quality Estimation through Spectral Vegetation Indices in Pear Orchards. *Remote Sens.* **2015**, *7*, 9886–9903.

1. Introduction

In capital-intensive horticultural cropping systems, estimating production or the production potential is essential in scheduling management decisions (*i.e.*, fruit thinning, harvest, *etc.*). One of the difficulties, however, is the variable influence of contributing factors on fruit yield and quality during different phenological stages (review by [1]). For example, water deficiencies during Stage I or III of fruit development—cell division and fruit thickening stage—will decrease yield, while a moderate deficiency during Stage II of fruit development—cell expansion—has no effect on yield [2,3]. Traditional *in situ* measurements of production variables and biophysical variables are time consuming and labor intensive. This results in limited samples and repetitions, which are insufficient to account for the high spatial and temporal variability within and between orchards [4,5]. It is yet well acknowledged that remote sensing can provide non-destructive, time efficient and cost beneficial alternatives for horticulture [6–8].

The application of remote sensing for crop yield estimation was mostly developed for annual crops [9–11]. For perennials, the estimation of production properties through remote sensing was previously investigated for different fruit crops, such as citrus [12,13], apple [5,14], peach [15], olives [15] and grapevines [16]. In these studies, the focus lay mostly on the estimation of overall yield, as higher yields were the main interest. In recent years, however, the focus in pear production systems shifted more towards quality-related production characteristics, because of the willingness to pay more for better quality fruit [17]. Although the research on quality estimation was primarily done post-harvest through proximal sensing [18], several studies have estimated qualitative traits through remote sensing imagery at time of harvest [12,15,19] or during specific periods within the growing season [13,16]. However, these studies were mostly based on single-image acquisitions and did not account for the variable nature of the growing season. Because the relationship between spectral measurements and production variables could vary between different phenological stages [13], the use of different vegetation indices during different growing stages would be required [20]. To optimize the scheduling of remote sensing missions and to monitor the production potential throughout the growing season, the temporal profile of the association between spectral information and production variables requires further investigation.

The primary goal of this study was to investigate the potential of remote sensing technology for estimating both production quality and quantity in pear orchards. The temporal variability of this relationship throughout the growing season—*i.e.*, optimal moments for yield and fruit quality monitoring—was explored for two orchards with different management and irrigation setups.

2. Materials and Methods

2.1. Study Area

The irrigated orchard, planted with Conference pear trees (*Pyrus communis* L. cv. "Conference") on Quince C rootstock, was situated in Bierbeek, Belgium (50°49′34.59″N, 4°47′42.83″E). The 2.5 m high trees were planted in a 3.5 by 1 meter grid in 2000 and were trained in a V-system with four fruiting branches on one central stem [21]. A side view of the irrigated orchard is shown in Figure 1A. The trees received 100% of the reference evapotranspiration (ETo) [22] throughout the growing season, except during Stage II of fruit development, characterized mostly by vegetative growth [3,23]. During this period, two irrigation treatments were applied. More information on the irrigation treatment can be found in Van Beek *et al.* [24]. Four plots of four trees each were selected on fixed intervals (±30 m) within four rows and monitored throughout the 2011, 2012 and 2013 growing seasons (48 plots).

(A)   (B)

(C)   (D)

Figure 1. Side view of V-shaped training system used in the irrigated orchard (**A**); side view of Spindle bush system in the rainfed orchard (**B**); spectral measurement setup (**C**); Top view of V-system in the irrigated orchard (**D**).

The non-irrigated or rainfed orchard, situated in Kerkom, Belgium (50°46′24.25″N, 5°09′27.05″E), was planted with Conference pear trees on Quince A rootstock in 2000. The 3.5 m high trees were planted in a 3.75 by 1.75 m grid and trained in a Spindle bush system [21]. A side view of the rainfed orchard is shown in Figure 1B. Two adjacent rows were selected and each row was divided into eight plots of four trees. Root pruning was applied on one side of the stem in the beginning of the growing season. This treatment was alternated between sides of the stem for subsequent growing seasons. In each row, a root-pruned plot was alternated with a non-treated plot. In 2011, only one row was monitored in the rainfed orchard (40 plots).

2.2. Ground Measurements

2.2.1. Fruit Yield and Quality

During the 2011, 2012 and 2013 growing seasons, harvest was carried out on Day of Year (DOY) 230 (243), 249 (250) and 253 (261) in the irrigated orchard (rainfed orchard), respectively. Total yield and number of fruits was determined on four trees per plot and averaged.

Fruit quality was determined three months after harvest with storage at $-0.5\,^{\circ}C$ in a cooling cell without controlled atmosphere. The green background color was determined with a Konia Minolta chromameter through chroma and hue values at the shadow side of the fruits (*i.e.*, the side that faces away from the sun) [25]. Chroma indicates the degree of departure from gray or white towards the pure color and is a measure of brightness, while hue angle quantifies color from red–green (0°–180°). Fruit firmness was measured with a penetrometer (0.5 cm^2 cylinder) after removal of the skin, while Total Soluble Solids (TSS, °brix) was determined with a hand-held refractometer. All fruit quality variables were determined on 60 fruits per plot and averaged per plot.

2.2.2. Spectral Measurements

Throughout the 2011, 2012 and 2013 growing seasons, canopy reflectance measurements were collected on cloud-free days using a full range (350–2500 nm) HR-1024 spectroradiometer (Spectra Vista Corporation, New York, NY, USA). The canopy spectra were taken from an elevated position between the rows at an average height of one meter above the top of the canopy (25° field of view). The experimental setup is shown in Figure 1C. Within this field of view, some within-canopy shadow and background will always be present (Figure 1D). However, all measurements were taken after full canopy disclosure to negate fractional cover differences and to minimize the effect of noise from shadow and/or background inclusion. Between plots, instruments were calibrated with a Spectralon reference panel. For each plot, 5–8 sunlit canopy spectra were taken and averaged per plot. To minimize differences with regards to solar geometry and illumination, all measurements were performed within 1.5 h from local solar noon. All spectra were smoothed using a 2nd order Savitsky-Golay filter with a window size of 21 nm [26].

2.2.3. Environmental Data

Daily precipitation (mm/day) and ETo (mm/day) were recorded and calculated at monitoring stations located 10 and 5 km from the irrigated and rainfed orchard, respectively (Portal of the Flemish Water managers, www.waterinfo.be (visited on 27 February 2014)). Average daily amount of rain deficit (or surplus) was calculated based on cumulative differences of precipitation and ETo [27,28]. To account for data gaps and measurement errors, the precipitation and ETo data from both monitoring stations was averaged on a daily basis prior to the calculation of cumulative available water.

2.3. Data Analysis

The spectral measurements (Section 2.2.2) were related to yield and quality variables (Section 2.2.1) through vegetation indices. The vegetation indices were chosen because of their proven relationship with water status and plant health in various agricultural crops and orchards. Moreover, the vegetation indices were associated either directly or indirectly with fruit yield and quality in horticultural crops [12,13,15].

The Normalized Difference Water Index (NDWI; Equation (1)) [29], was applied because of the association with canopy water status [29]. This resulted in a direct correlation between canopy water status and production variables or an indirect correlation between NDWI values and fruit yield and quality [12,16].

The Red-edge Normalized Difference Vegetation Index (ReNDVI; Equation (2)) [24], a normalized difference ratio between the NIR (Near-Infrared; 770–895 nm) and Red-edge (705–745 nm), was applied as it was previously related to water status (*i.e.*, stem water potential) and plant health in irrigated and rainfed pear orchards [24]. This association could provide significant correlations with fruit yield and quality [1,22]. The spectral bands used for ReNDVI were calculated based on the WorldView-2 spectral response function [30,31], similar to [24].

The Photochemical Reflectance Index (PRI; Equation (3)) [32] was applied because of the association with plant photosynthetic activity and water status and the proven relationship with fruit yield and quality in horticulture [13,19,33].

$$\text{NDWI} = (R_{860} - R_{1240})/(R_{860} + R_{1240}) \qquad (1)$$

$$\text{ReNDVI} = (R_{\text{Near Infrared}} - R_{\text{Red-edge}})/(R_{\text{Near Infrared}} + R_{\text{Red-edge}}) \qquad (2)$$

$$\text{PRI} = (R_{531} - R_{570})/(R_{531} + R_{570}) \qquad (3)$$

with R_x the reflectance at wavelength or band x.

The temporal variation of the correlation between spectral information and production variables (Section 2.2.1) was investigated. The correlation was analyzed at four key moments in the growing season coinciding with phenological stages of fruit development as specified through the BBCH code (Biologische Bundesanstalt, Bundessortenamt und CHemische Industrie) [34]. The considered phenological stages were fruitlet stage (± 90 days before harvest or BBCH 71–72), end of fruit fall (± 60 days before harvest or BBCH 73), fruit ripening (± 30 days before harvest or BBCH 81) and harvest stage (BBCH 87). For each fruit development stage, the nearest spectral measurement was selected. Only measurements prior to harvest were used because of the significant change of canopy reflectance after harvest. The strength of correlation between vegetation indices and production variables was determined with the Pearson correlation coefficient (r).

3. Results

3.1. Environmental Conditions

The gradient of cumulative rain deficit from 2011 to 2013 (Section 2.2.3), shown in Figure 2, highlights the differences between the monitored growing seasons. Note that in 2011, a dry spring (DOY 150–200) caused significant rain deficiencies, which could have affected fruit cell division. Oppositely, in 2012 a wet spring and summer caused rain surplus throughout the fruit cell division and vegetative growth period (DOY 100–200 or 150–50 days before harvest). In 2013, a rain surplus was present until 100 days before harvest (DOY 150) or the beginning of Stage II of fruit development, which is mostly associated with vegetative growth [3]. Subsequently, rain deficit steadily decreased towards the harvest period. Overall, the yearly precipitation was below the average precipitation of the last decade (*i.e.*, 622 ± 100 mm) in 2011 (546 mm) and 2013 (572 mm). In 2012, the yearly precipitation (711 mm) was above the 10-year-average.

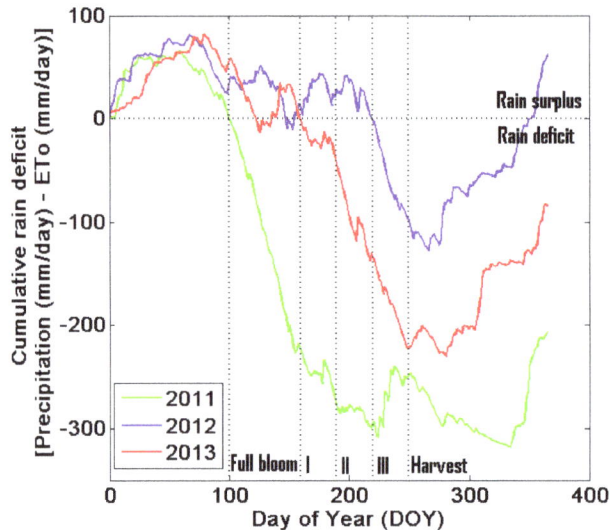

Figure 2. Cumulative rain deficit based on precipitation (mm/day) and ETo (mm/day) from 2011 to 2013 (Section 2.2.3). Vertical dotted lines indicate the approximate dates of full bloom, fruitlet stage (**I**), end of fruit fall (**II**), fruit ripening (**III**) and harvest.

3.2. Fruit Yield and Fruit Quality

An overview of the quantitative—total yield per tree and number of fruits per tree—and qualitative production variables—fruit firmness, total soluble solids, chroma and hue—is given in Table 1. Overall, the production in the irrigated orchard was more stable throughout the three growing seasons compared to the rainfed orchard. This was visible in the number of fruits per tree, the standard deviation of yield (kg/tree) and the relative differences between growing seasons. The rainfed orchard presented lower yields per tree in a dry season (2011; Figure 2), while the irrigated orchard had a more stable yield (*i.e.*, number of fruits) and improved fruit quality.

Table 1. Overview of quantitative (*i.e.*, total yield per tree and number of fruits per tree) and qualitative production variables after storage (*i.e.*, firmness, total soluble solids (TSS), chroma and hue) in the irrigated and rainfed orchard for 2011–2013 (±standard deviation). Values were averaged over 16 plots (eight plots in 2011 for the rainfed orchard).

Location	Year	Total Yield (kg/Tree)	Number of Fruits per Tree	Firmness (kg/0.5 cm^2)	TSS (°Brix)	Chroma (°)	Hue (°)
Irrigated Orchard	2011	28.4 (±3.1)	159 (±16)	5.7 (±0.4)	11.7 (±0.6)	41.5 (±1.5)	111.0 (±0.8)
	2012	17.5 (±3.0)	108 (±27)	7.0 (±0.2)	12.5 (±0.4)	41.2 (±0.6)	109.9 (±0.4)
	2013	19.1 (±2.9)	140 (±33)	7.3 (±0.3)	13.1 (±0.6)	38.3 (±0.9)	109.1 (±0.8)
Rainfed Orchard	2011	15.1 (±5.3)	90 (±32)	5.8 (±0.1)	13.2 (±0.2)	40.6 (±1.3)	108.1 (±1.7)
	2012	16.1 (±2.9)	88 (±18)	6.8 (±0.3)	12.3 (±0.3)	41.1 (±0.7)	109.7 (±0.4)
	2013	24.6 (±3.3)	170 (±36)	6.7 (±0.3)	12.9 (±0.5)	38.1 (±0.9)	107.7 (±1.5)

3.3. Production versus Spectral Measurements

The temporal profile of the measured vegetation indices is shown in Figure 3. Overall, the measured vegetation indices were less variable in the irrigated orchard throughout the different growing seasons. This was visible through the smaller standard deviations between measured plots and the smaller differences between growing seasons.

The temporal change of the correlation between spectral information (Section 2.3) and production variables (Section 2.2.1) was investigated at four moments in the growing season (*i.e.*, fruitlet stage, end of fruit fall, fruit ripening and harvest). The results are presented in Table 2. The relationship between spectral vegetation indices and a selection of production variables—providing a complete set of vegetation indices, phenological stages and production variables—is highlighted for the irrigated orchard in Figure 4 and for the rainfed orchard in Figure 5.

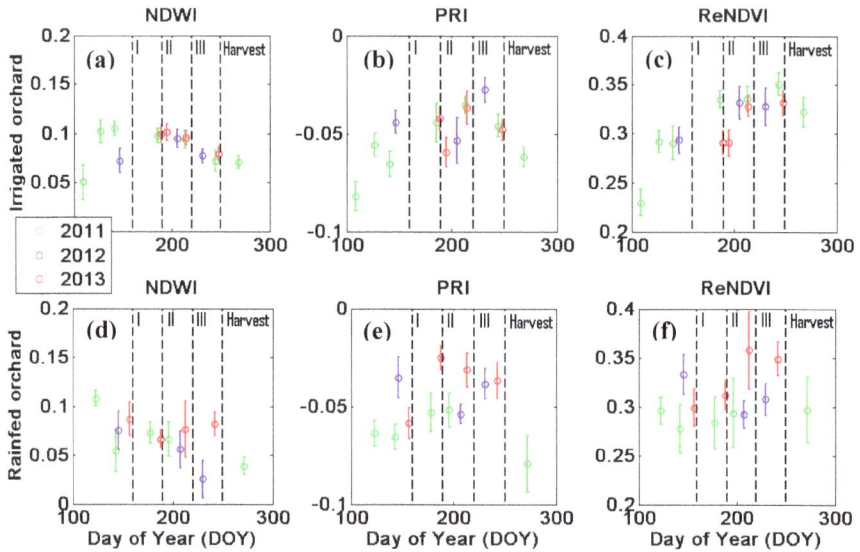

Figure 3. Profile of the Normalized Difference Water Index (NDWI; (**a**); the Photochemical Reflectance Index (PRI; (**b**) and the Red-edge Normalized Difference Vegetation Index (ReNDVI; (**c**) for the irrigated orchard throughout each growing season and profile of NDWI (**d**), PRI (**e**) and ReNDVI (**f**) for the rainfed orchard throughout each growing season. Bars represent the standard deviation between all the measured plots and vertical dashed lines indicate approximate dates of phenological stages of fruit development, namely fruitlet stage (**I**), end of fruit fall (**II**), fruit ripening (**III**) and harvest.

In the irrigated orchard, the correlation between vegetation indices and production variables was dependent on the phenological stage. Spectral indices were associated to production variables in the beginning of the growing season (fruitlet stage) and towards the harvest. For almost all production variables, a significant drop was noticeable at the end of fruit fall (± 60 days before harvest) compared to the rest of the growing season. This was illustrated between TSS and PRI values at the end of fruit fall (Figure 4c) and during fruit ripening (Figure 4d). NDWI values showed a positive correlation with production quantity (*i.e.*, total yield and number of fruits per tree) and a negative correlation with quality-related production variables (*i.e.*, firmness and TSS). Conversely, PRI and ReNDVI values displayed a negative correlation with production quantity and a negative with production quality. This was illustrated for both total yield (Figure 4a) and firmness (Figure 4b) combined with NDWI values at harvest. Furthermore, spectral indicators for color characteristics showed a similar gradient to quantity related variables, with a positive correlation with NDWI values and a negative correlation with PRI and ReNDVI values.

Table 2. Pearson correlation coefficient (r) values between production variables (*i.e.*, Total Yield, Number of fruits per tree, Fruit Firmness, Total Soluble Solids (TSS), Chroma and Hue) and the Normalized Difference Water Index (NDWI), the Photochemical Reflectance Index (PRI) and the Red-Edge Normalized Difference Vegetation Index (ReNDVI) respectively, for the irrigated and rainfed orchard throughout the growing season. The correlation was considered at four phenological stages in fruit development, namely fruitlet stage ($\pm$90 days before harvest), end of fruit fall ($\pm$60 days before harvest), fruit ripening ($\pm$30 days before harvest) and harvest. For each phenological stage the closest spectral measurements prior to harvest was chosen. Symbols indicated significance of correlation. Bold values point out correlations that are depicted in scatter plots in the following paragraphs.

	Total Yield (kg/tree)				Number of Fruits per Tree			
Phenological stage	Fruitlet	End of fruit fall	Fruit ripening	Harvest	Fruitlet	End of fruit fall	Fruit ripening	Harvest
Irrigated Orchard NDWI	0.56 **	0.19	0.22	**0.73****	0.53 **	0.11	0.06	0.47 **
PRI	−0.70 **	−0.02	−0.41 **	−0.50 **	−0.33 *	0.01	−0.05	−0.49 **
ReNDVI	0.12	−0.09	−0.30 *	−0.37 **	0.21	−0.26	−0.13	−0.31 *
Rainfed Orchard NDWI	0.48 **	0.12	0.59 **	0.59 **	0.43 **	0.04	0.64 **	0.64 **
PRI	−0.18	0.70 **	0.28	0.28	−0.31 *	**0.67 **	0.18	0.18
ReNDVI	−0.02	**0.66 **	0.65 **	0.65 **	−0.18	0.60 **	0.56 **	0.56 **
	*Significance at p < 0.05				**Significance at p < 0.001			

	Firmness (kg/0.5 cm²)				TSS (°brix)			
Phenological stage	Fruitlet	End of fruit fall	Fruit ripening	Harvest	Fruitlet	End of fruit fall	Fruit ripening	Harvest
Irrigated Orchard NDWI	−0.37 **	−0.12	−0.18	−0.69 **	−0.22	−0.11	−0.24	−0.62 **
PRI	0.79 **	−0.13	0.45 **	0.25	0.53 **	−0.06	**0.59 **	0.16
ReNDVI	0.06	−0.21	0.21	0.24	−0.22	−0.21	0.26	0.24
Rainfed Orchard NDWI	0.23	0.07	−0.25	−0.25	−0.02	0.04	**0.59 **	0.59 **
PRI	0.34 *	−0.01	0.21	0.21	−0.63 **	0.37 *	−0.13	−0.13
ReNDVI	0.35 *	0.15	0.07	0.07	−0.58 **	0.23	0.31	0.31
	*Significance at p < 0.05				**Significance at p < 0.001			

	Chroma (°)				Hue (°)			
Phenological stage	Fruitlet	End of fruit fall	Fruit ripening	Harvest	Fruitlet	End of fruit fall	Fruit ripening	Harvest
Irrigated Orchard NDWI	−0.25	−0.07	0.03	0.29 *	0.23	0.12	0.35 *	0.60 **
PRI	−0.47 **	0.18	−0.55 **	0.19	−0.62 **	0.14	−0.55 **	−0.15
ReNDVI	0.11	**0.51 **	−0.21	−0.20	−0.03	0.25	−0.38 **	−0.30 *
Rainfed Orchard NDWI	−0.44 **	−0.44 **	−0.66 **	−0.66 **	−0.19	−0.09	−0.53 **	−0.53 **
PRI	0.53 **	−0.71 **	−0.26	−0.26	**0.59 **	−0.53 **	0.06	0.06
ReNDVI	0.40 *	**0.73 **	−0.79 **	−0.79 **	0.58 **	−0.45 **	−0.35 **	−0.35 *
	*Significance at p < 0.05				**Significance at p < 0.001			

Similar to the irrigated orchard, the correlation between vegetation indices and production variables in the rainfed orchard was not constant throughout the growing season. For PRI values, the end of fruit fall ($\pm$60 days before harvest) showed significantly higher correlations ($r > 0.6$; $p < 0.001$) with quantity-related production variables compared to the rest of the growing season (Figure 5c). Similarly, ReNDVI values at the end of fruit fall were significantly correlated with quantity-related production variables (Figure 5f). Conversely to PRI values, the remainder of the growing season also presented high correlation coefficients ($r > 0.56$; $p < 0.001$). NDWI values were more related to both quantity and quality-related production variables towards the end of the growing season ($|r| \approx 0.6$; $p < 0.001$). This is illustrated for TSS and NDWI values at fruitlet (Figure 5a) and fruit ripening stages (Figure 5b).

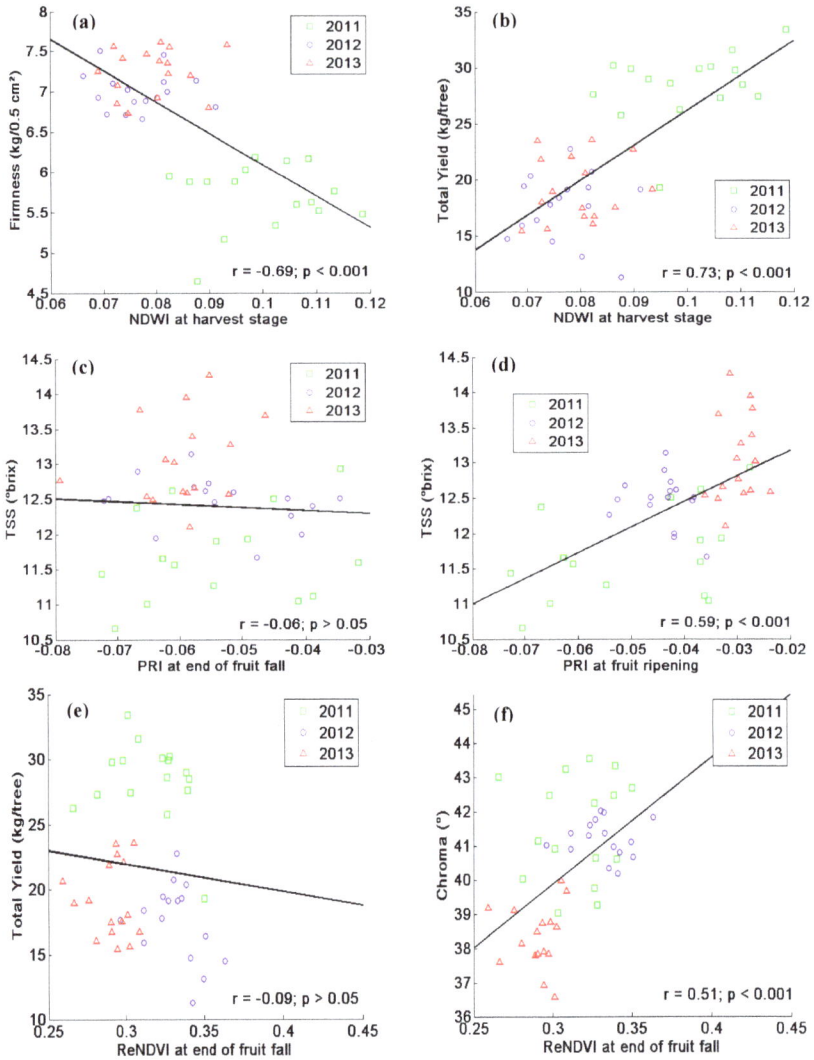

Figure 4. Scatter plots at various stages in the growing season between spectral vegetation indices, namely the Normalized Difference Water Index (NDWI; **a,b**), the Photochemical Reflectance Index (PRI; **c,d**) and the Red-edge Normalized Difference Vegetation Index (ReNDVI; **e,f**) and production variables, namely Fruit Firmness (a), Total Yield (b,f), Total Soluble Solids (TSS; c,d) and Chroma (e) in the irrigated orchard. All points were labeled for growing season.

In the rainfed orchard, spectral indices were more associated with color variables compared to firmness and TSS values. This is illustrated for PRI values at fruitlet stage and hue data (Figure 5d) and ReNDVI values at the end of fruit fall and chroma data (Figure 5e). However, in contrast with the irrigated orchard (Figure 4e), the correlation between spectral indices and color variables—chroma and hue—presented the reverse gradient compared to yield (Figure 5d,e).

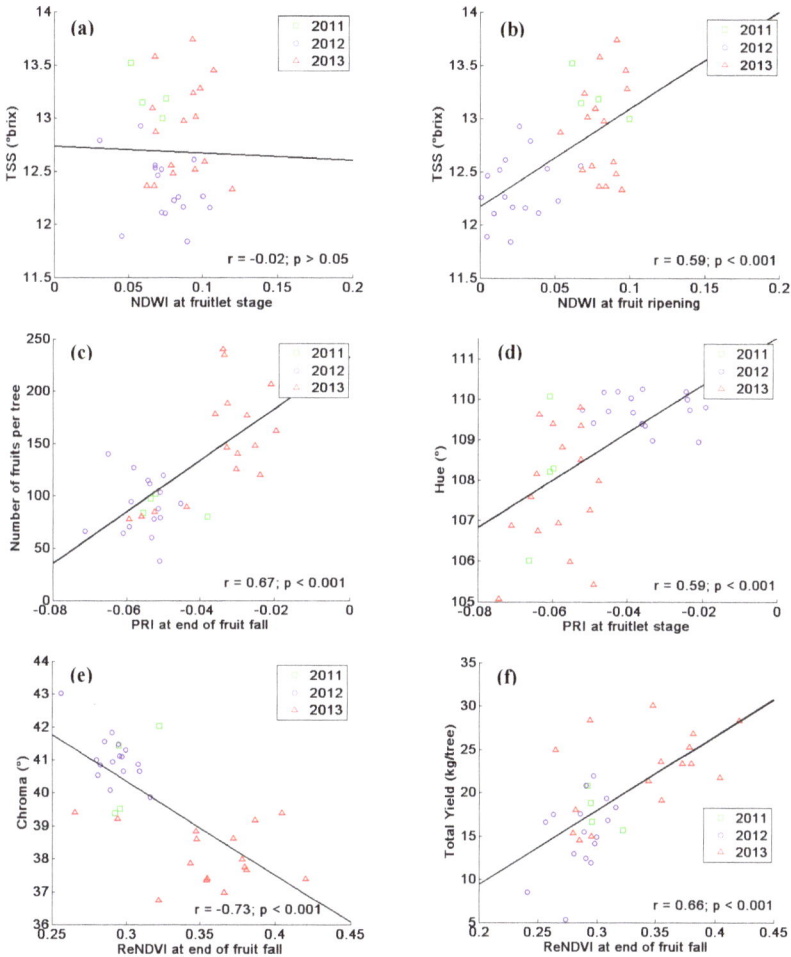

Figure 5. Scatter plots at various stages in the growing season between spectral vegetation indices, namely the Normalized Difference Water Index (NDWI; **a,b**), the Photochemical Reflectance Index (PRI; **c,d**) and the Red-edge Normalized Difference Vegetation Index (ReNDVI; **e,f**) and production variables, namely Total Soluble Solids (TSS; a,b); Number of fruits per tree (c); Hue (d); Chroma (e) and Total Yield (f) in the rainfed orchard. All points were labeled for growing season.

4. Discussion

4.1. Production versus Spectral Measurements

In general, Figure 2 and Table 1 illustrated the diversity of the growing conditions during the study. In these diverse conditions, remote sensing information was found to be associated with production variables and could provide agricultural managers with a reliable estimation of quantitative ($|r| > 0.6$; $p < 0.001$) and qualitative production variables ($|r| > 0.7$; $p < 0.001$) prior to harvest. Upon comparison to previous studies, the r values between vegetation indices at harvest and quantity-related production variables in this study were similar or higher. For instance, $|r|$ values between 0.3 and 0.7 (Table 2) compared to $|r|$ values of 0.6 for water related indices in vineyards [16] and between 0.2 and 0.6 for PRI in citrus orchards [13]. With regards to quality-related production variables, the association with vegetation indices yielded comparable r values as other studies. An $|r|$ value of 0.41 was found between TSS and PRI values 45 days before harvest (*i.e.*, between end of fruit fall and fruit ripening) in citrus orchards [19]. In vineyards, TSS was associated to water content related indices at harvest ($|r|$ value of 0.5) [16]. Similarly to Serrano *et al.* [16], this study showed that water related indices (*i.e.*, NDWI), plant health (*i.e.*, ReNDVI) and plant photosynthetic activity indices (*i.e.*, PRI) were related to total yield per tree and to fruit quality (Table 2). However, the choice of vegetation index was important, as the temporal dependence of these correlations was variable for different vegetation indices and between orchards (Figures 4 and 5 and Table 2).

The differences between the irrigated and rainfed orchard could be attributed to the differences in water availability at critical stages in the growing season (Figure 2), but could also be the result of differences in vigor for the different rootstocks [35] and training systems. In the rainfed orchard, the relationships between vegetation indices and production variables were more variable throughout the growing season. The absence of irrigation during dry periods possibly increased the influence of environmental conditions on TSS and fruit firmness (*i.e.*, water availability [36], amount of sunlight [37] and daily temperature [38]) and resulted in a variable correlation throughout the growing season (Figure 5c,b). The correlation between ReNDVI values and production variables was highly variable between both orchards (Figures 4e and 5f). This might be caused by the underlying relationship with stem water potential [24]. As a result of the deficit irrigation treatment in the irrigated orchard, large stem water potential differences were maintained during the end of fruit fall without significant fruit yield and fruit quality differences because of the ability to irrigate in later stages of fruit development. In the rainfed orchard, the potential stem water differences caused by the root pruning treatment would result in increased water deficiency [39] which in turn led to decreased fruit yield and improved fruit quality [27]. A similar trend was also visible for PRI values because of the association with stem water potential [40].

The correlation between vegetation indices and production variables was more stable throughout the growing season in the irrigated orchard, because of the possibility to irrigate during dry periods. This was also visible for the spectral vegetation indices (Figure 3), which were more stable towards the end of the growing season compared to the rainfed orchard. During the deficit irrigation period ($\approx$ end of fruit fall), large differences were achieved without significant fruit yield and fruit quality differences because of the ability to irrigate in later stages of fruit development. Therefore, the optimal period for remote sensing measurements in irrigated orchards would be before or after the vegetative growth period (*i.e.*, less than 30 days before harvest; Figure 5a,b,d), as small spectral differences would not result in variable fruit yield and quality in the vegetative growth period (Figure 5c). Measurements in the rainfed orchard at the end of fruit fall—associated with minimized fruit growth and more vegetative growth [3,23]—provided a good estimation of production quantity through ReNDVI (r = 0.66; $p < 0.001$; Figure 5f) and PRI (r = 0.67; $p < 0.001$; Figure 5c). Oppositely, water availability differences had a larger impact on the total production towards the end of the growing season, which resulted in higher $|r|$ values between total yield and NDWI (r $\approx$ 0.6; $p < 0.001$; Table 2).

4.2. Potential and Limitations

With the use of remote sensing, the estimation or prediction of production and visualization of optimal monitoring periods could be determined (Table 2 and Figures 4 and 5). This would allow remote sensing to visualize the large spatial variability (>20 kg/tree) [4,5] present within each orchard and optimize and schedule management procedures—deficit irrigation [27], summer pruning [39], root pruning [39], fruit thinning [41], fertigation [42] and harvest—throughout the growing season to improve production quality and yield [7]. However, several limitations remain with the use of remote sensing for the estimation of the orchards' production potential.

Although the results indicated a good correlation between spectral measurements and production variables, this relationship was dependent on the growing season (Figures 4 and 5). This was the result of highly variable weather conditions (Figure 2). Moreover, the alternate bearing tendency of some horticultural crops could cause large differences in subsequent growing seasons [13]. In this study, some alternate bearing was present (Table 1), especially in the rainfed orchard. However, the effect was smaller compared to other studies [13] because of the parthenocarpic tendency of Conference pears and the lack of return bloom inhibition by seeds [43]. Larger time series could provide more stable estimations of fruit yield and fruit quality because of the link between climatic differences and production potential. Moreover, larger time series would also provide information on the reliability of remote sensing data during each phenological stage in the growing season.

Conversely to annual crops [9–11], the significant correlations between vegetation indices and production variables were not the result of a direct relation between above ground biomass and crop yield [44]. The present season's growing conditions have a smaller impact compared to annual crops, as the amount of fruit buds (i.e., production potential) is influenced by crop load [45] and the plants' water status [27,39,46] of the previous growing season [1]. The correlation between production variables and vegetation indices most likely stemmed from the association with vegetative growth and the relationship between vegetative growth, water status and production [14,47,48]. Trees with more vegetative growth produced less flower buds (i.e., decreased number of fruits and total yields), as a result of the considerable consumption of water from excessive vegetative growth [47]. The spectral vegetation indices provided an overall indication of plant health—plant vigor [24], water content [24,29,32], photosynthetic efficiency [32]—which in turn was related to production variables. Because of the indirect nature of this relationship, several periods within the growing season showed insignificant correlation between vegetation indices and production variables (Figures 4a and 5a).

One of the difficulties with production estimation in orchards is the trade-off between fruit yield and fruit quality [16,28,36,42,49]. This link between fruit quality and quantity was also visible in the irrigated orchard (Table 1; Figure 4a,b; Figure 5e,f). For PRI values, a positive correlation was present with fruit firmness and TSS and a negative correlation was found with total yield, similarly to Serrano et al. [16]. Overall, a healthier tree in the irrigated orchard (i.e., higher PRI values) produced better quality fruit, while increased water availability (i.e., higher NDWI values) increased fruit yield. In the rainfed orchard, a healthier tree (i.e., higher PRI values) and increased water availability (i.e., higher NDWI values) both led to the production of more fruits with lower fruit quality (i.e., yellowing of fruit) [36].

In this study, the relationship between vegetation indices and production variables was shown to differ between irrigated and rainfed orchards (Figures 4 and 5 and Table 2), as a result of water availability and rootstock differences [35]. Although this might obstruct practical use of prediction models over larger areas, previous studies in orchard crops could distinguish irrigated from rainfed orchards [50]. As a result, the analysis could be adjusted based on these methodologies or through cooperation with fruit growers. Ultimately, the production estimates from remote sensing could only provide an indication of the production potential, as extreme conditions or circumstances—storms, hail or bird damage—could damage the crops. On the other hand, with the incorporation of environmental conditions into crop models, further improvement of horticultural production estimation and management should be achievable.

5. Conclusions

Remote sensing provides an alternative to time consuming, labor intensive and destructive *in situ* measurements required for yield and quality monitoring and estimation. In this study, an irrigated and a rainfed orchard were monitored with hyperspectral sensors through three subsequent growing seasons, demonstrating the potential of spectral measurements for the prediction of quality—fruit firmness and total soluble solids—and quantity-related production properties—total yield and amount of fruits per tree—throughout the growing season.

The results illustrated an association between vegetation indices—the Normalized Difference Water Index (NDWI), the Photochemical Reflectance Index (PRI) and Red-edge Normalized Difference Vegetation Index (ReNDVI)—and both fruit yield and fruit quality variables ($|r| > 0.6$; $p < 0.001$). However, the relationship between spectral indicators and production variables was variable throughout the growing season and between orchards. This temporal dependency demonstrated the usefulness of remote sensing and the necessity of optimized scheduling and interpretation of the results. In the rainfed orchard, NDWI values at harvest showed a positive correlation with yield ($r \approx 0.6$; $p < 0.001$), while PRI and ReNDVI values at the end of fruit fall (± 60 days before harvest) were strongly related to yield ($r > 0.6$; $p < 0.001$). In the irrigated orchard, PRI values near harvest showed a positive correlation with fruit firmness and TSS ($r \approx 0.5$; $p < 0.001$), while NDWI values showed the reverse gradient ($r \approx -0.6$; $p < 0.001$) and ReNDVI values showed no significant correlation. At the end of fruit fall—characterized by vegetative growth (± 60 days before harvest)—vegetation index values in the irrigated orchard were not correlated with yield and fruit quality.

Despite diverse conditions, remote sensing technology was able to correlate with production variables and could provide fruit growers with a reliable estimation of their production quantity ($|r| > 0.7$; $p < 0.001$) and quality ($|r| > 0.7$; $p < 0.001$) for several periods in the growing season. The results in this study highlighted the necessity of the careful use and selection of vegetation indices and monitoring times. Overall, these indices could enable managers to predict fruit yield and quality several months prior to harvest, allowing for optimized scheduling of management processes, such as deficit irrigation, hand thinning, fertigation and fruit harvest.

Acknowledgments: This work was supported by the Agency for Innovation by Science and Technology in Flanders (IWT-Vlaanderen). The research was funded through a project in collaboration with the Soil Service of Belgium (BDB) and the Research Center for Fruit (Proefcentrum Fruitteelt in Sint-Truiden). The authors would like to thank fruit grower Jan Vandervelpen for allowing research in his orchard and making this work possible.

Author Contributions: Experimental design: Jonathan Van Beek, Tom Deckers, Wim Verjans, Pieter Janssens and Pol Coppin; Data collection and processing: Jonathan Van Beek, Laurent Tits and Wim Verjans; Statistical analysis and interpretation: Jonathan Van Beek, Laurent Tits, Ben Somers, Tom Deckers and Pieter Janssens; Manuscript preparation, writing and paper review: Jonathan Van Beek, Laurent Tits, Ben Somers, Tom Deckers, Wim Verjans, Dany Bylemans, Pieter Janssens and Pol Coppin; Project coordination: Jonathan Van Beek, Tom Deckers, Dany Bylemans, Pieter Janssens and Pol Coppin.

Conflicts of Interest: The authors declare no conflict of interest.

References

1.   Webster, A.D. Factors influencing the flowering, fruit set and fruit growth of European pears. In *Acta Horticultura 569*; ISHS: Leuven, Belgium, 2002; pp. 699–709.
2.   Goodwin, I.; Boland, A.-M. Scheduling deficit irrigation of fruit trees for optimizing water use efficiency. In *Deficit Irrigation Practices. Water Reports Publication n. 22*; Food and Agriculture Organization (FAO): Rome, Italy, 2002; pp. 67–78.
3.   Mitchell, P.D.; Chalmers, D.J.; Jerie, P.H.; Burge, G. The use of initial withholding of irrigation and tree spacing to enhance the effect of regulated deficit irrigation on pear trees. *J. Am. Soc. Hortic. Sci.* **1986**, *111*, 858–861.
4.   Perry, E.M.; Dezzani, R.J.; Seavert, C.F.; Pierce, F.J. Spatial variation in tree characteristics and yield in a pear orchard. *Precis. Agric.* **2009**, *11*, 42–60.
5.   Aggelopoulou, K.D.; Wulfsohn, D.; Fountas, S.; Gemtos, T.A.; Nanos, G.D.; Blackmore, S. Spatial variation in yield and quality in a small apple orchard. *Precis. Agric.* **2009**, *11*, 538–556.
6.   Dorigo, W.A.; Zurita-Milla, R.; de Wit, A.J.W.; Brazile, J.; Singh, R.; Schaepman, M.E. A review on reflective remote sensing and data assimilation techniques for enhanced agroecosystem modeling. *Int. J. Appl. Earth Obs. Geoinf.* **2007**, *9*, 165–193.
7.   Usha, K.; Singh, B. Potential applications of remote sensing in horticulture—A review. *Sci. Hortic.* **2013**, *153*, 71–83.
8.   Van Beek, J.; Tits, L.; Somers, B.; Deckers, T.; Janssens, P.; Coppin, P. Reducing background effects in orchards through spectral vegetation index correction. *Int. J. Appl. Earth Obs. Geoinf.* **2014**, *34*, 167–177.
9.   Zarco-Tejada, P.J.; Ustin, S.L.; Whiting, M.L. Temporal and spatial relationships between within-field yield variability in cotton and high-spatial hyperspectral remote sensing imagery. *Agron. J.* **2005**, *97*, 641–653.
10.   Thenkabail, P.S.; Ward, A.D.; Lyon, J.G. Landsat-5 thematic mapper models of soybean and corn crop characteristics. *Int. J. Remote Sens.* **1994**, *15*, 49–61.
11.   Thenkabail, P.S.; Ward, A.D.; Lyon, J.G.; Merry, C.J. Thematic mapper vegetation indices for determining soybean and corn growth parameters. *Photogramm. Eng. Remote Sens.* **1994**, *60*, 437–442.
12.   Somers, B.; Delalieux, S.; Verstraeten, W.W.; van den Eynde, A.; Barry, G.H.; Coppin, P. The contribution of the fruit component to the hyperspectral citrus canopy signal. *Photogramm. Eng. Remote Sens.* **2010**, *76*, 37–47.

13. Ye, X.; Sakai, K.; Manago, M.; Asada, S.; Sasao, A. Prediction of citrus yield from airborne hyperspectral imagery. *Precis. Agric.* **2007**, *8*, 111–125.

14. Best, S.; Salazar, F.; Bastías, R.; Leon, L. Crop load estimation model to optimize yield-quality ratio in Apple Orchards, Malus Domestica Borkh, Var. Royal Gala. *J. Inf. Technol. Agric.* **2008**, *3*, 11–18.

15. Sepulcre-Canto, G.; Zarco-Tejada, P.; Jimenez-Munoz, J.; Sobrino, J.; Soriano, M.; Fereres, E.; Vega, V.; Pastor, M. Monitoring yield and fruit quality parameters in open-canopy tree crops under water stress. Implications for ASTER. *Remote Sens. Environ.* **2007**, *107*, 455–470.

16. Serrano, L.; González-Flor, C.; Gorchs, G. Assessment of grape yield and composition using the reflectance based Water Index in Mediterranean rainfed vineyards. *Remote Sens. Environ.* **2012**, *118*, 249–258.

17. Gallardo, R.K.; Kupferman, E.; Colonna, A. Willingness to pay for optimal "Anjou" pear quality. *Hortic. Sci.* **2011**, *46*, 452–456.

18. Nicolaï, B.M.; Beullens, K.; Bobelyn, E.; Peirs, A.; Saeys, W.; Theron, K.I.; Lammertyn, J. Nondestructive measurement of fruit and vegetable quality by means of NIR spectroscopy: A review. *Postharvest Biol. Technol.* **2007**, *46*, 99–118.

19. Suárez, L.; Zarco-Tejada, P.J.; González-Dugo, V.; Berni, J.A.J.; Sagardoy, R.; Morales, F.; Fereres, E. Detecting water stress effects on fruit quality in orchards with time-series PRI airborne imagery. *Remote Sens. Environ.* **2010**, *114*, 286–298.

20. Hatfield, J.L.; Prueger, J.H. Value of using different vegetative indices to quantify agricultural crop characteristics at different growth stages under varying management practices. *Remote Sens.* **2010**, *2*, 562–578.

21. Sansavini, S.; Musacchi, S. Canopy architecture, training and pruning in the modern European pear orchards: An overview. *Acta Hortic.* **1994**, *367*, 152–172.

22. Allen, R.G.; Pereira, L.S.; Raes, D.; Smith, M. *Crop Evapotranspiration-Guidelines for Computing Crop Water Requirements-FAO Irrigation and Drainage Paper 56*; FAO: Quebec City, QC, Canada, 1998.

23. Mitchell, P.D.; Jerie, P.H.; Chalmers, D.J. Effects of regulated water deficits on pear tree growth, flowering, fruit growth, and yield. *J. Am. Soc. Hortic. Sci.* **1984**, *109*, 604–606.

24. Van Beek, J.; Tits, L.; Somers, B.; Coppin, P. Stem water potential monitoring in pear orchards through WorldView-2 multispectral imagery. *Remote Sens.* **2013**, *5*, 6647–6666.

25. McGuire, R.G. Reporting of objective colour measurements. *Hortic. Sci.* **1992**, *27*, 1254–1255.

26. Savitsky, A.; Golay, M.J.E. Smoothing and differentiation of data by simplified least squares procedures. *Anal. Chem.* **1964**, *36*, 1627–1639.

27. Janssens, P.; Deckers, T.; Elsen, F.; Elsen, A.; Schoofs, H.; Verjans, W.; Vandendriessche, H. Sensitivity of root pruned "Conference" pear to water deficit in a temperate climate. *Agric. Water Manag.* **2011**, *99*, 58–66.

28. Alcobendas, R.; Mirás-Avalos, J.M.; Alarcón, J.J.; Pedrero, F.; Nicolás, E. Combined effects of irrigation, crop load and fruit position on size, color and firmness of fruits in an extra-early cultivar of peach. *Sci. Hortic.* **2012**, *142*, 128–135.

29. Gao, B. NDWI a normalized difference water index for remote sensing of vegetation liquid water from space. *Remote Sens. Environ.* **1996**, *58*, 257–266.

30. Updike, T.; Comp, C. *Radiometric Use of WorldView-2 Imagery Technical Note*; DigitalGlobe: Longmont, CO, USA, 2010.

31. Van Leeuwen, W.J.D.; Orr, B.J.; Marsh, S.E.; Herrmann, S.M. Multi-sensor NDVI data continuity: Uncertainties and implications for vegetation monitoring applications. *Remote Sens. Environ.* **2006**, *100*, 67–81.

32. Gamon, A.; Serrano, L.; Surfus, S. The photochemical reflectance index: An optical indicator of photosynthetic radiation use efficiency across species, functional types, and nutrient levels. *Oecologia* **1997**, *112*, 492–501.

33. Stagakis, S.; González-Dugo, V.; Cid, P.; Guillén-Climent, M.L.; Zarco-Tejada, P.J. Monitoring water stress and fruit quality in an orange orchard under regulated deficit irrigation using narrow-band structural and physiological remote sensing indices. *ISPRS J. Photogramm. Remote Sens.* **2012**, *71*, 47–61.

34. Meier, U.; Graf, H.; Hack, H.; Hess, M.; Kennel, W.; Klose, R.; Mappes, D.; Seipp, D.; Stauss, R.; Streif, J.; *et al.* Phaenologische entwicklungsstadien des Kernobstes (Malus domestica Borkh. und Pyrus communis L.), des Steinobstes (Prunus-Arten), der Johannisbeere (Ribes-Arten) und der Erdbeere (Fragaria x ananassa Duch.). *Nachrichtenblatt Dtsch. Pflanzenschutzdienstes* **1994**, *46*, 141–153.

35. Massai, R.; Loreti, F.; Fei, C.; Legnose, S. Growth and yield of "Conference" pears grafted on quince and pear rootstocks. *Acta Hortic.* **2008**, 617–624.

36. Mpelasoka, B.S.; Behboudian, M.H.; Green, S.R. Water use, yield and fruit quality of lysimeter-grown apple trees: Responses to deficit irrigation and to crop load. *Irrig. Sci.* **2001**, *20*, 107–113.

37. Barrit, B.H.; Konishi, B.S.; Drake, S.R.; Rom, C.R. Influence of sunlight level and rootstock on apple fruit quality. *Acta Hortic.* **1997**, *451*, 569–577.

38. Oke, A.M.C.C.; Tapper, N.J.; Barlow, E.W.R. Within-vineyard variability in grape quality and yield and its relationship to the vineyard environment. *Acta Hortic.* **2007**, *754*, 507–514.

39. Asın, L.; Alegre, S.; Montserrat, R. Effect deficit irrigation, sumer pruning and root pruning on shoot growth, yield, and return bloom, in a "Blanquilla" pear orchard. *Sci. Hortic.* **2007**, *113*, 142–148.

40. Suárez, L.; Zarco-Tejada, P.J.; Sepulcre-Cantó, G.; Pérez-Priego, O.; Miller, J.R.; Jiménez-Muñoz, J.C.; Sobrino, J. Assessing canopy PRI for water stress detection with diurnal airborne imagery. *Remote Sens. Environ.* **2008**, *112*, 560–575.

41. Treder, W. Crop loading studies with "Jonagold" Apple tree. *J. Fruit Ornam. Plant Res.* **2010**, *18*, 59–69.

42. Hudina, M.; Štampar, F. The correlation of the pear (Pyrus communis L.) cv. "Williams" yield quality to the foliar nutrition and water regime. *Acta Agric. Slov.* **2005**, *85*, 179–185.

43. Chan, B.G.; Cain, J.C. The effect of seed formation on subsequent flowering in apple. *Proc. Am. Soc. Hortic. Sci.* **1967**, *91*, 63–68.

44. Thenkabail, P.S.; Smith, R.B.; de Pauw, E. Evaluation of narrowband and broadband vegetation indices for determining optimal hyperspectral wavebands for agricultural crop characterization. *Photogramm. Eng. Remote Sens.* **2002**, *68*, 607–621.

45. Webster, A.D.; Spencer, J.E. New strategies for the chemical thinning of apple And, (Malus domestica Borkh.) cultivars Queen Cox and Royal Gala. *J. Hortic. Sci. Biotechnol.* **1999**, *74*, 337–346.

46. Marsal, J.; Arbone, A.; Rufat, J.; Girona, J. Regulated deficit irrigation and rectification of irrigation scheduling in young pear trees: An evaluation based on vegetative and productive response. *Eur. J. Agronomy* **2002**, *17*, 111–122.

47. Wu, Y.; Zhao, Z.; Wang, W.; Ma, Y.; Huang, X. Yield and growth of mature pear trees under water deficit during slow fruit growth stages in sparse planting orchard. *Sci. Hortic.* **2013**, *164*, 189–195.

48. Jung, S.-K.; Choi, H.-S. Light penetration, growth, and fruit productivity in "Fuji" apple trees trained to four growing systems. *Sci. Hortic.* **2010**, *125*, 672–678.

49. Naor, A. Irrigation and crop load influence fruit size and water relations in field-grown "Spadona" pear. *J. Am. Soc. Hortic. Sci.* **2001**, *126*, 252–255.

50. Sepulcre-Cantó, G.; Zarco-Tejada, P.J.; Sobrino, J.A.; Berni, J.A.J.; Jiménez-Muñoz, J.C.; Gastellu-Etchegorry, J.P. Discriminating irrigated and rainfed olive orchards with thermal ASTER imagery and DART 3D simulation. *Agric. For. Meteorol.* **2009**, *149*, 962–975.

Comparison of NDVIs from GOCI and MODIS Data towards Improved Assessment of Crop Temporal Dynamics in the Case of Paddy Rice

Jong-Min Yeom and Hyun-Ok Kim

Abstract: The monitoring of crop development can benefit from the increased frequency of observation provided by modern geostationary satellites. This paper describes a four-year testing period from 2010 to 2014, during which satellite images from the world's first Geostationary Ocean Color Imager (GOCI) were used for spectral analyses of paddy rice in South Korea. A vegetation index was calculated from GOCI data based on the bidirectional reflectance distribution function (BRDF)-adjusted reflectance, which was then used to visually analyze the seasonal crop dynamics. These vegetation indices were then compared with those calculated using the Moderate-resolution Imaging Spectroradiometer (MODIS)-normalized difference vegetation index (NDVI) based on Nadir BRDF-adjusted reflectance. The results show clear advantages of GOCI, which provided four times better temporal resolution than the combined MODIS sensors, interpreting subtle characteristics of the vegetation development. Particularly in the rainy season, when data acquisition under clear weather conditions was very limited, it was possible to find cloudless pixels within the study sites by compiling GOCI images obtained from eight acquisition periods per day, from which the vegetation index could be calculated. In this study, ground spectral measurements from CROPSCAN were also compared with satellite-based vegetation products, despite their different index magnitude, according to systematic discrepancy, showing a similar crop development pattern to the GOCI products. Consequently, we conclude that the very high temporal resolution of GOCI is very beneficial for monitoring crop development, and has potential for providing improved information on phenology.

Reprinted from *Remote Sens*. Cite as: Yeom, J.-M.; Kim, H.-O. Comparison of NDVIs from GOCI and MODIS Data towards Improved Assessment of Crop Temporal Dynamics in the Case of Paddy Rice. *Remote Sens*. **2015**, *7*, 11326–11343.

1. Introduction

Phenological changes in land surface vegetation, which are closely related to boundary-layer atmospheric dynamics, have been increasingly seen as important signals of year-to-year climate variations or even global environmental changes [1–4]. The time series of wide-field-of-view sensors such as the Advanced Very High

Resolution Radiometer (AVHRR), Medium Resolution Imaging Spectrometer (MERIS), Moderate-resolution Imaging Spectroradiometer (MODIS), and SPOT VEGETATION have proven appropriate for phenology detection from multi-temporal vegetation indices [5–10]. Particularly for crop monitoring, the MODIS multi-year time-series analysis may make a significant contribution to providing temporal dynamics on rice cropping systems, as well as determining the spatial distribution of rice phenology [11–14]. Furthermore, the temporal information of crops from low resolution satellite imagery is useful for mapping different vegetation and crop types [15], and assessing yield and production [16].

When observing reflected solar spectral radiation from vegetation on the land surface using an optical sensor, cloud cover can prevent the accurate collection of surface physical characteristics. It is impossible to obtain surface spectral information from optical satellites over a cloudy area because the wavelength of the reflected solar spectrum cannot penetrate the cloudy area. Therefore, it is important to secure timely surface information from optical sensors under severe weather conditions. To overcome the limitations of polar orbiting reflective wavelength sensors for interpreting vegetation development, various temporal smoothing techniques such as Fourier harmonics, threshold methods, and curve-fitting methods have been suggested to fill or smooth noise and sparse greenness observations from satellite images [17–24]. Although these techniques are effective for dealing with sporadic missing data, using them for long-term missing data during the cloudy monsoon period of crop growth may produce detrimental results. Therefore, it is important to use high-temporal-resolution satellite images to obtain meaningful information. The combined MODIS observation characteristics from the Terra and Aqua satellites have been optimized to estimate vegetation phenology under normal weather conditions. However, during the monsoon rainy season (called Jang-Ma in Korea) between June and August, the high level of cloud cover makes it difficult to acquire timely surface information from MODIS observations.

The objective of this study was to calculate vegetation index profiles for two points using data from the first Korean geostationary orbit satellite, the Geostationary Ocean Color Imager (GOCI) launched successfully on 27 June 2010. GOCI was designed to detect, monitor, and predict regional ocean phenomena around Korea but is equipped with eight spectral bands (six visible, two near infrared). So, there is great interest in its terrestrial application because of its high temporal resolution as well as its vegetation-sensitive multispectral bands. The high temporal resolution of GOCI allows for eight acquisitions of imagery during the daytime and it is four times better than the MODIS observation system combining Terra and Aqua. The frequent observation characteristics of GOCI are therefore expected to provide more reliable information on crop temporal dynamics. We compared for two sample sites four-year GOCI data with corresponding MODIS image data to detect spectral signals according to crop growth and development.

2. Materials and Methods

2.1. Study Area

In this study, two paddy rice areas were selected; one was located in Kyehwa and corresponds to a GOCI pixel with coordinates of 35°46′37N and 126°41′03E (Figure 1b). The other was in Kimjae and corresponds to a GOCI pixel with coordinates of 35°44′59N and 126°52′15E (Figure 1c). These paddy areas were included in the monitoring site for the rice yield estimation by the Korea Agricultural Research & Extension Services. The study site at Kimjae represents the double cropping of barley and early maturing rice cultivars, and the site in Kyehwa represents the most popular paddy rice agriculture with an intermediate-late-maturing rice cultivar. The early maturing rice cultivars are generally transplanted a little later than intermediate-late-maturing species, around the middle of June, and harvested at the end of September or the beginning of October. The intermediate-late rice cultivars are transplanted from the end of May until the beginning of June and harvested around the middle of October. As these study sites are relatively homogeneous, despite the small paddy units, the temporal dynamics of different crops should be recognizable in the daily satellite image data analysis.

2.2. Satellite Data Used in the Present Study

We compared two sets of optical earth observation satellite data with the same spatial resolution of 500 m; one was from a geostationary (GOCI) and the other from a sun-synchronous satellite (MODIS). GOCI is limited to a 2500×2500 km^2 field of view (FOV) centered with respect to the Korean Peninsula and its eight multispectral bands cover visible and near-infrared (NIR) spectral wavelengths (Table 1). In addition, its geometric accuracy is better than 0.4 pixels. The GOCI viewing zenith angle (VZA) ranges from 32.38° to 63.74°. The GOCI VZA for the study areas was 48.47 (Figure 1b) and 48.46 (Figure 1c). We used the fifth and eighth GOCI bands for calculating the normalized difference vegetation index (NDVI). For comparison, MODIS NDVI products were applied as a reference. This study analyzed the data for the years 2011 to 2014.

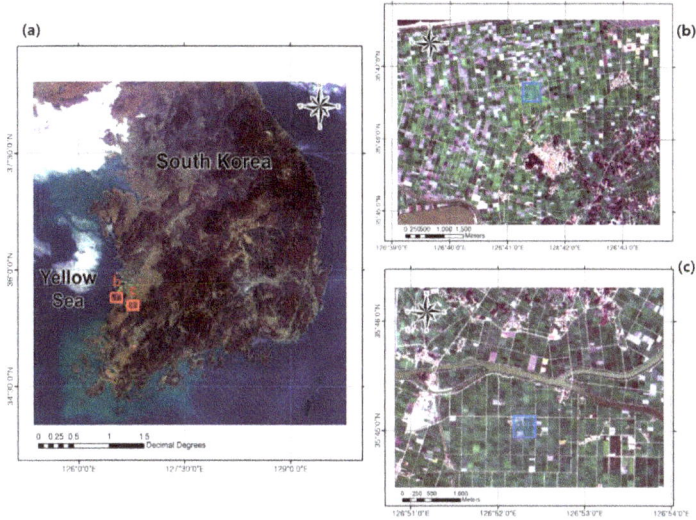

Figure 1. Study area. (**a**) Red Green Blue (RGB) color composite image from the Geostationary Ocean Color Imager (GOCI) acquired on 1 April 2011. The red rectangles, (**b**) and (**c**), in (a) are shown in (b), and (c), respectively, giving detailed views using high-resolution RapidEye multispectral data obtained on 5 August 2011 (b), and 11 October 2011 (c). The blue rectangles in (b), and (c) are geometrically matched with corresponding satellite observation pixels.

Table 1. Detailed characteristics of the GOCI and MODIS sensors used for estimating land-surface products.

Satellite Sensor	Orbit Type	Altitude	Wavelength	Spatial Resolution
GOCI	Geo-synchronous	≈36,000 km	B1: 402–422 nm B2: 433–453 nm B3: 480–500 nm B4: 545–565 nm B5: 650–670 nm B6: 675–685 nm B7: 735–755 nm B8: 845–885 nm	Approximately 500 m over South Korea area (≈390 m at nadir)
MODIS	Sun-synchronous	≈705 km	B1: 620–670 nm B2: 841–876 nm B3: 459–479 nm B4: 545–565 nm B5: 1230–1250 nm B6: 1628–1652 nm B7: 2105–2155 nm	500 m at nadir

Figure 2 shows the spectral response functions of MODIS (in blue) from MODIS Characterization Support Team and GOCI (in red) from Korea Institute of Ocean Science & Technology (KIOST); the straight and dashed lines in the two colors shown correspond to red and NIR wavelengths, respectively. The spectral response

121

functions (SRFs) shown in Figure 2 for the red and NIR frequencies were slightly different because GOCI was designed to observe ocean products such as chlorophyll. The GOCI visible red band SRF is narrower than that for MODIS because its original band purpose was as a baseline for fluorescence, chlorophyll, and suspended sediment. In this study, interpreting the effect of different SRFs was beyond the scope of our research, requiring sensor calibration with atmospheric constituents and ground spectral information for an accurate reading of the spectral vegetation index from different sensors. We assumed the MODIS spectral bands as a reference and compared them with GOCI land products to determine the feasibility of GOCI land application.

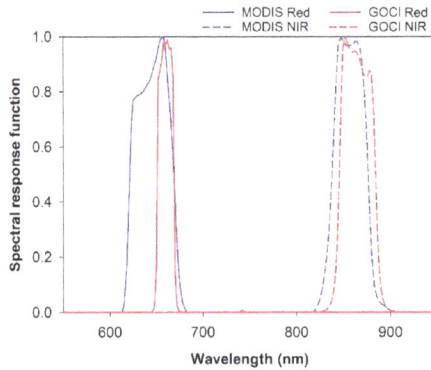

Figure 2. Spectral response function (SRF) of Moderate resolution Imaging Spectroradiometer (MODIS) (in blue) and GOCI (in red). The straight and dashed lines correspond to visible and near-infrared wavelengths, respectively.

2.3. Ground Measurements Using a Multispectral Radiometer

In this study, ground measurements were performed using the multispectral radiometer (MSR) to evaluate satellite-based vegetation profiles for comparative analysis. The CROPSCAN MSR16 used in this study was equipped with 16 spectral sensor bands in the 450–1750 nm region. When measuring ground spectral information on rice paddy with CROPSCAN, we observed three different points within selected blue rectangle areas in Figure 1b and 1c, and then averaged the tree points of spectral measurements to reflect spatial representation of chosen rice paddy. The blue rectangles (500 × 500 m) in Figure 1b,c are geometrically matched with corresponding satellite observation pixels for comparison. Field measurements were carried out from June to October 2014. To obtain the crop development characteristics of the paddy rice, measurements were made on eight dates based on the cultivation schedule, including transplantation and harvest. Table 2 lists paddy rice development during the growing season over Kyehwa (Figure 1b) and Kimjae (Figure 1c).

Table 2. Time-series photographs of paddy rice in the Kyehwa and Kimjae areas.

Date	Kyehwa	Kimjae	Status
06/13			After transplantation
06/26			Growing season
07/15			Growing season
07/22			Growing season
08/11			Earing season

Table 2. *Cont.*

Date	Kyehwa	Kimjae	Status
08/26			Earing season
09/15			Heading stage
10/03			Harvest

2.4. Satellite Data Pre-Processing

For the GOCI satellite image, further pre-processing, including conversion of digital numbers (DN) to radiance, cloud masking, and atmospheric correction, was performed to calculate surface reflectance. To undertake cloud masking, a threshold method was adopted [25]. The look-up table (LUT) from the Second Simulation of a Satellite Signal in the Solar Spectral (6S) atmospheric correction model was used for calculating the GOCI surface reflectance [26–28]. The 6S radiative transfer model is advantageous for atmospheric correction because it is flexible in applying particular regional characteristics (e.g., topography, land type, or atmospheric condition) and sensor properties (e.g., band width or spectral response function of each band) [29]. The LUT is preliminarily constructed to invert 6S radiative transfer model for calculating the surface reflectance. When simulating 6S modeling for GOCI, atmospheric products such as aerosol optical thickness, aerosol type, ozone, and water vapor were acquired from MODIS atmospheric products (MOD04, MOD05, and MOD07) from NASA's Earth Observing System Data an Information System (EOSDIS). When using MODIS atmospheric products, which did not fully cover the

GOCI observation times, we assumed that the daily variation in the atmospheric constituents from the MODIS atmospheric products was low. When comparing ground station particulate matter ($PM_{2.5}$), we found that the overall root mean square error (RMSE) of the aerosol optical depth (AOD) was 0.123 [30]; it follows that the expected error in the surface reflectance using the MODIS daily AOD will be less than 3% in the 6S radiative transfer model. When MODIS products were unavailable (mainly due to cloud contamination), we substituted the aerosol optical thickness based on COMS MI [31] for the MODIS aerosol optical thickness. In this study, for the MODIS satellite image, the MODIS atmospheric corrected reflectance (MOD09GA, MYD09GA, collection 5) from NASA's EOSDIS was used to estimate the normalized NDVI products. For geometric matching, we applied the nearest-neighbor method to the GOCI and MODIS data by resampling different projected images. Since reflectance measurements from satellite data are affected by the surface anisotropy, the semi-empirical bidirectional reflectance distribution function (BRDF) model was applied to normalize surface reflectance from GOCI and MODIS images.

2.5. BRDF Modeling and Calculation of Vegetation Index

We applied the BRDF model based on Ross-thick/Li-sparse reciprocal (RTLSR) kernels to estimate the normalized reflectance [32–34] and correct surface anisotropy effects. Surface reflectance data from GOCI and MODIS were used in the BRDF model to calculate the GOCI and MODIS BRDF-adjusted reflectance, respectively [35–37]. The BRDF model kernel coefficients were estimated independently for each gridded pixel location using available cloud-cleared observations for a 16-day composite period to estimate daily rolling products [34,35]. In other words, the cloud-free surface reflectance during the 16-day composite period was assembled to simulate the BRDF model, and then the estimated kernel coefficients were utilized to retrieve the angle-adjusted reflectance. In this study, BRDF-adjusted reflectances from GOCI and MODIS were estimated using a daily rolling strategy over a 16-day composite period to interpret more subtle characteristics of the phenology [36,38]. The BAR products, which were less sensitive to variations in the sun and viewing geometry, were used to estimate daily NDVI products using the following equation:

$$NDVI_{BAR} = \frac{NIR_{BAR} - red_{BAR}}{NIR_{BAR} + red_{BAR}} \tag{1}$$

where $NDVI_{BAR}$ is the vegetation index based on BRDF-adjusted surface reflectance, and NIR_{BAR} and red_{BAR} represent the BRDF-corrected surface NIR and red bands, respectively.

Lastly, the 10-day NDVI maximum value composite (MVC) is also estimated for comparing the GOCI BAR NDVIs with the GOCI 10-day MVC NDVI. The 10-day NDVI MVC method has been recommended in many cases to minimize the effect of cloud contamination on optical sensors [39] because the highest NDVI value during the 10-day period is retained under the assumption that it represents the NDVI value least affected by the presence of clouds, smoke, haze, snow, and ice during the composite period.

3. Results and Discussion

3.1. Spectral Analysis of Crop Temporal Dynamics

The temporal changes in BAR NDVIs in the four-year GOCI data were compared with those in the corresponding MODIS NBAR NDVIs data for two rice paddies, shown in Figure 1b,c. As Figure 3 shows, the annual NDVI changes correspond well with the crop development of the intermediate-late-maturing rice paddy (Figure 3a), and early maturing rice paddy (Figure 3b).

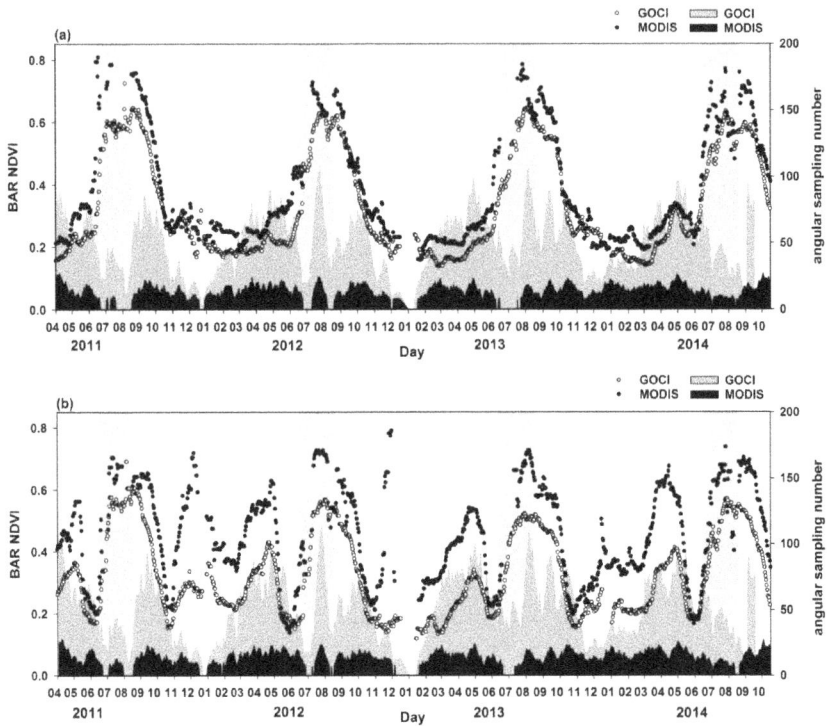

Figure 3. Year-to-year variation in crop seasonal dynamics using GOCI- and MODIS-based vegetation indices. The open and solid circles show the GOCI BAR NDVI and MODIS NBAR NDVI, respectively. The gray and black histograms show the number of GOCI and MODIS angular samples, respectively. (**a**) Intermediate-late-maturing rice paddy, and (**b**) Early maturing rice paddy. The four light gray areas from middle June to middle August are rainy summer seasons in South Korea.

For the intermediate-late-maturing rice paddy in Figure 3a, compared with the NBAR NDVIs derived from MODIS (solid circles), the GOCI BAR NDVI (open circles) better reflects the annual tendency with less scattering from general crop seasonal dynamics. The advantages of the GOCI are particularly during the summer from June to August, when the weather conditions are very changeable, and rain can persist for long periods. Whereas MODIS resulted in intermittent NDVI values during the long rainy periods (light gray areas in Figure 3) usually between middle June and early August, GOCI provided increasing NDVI values, which appear reasonable for the growing season of paddy rice. In addition, as shown in Figure 3a, the single crop development patterns from the GOCI BAR NDVIs and MODIS NBAR NDVI were similar, but exhibited more discontinuous crop signal transitions of MODIS during the summer from June to August. For the early maturing rice paddy, double cropping spectral patterns were detected (see Figure 3b) for both GOCI (open circles) and MODIS (solid circles). Whereas the GOCI- and MODIS-based vegetation index profiles show similar patterns under benign weather conditions with a high number of angular samples, there is a clear difference between the GOCI- and MODIS-based spectral dynamic patterns during the rainy summer and snowy winter season.

The quality of BRDF modeling for normalized reflectance is dependent on acquiring at least seven cloud-free observations of each gridded pixel during the 16-day composite period [36]. In Figure 3, the number of cloud-free observations for BRDF modeling is depicted in histograms (gray shows GOCI acquisition and black is MODIS) to ensure the full inversion BRDF parameters required for obtaining reliable surface estimations. If only one to six clear observations are available during the 16-day composite period, then angular sampling numbers of fewer than six were replaced with zero to clearly identify whether full inversion BRDF modeling was applicable. The results from the two study areas show that MODIS (black histogram) did not perform the full inversion with Equation (1) during the rainy season because MODIS from Terra and Aqua can only make two observations over a pixel location. In contrast, GOCI displays exhibited increasing NDVI values during the cloudy summer periods, which appear to be reasonable for the growing seasons (from July to August) of crop areas. As GOCI offers eight multispectral images every day during the daytime (from 9 a.m. to 4 p.m.), the intuitive multi-temporal NDVI can be estimated from sufficient cloud-free observations despite the rainy season.

Given the steady margins of the absolute difference between MODIS NBAR NDVIs and GOCI BAR NDVIs under benign weather condition shown in Figure 3, Figure 4 makes one-to-one comparisons of the NDVI, NIR, and Red bands to identify different characteristics of MODIS and GOCI. For both rice paddy areas, GOCI BAR NDVI gave lower values than MODIS, implying that the different SRF of the red band described in Figure 2 might cause the steady margin difference. In Figure 2, NIR SRF had a similar function, but the red band of GOCI has a narrower SRF than

MODIS. Therefore, we think that the BAR red band of GOCI gave higher reflectance resulting in lower NDVI values in Figure 4. We inferred that the higher red band RMSE between GOCI and MODIS would cause the higher NDVI RMSE due to SRF different in Figure 4b.

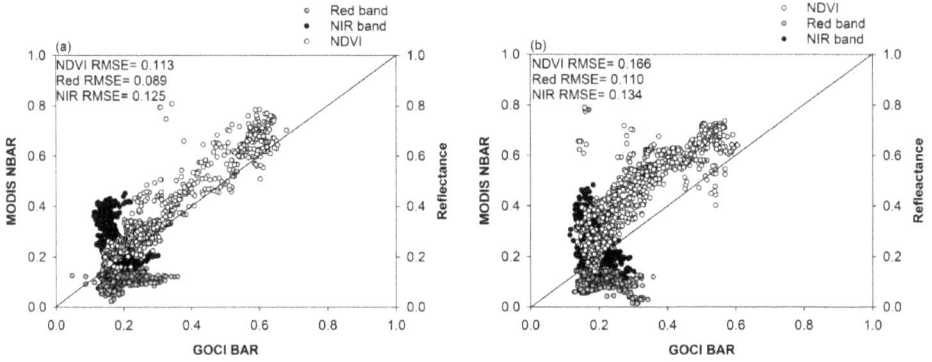

Figure 4. Scatterplots of the GOCI and MODIS vegetation products. The open circles show the BAR NDVI, and the solid and gray circles show the BAR NIR and BAR Red bands, respectively. (**a**) Intermediate-late maturing rice paddy, and (**b**) early intermediate-late maturing rice paddy.

Figure 5 compares two NDVI datasets: one is the instantaneous NDVI measurement from atmospherically corrected reflectance (open circles), and the other is the equivalent processed BAR NDVI (solid circles) for the two crop areas.

For the intermediate-late-maturing rice paddy, the BRDF-adjusted NDVI (solid circles) from GOCI (Figure 5a) and MODIS (Figure 5b) show annual spectral change that corresponds well to the development of vegetation. However, the instantaneous measurements of NDVI from both GOCI and MODIS (open circles) are scattered mostly with the BAR NDVI as the center because of BRDF effects and cloud contamination. For the early maturing rice paddy in Figure 5c,d, similar patterns are shown. In Figure 5, the maximum value of the instantaneous NDVI measurements among the daily values is described alongside the crop dynamics of the study area.

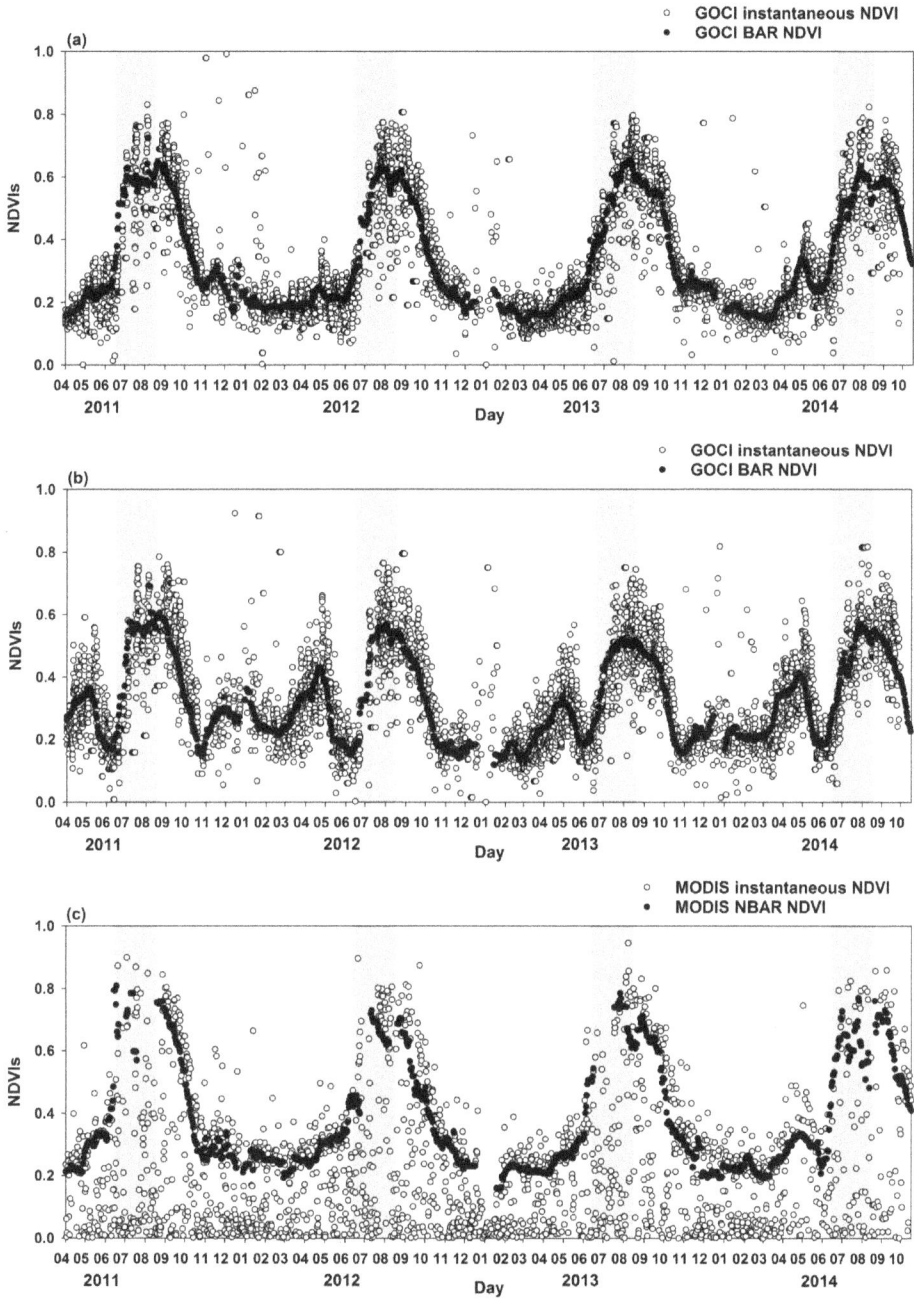

Figure 5. *Cont.*

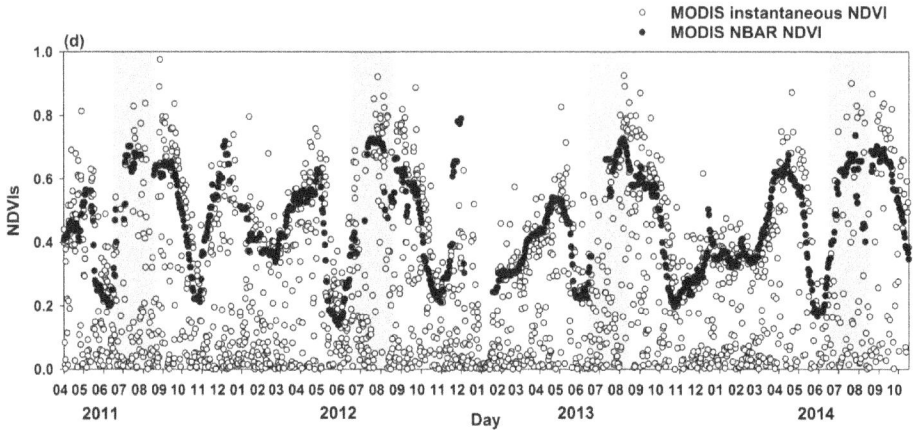

Figure 5. Comparisons of the temporal NDVI variation derived from GOCI and MODIS. The GOCI NDVI profiles for BAR (solid circles) and instantaneous measurement of NDVI (open circles) over (**a**) intermediate-late maturing paddy rice, and (**b**) early maturing paddy rice; MODIS NBAR (solid circles) and instantaneous measurement of NDVI (open circles) profiles over (**c**) intermediate-late-maturing paddy rice, and (**d**) early maturing paddy rice. The four light gray areas from middle June to middle August are rainy summer seasons in South Korea.

In this study, we also compared the GOCI BAR NDVIs with the GOCI 10-day MVC NDVI based on a daily rolling strategy to determine efficient methods for interpreting intuitive crop dynamics (see Figure 6). The days used to represent these products were the center day of the time window; therefore, matched center data are used for comparison. As shown in Figure 6a, MVC values above 1 were considered outliers, indicating that the MVC method could be used to reveal the limitations associated with minimizing the effect of cloud contamination. However, the crop temporal dynamics of the GOCI 10-day MVC NDVI (open gray circles) did not describe the general phenology pattern, which still remained scattered (Figure 6b). Although scatterplots of the 10-day MVC NDVI displayed better a crop seasonal dynamic pattern compared with the instantaneous measurement NDVI, it was still insufficient with regards to obtaining detailed crop growth and development information, such as the time required for the onset of green-up, the maximum rate of green-up, and time-integrated NDVI as a measure of net primary productivity. For the early maturing rice paddy area, similar characteristics are seen in Figure 6b. Figure 6b shows that the spectral features during the crop-growing season and agricultural off-season are captured in GOCI BAR NDVI, but the crop signal dynamics are not described in the GOCI 10-day MVC NDVI.

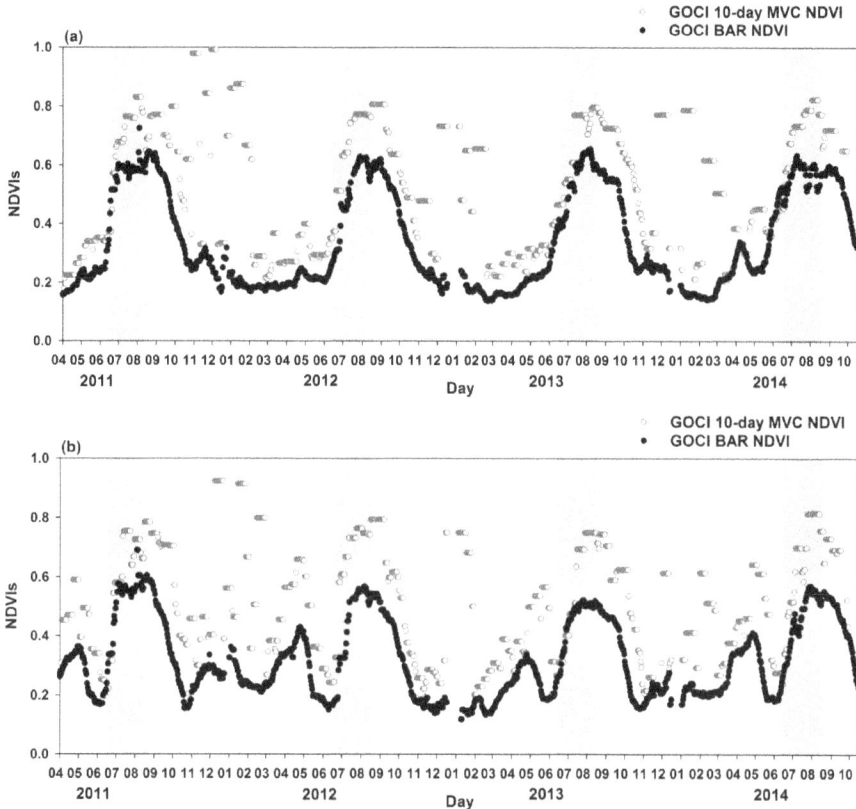

Figure 6. Comparison of the BAR NDVI and 10-day MVC NDVI from GOCI. Temporal variation in BAR NDVI (solid circles) and 10-day MVC NDVI (open gray circles) over intermediate-late-maturing paddy rice (**a**) and early paddy rice (**b**). The 4 number of light gray areas from middle June to middle August are rainy summer seasons in South Korea.

Finally, we compared temporal BRDF-adjusted NDVIs from GOCI and MODIS with field measurement data from CROPSCAN gathered in 2014. Figure 7 shows the comparison of CROPSCAN measured NDVI, interpolated using the cubic spline function (triangle points for measurement and dashed line for interpolated values) and multi-temporal satellite-based NDVIs (open circles for GOCI; solid circles for MODIS) for intermediate-late-maturing rice paddy (Figure 7a) and early-maturing rice paddy (Figure 7b). The field measured NDVIs clearly exhibit higher values than the values based on satellite data values. This may be explained by considering that the CROPSCAN-measured NDVI values represent the coverage of rice planted on a paddy unit whereas the moderate spatial resolution satellite data-based NDVI values include other types of land cover, such as farm roads, vinyl greenhouses, and

131

artificial structures, in its pixel. A comparison without considering the mixed land cover in the MODIS and GOCI data prevents exact validation of the crop temporal dynamics based on the satellite data. However, when interpreted visually, the GOCI BAR NDVI multispectral changes during the growing season appear to better match with the crop development dynamics in the field than the MODIS data. GOCI has similar vegetation trajectory patterns with a constant margin as the CROPSCAN measurement, from date of maximum growth to senescence. However, BRDF adjusted NDVI profiles appeared as shifted to the right side when comparing with the CROPSCAN measurements. It would be caused by the 16-day composite method for simulating BRDF model. BRDF adjusted NDVI might be less sensitive for real time change due to temporal composite than ground measured NDVI representing the immediate reaction of targets.

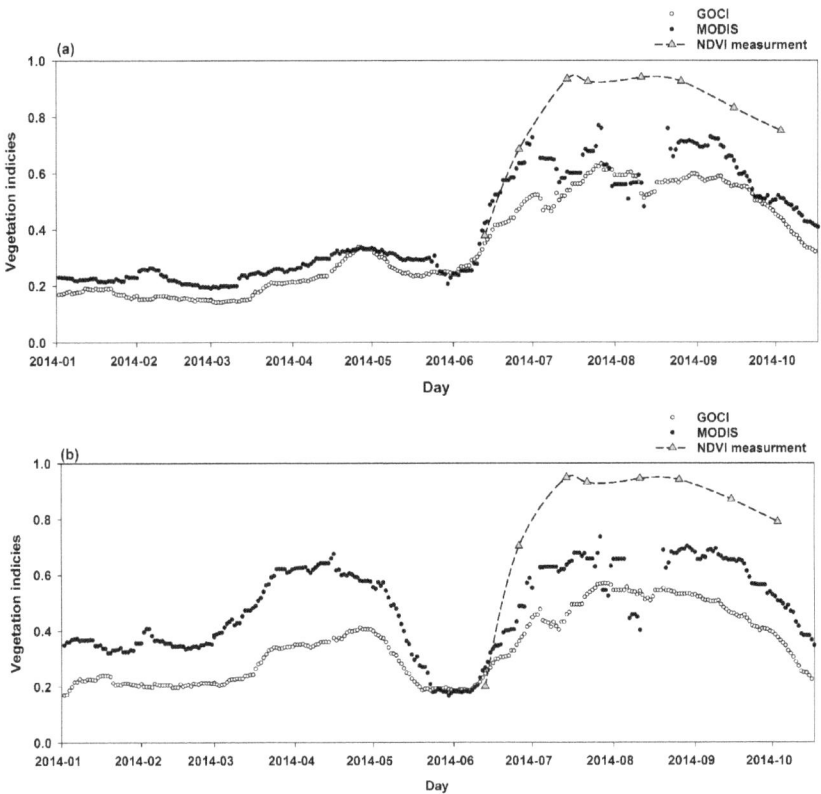

Figure 7. Comparison of temporal BRDF-adjusted NDVI from GOCI (solid circles) and MODIS (open circles) with CROPSCAN measurements (solid triangles) over rice paddy with (**a**) intermediate-late maturing and (**b**) early-maturing rice cultivar during 2014. The dashed line over scan measurements is interpolated NDVI using the cubic spline function.

3.2. Discussion

Our four-year GOCI BAR NDVI analysis showed the significant benefit of high temporal resolution for monitoring crop development. However, there were a number of limitations to this work. First, the wavelengths of the red and NIR bands in GOCI and MODIS used for NDVI calculation differed slightly. In principle, for interpreting the different SRFs effects, inter-calibration needs to be performed using stable, homogenous, and less anisotropic natural targets [40]. However, the previous studies revealed the difficulty to compare biophysical products even if derived from the same sensor [41]. So, we performed the one-to-one comparison of the Red, NIR, and NDVI products in order to complement the SRFs difference between GOCI and MODIS and found that the different SRF of the red band might cause the steady margin difference. The interesting fact in the one-to-one comparison was that the BRDF adjusted reflectance showed mostly the less relations, while the NDVIs were well correlated.

Second, there were only two sample sites, and each sample site corresponded to one satellite image pixel. As the main purpose of our study is rapidly to test the benefits of GOCI data with a very high temporal resolution for extracting reliable crop temporal dynamics, we focused on selecting representative study sites instead of quantitative number of study sites. We very carefully chose those two rice paddy sample sites, which are homogeneous despite of small paddy units and covered by the monitoring site for the rice yield estimation by Korea Agricultural Research & Extension Services.

Third, for the comparison of the NDVI calculated using moderate-resolution satellite data with the field measurements, it is necessary to consider the mixed-pixel problem. Because the paddy units in South Korea are relatively very small, it is very difficult to observe a non-mixed spectral value for rice paddies on the moderate spatial resolution GOCI data. The challenges of insufficient spatial resolution were also mentioned in many other studies [42].

4. Conclusion

We investigated the applicability of high-temporal-resolution GOCI satellite data for monitoring crop development. We found that the high temporal resolution of GOCI is advantageous for simulating full inversion BRDF modeling and detecting crop temporal dynamics, which is useful in crop phenology analysis, particularly during the rainy season. In general, GOCI and MODIS displayed similar temporal variation in NDVI under benign weather conditions, because they can secure enough cloud-free observations for full inversion BRDF modeling. During the monsoon season, however, with its long periods of rain and many cloudy days, GOCI was found to be more useful for extracting cloudless or less cloudy areas by arraying its eight images to calculate representative daily data. We also found that the GOCI BAR

NDVI was more useful for crop signal monitoring than the widely used MVC NDVI. Lastly, we compared the multi-year NDVI profiles derived from GOCI and MODIS data with field measurements and visually verified the similar crop development patterns between satellite data and field measurement, despite of their different index magnitude.

So, we could conclude that GOCI's very high-temporal-resolution originally desired for ocean color monitoring is also very applicable for terrestrial monitoring. For the GOCI BAR NDVI, it would be useful to calculate the crop temporal dynamics in greater detail, including the time required for the onset of green-up, maximum rate of green-up, and time-integrated NDVI as a measure of net primary productivity.

We expect that stable vegetation profiles derived from high-temporal-resolution GOCI data will be useful for analyzing crop phenology, as well as phonological parameters reflecting the exact field conditions, which will be the subject of future study. To ensure the GOCI application for land areas, the future study will (1) expand the spatial coverage at a regional or continental scale to show the spatial representativeness, (2) verify the spectral values derived from GOCI and MODIS with ground based spectral measurements to model the real crop development, and lastly (3) simulate the rice yield using GOCI BAR NDVIs and verify its applicability.

Acknowledgments: We thank the Korea Institute of Ocean Science & Technology (KIOST) for providing GOCI data.

Author Contributions: All authors assisted in the analysis and editing of the paper. Jong-Min Yeom, the main author, processed the data, developed the methodology, analyzed the results, and wrote the manuscript. Hyun-Ok Kim, the corresponding author reviewed the paper and supported the analysis and interpretation of the results.

Conflicts of Interest: The authors declare no conflict of interest.

References

1. Reed, B.; Brown, J.F.; Vanderzee, D.; Loveland, T.R.; Merchant, J.W.; Ohlen, D.O. Measuring phenological variability from satellite imagery. *J. Veg. Sci.* **1994**, *5*, 703–714.
2. Schwartz, M.D.; Ahas, R.; Aasa, A. Onset of spring starting earlier across the Northern hemisphere. *Int. J. Climatol.* **2006**, *22*, 343–351.
3. Cleland, E.E.; Chuine, I.; Menzel, A.; Nooney, H.A.; Schwartz, M.D. Shifting plant phenology in response to global change. *Trends Ecol. Evol.* **2007**, *22*, 357–365.
4. Shuai, Y.; Schaaf, C.B.; Zhang, X.; Strahler, A.; Roy, D.; Morisette, J.; Wang, Z.; Nightingale, J.; Nickeson, J.; Richardson, A.D.; *et al.* Daily MODIS 500 m reflectance anisotropy direct broadcast (DB) products for monitoring vegetation phenology dynamics. *Int. J. Remote Sens.* **2013**, *34*, 5997–6016.
5. John, F.H.; Robert, W.J.; Bethany, A.B.; John, F.M. Extracting phenological signals from multiyear AVHRR NDVI time series: Framework for applying high-order annual splines with roughness damping. *IEEE Trans. Geosci. Remote Sens.* **2007**, *45*, 3264–3276.

6. Ganguly, S.; Friedl, M.; Tan, B.; Zhang, X.; Verma, M. Land surface phenology from MODIS: characterization of the Collection 5 global land cover dynamics product. *Remote Sens. Environ.* **2010**, *114*, 1805–1816.

7. Jin, J.; Jiang, H.; Zhang, X.; Wang, Y. Characterizing spatial-temporal variations in Vegetation phenology over the North-South transect of Northeast Asia based upon the MERIS terrestrial chlorophyll index. *Terr. Atmos. Ocean. Sci.* **2012**, *23*, 413–424.

8. Vintrou, E.; Bégué, A.; Baron, C.; Saad, A.; Seen, D.L.; Traoré, S.B. A comparative study on satellite- and model-based crop phenology in West Africa. *Remote Sens.* **2014**, *6*, 1367–1389.

9. Delbart, N.; Toan, T.L.; Kergoat, L.; Fedotova, V. Remote sensing of spring phenology in boreal regions: A free of snow-effect method using NOAA-AVHRR and SPOT-VGT data (1982–2004). *Remote Sens. Environ.* **2006**, *101*, 52–62.

10. Atzberger, C.; Eilers, P.H.C. A time series for monitoring vegetation activity and phenology at 10-daily time steps covering large parts of South America. *Int. J. Digit. Earth* **2011**, *4*, 365–386.

11. Boschetti, M.; Nutini, F.; Manforn, G.; Brivio, P.A.; Nelson, A. Comparative analysis of normalized difference spectral indices derived from MODIS for detecting surface water in flooded rice cropping systems. *PloS ONE.* **2014**, *9*, 1–21.

12. Boschetti, M.; Stroppiana, D.; Brivio, P.A.; Bocchi, S. Multi-year monitoring of rice crop phenology through time series analysis of MODIS images. *Int. J. Remote Sens.* **2009**, *30*, 4643–4662.

13. Sakamoto, T.; Nguyen, N.V.; Ohno, H.; Ishitsuka, N.; Yokozawa, M. Spatio-temporal distribution of rice phenology and cropping systems in the Mekong Delta with spectral reference to the seasonal water flow of the Mekong and Bassac rivers. *Remote Sens. Environ.* **2006**, *100*, 1–16.

14. Sari, D.K.; Ismullah, I.H.; Sulasdi, W.N.; Harto, A.B. Detecting rice phenology in paddy fields with complex cropping pattern using time series MODIS data. *ITB J. Sci.* **2010**, *42*, 91–106.

15. Friedl, M.A.; Mclver, D.K.; Hodges, J.C.F.; Zhang, X.Y.; Muchoney, D.; Strahler, A.H.; Woodcock, C.E.; Gopal, S.; Schneider, A.; Cooper, A.; *et al.* Global land cover mapping from MODIS: Algorithms and early results. *Remote Sens. Environ.* **2002**, *83*, 287–302.

16. Rembold, F.; Atzberger, C.; Savin, I.; Rojas, O. Using low resolution satellite imagery for yield prediction and yield anomaly detection. *Remote Sens.* **2013**, *5*, 5572–5573.

17. Viovy, N.; Arino, O.; Belward, A.S. The best index slope extraction (BISE): A method for reducing noise in NDVI time-series. *Int. J. Remote Sen.* **1992**, *13*, 1585–1590.

18. Roerink, G.J.; Menenti, M.; Verhoef, W. Reconstructing cloud free NDVI composites using Fourier analysis of time series. *Int. J. Remote Sen.* **2000**, *21*, 1911–1917.

19. Jakubauskas, M.E.; Legates, D.R.; Kastens, J.H. Harmonic analysis of time-series AVHRR NDVI data. *Photogramm. Eng. Remote Sens.* **2001**, *67*, 461–470.

20. Jonsson, P.; Eklundh, L. Seasonality extraction by function fitting to time-series of satellite sensor data. *IEEE Trans. Geosci. Remote Sens.* **2002**, *40*, 1824–1832.

21. Beck, P.S.A.; Atzberger, C.; Høgda, K.A.; Johansen, B.; Skidmore, A.K. Improved monitoring of vegetation dynamics at very high latitudes: A new method using MODIS NDVI. *Remote Sens. Environ.* **2006**, *100*, 321–334.

22. Chen, J.M.; Deng, F.; Chen, M.Z. Locally adjusted cubic-spline capping for reconstructing seasonal trajectories of a satellite-derived surface parameter. *IEEE Trans. Geosci. Remote Sens.* **2006**, *44*, 2230–2238.

23. Hird, J.N.; McDermid, G.J. Noise reduction of NDVI time series: An empirical comparison of selected techniques. *Remote Sens. Environ.* **2009**, *113*, 248–258.

24. Atkinson, P.M.; Jeganathan, C.; Dash, J.; Atzberger, C. Inter-comparison of four models for smoothing satellite sensor time-series data to estimate vegetation phenology. *Remote Sens. Environ.* **2012**, *123*, 400–417.

25. Saunders, R.W.; Kriebel, K.T. An improved method for detecting clear sky and cloudy radiances from AVHRR data. *Int. J. Remote Sens.* **1988**, *9*, 123–150.

26. Vermote, E.F.; Tanre, D.; Deuze, J.L.; Herman, M.; Morcette, J.J. Second simulation of the satellite signal in the solar spectrum, 6S: An overview. *IEEE Trans. Geosci. Remote Sens.* **1997**, *35*, 675–686.

27. Wang, M. An efficient method for multiple radiative transfer computations and the lookup table generation. *J. Quant. Spectrosc. Radiat. Transf.* **2003**, *78*, 471–480.

28. Lyapustin, A.; Martonchik, J.; Wang, Y.; Laszlo, I.; Korkin, S. Multiangle implementation of atmospheric correction (MAIAC): 1. Radiative transfer basis and look-up tables. *J. Geophys. Res.* **2011**.

29. Lee, C.S.; Yeom, J.M.; Lee, H.L.; Kim, J.J.; Han, K.S. Sensitivity analysis of 6S-based look-up table for surface reflectance retrieval. *Asia-Pac. J. Atmos. Sci.* **2015**, *51*, 91–101.

30. Green, M.; Kondragunta, S.; Xu, P.C.; Xu, C. Comparison of GOES and MODIS aerosol optical depth (AOD) to aerosol robotic network (AERONET) AOD and IMPROVE PM2.5 mass at Bondville, Illinois. *J. Air Waste Manage. Assoc.* **2009**, *59*, 1082–1091.

31. Kim, Y.J.; Ahn, M.H.; Sohn, B.J.; Lim, H.S. Retrieving aerosol optical depth using visible and mid-IR channels from geostationary satellite MTSAT-1R. *Int. J. Remote Sens.* **2008**, *29*, 6181–6192.

32. Roujean, J.L.; Leroy, M.; Deschamps, P.Y. A bidirectional reflectance model of the Earth's surface for the correction of remote sensing data. *J. Geophys. Res.* **1992**, *97*, 20455–20468.

33. Wanner, W.; Li, X.; Strahler, A.H. On the derivation of kernels for kernel-driven models of bidirectional reflectance. *J. Geophys. Res.* **1995**, *100*, 21077–21089.

34. Schaaf, C.B.; Gao, F.; Strahler, A.H.; Lucht, W.; Li, X.; Tsang, T.; Strugnell, N.C.; Zhang, X.; Jin, Y.; Muller, J.P.; *et al.* First operational BRDF, albedo and nadir Reflectance Products from MODIS. *Remote Sens. Environ.* **2002**, *83*, 135–148.

35. Lucht, W.; Schaaf, C.B.; Strahler, A.H. An Algorithm for the retrieval of albedo from space using semiempirical BRDF Models. *IEEE Trans. Geosci. Remote Sens.* **2000**, *38*, 977–998.

36. Shuai, Y.; Schaaf, C.B.; Strahler, A.H.; Liu, J.; Jiao, Z. Quality assessment of BRDF/Albedo retrievals in MODIS operational system. *Geophys. Res. Lett.* **2008**, *35*, L05407.

37. Yeom, J.M.; Kim, H.O. Feasibility of using Geostationary Ocean Color Imager (GOCI) data for land applications after atmospheric correction and bidirectional reflectance distribution function modelling. *Int. J. Remote Sens.* **2013**, *34*, 7329–7339.

38. Ju, J.; Roy, D.; Shuai, Y.; Schaaf, C.B. Development of an approach for generation of temporally complete daily nadir MODIS reflectance time series. *Remote Sens. Environ.* **2010**, *114*, 1–20.

39. Holben, B.N. Characteristics of maximum-value composite images from temporal AVHRR data. *Int. J. Remote Sens.* **1986**, *7*, 1417–1434.

40. Chander, G.; Hewison, T.J.; Fox, N.; Wu, X.; Xiong, X.; Blackwell, W.J. Overview of intercalibration of satellite instruments. *IEEE Trans. Geosci. Remote Sens.* **2013**, *51*, 1056–1080.

41. Meroni, M.; Atzberger, C.; Vancutsem, C.; Gobron, N.; Baret, F.; Lacaze, R.; Eerens, H.; Leo, O. Evaluation of agreement between space remote sensing SPOT-VEGETATION fAPAR time series. *IEEE Trans. Geosci. Remote Sens.* **2013**, *51*, 1951–1932.

42. Ozdogan, M. The spatial distribution of crop types from MODIS data: Temporal unmixing using independent component analysis. *Remote Sens. Environ.* **2010**, *114*, 1190–1204.

Exploring the Vertical Distribution of Structural Parameters and Light Radiation in Rice Canopies by the Coupling Model and Remote Sensing

Yongjiu Guo, Ling Zhang, Yehui Qin, Yan Zhu, Weixing Cao and Yongchao Tian

Abstract: Canopy structural parameters and light radiation are important for evaluating the light use efficiency and grain yield of crops. Their spatial variation within canopies and temporal variation over growth stages could be simulated using dynamic models with strong application and predictability. Based on an optimized canopy structure vertical distribution model and the Beer-Lambert law combined with hyperspectral remote sensing (RS) technology, we established a new dynamic model for simulating leaf area index (LAI), leaf angle (LA) distribution and light radiation at different vertical heights and growth stages. The model was validated by measuring LAI, LA and light radiation in different leaf layers at different growth stages of two different types of rice (*Oryza sativa L.*), *i.e.*, *japonica* (Wuxiangjing14) and *indica* (Shanyou63). The results show that the simulated values were in good agreement with the observed values, with an average RRMSE (relative root mean squared error) between simulated and observed LAI and LA values of 14.75% and 21.78%, respectively. The RRMSE values for simulated photosynthetic active radiation (PAR) transmittance and interception rates were 14.25% and 9.22% for Wuxiangjing14 and 15.71% and 4.40% for Shanyou63, respectively. In addition, the corresponding RRMSE values for red (R), green (G) and blue (B) radiation transmittance and interception rates were 16.34%, 15.96% and 15.36% for Wuxiangjing14 and 5.75%, 8.23% and 5.03% for Shanyou63, respectively. The results indicate that the model performed well for different rice cultivars and under different cultivation conditions.

Reprinted from *Remote Sens.* Cite as: Guo, Y.; Zhang, L.; Qin, Y.; Zhu, Y.; Cao, W.; Tian, Y. Exploring the Vertical Distribution of Structural Parameters and Light Radiation in Rice Canopies by the Coupling Model and Remote Sensing. *Remote Sens.* **2015**, *7*, 5203–5221.

1. Introduction

Crop canopy structure depends on the crop's genetic characteristics and its physiological and biochemical processes, as well as its planting pattern and growth status. As crop canopies represent an integrated photosynthetic and matter production system, the structure of a crop canopy plays an important role in its function [1].

Canopy structure, which is directly related to the ability of within-canopy light interception, is the most important factor influencing light radiation distribution and light use efficiency [2]. Leaf angle (LA) and leaf area index (LAI) distribution are the principle factors that determine light radiance distribution and leaf physiological characteristic in crop canopies [3,4]. Employing crop varieties with compact plant types is beneficial for improving middle and bottom canopy light conditions, which enhances light use efficiency. Developing quantitative models of crop canopy structure is an important aspect of light energy use and balance research [5,6].

The simulation of canopy light distribution is based on the simulation of canopy structure parameters, such as the number of leaves characterized by the LAI and the distribution of leaves characterized by the leaf inclination angle. Since Monsi and Saeki [7] first applied the Beer–Lambert law describing the random distribution of light in a medium to predict light transmission in the plant canopy, many studies have focused on modeling canopy light transmission in crops, such as wheat, rice, maize and cotton [8–10]. The most classical approach is to utilize the extinction coefficient (K) and cumulative LAI values to simulate vertical light distribution in the crop canopy, when the larger LAI, the larger K under the same conditions and the canopy intercepted more sunlight. K values are affected by many other factors, such as structural parameters and solar elevation angle [11]. Some researchers calculated extinction coefficients through the function of canopy projected area (G) [12], while Nilson [13] and Ross [14] simulated the G function by determining the probability density function of the leaf angle, which downplays the relationship between canopy structure and K. Campbell [15] proposed a method that uses an elliptic function to describe leaf angle probability density, while Verhoef [16] tried to use a linear combination of trigonometric functions to describe the probability density distribution of leaf angle. The rapid development of three dimensional (3D) graphics technology and virtual simulation technology has led to the development of a plant canopy radiation distribution model for accurate simulation within the 3D space using irradiance and ray tracing techniques [17,18]. To date, few studies have focused on the characteristics of radiation transfer in the canopy and on modeling its spatial and temporal distribution. Several published studies have neglected the inhomogeneity of crop canopy level distribution and the differences in different wavelength radiation [8,10]. Moreover, the derivation of developing models is particularly complicated, and it is difficult to obtain input parameters, especially when simulating the intensity of radiation at different canopy heights. These problems have restricted the simulation accuracy of crop canopy light distribution and photosynthetic production [2]. In rice, two questions arise in this respect: first, how can Lambert-Beer's law be expanded to account for radiation at different canopy heights of rice, and second, where do radiative transfer differences occur between different wavelength radiations.

To this end, dedicated experiments were established, which contained two rice varieties of high yield with different morphologies and varied planting densities and nitrogen rates. The objective of this study was to develop a novel dynamic model for simulating the spatial and temporal distribution characteristics of LAI, LA and light radiation by the coupling model and remote sensing. As a precise, nondestructive and rapid method, hyperspectral remote sensing was used to estimate agronomic parameters, such as LAI [11,19,20]. This anticipated outcome would help to improve the understanding of yield formation and to identify key structural parameters for rice breeding programs.

2. Materials and Methods

2.1. Experiments

Two rice (*Oryza sativa* L.) field experiments were conducted to test the performance of the dynamic models developed in this study. These experiments involved different cultivars, planting densities and nitrogen fertilization rates during different years (Table 1), as described below.

Table 1. Treatment and sampling information. N1, Nitrogen Rate 1.

Experiment	Year	Site Location	Cultivar	Nitrogen Rate $(kg \cdot ha^{-2})$	Planting Density	Sampling Date	Planting Data
1	2012	Rugao 120°19'E 32°14'N	Wuxiangjing14 (WXJ14, V1) Shanyou63 (SY63, V2)	150 (N1) 250 (N2) 350 (N4)	22.2 plants/m² (D1) 13.3 plants/m² (D2)	7/22, 8/6, 8/19 8/30, 9/15, 9/24	6/18
2	2013	Rugao 120°19'E 32°14'N	Wuxiangjing14 (WXJ14, V1) Shanyou63 (SY63, V2)	150 (N1) 300 (N3)	22.2 plants/m² (D1) 13.3 plants/m² (D2)	7/9, 7/19, 7/30 8/8, 8/18, 8/30 9/10, 9/17	6/21

Experiment 1 was conducted in 2012 at the Rugao experiment station of the national engineering and technology center for information agriculture, Nantong city, China (120°19'E, 32°14'N). Two rice cultivars, *japonica* (Wuxiangjing14, V1, with inclined leaves) and *indica* (Shanyou63, V2, with erect leaves), were sown on 18 May and transplanted on 18 June with row and plant spacings of 30 cm and 15 cm (D1) and 50 cm and 15 cm (D2), respectively. The plot area was 42 m², and the plots were 7 m long and 6 m wide. Three nitrogen (N) fertilization rates (150 (N1), 250 (N2) and 350 (N4) kg N·ha⁻¹) were applied in the form of urea at a rate of 50% at preplanting, 10% at tillering, 20% at panicle initiation and 20% at spikelet initiation. For all treatments, 135 kg·ha⁻¹ P_2O_5 (as monocalcium phosphate $(Ca(H_2PO_4)_2)$) and 190 kg·ha⁻¹ K_2O (as KCl) were applied prior to transplanting. The experiment

140

employed a two-way factorial arrangement of treatments within a randomized complete block design with three replications.

Experiment 2 was conducted in 2013 using the same cultivars and location as Experiment 1. The cultivars were sown on 19 May and transplanted on 21 June with row and plant spacings of 30 cm and 15 cm (D1) and 50 cm and 15 cm (D2), respectively. The plot area was 30 m^2, with plots 6 m long and 5 m wide. Two N rates (150 (N1) and 300 (N3) kg N·ha^{-1}) were applied in the form of urea at a rate of 40% at preplanting, 10% at tillering, 20% at panicle initiation and 30% at spikelet initiation. Phosphate and potassium fertilizers were applied as described for Experiment 1.

In these experiments, different N rates and application times in different experiments were conducted according to basal soil fertility, the growth status of rice and the weather during various growth periods. Pest management and other management procedures follow.

2.2. Spectral Measurements of Canopy Leaves

Canopy reflectance values were acquired with A FieldSpec 3 (Analytical Spectral Devices, Boulder, CO, USA). This instrument recorded reflectance between 350 and 1000 nm, with a sampling interval of 1.40 nm and a resolution of 3 nm, and reflectance between 1000 and 2500 nm, with a sampling interval of 2 nm and a resolution of 10 nm. The spectroradiometer, with a 25° field of view (FOV), was positioned 1 m above the rice canopy. The radiance was measured at five positions within each plot to cover the entire plot and to characterize variability. Three scans were performed for each position and averaged to produce the final canopy spectra, and the average of five positions was used as the measurement for the plot. All spectral measurements were performed during cloud-free periods at midday (between 10:00 and 14:00). A white Spectralon reference panel (Labsphere, North Sutton, NH, USA) was used under the same illumination conditions to convert the spectral radiance measurements to reflectance.

2.3. Irradiance Measurements

The spectroradiometer (FieldSpec 3) and its cosine correctors were used to measure the vertical downward and upward irradiance (W·m^{-2}·nm^{-1}) at three different rice canopy observation depths (rice canopy depth was averaged and divided into three layers; Figure 1). The irradiance was measured at five positions within each plot. Three scans were performed for each position and averaged to produce the final canopy irradiance, and the average of five positions was used as the measurement for the plot. The transmittance and interception of each canopy layer were calculated as follows [21]:

Interception of the i-th layer:

$$I_i = \frac{(E_0\downarrow - E_0\uparrow - E_i\downarrow)}{E_0\downarrow}$$

Transmittance of the i-th layer:

$$T_i = \frac{E_i\downarrow}{E_0\downarrow}$$

where $E_0\downarrow$ is incident solar radiation of the top canopy (W·m^{-2}·nm^{-1}), $E_0\uparrow$ is reflected solar radiation of the top canopy (W·m^{-2}·nm^{-1}) and $E_i\downarrow$ is incident solar radiation of the i-th layer (W·m^{-2}·nm^{-1}).

Figure 1. Sketch map for measuring irradiance in the rice canopy.

2.4. Determination of Agronomic Parameters

During the 2012 and 2013 growing seasons, repeated destructive samplings were carried out in each plot. After each measurement of canopy spectra and irradiance, three plants from each experimental plot were randomly selected to determine leaf area and leaf angle. For each sample, the plants were equally divided into three layers in the direction from the ground to the canopy top. The green leaves from each layer were separated from the stems and immediately scanned using an LI-3000A. Then, the leaf area (L) of each layer was obtained, and the LAI for each layer and each plot (the sum of different layers) was calculated. The LAI represents the product of the number of plants per square area surveyed in a field and the L. The L of 10 randomly-selected rice stems from each experimental plot was measured using an inclinometer, the leaf angle was averagely divided into 6 parts in the range of 0–90° in the natural state, and the curved leaves were divided into 3 parts for testing, which recorded the horizontal as 0°, measuring the leaf area of each leaf piece at the same time.

2.5. Meteorological Data

Meteorological data were collected by an automatic meteorological station (Dynamet, Bellevue, WA, USA) installed in the experimental field. The average temperature (°C) and photosynthetic active radiation energy (kW·m^{-2}) value per hour were recorded.

2.6. Calculation

2.6.1. Growing Degree Day Calculation

GDD (growing degree days) values were calculated as follows [22]:

$$GDD = \sum (T_i - T_b)$$

where T_i is the average temperature throughout the experiment and T_b is the base temperature, which is usually set to 12 °C for rice.

Fitting analysis of the test data was performed using SPSS and Origin statistics software. Programming calculation and drawing were performed using MATLAB 7.11. Relative root mean square error (RRMSE) was used to calculate the fitness between the estimated and observed values [23] and to evaluate the overall performance of the model.

2.6.2. Spectral Index Calculation

In this study, three types of two-band indices were calculated: (1) normalized difference index (ND); (2) simple ratio index (SR); and (3) difference index (DI). These values were calculated using the original reflectance and the first derivative values from all available two-waveband ($\lambda 1$ and $\lambda 2$) combinations in the 350–2500 nm region to select the best two-band indices or the effective two-band combination regions for model parameter estimating. SR, ND and DI values were calculated as follows [24]:

$$SR\ (R_{\lambda 1}, R_{\lambda 2}) = \frac{R_{\lambda 1}}{R_{\lambda 2}}$$

$$\text{ND } (R_{\lambda 1}, R_{\lambda 2}) = \frac{|R_{\lambda 1} - R_{\lambda 2}|}{R_{\lambda 1} + R_{\lambda 2}}$$

$$\text{DI } (R_{\lambda 1}, R_{\lambda 2}) = R_{\lambda 1} - R_{\lambda 2}$$

2.6.3. Leaf Area Index Vertical Distribution Model

The logistic model can be used to simulate the vertical distribution of rice leaf area index quite well, but it cannot be used to simulate dynamic changes in LAI vertical distribution during different growth stages [25]. In this study, using the canopy depth (h) parameter to modify the logistic model, a dynamic simulation model of the downward accumulation leaf area index in rice was built. The model is expressed as follows:

$$LAI_{(h)} = LAI - \frac{LAI}{1 + b \times e^{-c\,(1-h)}} \tag{1}$$

where h is the relative depth of top-downward accumulation in the rice canopy (whole canopy depth is set to 1, $h \leqslant 1$), LAI is the leaf area index of the whole canopy and b, c are two structural adjustment parameters. Different b and c values represent different vertical structures, and greater b and c values mean less and more proportions of leaves under rice canopy, respectively. Thus, a vertical distribution dynamic model of rice canopy LAI was developed.

2.6.4. Leaf Angle Distribution Simulation Model

The leaf angle distribution function (LADF) is an important factor in describing canopy structure. In the radiation transfer theory (RT), leaf inclination angles and their distribution are key functions for solving RT problems in vegetative canopies. The ellipsoidal function [15] is based on the assumption that the canopy leaf area has inclination angles distributed parallel to the surface of a prolate or oblate ellipsoid. This function is described as:

$$f(\theta) = \frac{2(ELADP)^3 \sin\theta}{\Lambda[\cos^2\theta + (ELADP)^2 \sin^2\theta]^2} \tag{2}$$

where $ELADP$ (the ellipsoidal leaf angle distribution parameter) is the ratio of the horizontal semi-axis length (l) to the vertical semi-axis length (a) of an ellipsoid, $i.e.$, $ELADP = l/a$, where θ is the leaf inclination angle and Λ is a parameter defined by $ELADP$.

In this study, the canopy relative depth parameter h was added to the ellipsoid function model. Assuming that the canopy of downward accumulation relative depth h is an ellipsoid-structured layer, all values conform to the Campbell ellipsoid distribution function. We can calculate the horizontal semi-axis length (l) as $l \times \sqrt{1-(1-h)^2}$ using the elliptic equations and the vertical semi-axis length (a) as $a \cdot h$; the $ELADP_h$. The new ellipsoidal functions are written as follows:

$$ELADPh = \frac{lh}{ah} = \frac{l \times \sqrt{1-(1-h)^2}}{a \cdot h} = \frac{\sqrt{1-(1-h)^2}}{h} \cdot ELADP \tag{3}$$

$$f(\theta) = \frac{2(ELADP_h)^3 \sin\theta}{\Lambda[\cos^2\theta + (ELADP_h)^2 \sin^2\theta]^2} \tag{4}$$

When $ELADP = 1$, the ellipsoidal distribution becomes spherical, and $\Lambda = 2$. For vertical distributions, $ELADP_h < 1$, and therefore:

$$\Lambda = ELADP_h + \frac{\sin^{-1}\varepsilon}{\varepsilon}$$

with:

$$\varepsilon = \sqrt{1 - \left[\frac{a \times h}{l \times \sqrt{1-(1-h)^2}}\right]^2}$$

Finally, when $ELADP_h > 1$, the distribution is horizontal and:

$$\Lambda = ELADP_h + \frac{\ln(1+\varepsilon)/(1-\varepsilon)}{2 \cdot \varepsilon \cdot ELADP_h}$$

with:

$$\varepsilon = \sqrt{1 - \left[\frac{l \times \sqrt{1-(1-h)^2}}{a \times h}\right]^2}$$

2.6.5. Rice Canopy Radiation Vertical Distribution Model

Nilson and Ross's algorithm was employed, assuming uniform leaf azimuthal distribution [13,14]. The extinction coefficient K is calculated from the mean projection of unit leaf area on the plane perpendicular to the beam direction as follows:

$$K = \frac{G}{\cos\phi} \tag{5}$$

with:

$$G = \int_0^{\frac{\pi}{2}} A(\theta, \phi) f(\theta) d\theta$$

and:

$$A(\theta, \phi) = \begin{cases} \cos\theta\cos\phi & \theta + \phi \leqslant \frac{\pi}{2} \\ \cos\theta\cos\phi[1 + \frac{2}{\pi}(\tan\delta - \delta)] & \theta + \phi > \frac{\pi}{2} \end{cases}$$

$$\delta = \cos^{-1}(\cot\theta\cot\phi)$$

Then, based on the Beer-Lambert law [7], the solar transmittance at the top-downward relative depth h can be calculated using the canopy structure parameters model and the extinction coefficient model; the equation is as follows:

$$T_{(i,h)} = e^{-LAI_{(h)} \times K_{(i,h)}} \tag{6}$$

where $T_{(i, h)}$, $k_{(i, h)}$ is the transmittance and extinction coefficient of wavelength light i at the top-downward relative depth h. $LAI(h)$ is LAI at the top-downward relative depth h.

Finally, the interception corresponding to transmittance is calculated as follows [26]:

$$f_{APAR} = 0.95 \times (1 - e^{-K \times LAI}) \tag{7}$$

3. Results

3.1. Determination Partial Model Parameters Using RS and GDD

In the LAI vertical distribution model (Formula 1), LAI was estimated by a hyperspectral vegetation index. The best two-band ND, SR and difference VIs for estimating LAI is plotted in Figure 2. LAI had a linear relationship with $DI(R_{800}, R_{750})$ and an exponential relation with $ND(R_{930}, R_{730})$ and $SR(R_{730}, R_{930})$. The differential spectral index $DI(R_{800}, R_{750})$ could best predict the LAI values of different rice varieties (*japonica* and *indica*) under different cultivation conditions (Figure 2). Moreover, *ELADP* values (Equation (4)) under different experimental conditions were estimated by using the measured canopy radiation transmission values and a nonlinear least squares fitting method. The quantitative relationships between *ELADP* and spectral indices (Figure 3) were also analyzed. The results show that the parameter *ELADP* could be successfully estimated by the hyperspectral vegetation index. However, different rice varieties had different optimal vegetation indices; the differential index $DI(R_{945}, R_{915})$ and normalized differential index $ND(R_{700}, R_{525})$ were used to develop corresponding monitoring models for *ELADP* estimation in *japonica* and *indica* rice, respectively (Figure 3).

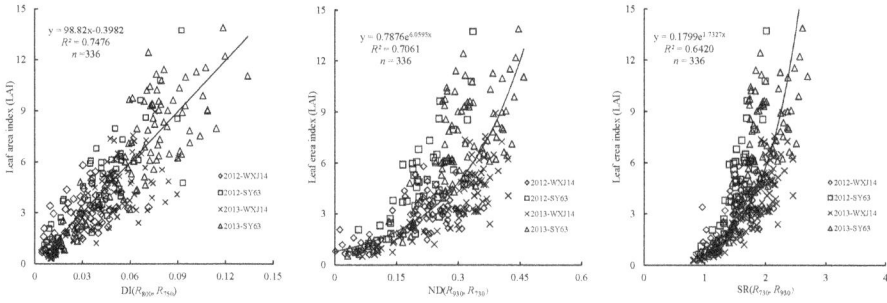

Figure 2. Relationship between the canopy leaf area index and spectral index in rice.

In addition, using data from two years of experiments to fit the LAI vertical distribution model, b was found to have significant correlation with c (Figure 4A); the parameters b and c could successfully be simulated by GDD (Figure 4). Canopy structure changes over time, and different canopy structure corresponds to varied b and c values; thus, the dynamic changes of the canopy structure can be calculated by GDD values.

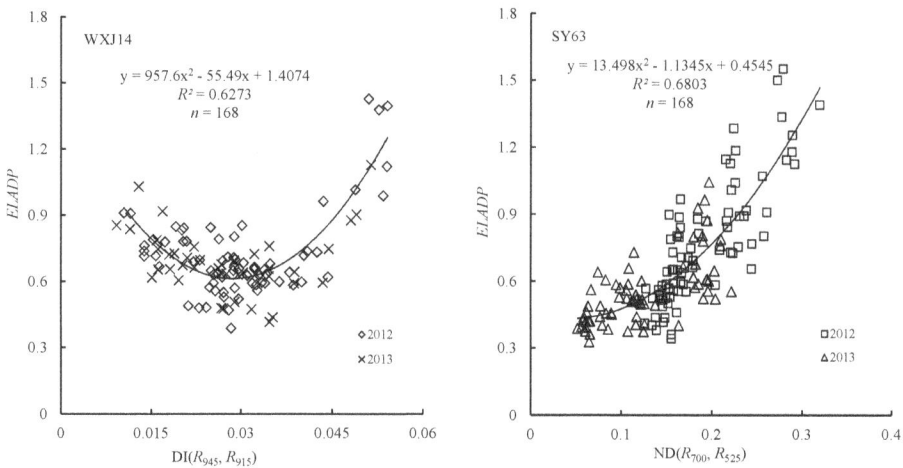

Figure 3. Relationship between the ellipsoidal leaf angle distribution parameter (*ELADP*) and the spectral index in rice. DI, difference index; ND, normalized difference index.

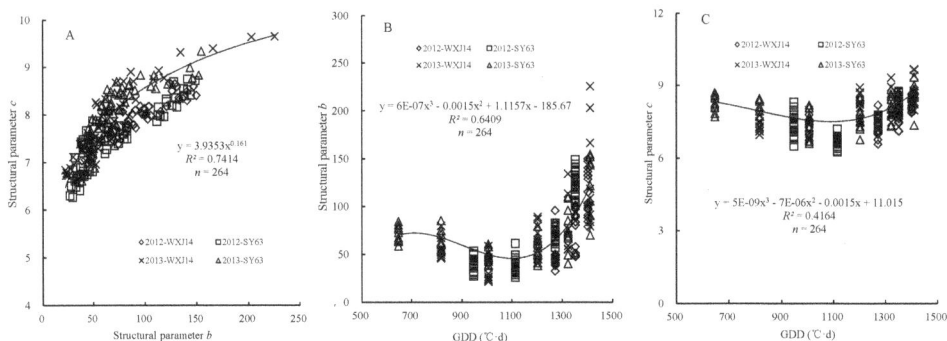

Figure 4. Relationship between rice canopy structure parameters b and c (**A**), growing degree days (*GDD*) and b (**B**), c (**C**).

3.2. Radiation Transfer of Different Light Qualities (Wavelengths)

Green vegetable organs have selective absorption characteristics for solar radiation. The quantitative relationship between the interception and transmission of blue (B), green (G) and red (R) radiation and photosynthetic active radiation (PAR) were analyzed based on the experimental data. The results show that good correlation existed between these values despite the use of different rice varieties, cultivation conditions and canopy heights (Figure 5). Thus, the radiation transmittance of different wavelengths (light quality) at relative depth h of the rice canopy could be calculated based on its relationship to PAR transmittance.

3.3. Validation of LAI Vertical Distribution Model

The LAI of different canopy height layers in different treatments and growth stages was simulated using the developed vertical LAI distribution model based on 2012 and 2013 experimental conditions. The result show (Figure 6; 2013, for example) that the simulated values closely matched the measured values and complied with the laws of different cultivation treatments and different growth stages. The *RRMSE* values derived from model simulation of the first, second and third layers of the rice canopy were 20.81%, 14.89% and 15.21% in 2012 and 18.97%, 10.53% and 10.48% in 2013, respectively, which indicates that the model could successfully simulate LAI in different rice canopy layers. The total mean *RRMSE* value of the model simulating different canopy layers of rice LAI was 14.75%. Better simulation results could be obtained under normal planting density conditions (*RRMSE* of 10.21%) than under lower planting density conditions (*RRMSE* of 16.26%; Figure 7). Large deviation existed between the simulated and measured LAI values in the group of V2D2 due to the measurement errors for SY63 (*indica* rice) that was planted in large row spacing and grew with inclined leaves. On the whole, this model is suitable for simulating the spatial and temporal distribution characteristics of canopy LAI in rice.

148

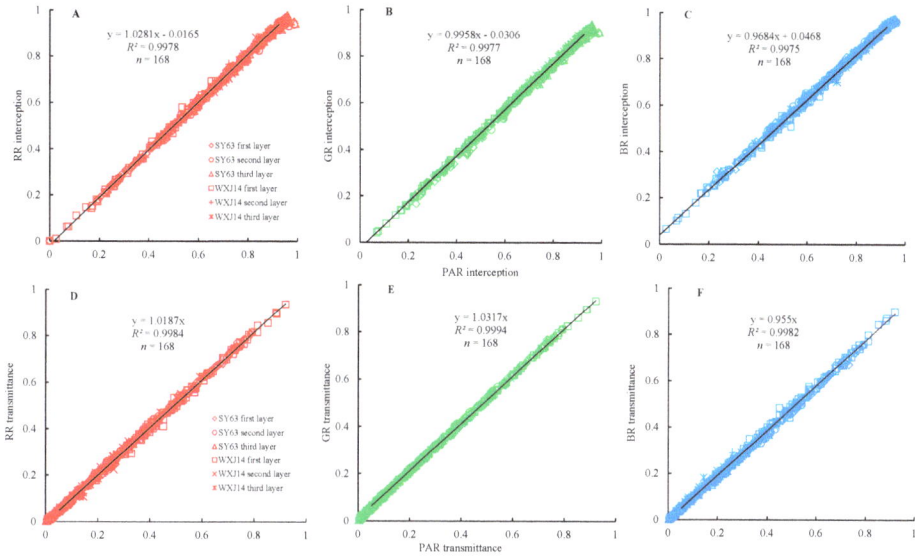

Figure 5. Relationship between transmittance and interception of different wavelengths of light and photosynthetic active radiation (PAR). (**A**) Red radiation interception; (**B**) green radiation interception; (**C**) blue radiation interception; (**D**) red radiation transmittance; (**E**) green radiation transmittance; (**F**) bule radiation transmittance.

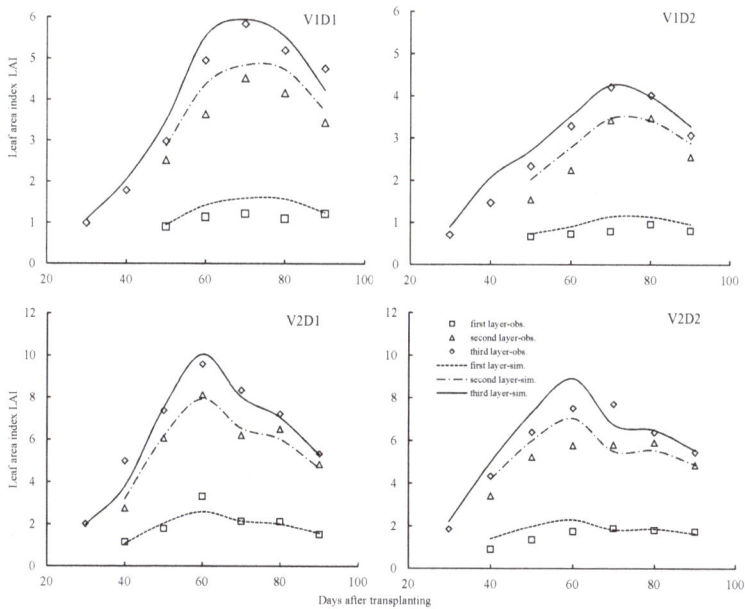

Figure 6. Comparison of simulated and observed downward cumulative LAI in rice, 2013.

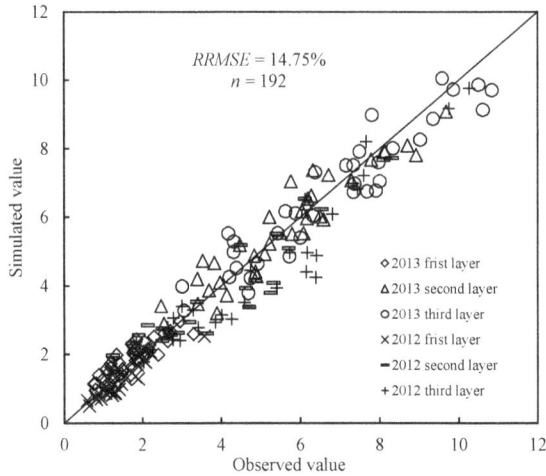

Figure 7. Comparison of simulated and observed downward cumulative LAI in rice, 2012 and 2013.

3.4. Simulation Analysis of Leaf Angle Distribution

The developed LA distribution function model was used to simulate LA at h height within the rice canopy during different growth stages. The results show that there were large differences in leaf angle distribution in the *japonica versus indica* rice varieties at different growth stages; however, these varieties exhibited similar patterns of spatial distribution of LA within the crop canopy. Overall, the leaf angles at the top of the canopy were smaller than those in the lower canopy for both rice varieties. For leaves in the upper part of the canopy, the LA values were within 20°–50° for Wuxiangjing14 (*japonica* rice) and 20°–40° for Shanyou63 (*indica* rice). For leaves in the lower part of the canopy, the LA values were within 50°–90° for Wuxiangjing14 and 40°–90° for Shanyou63 (Figure 8; N1D1 treatment in 2013, for example). The *RRMSE* value for simulated rice canopy LA was 21.78%, which indicates that the model developed in this study has a good prediction ability for rice LA (Figure 9).

3.5. Validation of Rice Canopy Light Radiation Vertical Distribution Model

The transmittance and interception of two rice varieties under different growth stages and treatment conditions (Figure 10) were simulated using the developed rice canopy optical light radiation vertical distribution model, and the corresponding measured values were used to validate the model. The results show that for the two rice varieties examined, the simulated and measured values of transmittance and optical radiation interception rates were consistent. Over the entire growth period, the mean *RRMSE* values of rice canopy PAR transmittance and interception were 13.62% and 7.61% for Wuxiangjing14 and 18.69% and 3.61% for Shanyou63, respectively. The average *RRMSE* values for different qualities of light radiation (R, G and B) at different growth stages were 16.34%, 15.96% and 15.36% (transmittance) and 5.75%, 8.23% and 5.03% (interception), respectively.

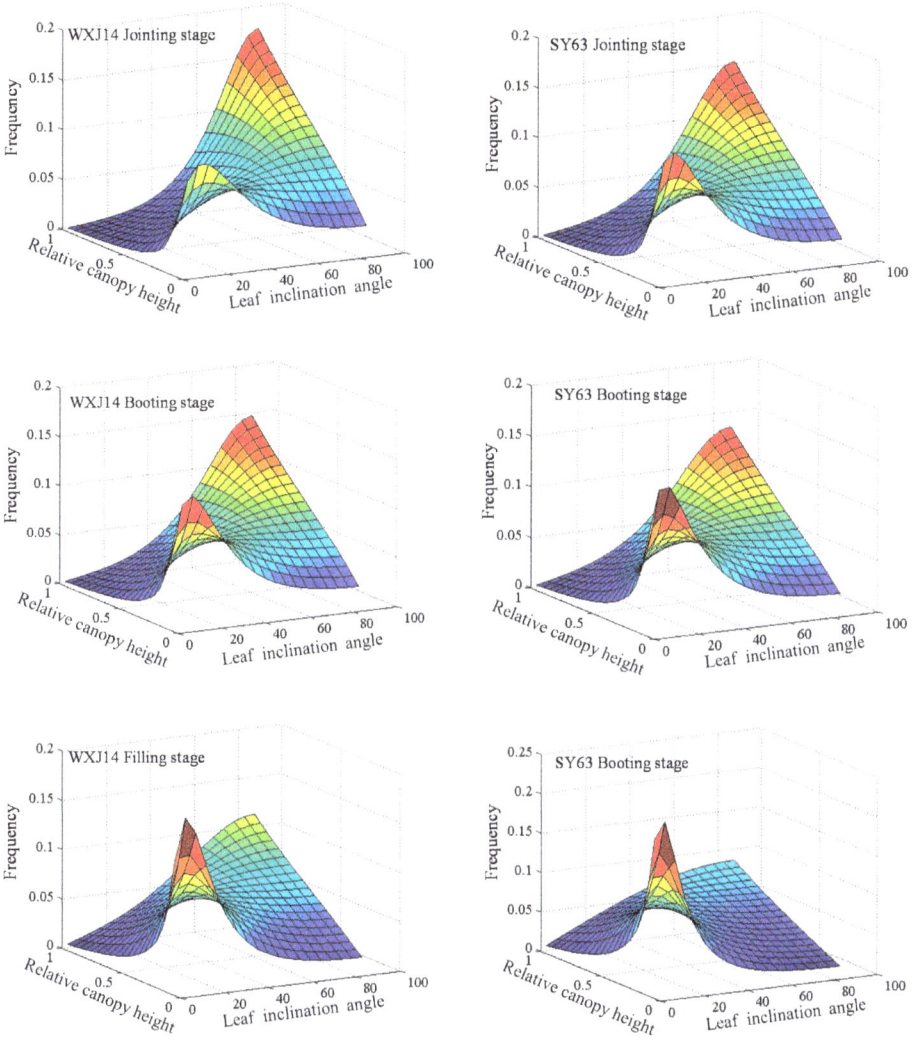

Figure 8. Frequency of leaf angle distribution of different canopy layers of rice at the jointing, booting and filling stage.

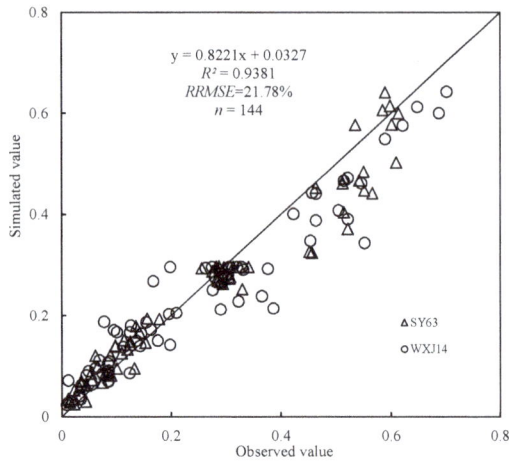

Figure 9. Comparison of simulated and measured leaf inclination angle distribution in the rice canopy.

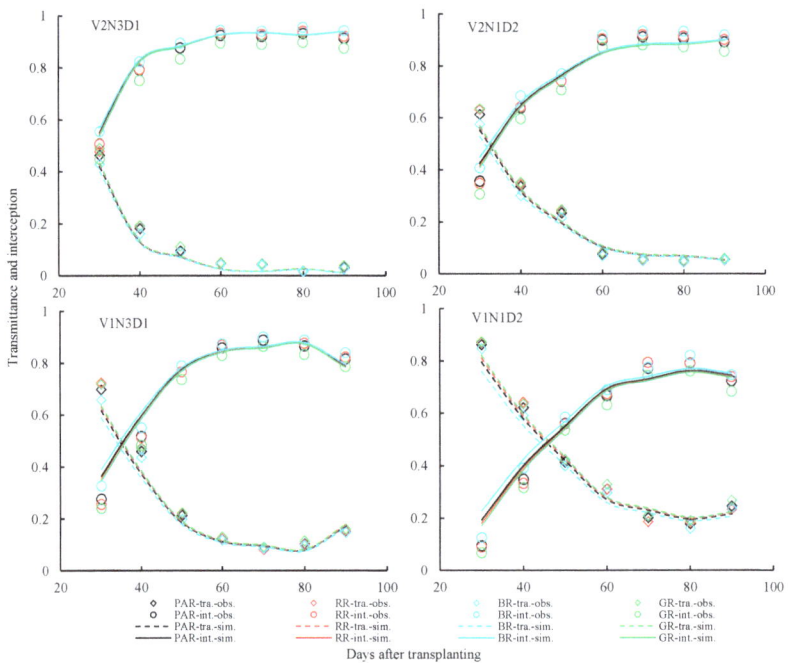

Figure 10. Comparison of simulated and observed radiation transmittance at different growth stages and under different nitrogen rates and planting densities for two rice varieties. V1, Wuxiangjing14; V2, Shanyou63.

The simulation results for canopy layers at different heights show that PAR transmittance and interception changed quickly in the upper part of the rice canopy, exhibiting an approximately linear trend, which began to slow down when the relative depth (h) reached approximately 0.4. There was little change when the h reached 0.7 (Figure 11), which indicates that most of the solar radiation was absorbed by the top 70% of the rice canopy. Validation results using the measured optical radiation data from different canopy layers show that the average RRMSE values in the first, second and third layers of the rice canopy were 8.49%, 16.98% and 26.06% (PAR transmittance) and 10.88%, 4.39% and 4.76% (PAR interception), respectively. For radiation of different light qualities (R, G and B), the RRMSE values were 23.58%, 23.77% and 26.92% (transmittance) and 6.73%, 8.02% and 6.88% (interception), respectively (Figure 12). These results demonstrate that the model performed well in simulating the spatial and temporal distribution of optical radiation in the rice canopy.

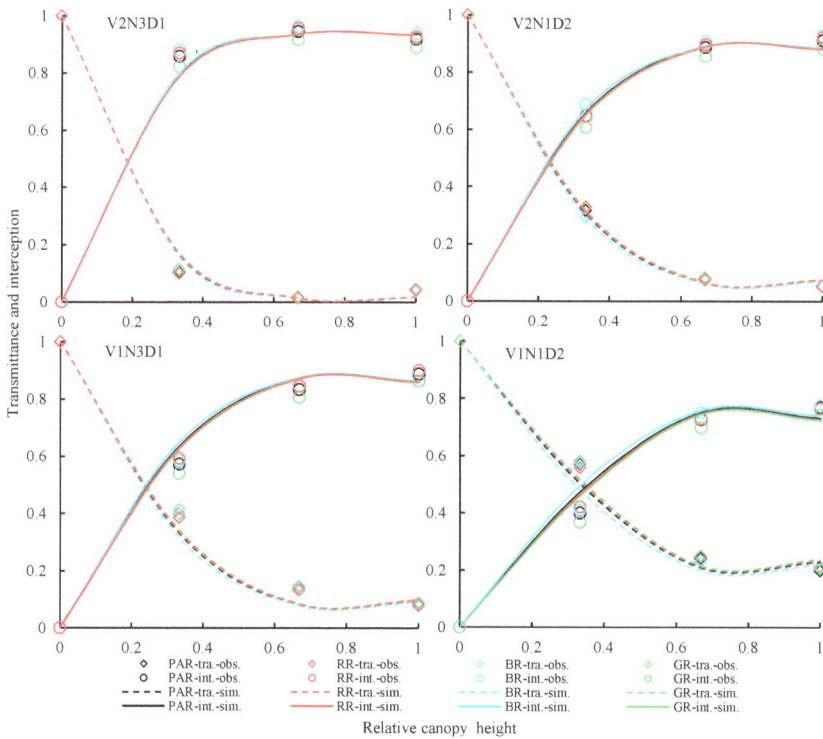

Figure 11. Comparison of simulated and observed radiation transmittance and interception in different relative canopy depths under different nitrogen rates and planting densities for two rice varieties, heading stage; V1, Wuxiangjing14; V2, Shanyou63.

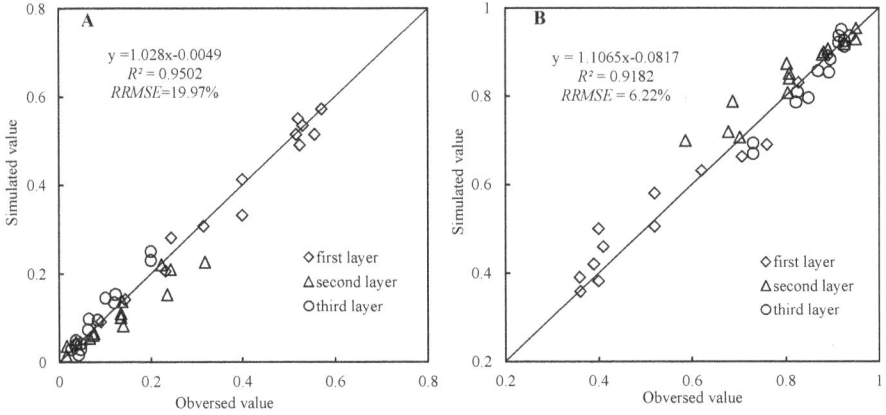

Figure 12. Comparison of simulated and observed PAR transmittance and interception values. (**A**) Transmittance and (**B**) interception.

4. Discussion

LAI and LA distributions greatly influence the absorption and transmission of solar radiation in plant populations [27,28]. Hu *et al.* [27] simulated the canopy structure index based on the plant-type factor, but the plant-type factor was difficult to obtain, and its continuous changes throughout the growth period could not be simulated. Crop LAI was successfully estimated by RS technology [20,29]. However, it is difficult to invert crop LAI in the vertical direction using the RS method. Therefore, in the current study, we estimated rice canopy LAI values in different growth stages using hyperspectral vegetation indices, and we developed a canopy LAI vertical distribution model based on the logistic equation, to which we added the canopy depth factor h. In addition, the results of this study show that the model adjustment parameters b and c could be successfully simulated by *GDD*. Thus, we successfully performed dynamic simulation of rice canopy LAI in different canopy layers and at different growth stages.

The Campbell ellipsoidal distribution function is the most commonly-used function in the study of LA distribution simulation [15]. This model is highly versatile for different plant vegetation types [12], but it cannot simulate the temporal and spatial variation of vegetation canopy LA distribution [15]. In the study, the Campbell ellipsoidal distribution function was improved to simulate the probability of LA distribution at different canopy height levels using the canopy vertical depth parameter h. Moreover, this study shows that rice canopy hyperspectral vegetation indices $ND(R_{700}, R_{525})$ and $DI (R_{945}, R_{915})$ had good correlations with the input parameter *ELADP* of the LA distribution model for both Shanyou63 and Wuxiangjing14 rice; this input parameter was estimated by hyperspectral RS, which

154

helped us obtain the model parameters. The results show that the model has good descriptive and predictive value for modeling the spatial and temporal distribution of rice canopy LA.

Mathematical functions and three-dimensional structures of computer simulation modeling methods are commonly used to simulate the light radiation distribution in the crop canopy [30–32]. The Beer-Lambert law proposed by Monsi and Saeki [7] is one of the most popular classical models for simulating light radiation distribution. Some studies have employed ray tracing techniques [33], radiosity [34] and other methods to simulate canopy light distribution. These models can accurately simulate light radiation intensity at any point within a canopy, but they require huge amounts of calculations, and input parameters are also difficult to obtain, which limits their application. Therefore, by combining the Beer-Lambert law and the hyperspectral RS technique, we developed a dynamic model of the vertical distribution of light in a rice canopy. This simplified model, with improved input parameter acquisition, provided good simulations of light radiation transmission and interception in canopy layers at different heights and growth stages. This model does not require the input of various agronomic parameters that are difficult to obtain, and it requires a small amount of computation and yields highly accurate results. In addition, two rice varieties (*japonica* and *indica*) with compact and loose canopy structures were studied in this paper. Different N treatments induced the significant differences in LAI and LA. The good performance of the newly proposed model suggested its strong applicability and predictability to simulate spatial and temporal distribution characteristics of LAI, LA and light radiation in rice.

Crop canopy leaves have selective absorption characteristics for different wavelengths of light radiation, and the transmittance and interception of different wavelengths of light radiation in the crop canopy differ; R and B radiation exhibit the minimum and maximum differences with PAR, respectively [35]. This observation was confirmed in the current study. Previous studies have primarily focused on PAR transport simulation and utilization efficiency, while studying the temporal and spatial distribution of different wavelengths of light radiation and its simulation models is necessary for improving radiation use efficiency in crop production [36]. Thus, we analyzed the relationships between different wavelengths of light (R, G, B) radiation and PAR, and we simulated the radiation transmission characteristics of different wavelengths of light radiation in canopy layers at different heights and at different growth stages.

The vertical distribution of canopy LAI is uneven, and stems have a greater impact on light radiation in the lower part of the canopy. The impact of the stem on light transmission was not considered in the current model, which leads to relatively low accuracy of simulation in the lower part of the canopy. Therefore, the stem factor should be considered in further studies by separating LAI from the stem area index

(SAI) [37]. In addition, the influence of the physiological and biochemical parameters of rice leaves on differential light radiation transfer requires further study.

5. Conclusions

In this study, we developed a dynamic model of the vertical distribution of light radiation in the rice canopy based on the Beer-Lambert law coupled with hyperspectral RS technology. Based on the logistic equation and the Campbell ellipsoidal distribution model, we introduced the canopy height parameter h, assuming that the canopy of downward accumulation relative depth h is an ellipsoid-structured layer, and combined RS inversion of model input parameters to develop LAI and LA vertical distribution dynamic models. The extinction coefficients of photosynthetic active radiation under different experimental conditions were simulated by the G function, and then, the relationship between radiation transmission of PAR and different wavelengths radiation (R, G, B) were quantified, which enabled us to simulate the spatial and temporal distribution of different wavelengths' radiation in the rice canopy. This model represents a new tool for simulating crop canopy photosynthetic production and light use efficiency and for evaluating the photosynthetic efficiency of different rice plant types.

Acknowledgments: This research was supported by the National Natural Science Foundation of China (31371535), the National 863 High-tech Program (2013AA102301 and 2011AA100703), the Special Fund for Agro-scientific Research in the Public Interest (201303109), the Science and Technology Support Program of China (2013BAD20B05), the Science and Technology Support Program of Jiangsu (BE2011351 and BE2012302), the Jiangsu Collaborative Innovation Center for Modern Crop Production and the Priority Academic Program Development of Jiangsu Higher Education Institutions, China.

Author Contributions: Yongjiu Guo designed the experiments, measured field spectral data, analyzed the data, interpreted the results and wrote the manuscript. Ling Zhang and Yehui Qin carried out the spectral measurements during field work and assisted with the results interpretation. The idea for the study and the basic structure of the manuscript were developed by Yan Zhu, Weixing Cao and Yongchao Tian. Yongchao Tian proposed the research design and manuscript revision.

Conflicts of Interest: The authors declare no conflict of interest.

References

1. Hirose, T. Development of the monsi-saeki theory on canopy structure and function. *Ann. Bot.* **2005**, *95*, 483–494.
2. Yu, Q.; Wang, T.D.; Liu, J.D.; Sun, F. A mathematical study on crop architecture and canopy photosynthesis I. Model. *Sci. Agric. Sin.* **1998**, *24*, 7–15.
3. Sassenrath-Cole, G.F. Dependence of canopy light distribution on leaf and canopy structure for two cotton (gossypium) species. *Agric. For. Meteorol.* **1995**, *77*, 55–72.
4. Stewart, D.W.; Costa, C.; Dwyer, L.M.; Smithc, D.L.; Hamiltona, R.I.; Ma, B.L. Canopy structure, light interception, and photosynthesis in maize. *Agron. J.* **2003**, *95*, 1465–1474.

5. Tappeiner, U.; Cernusca, A. Model simulation if spatial distribution of photosynthesis in structurally differing plant communities in the central Caucasus. *Ecol. Model.* **1998**, *24*, 272–279.

6. Wang, W.M.; Li, Z.L.; Su, H.B. Comparison of leaf angle distribution functions: Effects on extinction coefficient and fraction of sunlit foliage. *Agric. For. Meteorol.* **2007**, *143*, 106–122.

7. Monsi, M.; Saeki, T. Uber den lichtfaktor in den pflanzengesellschaften und seine bedeutung fur die stoffproduktion. *Jpn. J. Bot.* **1953**, *14*, 22–52.

8. Li, Y.D.; Tang, L.; Zhang, Y.P.; Liu, L.L.; Cao, W.X.; Zhu, Y. Spatiotemporal distribution of photosynthetically active radiation in rice canopy. *Chin. J. Appl. Ecol.* **2010**, *21*, 952–958.

9. Liu, R.Y.; Huang, W.J.; Ren, H.Z.; Yang, G.J.; Xie, D.H.; Wang, J. Photosynthetically active radiation vertical distribution model in maize canopy. *Trans. Chin. Soc. Agric. Eng.* **2011**, *27*, 115–121.

10. Tang, L.; Li, Y.D.; Zhang, Y.P.; Zhu, X.C.; Liu, X.J.; Cao, W.X.; Zhu, Y. Simulation of canopy light distribution and application in rice. *Rice Sci.* **2011**, *4*, 427–434.

11. Wang, F.M.; Huang, J.F.; Tang, Y.L.; Wang, X.Z. New vegetation index and its application in estimating leaf area index of rice. *Rice Sci.* **2007**, *14*, 195–203.

12. Campbell, G.S. Extinction coefficients for radiation in plant canopies calculated using an ellipsoidal inclination angle distribution. *Agric. For. Meteorol.* **1986**, *36*, 317–321.

13. Nilson, T. A theoretical analysis of the frequency of gaps in plant stands. *Agric. For. Meteorol.* **1971**, *8*, 25–38.

14. Ross, J. *The Radiation Regime and Architecture of Plant Stands*; Springer Science & Business Media: Boston, MA, USA, 1981.

15. Campbell, G.S. Derivation of an angle density function for canopies with ellipsoidal leaf angle distributions. *Agric. For. Meteorol.* **1990**, *49*, 173–176.

16. Verhoef, W. Theory of Radiative Transfer Models Applied in Optical Remote Sensing of Vegetation Canopies. Ph.D. Thesis, Wageningen Agricultural University, Wageningen, The Netherlands, 1998.

17. Wang, X.Y.; Guo, Y.; Li, B.G. Evaluating a three dimensional model of diffuse radiation in maize canopies. *Int. J. Biometeorol.* **2006**, *50*, 349–357.

18. Zheng, B.Y.; Shi, L.J.; Ma, Y.T. Comparison of architecture among different cultivars of hybrid rice using a spatial light model based on 3d digitizing. *Funct. Plant Biol.* **2008**, *35*, 900–910.

19. Tian, Y.C.; Yang, J.; Yao, X.; Zhu, Y.; Cao, W.X. Quantitative relationships between hyperspectral vegetation indices and leaf area index of rice. *Chin. J. Appl. Ecol.* **2009**, *20*, 1685–1690.

20. Viña, A.; Gitelson, A.A.; Nguy-Robertson, A.L.; Peng, Y. Comparison of different vegetation indices for the remote assessment of green leaf area index of crops. *Remote Sens. Environ.* **2011**, *115*, 3468–3478.

21. Gallo, K.P.; Daughtry, C.S.T. Techniques for measuring intercepted and absorbed photosynthetically active radiation in corn canopies. *Agron. J.* **1986**, *78*, 752–756.

22. McMaster, G.S.; Wilhelm, W.W. Growing degree-days: One equation, two interpretations. *Agric. For. Meteorol.* **1997**, *87*, 291–300.

23. Wallach, D.; David, M.; Jones, J.W. *Working with Dynamic Crop Models*; Elsevier BV: Boston, MA, USA, 2006.

24. Tian, Y.C.; Yao, X.; Yang, J.; Cao, W.X.; Hannaway, D.B.; Zhu, Y. Assessing newly developed and published vegetation indices for estimating rice leaf nitrogen concentration with ground- and space-based hyperspectral reflectance. *Field Crops Res.* **2011**, *120*, 299–310.

25. Li, Y.D.; Tang, L.; Zhang, Y.P.; Zhu, X.C.; Cao, W.X.; Zhu, Y. Relationship of par interception of canopy to leaf area and yield in rice. *Sci. Agric. Sin.* **2010**, *43*, 3296–3305.

26. Ruimy, A.; Kergoat, L.; Bondeau, A. Comparing global models of terrestrial net primary productivity (NPP): Analysis of differences in light absorption and light-use efficiency. *Glob. Chang. Biol.* **1999**, *5*, 56–64.

27. Hu, N.; Yao, K.M.; Zhang, X.C.; Lv, C.G. Effect and simulation of plant type on canopy structure and radiation transmission in rice. *Rice Sci.* **2011**, *25*, 535–543.

28. Zhou, Y.H.; Xiang, Y.Q.; Lin, Z.H. Radiation interception by erectophile maize colony. *Chin. J. Appl. Ecol.* **1997**, *8*, 21–25.

29. Liu, Q.; Liang, S.; Xiao, Z.; Fang, H. Retrieval of leaf area index using temporal, spectral, and angular information from multiple satellite data. *Remote Sens. Environ.* **2014**, *145*, 25–37.

30. Wang, X.P.; Guo, Y.; Wang, X.Y. Estimating photosynthetically active radiation distribution in maize canopies by a three-dimensional incident radiation model. *Funct. Plant Biol.* **2008**, *35*, 867–876.

31. Evers, J.B.; Vos, J.; Yin, X.; Romero, P.; van der Putten, P.E.; Struik, P.C. Simulation of wheat growth and development based on organ-level photosynthesis and assimilate allocation. *J. Exp. Bot.* **2010**, *61*, 2203–2216.

32. Sarlikioti, V.; de Visser, P.H.; Marcelis, L.F. Exploring the spatial distribution of light interception and photosynthesis of canopies by means of a functional-structural plant model. *Ann. Bot.* **2011**, *107*, 875–883.

33. Ross, J.K.; Marshak, A.L. Calculation of canopy bidirectional reflectance using the Monte Carlo method. *Remote Sens. Environ.* **1988**, *24*, 213–225.

34. Chelle, M.; Andrieu, B. The nested radiosity model for the distribution of light within plant canopies. *Ecol. Model.* **1998**, *111*, 75–91.

35. Awad, M.A.; Wsgenmakers, P.S.; de Jager, A. Effect of light on flavonoid and chlorugenic acid levels in the skin of "jonagold" apples. *Sci. Hortic.* **2001**, *88*, 289–298.

36. Yang, C.M.; Yang, L.Z.; Wei, C.L.; Ding, C.C. Canopy spectral characteristics of different rice varieties. *Chin. J. Appl. Ecol.* **2002**, *13*, 689–692.

37. Fang, H.; Li, W.; Wei, S.; Jiang, C. Seasonal variation of leaf area index (LAI) over paddy rice fields in NE China: Intercomparison of destructive sampling, LAI-2200, digital hemispherical photography (DHP), and AccuPAR methods. *Agric. For. Meteorol.* **2014**, *198–199*, 126–141.

The Impact of Sunlight Conditions on the Consistency of Vegetation Indices in Croplands—Effective Usage of Vegetation Indices from Continuous Ground-Based Spectral Measurements

Mitsunori Ishihara, Yoshio Inoue, Keisuke Ono, Mariko Shimizu and Shoji Matsuura

Abstract: A ground-based network of spectral observations is useful for ecosystem monitoring and validation of satellite data. However, these observations contain inherent uncertainties due to the change of sunlight conditions. This study investigated the impact of changing solar zenith angles and diffuse/direct light conditions on the consistency of vegetation indices (normalized difference vegetation index (NDVI) and green-red vegetation index (GRVI)) derived from ground-based spectral measurements in three different types of cropland (paddy field, upland field, cultivated grassland) in Japan. In general, the vegetation indices decreased with decreasing solar zenith angle. This response was affected significantly by the growth stage and diffuse/direct light conditions. The decreasing response of the NDVI to the decreasing solar zenith angle was high during the middle growth stage ($0.4 <$ NDVI < 0.8). On the other hand, a similar response of the GRVI was evident except in the early growth stage (GRVI < 0). The response of vegetation indices to the solar zenith angle was evident under clear sky conditions but almost negligible under cloudy sky conditions. At large solar zenith angles, neither the NDVI nor the GRVI were affected by diffuse/direct light conditions in any growth stage. These experimental results were supported well by the results of simulations based on a physically-based canopy reflectance model (PROSAIL). Systematic selection of the data from continuous diurnal spectral measurements in consideration of the solar light conditions would be effective for accurate and consistent assessment of the canopy structure and functioning.

Reprinted from *Remote Sens*. Cite as: Ishihara, M.; Inoue, Y.; Ono, K.; Shimizu, M.; Matsuura, S. The Impact of Sunlight Conditions on the Consistency of Vegetation Indices in Croplands—Effective Usage of Vegetation Indices from Continuous Ground-Based Spectral Measurements. *Remote Sens.* **2015**, *7*, 14079–14098.

1. Introduction

Timely and large-scale observations of agroecosystems by remote sensing are crucial for food and environment security [1–4]. In many agroecosystem applications, high spatial and temporal resolutions are required at the same time. In particular, in many Asian countries, high spatial resolution is critical because agricultural fields are small and land use is mosaic. For example, high-spatial-resolution optical satellites are used for mapping of the protein content and the full ripe stage of rice in a large number of individual fields [5,6]. However, despite the constellation of satellites, the probability of image acquisition at specific target periods is still unsatisfactory for timely mapping. Low-spatial-resolution optical satellite sensors, such as the Terra and Aqua Moderate Resolution Imaging Spectroradiometer (MODIS), SPOT-VEGETATION (SPOT-VGT), and NOAA Advanced Very High Resolution Radiometer (AVHRR), can make global and regional observations with high temporal frequency, but their spatial resolutions range from 250 m to 1000 m. On the other hand, high-spatial-resolution optical satellite sensors, such as SPOT, RapidEye, and WorldView can observe the land surface with spatial resolutions of 2–15 m, but their temporal frequency is low. Additionally, medium-spatial-resolution optical satellite sensors, such as HJ-1A/B, with spatial resolutions of 30 m can observe the same position at temporal intervals of four days, but their spatial resolution is insufficient for monitoring agricultural fields in many Asian countries [7].

Under these circumstances, a ground-based network of spectral measurements would be important in ecosystem monitoring as an addition to synthetic aperture radar (SAR) satellites and drone-based remote sensing to compensate for the limitations of optical satellite sensors. SAR sensors have good potential for crop monitoring because they are not affected by sky conditions [8–10]. Drone-based remote sensors can play unique roles due to their timely and flexible operation and super-high spatial resolution (~10 cm) [11,12]. A ground-based network of spectral observations has proved to be useful for ecosystem monitoring and validation of optical satellite data (EUROSPEC [13], Spectral Network (SpecNet) [14], and Phenological Eyes Network (PEN) [15]). Ground-based sensors automatically acquire spectral reflectance, in addition to CO_2 flux, micrometeorological data, and digital images, at high temporal resolution (~30 min) [15,16]. Such datasets can be used to investigate the dynamic change of ecosystems in detail by making the most of the high temporal resolution and continuous measurements. For assessment of phenological changes, such as timing of leaf green-up and autumn coloring or crop status in agroecosystems, such as protein content and water stress, spectral reflectance or vegetation indices from optical satellite data can be validated directly using ground-based spectral measurements. For example, Motohka et al. [17] reported that phenological features observed in MODIS data were validated using ground-based spectral reflectance observations in a paddy field. Sakamoto *et al.* [18] proposed a monitoring method for crop status based on ground-based digital camera images.

However, these ground-based spectral reflectance observations do not ensure consistency due to the differences in the canopy structure, viewing geometry, and illumination. These changes can be expressed by a bidirectional reflectance distribution function (BRDF) [19], but determination of surface parameters for the BRDF is not easy. The canopy structure can change drastically according to the growth stage and vegetation type. The view zenith angle of ground-based sensors is usually fixed at $0°$ (nadir observation), but the solar zenith angle changes with the time of day and day of the year. Additionally, the diffuse/direct light ratio changes with the daily weather conditions. While the ground-based spectral reflectances are used for calibration or validation of satellite data [20,21], the changes caused by diurnal and seasonal variation of canopy structures and light conditions are often ignored. Cogliati *et al.* [22] reported that the normalized difference vegetation index (NDVI) from continuous ground-based measurement showed some diurnal change as affected by the photosynthetic photon flux density (PPFD). Rahman *et al.* [23] reported that NDVI from ground-based observations was affected by the solar zenith angle in a pasture site. However, this relation was examined using a dataset for a full-cover pasture canopy only on two days under clear sky conditions during the vegetative stage. Thus, a generalized relation throughout the growth season under various light conditions and/or in different types of vegetation is necessary.

The objectives of this study are (1) to examine the diurnal and seasonal fluctuations of vegetation indices derived from ground-based spectral measurements for three different types of cropland (paddy field, upland field, cultivated grassland); (2) to investigate the impact of changing solar zenith angles and diffuse/direct light conditions on the consistency of vegetation indices; and (3) to propose efficient usage of ground-based spectral data. In this study, we used the NDVI and green-red vegetation index (GRVI) as vegetation indices because these indices are widely used in remote sensing studies [20,24]. The NDVI has been used to estimate variations in vegetation conditions [25,26]. The GRVI is a new vegetation index and has been used to detect subtle vegetation changes (e.g., leaf fall due to a typhoon or mowing of plants) or differences among ecosystem types [20,27].

2. Materials and Methods

2.1. Study Sites

The datasets were acquired in three types of cropland at different locations in Japan: a paddy field in Mase, Tsukuba (36°03'14.3"N/140°01'36.9"E: rice, MSE), an upland field in Shinhidaka (42°24'41.4"N/142°28'16.6"E: maize, SHD), and a cultivated grassland in Nasushiobara (36°54'54.3"N/139°56'12.8"E: grass, NSS).

The details of each site are shown in Table 1. All three sites belong to AsiaFlux (http://asiaflux.net/), where fluxes of CO_2, sensible heat, and latent heat, in addition to basic micrometeorological and physiological data have been collected since 1999 at the rice site, 2007 at the maize site, and 2004 at the grass site [28–30].

Table 1. Details of the study sites.

	Rice	Maize	Grass
Site code	MSE	SHD	NSS
Position	36°03'14.3"N, 140°01'36.9"E	42°24'41.4"N, 142°28'16.6"E	36°54'54.3"N, 139°56'12.8"E
Elevation (m asl)	11	120–130	305
Mean annual air temperature (°C)	13.7	8.0	12.2
Mean annual precipitation (mm)	1200	1290	1561
Vegetation type	Paddy field	Upland field	Cultivated grassland
Dominant species	Rice (Oryza sativa L.; cultivar Koshihikari)	Maize (Zea mays L.)	Orchardgrass (Dactylis glomerata L.), Italian lyegrass (Lolium multiflorum Lam.)
Canopy height (m)	0–1.2	0–3.2	0–1.2
Annual maximum leaf area index ($m^2 \cdot m^{-2}$)	5.0	NA	NA
Height of sensor arm (m)	2.88	5.15	1.55
Data logger	CR3000	CR23X	CR23X
Observation year	2013	2013	2014
Growth stage	Transplanting: DOY 122 (2 May) Heading: DOY 204 (23 Jul.) Harvesting: DOY 249 (6 Sep.)	Budding: DOY 150 (30 May) Silking: Dot 208 (27 Jul.) Harvesting: DOY 261 (18 Sep.)	Second Harvesting: DOY 178 (27 Jun.) Third harvesting: DOY 239 (27 Aug.)

NA: not available.

2.2. Data and Analytical Methods

2.2.1. Multispectral Radiance Measurement

Measurements of multispectral radiation were obtained by using a four-channel sensor (SKR1850, Skye Instruments Ltd, Llandrindod Wells, UK) at each study site in 2013 and 2014. The average center wavelength (average full width at half maximum (FWHM)) of each spectral band for the three sites was 478.3 ± 1.5 (9.3 ± 0.2) nm (blue), 549.0 ± 0.6 (9.7 ± 0.1) nm (green), 657.7 ± 0.6 (21.7 ± 0.1) nm (red), and 827.9 ± 0.6 (37.4 ± 0.1) nm (near infrared: NIR). The center wavelength and bandwidth were slightly different among the three sites, but the standard deviations for both the center wavelength and bandwidth were small enough to assume that the wavebands of all sensors were identical. A set of two sensors, one directed upwards and the other downwards was attached to a horizontal arm to measure the spectral irradiance of incident light and the radiance of reflected light. The field of view (FOV) of the sensors was 180° in the upward direction with a removable diffusing cosine correction head, and 25° in the downward direction. The height of the sensor arms was 2.88 m at the rice site, 5.15 m at the maize site, and 1.55 m at the grass site above the ground (Table 1). All measurements from individual spectral channels were recorded by a data logger (CR3000 (rice site) and CR23X (maize and grass sites), Campbell Scientific, USA) at an interval of 10 min throughout the seasons. We used the spectral data from 09:00 to 16:00 local time for the period of the day of the year (DOY) 130 (May 10)–DOY 230 (August 18), in 2013 at the rice and maize sites, and the same period in 2014 at the grass site.

2.2.2. Vegetation Indices Based on Ground-Based Spectral Measurements

The NDVI and the GRVI were calculated from the ground-based radiometer data. The NDVI and the GRVI are defined as follows [20,24]:

$$\text{NDVI} = (\rho_{NIR} - \rho_{red}) / (\rho_{NIR} + \rho_{red}) \tag{1}$$

$$\text{GRVI} = (\rho_{green} - \rho_{red}) / (\rho_{green} + \rho_{red}) \tag{2}$$

where ρ_{NIR}, ρ_{red} and ρ_{green} are the reflectance factors in the NIR, red, and green regions, respectively. The sensor with the removable diffusing cosine correction head for incident light was calibrated for irradiance by a National Physical Laboratory UK reference standard lamp. However, the sensor for reflected light did not have an absolute calibration [13]. Therefore, instead of calculating the reflectance for each channel directly, the NDVI and the GRVI were determined from the following equations using the incident and reflected light intensity in each spectral band:

$$\text{NDVI} = [(Z_1 \times R_{NIR} / I_{NIR}) - (R_{red} / I_{red})] / [(Z_1 \times R_{NIR} / I_{NIR}) + (R_{red} / I_{red})] \tag{3}$$

$$\text{GRVI} = [(Z_2 \times R_{green} / I_{green}) - (R_{red} \times I_{red})] / [(Z_2 \times R_{green} / I_{green}) + (R_{red} / I_{red})] \tag{4}$$

where Z_1 is the sensitivity ratio of reflected NIR to red light; Z_2 is the sensitivity ratio of green to red light; R_{NIR}, R_{red}, and R_{green} are the reflected readings in the NIR, red, and green regions (nano ampere: nA), respectively; and I_{NIR}, I_{red}, and I_{green} are the incident ($\mu mol \cdot m^{-2} \cdot s^{-1}$) readings for the NIR, red, and green regions, respectively [31,32]. We used only the vegetation indices in the range from −1 to 1 to exclude abnormal data that were presumably caused by insufficient irradiance, rain, birds, insects, *etc.*

The solar zenith angle was calculated based on the geolocation of each site and the time of the spectral measurements. To investigate the influence of the diffuse/direct light conditions, we selected "clear sky" days and "cloudy sky" days by using the intensity and diurnal change of the global solar radiation measured by a pyranometer (rice and grass sites) and photosynthetically-active radiation (PAR) measured by a PAR sensor (maize site). A clear sky day was defined as a day with high radiation values and a smooth diurnal curve (see some examples in Figure 2). The PAR was proportional to the solar radiation and the ratio of PAR ($\mu mol \cdot m^{-2} \cdot s^{-1}$) to solar radiation ($W \cdot m^{-2}$) was 1.863. In contrast, a cloudy sky day was defined as a day with low incident radiation values throughout the daytime. We used these vegetation indices on the clear and cloudy sky days to analyze the effects of the solar zenith angle and diffuse/direct light conditions. The proportion of clear sky days was 19% at the rice site, 14% at the maize site, and 9% at the grass site.

2.2.3. A Radiative Transfer Model for Simulating Vegetation Indices

We used the PROSAIL radiative transfer model to simulate the influence of the solar zenith angle and diffuse/direct light condition on the vegetation indices [33]. The PROSAIL model is a combination of the canopy reflectance model SAIL [34,35] and the leaf reflectance model PROSPECT [36]. The model can simulate the canopy bidirectional reflectance in the 400–2500 nm wavelength region at 1 nm resolution under various biophysical conditions and/or measurement configurations. In this study, model parameters for the actual canopies in each experiment were not determined, so we used typical parameter values from the literature for the maize canopies, as shown in Table 2 [12,37,38]. Therefore, we assumed that the general relations between solar zenith angle, diffuse/direct light conditions (the ratio of diffuse light to total radiation), and leaf area index (LAI) could be investigated properly by simulations with these typical parameters.

Table 2. List of input parameters for the PROSAIL model.

Parameter	Value
Chlorophyll a and b content (Cab)	40
Carotenoid content (Car)	12.3
Brown pigment content (Cbrown)	0
Leaf water content (Cw)	0.015
Leaf dry matter content (Cm)	0.0055
Structure coefficient (N)	1.5
Leaf angle distribution (LIDF)	Spherical
Leaf area index (LAI)	0.1, 0.5, 1, 2, 3, 4, 5
Solar zenith angle (tts)	20, 30, 40, 50, 60
Observer zenith angle (tto)	0
Azimuth (psi)	0
Soil reflectance properties (psoil)	0.7

3. Results

3.1. The Effects of Solar Zenith Angle on Diurnal and Seasonal Change of Vegetation Indices

Figure 1 shows a time series of the NDVI and the GRVI with the same solar zenith angle, *i.e.*, 20°, 30°, 40°, 50°, and 60°, during the growing season. These data were extracted from the diurnal data so that the individual data had a similar solar zenith angle. Accordingly, the time of day for the individual data varied from morning (9:00 local time) to afternoon (16:00 local time). At the rice site (MSE), the dates for transplanting, heading, and harvesting were DOY 122 (2 May), DOY 204 (23 July), and DOY 249 (6 September), respectively (Table 1). At the maize site (SHD), the dates for budding, silking, and harvesting were DOY 150 (30 May), DOY 208 (27 July), and DOY 261 (18 September), respectively (Table 1). At the grass site (NSS), regular renovation of the grassland was conducted in 2012, and the second and third harvesting were on DOY 178 (27 June) and DOY 239 (27 August), respectively (Table 1). Overall, both the NDVI and the GRVI increased with plant growth at all sites. After reaching the maximum level, the NDVI remained nearly constant, whereas the GRVI gradually decreased. Most importantly, the difference in solar zenith angle caused some systematic changes in the seasonal pattern of both the NDVI and the GRVI. The influence of the solar zenith angle was slightly larger for the GRVI than for the NDVI.

Figure 2 shows the distinctive diurnal change of the NDVI, the GRVI and the solar radiation at the rice site. These figures show some selected days under clear and cloudy sky conditions during the early growth stage (NDVI < 0.4), middle growth stage (NDVI: 0.4–0.8), and late growth stage (NDVI > 0.8). Under clear sky conditions, both the NDVI and the GRVI showed significant diurnal changes during the middle and the late growth stages (Figure 2c,e), whereas the diurnal change was small during the early growth stage (Figure 2a). The NDVI and the GRVI showed minimum values from 11:00 to 12:00, when the solar radiation reached a maximum during the middle and late growth stages. The precipitous decrease of the NDVI and the GRVI showed around solar noon. On the other hand, under cloudy sky conditions, neither the NDVI nor the GRVI showed significant diurnal changes in spite of changes in the solar radiation throughout the growing season (Figure 2b,d,f).

165

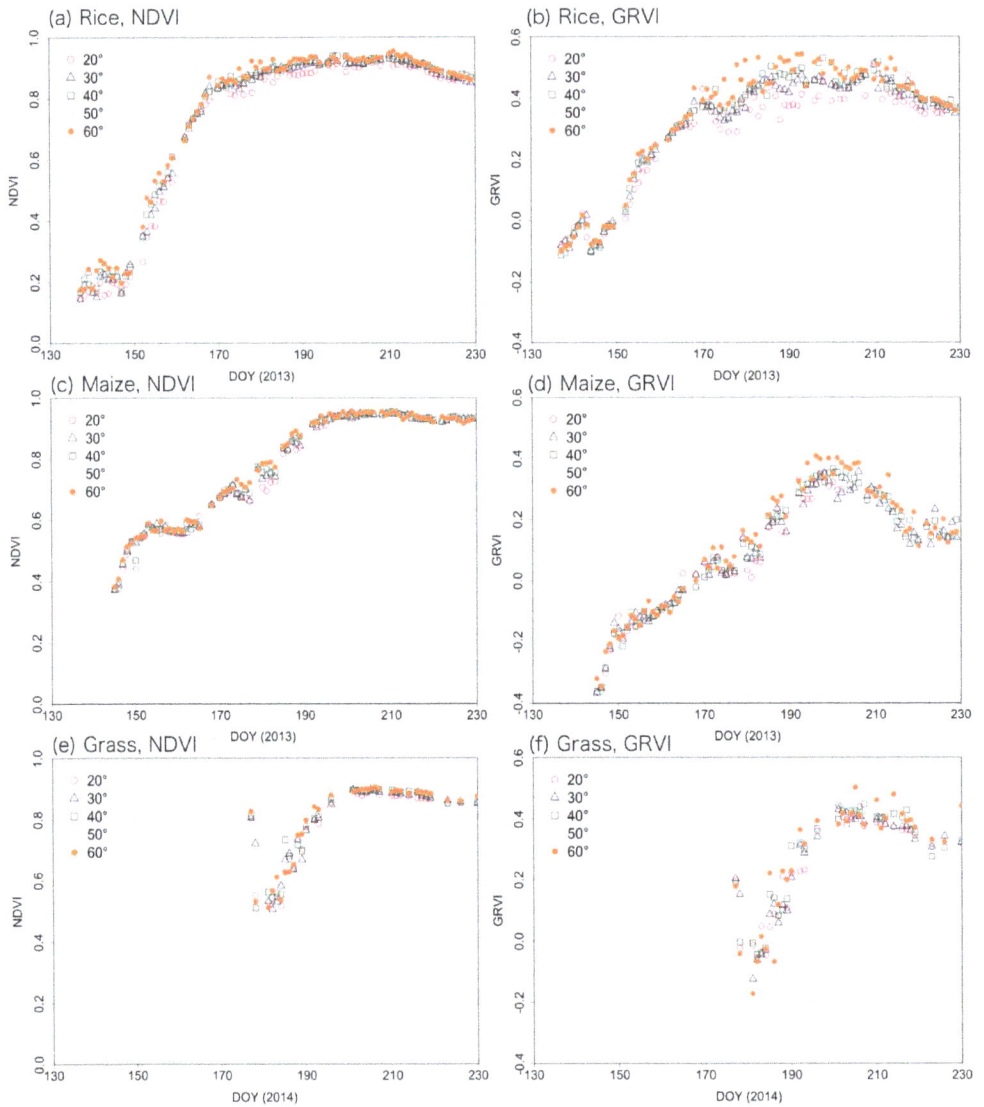

Figure 1. Time series of the NDVI (**left**) and the GRVI (**right**) with the same solar zenith angles (20°, 30°, 40°, 50°, and 60°) during the growing season at the rice (**a,b**), maize (**c,d**), and grass (**e,f**) sites.

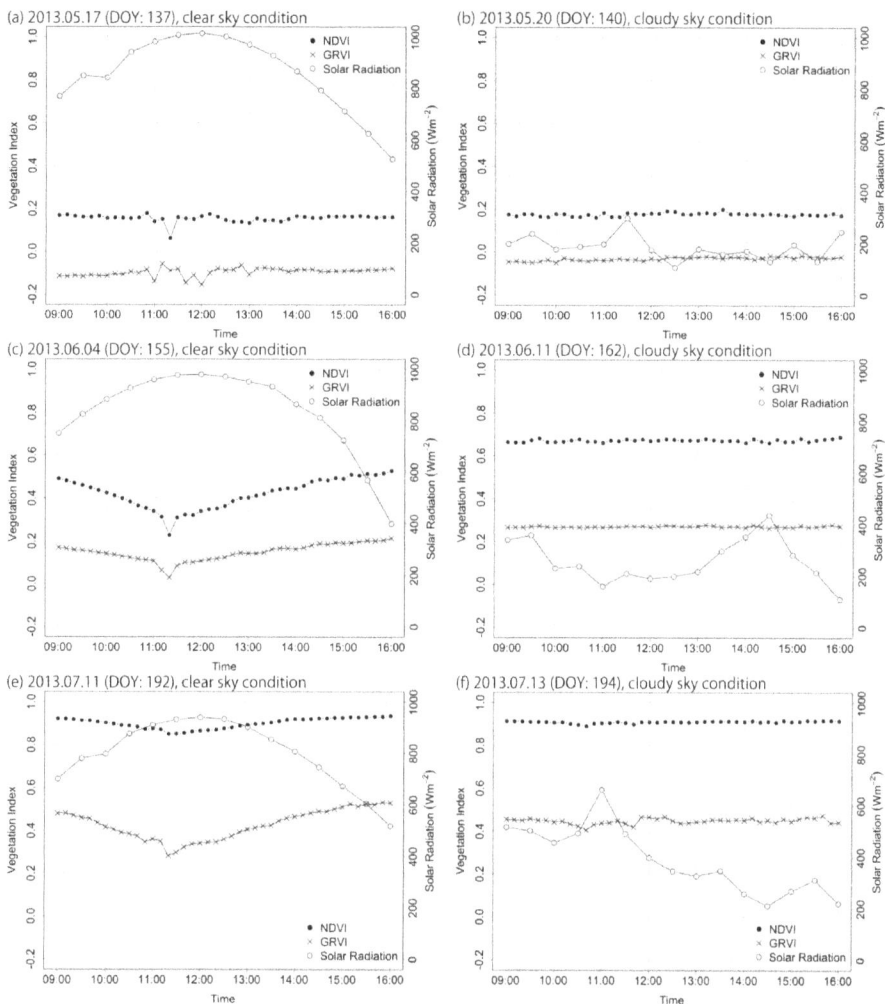

Figure 2. Distinctive diurnal changes of the NDVI and the GRVI under clear sky conditions (**left**) and cloudy sky conditions (**right**) at the rice site. Solar noon is between 11:36 and 11:45 local time. (**a**) DOY: 137; (**b**) 140; (**c**) 155; (**d**) 162; (**e**) 192; (**f**) 194.

Figure 3 shows the dependence of the NDVI and the GRVI on the solar zenith angle for selected days during the growing season. In this figure, data points are shown for sets of two days (clear and cloudy sky conditions) with almost the same crop conditions, in order to determine the effects of the diffuse/direct light conditions. The open symbols show the days under clear sky conditions and the closed symbols show the days under cloudy sky conditions. Overall, the NDVI was not affected by the solar zenith angle under cloudy sky conditions throughout the growing season. However, under clear sky conditions, the NDVI decreased significantly with decreasing solar zenith angle during the middle growth stage (NDVI: 0.4–0.8), whereas, even under clear sky conditions, the influence of the solar zenith angle on the NDVI was not clear during the early and late growth stages (NDVI < 0.4 and NDVI > 0.8). On the other hand, the GRVI decreased with decreasing solar zenith angle under clear sky conditions after the middle growth stage (GRVI > 0), whereas it was not affected by the solar zenith angle under either clear or cloudy sky conditions during the early growth stage (GRVI < 0). These responses of the vegetation indices to the change in solar zenith angle were much more significant at the rice site than at the other sites. In particular, on DOY 155, the NDVI decreased by more than 0.2 in response to a 25° decrease in solar zenith angle at the rice site. Meanwhile, there were some fluctuations in NDVI and the GRVI with the change of the solar zenith angle.

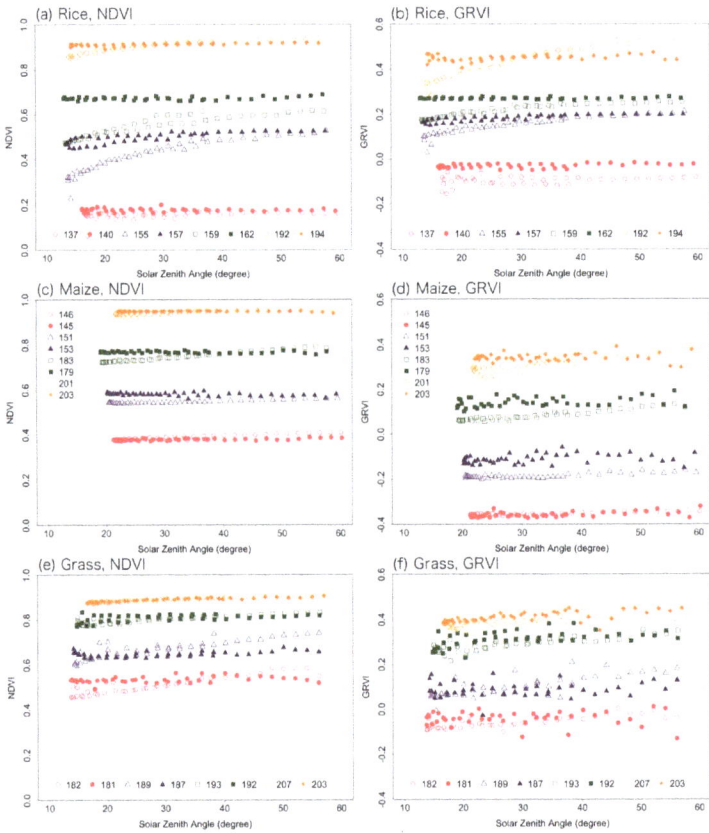

Figure 3. Relation between the NDVI (**left**), GRVI (**right**), and the solar zenith angles on selected days during the growing season at the rice (**a,b**), maize (**c,d**) and grass (**e,f**) sites. Open symbols show the days under clear sky conditions and closed symbols show the days under cloudy sky conditions.

Table 3 shows the statistical coefficients for the linear regression between the vegetation indices and the solar zenith angle on the selected days used in Figure 3. The coefficient of determination for linear regression was high under clear sky conditions. Under cloudy sky conditions, the variation of the vegetation indices with changing the solar zenith angle was small, and linear relationship was not significant. The slope of linear regression between NDVI and solar zenith angle varied in a range from 0.0019 to 0.0050 under clear sky conditions during the middle growth stage (NDVI: 0.4–0.8) except DOY 151 at the maize site. On the other hand, the slope of linear regression between GRVI and solar zenith angle varied in a range from 0.0018 to 0.0054 under clear sky conditions after the middle growth stage (GRVI > 0). The mean slope of the regression lines for NDVI was 0.0042 at the rice site, 0.0012 at the maize site, and 0.0029 at the grass site, respectively. Similarly, the mean slope of the regression lines for GRVI was 0.0034 at the rice site, 0.0025 at the maize site, and 0.0021 at the grass site, respectively. All slopes of both NDVI and GRVI under cloudy sky conditions were less than 0.001 except DOY 157 at rice site in NDVI and DOY192 at grass site in GRVI.

Table 3. The statistical coefficients for the linear regression between the vegetation indices and the solar zenith angle on the selected days during the growing season at the rice, maize, and grass site. The days under clear sky conditions is highlighted by gray color.

(a) NDVI

	Rice				Maize				Grass		
DOY	Slope	Intercept	R^2	DOY	Slope	Intercept	R^2	DOY	Slope	Intercept	R^2
137	0.0005	0.140	0.111	146	0.0009	0.357	0.799	182	0.0029	0.426	0.855
140	−0.00003	0.176	0.003	145	0.0002	0.371	0.295	181	0.0003	0.524	0.092
155	0.0050	0.274	0.847	151	0.0005	0.530	0.869	189	0.0029	0.585	0.775
157	0.0015	0.455	0.642	153	−0.0003	0.590	0.215	187	0.0004	0.634	0.207
159	0.0034	0.449	0.829	183	0.0019	0.686	0.987	193	0.0012	0.766	0.798
162	0.0001	0.669	0.047	179	−0.0001	0.772	0.102	192	0.0006	0.789	0.219
192	0.0019	0.844	0.856	201	0.0008	0.916	0.707	207	0.0008	0.865	0.897
194	0.0003	0.903	0.364	203	−0.00007	0.950	0.161	203	0.0007	0.867	0.831

(b) GRVI

	Rice				Maize				Grass		
DOY	Slope	Intercept	R^2	DOY	Slope	Intercept	R^2	DOY	Slope	Intercept	R^2
137	0.0004	−0.109	0.072	146	0.0005	−0.370	0.609	182	0.0016	−0.105	0.636
140	0.0003	−0.041	0.173	145	0.0006	−0.378	0.290	181	−0.0001	−0.039	0.002
155	0.0028	0.065	0.831	151	0.0007	−0.207	0.628	189	0.0031	0.010	0.603
157	0.0010	0.154	0.680	153	0.0005	−0.126	0.072	187	0.0003	0.076	0.014
159	0.0020	0.159	0.863	183	0.0018	0.023	0.921	193	0.0019	0.233	0.651
162	0.00004	0.269	0.019	179	0.0006	0.123	0.097	192	0.0014	0.272	0.242
192	0.0054	0.270	0.912	201	0.0032	0.204	0.749	207	0.0014	0.346	0.470
194	0.0004	0.438	0.115	203	0.0001	0.337	0.004	203	0.0013	0.371	0.412

3.2. Vegetation Indices Simulated Using the Radiative Transfer Model

Figure 4 shows the reflectance values simulated by the PROSAIL model for a range of solar zenith angles (10°, 20°, 30°, 40°, 50°, 60°) under different LAI (0.1, 0.5,

1, 2, 3, 4, 5). For all solar zenith angles, the reflectance decreased with increasing LAI in the visible to red-edge range (400–750 nm). In contrast, in the red-edge to NIR range (750–1000 nm), the reflectance increased consistently. The response of the reflectance to the solar zenith angle was weak for low LAI values (~0.1), but the reflectance in the visible range for higher LAI (0.5–5) showed a decreasing trend with increasing solar zenith angle. The reflectance in the NIR region for LAI values of 0.5–4 showed a decreasing trend for solar zenith angles from 10 to 40°, but an increasing trend from 40 to 60°. The reflectance of the NIR for the high LAI (5) showed a decreasing trend in the order of solar zenith angle in the full range of 10 to 60°. In summary, the spectral response to the change in solar zenith angle was largest in the red band (657.7 nm), followed by the green (549.0 nm), and the NIR (827.9 nm) bands.

Figure 5 shows the relations between the vegetation indices and the solar zenith angle simulated by the PROSAIL model for a range of solar zenith angles (10°, 20°, 30°, 40°, 50°, 60°) and LAI values (0.1, 0.5, 1, 2, 3, 4, 5). The NDVI decreased with decreasing solar zenith angle for all LAI values. Nevertheless, the response of the NDVI to the change in solar zenith angle was negligible for low and high LAI values. The response of the GRVI to solar zenith angle was similar to that of the NDVI. However, the response of the GRVI for high LAI values was much clearer than that of the NDVI, whereas the response for a low LAI value (0.1) was almost negligible as in the case of the NDVI.

Figure 6 shows the relations between the simulated vegetation indices and the solar zenith angles for different diffuse light ratios (40%, 60%, 80%, 100%). Under 100% diffuse light conditions, neither the NDVI nor the GRVI changed with changing solar zenith angle, irrespective of the LAI values. However, for lower diffuse light ratios (clear sky conditions), a response of the vegetation indices to the solar zenith angle was evident. Both the NDVI and the GRVI decreased with decreasing solar zenith angle. These responses were clearer at the middle LAI values, but were negligible for low and high LAI values in the case of the NDVI and for low LAI values in the case of the GRVI.

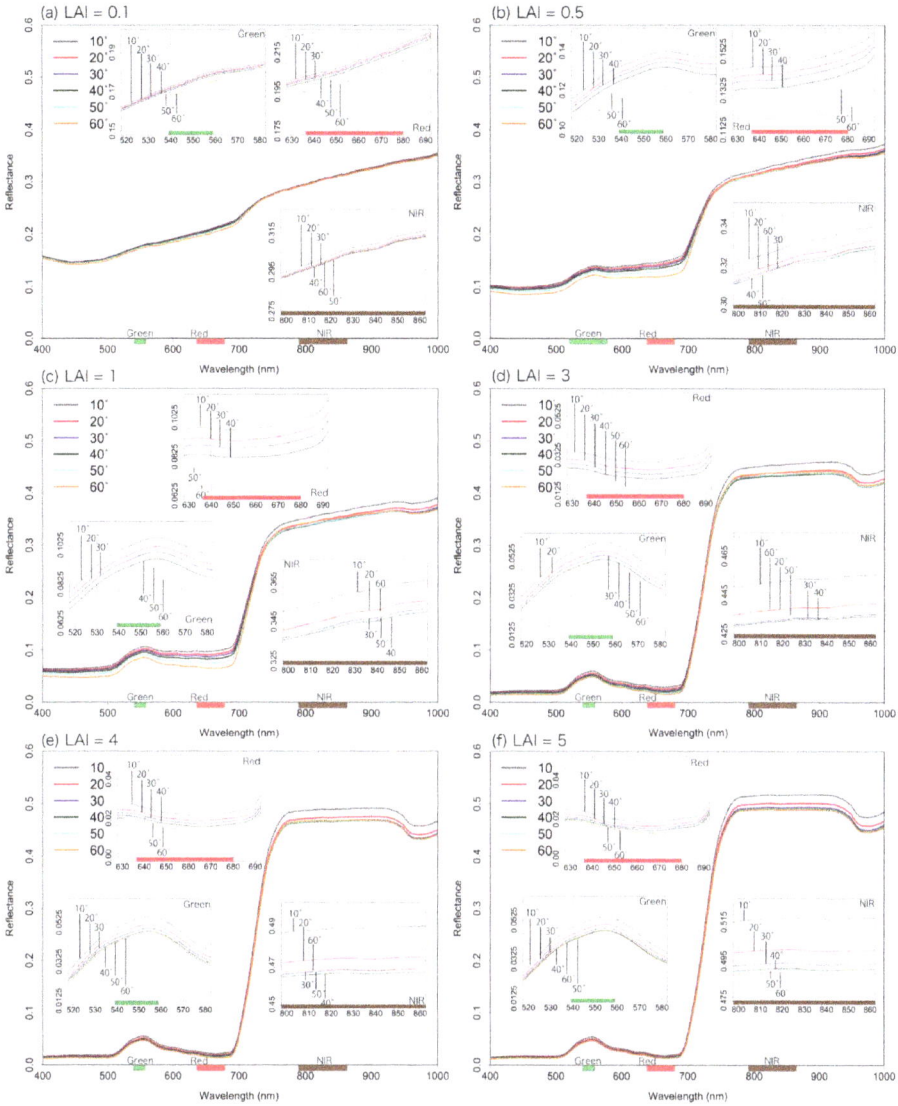

Figure 4. Reflectance spectra simulated using the PROSAIL model for a range of solar zenith angles (10°, 20°, 30°, 40°, 50°, 60°) and LAI values ((**a**) 0.1, (**b**) 0.5, (**c**) 1, (**d**) 3, (**e**) 4, (**f**) 5).

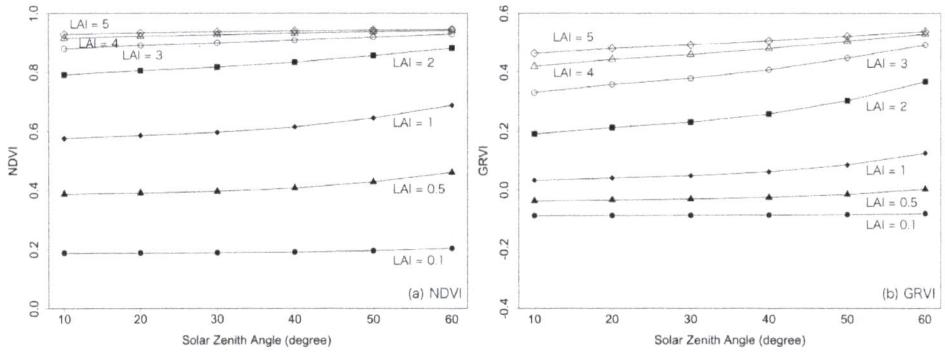

Figure 5. Relation between vegetation indices ((**a**) NDVI and (**b**) GRVI) and solar zenith angles simulated by the PROSAIL model for a range of LAI values (0.1, 0.5, 1, 2, 3, 4, 5).

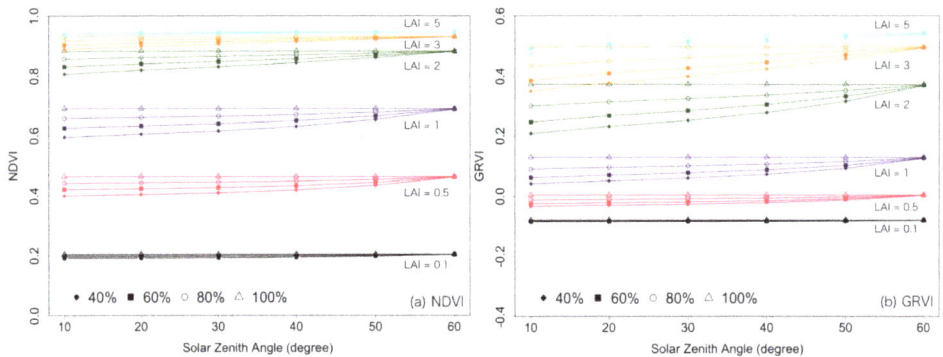

Figure 6. Relation between vegetation indices ((**a**) NDVI and (**b**) GRVI) and solar zenith angles simulated by the PROSAIL model for a range of LAI values (0.1, 0.5, 1, 2, 3, 4, 5) and different ratios of diffuse light (40%, 60%, 80%, 100%).

4. Discussion

4.1. Influence of the Solar Zenith Angle on the Change in Vegetation Indices

The values of the vegetation indices for a vegetation canopy fluctuated in response to the solar zenith angle. The values were not consistent even during a day due to the change in solar radiation and solar zenith angle (Figures 1 and 2). The precipitous decrease of the NDVI and the GRVI around solar noon may be attributable to the hot spot phenomenon [39]. In general, the vegetation indices decreased with decreasing solar zenith angle (Figure 3). This response was affected significantly by the growth stage and diffuse/direct light conditions. The decreasing response of the NDVI to decreasing solar zenith angle was high during the middle growth stage (0.4 < NDVI < 0.8). The decrease ratio of NDVI by decreasing solar zenith angle was in the range from 0.0019 to 0.0050 under clear sky conditions in this growth stage and the NDVI value decreased within the range from 0.0057 to 0.15 with decreasing solar zenith angle from 50 to 20 (Table 3). On the other hand, a similar response of the GRVI was evident, except for the early growth stage (GRVI < 0). The decrease in ratio of the GRVI by the decreasing solar zenith angle was from the range of 0.0018 to 0.0054 under clear sky conditions in this growth stage and the GRVI value decreased within the range from 0.0054 to 0.162 with decreasing solar zenith angle from 50 to 20. The response of the vegetation indices to the solar zenith angle was also affected by the diffuse/direct light conditions. The change in the vegetation indices in response to the solar zenith angle was evident under clear sky conditions but almost negligible in cloudy sky conditions. Under cloudy sky conditions, the variation of the vegetation indices by change of the solar zenith angle was small, and the clear linear relationship was not found. A part of the fluctuations of NDVI and the GRVI observed in Figure 3 would be attributable to the interaction of solar azimuth angle with crop row orientation, although the other parts might have been caused by other environmental factors such as rain and birds [1].

Rahman *et al.* [23] reported that the NDVI determined by ground-based observations decreased with decreasing solar zenith angle at a pasture site. These results indicated a similar tendency to those obtained in the present study. However, the previous study used a dataset for only two days under conditions of vegetation cover and did not show results throughout the growth season. Furthermore, because the observation of radiation was conducted only under clear sky conditions, the relations between the NDVI and the solar zenith angle were not investigated under cloudy sky conditions. In this study, we compared the influences of various growth stages and diffuse/direct light conditions on vegetation indices by using continuous ground-based measurements.

These experimental results were well supported by the results of simulations based on the physically-based canopy reflectance model (PROSAIL) (Figures 5 and 6). First, the effect of the growth stage (as represented by LAI) on the sensitivity of the NDVI and the GRVI to the solar zenith angle was assessed quantitatively. The results agreed well with the experimental results, in which the sensitivity of the NDVI to LAI was evident during the middle growth stage but low during the early and late growth stages. The sensitivity of the GRVI was similar to that of the NDVI, but negligible only during the early growth stage. Second, the simulation results concerning the effect of light conditions (ratio of diffuse light) on the sensitivity of the NDVI and the GRVI to the solar zenith angle also agreed well with the experimental results. The response of the NDVI and the GRVI to the difference in solar zenith angle was evident under clear sky conditions (less diffuse light), but negligible under cloudy sky conditions, irrespective of the growth stage (LAI). Nevertheless, the sensitivity of these responses was significant during the middle growth stage, but weak during the early and late growth stages for the NDVI and during the early growth stage for the GRVI.

The relation between the vegetation indices and the solar zenith angle was caused by the response of the reflectance to the solar zenith angle (Figure 4). In general, the canopy reflectance is affected by view/illumination geometry as well as the canopy structure and optical properties of leaves and soils [1,33,34]. The variations in solar zenith angle alter both the optical thickness of a canopy and the illuminated components of vegetation/background [40]. In most crop canopies, the canopy reflectance is determined mainly by the soil reflectance in small LAI conditions (~1) and by the vegetation reflectance in large LAI conditions (4~). Accordingly, the response of canopy reflectance to the solar zenith angle is determined by the interactive effects of the canopy structure (LAI and leaf angle distribution) in the direction of sun-beam as well as the BRDF of soil surface. Similarly, the small influence of solar zenith angle under diffuse light is explained by the isotropic illumination conditions, *i.e.*, stability of the optical thickness and the illuminated components.

4.2. Response of Vegetation Indices to Solar Zenith Angle and Diffuse/Direct Light Conditions in Different Vegetation Types

The overall relation between the vegetation indices and the solar zenith angles was similar for all three vegetation types. However, their responses were affected by the differences in canopy structure and the growth pattern for each vegetation type. The vegetation indices decreased with decreasing solar zenith angle for all vegetation types, but the sensitivity of the response was somewhat different across the three crops (Figure 3 and Table 3). In particular, the sensitivity of the response was much higher for the paddy field than for the other types. For the paddy field, the decrease in the NDVI was remarkable for solar zenith angles smaller than 30°. In contrast, the reduction rate for the GRVI with decreasing solar zenith angle was not affected by the vegetation type.

The relation between the NDVI and the solar zenith angle in the simulation was similar to the experimental results for upland field and cultivated grassland, whereas the experimental results for the paddy field showed a higher response than the simulation results (Figures 5 and 6). This difference may be attributable to the unique ground surface condition in paddy fields. The soil surface of paddy fields is under flooded conditions during the majority of the growing period. All selected days used in Figure 3 were under flooded conditions. Under such conditions, radiation in the NIR region is absorbed by the background water, and so the reflectance in the NIR region would be decreased [41]. Especially during the early growth stage when the rice canopy is not closed, the effects of the water surface on the reflectance in the

NIR region can be more significant than in other growth stages. Accordingly, under these conditions, the reflectance in the NIR region decreases when the solar zenith angle is small, whereas the reflected radiation would increase for high solar zenith angles. Our experimental and simulation results suggest that the higher sensitivity of vegetation indices to the solar zenith angle in paddy fields would be caused by the unique flooded conditions beneath the rice plants.

On the other hand, the relation between the NDVI and the solar zenith angle in upland field and cultivated grassland was slightly different from that for the paddy field. In cultivated grassland, the decrease in the NDVI was not significant for small solar zenith angles. A grass canopy usually closes earlier than row crops, such as rice and maize, because of the broadcast sowing method. This may be the reason why the relation between the NDVI and the solar zenith angle for cultivated grassland was less sensitive than that for the other vegetation types. The relation between the NDVI and the solar zenith angle for the maize canopy was also less sensitive than that for the paddy field because of the difference of background surface condition (upland or flooded). The vegetation indices were also affected by the soil surface condition when the vegetation cover was small [40]. The soil-adjusted vegetation index (SAVI) have been used to minimize the effects of soil background [40]. Note that the relation between the NDVI and the solar zenith angle was affected to some extent by the background surface conditions and differences in canopy structure.

4.3. Effective Usage of Vegetation Indices Derived from Continuous Ground-Based Spectral Measurement

Ground-based spectral observations can provide detailed and accurate information on the dynamic change in structure and/or function of vegetation based on high temporal resolution data. In addition, these ground-based measurements can be used for absolute calibration of satellite or airborne images. However, these data are affected by the solar zenith angle and diffuse/direct light conditions at the time of measurement. Therefore, we have to take account of such fluctuations in the analysis of continuous measurements on the ground. However, in previous studies, instantaneous or mean values at a specified time were often used throughout the season [20,21]. In such practices, the effects of diurnal and seasonal changes of the solar zenith angle are ignored, so the actual vegetation parameters would not be estimated properly. To reduce these influences, it is desirable to use data obtained under identical measurement conditions for the solar zenith angle and the diffuse/direct light. In general, the number of usable data is restricted to satisfy such measurement conditions. According to our results (Figures 3 and 6), for larger solar zenith angles, the vegetation indices are not significantly affected by the diffuse/direct light conditions and growth stage. For example, for a solar zenith angle of 60°, neither the NDVI nor the GRVI are affected by the diffuse/direct light conditions in any growth stage. Therefore, our results suggest that using the selected data at the solar zenith angle of 60° would be effective for accurate assessment of the canopy structure and function based on continuous diurnal spectral measurements.

In Figure 3, we investigated the influence of solar zenith angle on vegetation indices based on the ground-based measurements on typical clear- and cloudy-sky days. However, in general, the distribution and optical thickness of clouds in the sky vary during a day or a season depending on the weather conditions. Therefore, we have examined similar relationships as in Figure 3 using the data on the other type of days, *i.e.*, those with some fluctuations between clear- and cloudy-sky conditions in a day. The result suggested that the effect of solar zenith angle on the vegetation indices under such days varied within the variation range of the two typical conditions depending on the diffuse/direct light ratio. These fluctuations are explained mainly by the change of the diffuse/direct light ratio as investigated in our simulation study (Figure 6). Accordingly, the possible influences of sky conditions (diffuse/direct light ratio) on the relationship between vegetation indices and solar zenith angle can be assessed by using instantaneous measurements of incident light obtained by spectral radiometers, pyranometers, or PAR sensors.

Ground-based spectral measurements are also used for validation of satellite observations [17,42]. In the case that the satellite data are validated by synchronized ground-based measurements, the solar zenith angle and the diffuse/direct light conditions are basically the same for the ground-based and the satellite observations. However, the viewing angle of satellites does not always agree with the ground-based sensors because the view zenith angle is usually fixed at 0° (nadir observation) for the ground-based sensors. Therefore, similar to the fluctuations caused by the solar zenith angle, the influence of this difference in viewing angle has to be considered because the vegetation indices would be affected by it [43].

In a wide range of experimental studies in the field, spectral data are measured periodically using a portable spectro-radiometer at some intervals, and it is assumed that the effect of the solar zenith angle is small [44,45]. These observations are usually observed midday under clear sky conditions. When the observation dates are close to each other, the difference in solar zenith angle can be negligible. However, when the observation dates are different to some extent, the solar zenith angles may not be comparable. Rahman *et al.* [23] proposed a method to correct the influence of the solar zenith angle on the NDVI by using the relation between the NDVI and the solar zenith angle. However, the applicability of the method may be limited because the data used was from a narrow and specific period. In addition, the relation between the vegetation indices and the solar zenith angle is affected by the growth stage and vegetation type (Figure 3). The continuous spectral measurement allows the selection of some preferable data from the diurnal data for specific purposes, although measurements are taken for a fixed point in the field. Generally, field measurements using a portable spectro-radiometer allow us to acquire the spatial average of spectral measurements over a range of different targets. Nevertheless, data acquisition at the same solar zenith angle would be useful to improve the seasonal consistency of vegetation indices. If multiple measurements at different times of day (different solar zenith angles) can be made, the observation data can be corrected to be more consistent based on the relation between the solar zenith angle and the vegetation indices.

5. Conclusions

In this study, we investigated the impact of the changing solar zenith angle and diffuse/direct light conditions on the consistency of vegetation indices (NDVI and GRVI) derived from ground-based spectral measurements in three kinds of croplands (paddy field, upland field, cultivated grassland). The vegetation indices showed some systematic changes in response to the solar zenith angle, the ratio of diffuse light, and the growth stage.

Our comprehensive analysis revealed the general effects of the growth stage and light conditions on the diurnal and seasonal fluctuations of vegetation indices. In general, the vegetation indices decrease with decreasing solar zenith angle. This response can be affected significantly by the growth stage and diffuse/direct light conditions. The decreasing response of the NDVI to decreasing solar zenith angle is high during the middle growth stage (0.4 < NDVI < 0.8). On the other hand, a similar response of the GRVI is evident, except for the early growth stage (GRVI < 0). The change in vegetation indices in response to the solar zenith angle is evident under clear sky conditions, but almost negligible under cloudy sky conditions irrespective of the growth stage. Furthermore, for larger solar zenith angles, the vegetation indices are not significantly affected by the diffuse/direct light conditions and growth stage. These experimental results are well supported by the simulation results based on a physically-based canopy reflectance model (PROSAIL). Basically, the vegetation indices decrease with decreasing solar zenith angle for all vegetation types, but the sensitivity of the response is somewhat different for the three crops. In particular, the sensitivity of the response is much higher for the paddy field than for the other types, and this could be attributable to the uniquely flooded conditions in paddy fields.

Systematic selection of data from continuous diurnal spectral measurements in consideration of the solar light conditions would be effective for accurate and consistent assessment of canopy structure and function. Necessary corrections for the influences of sky conditions on the relationship between vegetation indices and solar zenith angle can be made by using instantaneous measurements of incident light obtained by spectral radiometers, pyranometers, or PAR sensors. These results would provide useful insights into the consistency of vegetation indices obtained by various sensors and platforms.

Acknowledgments: We thank M. Mano (Chiba University) for the assistance in field measurements. This work was supported by JSPS KAKENHI Grant Number 26850164. We also thank the three anonymous reviewers for their valuable comments and suggestions.

Author Contributions: Yoshio Inoue conceived and designed the spectral sensing network and research plans. Mitsunori Ishihara collected the field data, and analyzed the data for this study. Keisuke Ono, Mariko Shimizu and Shoji Matsuura assisted with the field data collection. Mitsunori Ishihara and Yoshio Inoue jointly wrote the manuscript. All the authors contributed to editing and reviewing the manuscript.

Conflicts of Interest: The authors declare no conflict of interest.

References

1. Hatfield, J.L.; Gitelson, A.A.; Schepers, J.S.; Walthall, C.L. Application of spectral remote sensing for agronomic decisions. *Agron. J.* **2008**, *100*, 117–131.
2. Inoue, Y.; Sakaiya, E.; Zhu, Y.; Takahashi, W. Diagnostic mapping of canopy nitrogen content in rice based on hyperspectral measurements. *Remote Sens. Environ.* **2012**, *126*, 210–221.
3. Peng, Y.; Gitelson, A.; Sakamoto, T. Remote estimation of gross primary productivity in crops using MODIS 250 m data. *Remote Sens. Environ.* **2013**, *128*, 186–196.
4. Zhong, L.; Gong, P.; Biging, G.S. Efficient corn and soybean mapping with temporal extendability: A multi-year experiment using Landsat imagery. *Remote Sens. Environ.* **2014**, *140*, 1–13.
5. Asaka, D.; Shiga, H. Estimating rice grain protein contents with SPOT/HRV data acquired at maturing stage. *J. Remote Sens. Soc. Jpn.* **2003**, *23*, 451–457. (In Japanese with English summary)
6. Sakaiya, E.; Inoue, Y. Operational use of remote sensing for harvest management of rice. *J. Remote Sens. Soc. Jpn.* **2013**, *33*, 185–199. (In Japanese with English summary)
7. Wang, J.; Huang, J.; Zhang, K.; Li, X.; She, B.; Wei, C.; Gao, J.; Song, X. Rice fields mapping in fragmented area using multi-temporal HJ-1A/B CCD images. *Remote Sens.* **2015**, *7*, 3467–3488.
8. Inoue, Y.; Sakaiya, E. Relationship between X-band backscattering coefficients from high-resolution satellite SAR and biophysical variables in paddy rice. *Remote Sens. Lett.* **2013**, *4*, 288–295.
9. Inoue, Y.; Sakaiya, E.; Wang, C. Potential of X-band images from high-resolution satellite SAR sensors to assess growth and yield in paddy rice. *Remote Sens.* **2014**, *6*, 5995–6019.
10. Inoue, Y.; Sakaiya, E.; Wang, C. Capability of C-band backscattering coefficients from high-resolution satellite SAR sensors to assess biophysical variables in paddy rice. *Remote Sens. Environ.* **2014**, *140*, 257–266.
11. Hunt, E.R.; Hively, W.D.; Fujikawa, S.J.; Linden, D.S.; David, S.C.; Daughtry, S.T.; McCarty, G.W. Acquisition of NIR-green-blue digital photographs from unmanned aircraft for crop monitoring. *Remote Sens.* **2010**, *2*, 290–305.

12. Duan, S.B.; Li, Z.L.; Wu, H.; Tang, B.H.; Ma, L.; Zhao, E.; Li, C. Inversion of the PROSAIL model to estimate leaf area index of maize, potato, and sunflower fields from unmanned aerial vehicle hyperspectral data. *Int. J. Appl. Earth Obs. Geoinf.* **2014**, *26*, 12–20.

13. Balzarolo, M.; Anderson, K.; Nichol, C.; Rossini, M.; Vescovo, L.; Arriga, N.; Wohlfahrt, G.; Calvet, J.-C.; Carrara, A.; Cerasoli, S.; *et al.* Ground-based optical measurements at European flux sites: A review of methods, instruments and current controversies. *Sensors* **2011**, *11*, 7954–7981.

14. Gamon, J.A.; Rahman, A.F.; Dungan, J.L.; Schildhauer, M.; Huemmrich, K.F. Spectral Network (SpecNet): What is it and why do we need it? *Remote Sens. Environ.* **2006**, *103*, 227–235.

15. Nasahara, K.N.; Nagai, S. Review: Development of an *in situ* observation network for terrestrial ecological remote sensing: the Phenological Eyes Network (PEN). *Ecol. Res.* **2015**, *30*, 211–223.

16. Soudani, K.; Hmimina, G.; Delpierre, N.; Pontailler, J.Y.; Aubinet, M.; Bonal, D.; Caquet, B.; de Grandcourt, A.; Burban, B.; Flechard, C.; *et al.* Ground-based Network of NDVI measurements for tracking temporal dynamics of canopy structure and vegetation phenology in different biomes. *Remote Sens. Environ.* **2012**, *123*, 234–245.

17. Motohka, T.; Nasahara, K.N.; Miyata, A.; Mano, M.; Tsuchida, S. Evaluation of optical satellite remote sensing for rice paddy phenology in monsoon Asia using a continuous *in situ* dataset. *Int. J. Remote Sens.* **2009**, *30*, 4343–4357.

18. Sakamoto, T.; Gitelson, A.A.; Nguy-Robertson, A.L.; Arkebauer, T.J.; Wardlow, B.D.; Suyker, A.E.; Verma, S.B.; Shibayama, M. An alternative method using digital cameras for continuous monitoring of crop status. *Agric. For. Meteorol.* **2012**, *154–155*, 113–126.

19. Schaaf, C.B.; Gao, F.; Strahler, A.H.; Lucht, W.; Li, X.; Tsang, T.; Strugnell, N.C.; Zhang, X.; Jin, Y.; Muller, J.P.; *et al.* First operational BRDF, Albedo and Nadir reflectance products from MODIS. *Remote Sens. Environ.* **2002**, *83*, 135–148.

20. Motohka, T.; Nasahara, K.N.; Oguma, H.; Tsuchida, S. Applicability of green-red vegetation index for remote sensing of vegetation phenology. *Remote Sens.* **2010**, *2*, 2369–2387.

21. Nagai, S.; Saitoh, T.M.; Kobayashi, H.; Ishihara, M.; Suzuki, R.; Motohka, T.; Nasahara, K.N.; Muraoka, H. *In situ* examination of the relationship between various vegetation indices and canopy phenology in an evergreen coniferous forest, Japan. *Int. J. Remote Sens.* **2012**, *33*, 6202–6214.

22. Cogliati, S.; Rossini, M.; Julitta, T.; Meroni, M.; Schickling, A.; Burkart, A.; Pinto, F.; Rascher, U.; Colombo, R. Continuous and long-term measurements of reflectance and sun-induced chlorophyll fluorescence by using novel automated field spectroscopy systems. *Remote Sens. Environ.* **2015**, *164*, 270–281.

23. Rahman, M.M.; Lamb, D.W.; Stanley, J.N. The impact of solar illumination angle when using active optical sensing of NDVI to infer fAPAR in a pasture canopy. *Agric. For. Meteorol.* **2015**, *202*, 39–43.

24. Tucker, C.J. Red and photographic infrared linear combinations for monitoring vegetation. *Remote Sens. Environ.* **1979**, *8*, 127–150.

25. Baez-Gonzalez, A.D.; Kiniry, J.R.; Maas, S.J.; Tiscareno, M.L.; Macias, J.C.; Mendoza, J.L.; Richardso, C.W.; Salinas, J.G.; Manjarrez, J.R. Large-area maize yield forecasting using leaf area index based yield model. *Agron. J.* **2005**, *97*, 418–425.

26. Funk, C.; Budde, M. Phenologically-tuned MODIS NDVI-based production anomaly estimates for Zimbabwe. *Remote Sens. Environ.* **2009**, *113*, 115–125.

27. Nagai, S.; Ishii, R.; Suhaili, A.B.; Kobayashi, H.; Matsuoka, M.; Ichie, T.; Motohka, T.; Kendawang, J.J.; Suzuki, R. Usability of noise-free daily satellite-observed green-red vegetation index values for monitoring ecosystem changes in Borneo. *Int. J. Remote Sens.* **2014**, *35*, 7910–7926.

28. Nkongolo, N.V.; Hatano, R.; Kakembo, V. Diffusivity models and greenhouse gases fluxes from a forest, pasture, grassland and corn field in Northern Hokkaido, Japan. *Pedosphere* **2010**, *20*, 747–760.

29. Ono, K.; Maruyama, A.; Kuwagata, T.; Mano, M.; Takimoto, T.; Hayashi, K.; Hasegawa, T.; Miyata, A. Canopy-scale relationships between stomatal conductance and photosynthesis in irrigated rice. *Glob. Change Biol.* **2013**, *19*, 2209–2220.

30. Matsuura, S.; Miyata, A.; Mano, M.; Hojito, M.; Mori, A.; Kano, S.; Sasaki, H.; Kohyama, K.; Hatano, R. Seasonal carbon dynamics and the effects of manure application on carbon budget of a managed grassland in a temperate, humid region in Japan. *Grassl. Sci.* **2014**, *60*, 76–91.

31. Harris, A.; Gamon, J.A.; Pastorello, G.Z.; Wong, C.Y.S. Retrieval of the photochemical reflectance index for assessing xanthophyll cycle activity: A comparison of near-surface optical sensors. *Biogeosciences* **2014**, *11*, 6277–6292.

32. Application Notes Sensors for NDVI Calculations. Available online: http://www.skyeinstruments.com/wp-content/uploads/Application-Notes-for-NDVI.pdf (accessed on 17 August 2015).

33. Jacquemoud, S.; Verhoef, W.; Baret, F.; Bacour, C.; Zarco-Tejada, P.J.; Asner, G.P.; Francois, C.; Ustin, S.L. PROSPECT + SAIL models: A review of use for vegetation characterization. *Remote Sens. Environ.* **2009**, *113*, S56–S66.

34. Verhoef, W. Light scattering by leaf layers with application to canopy reflectance modeling: The SAIL model. *Remote Sens. Environ.* **1984**, *16*, 125–141.

35. Verhoef, W. Earth observation modeling based on layer scattering matrices. *Remote Sens. Environ.* **1985**, *17*, 165–178.

36. Jacquemoud, S.; Baret, F. PROSPECT: A model of leaf optical properties spectra. *Remote Sens. Environ.* **1990**, *34*, 75–91.

37. Jacquemoud, S.; Bacour, C.; Poilvé, H.; Frangi, J.-P. Comparison of four radiative transfer models to simulate plant canopies reflectance: direct and inverse mode. *Remote Sens. Environ.* **2000**, *74*, 471–481.

38. Wu, C.; Niu, Z.; Tang, Q.; Huang, W. Estimating chlorophyll content from hyperspectral vegetation indices: Modeling and validation. *Agric. For. Meteorol.* **2008**, *202*, 39–43.

39. Lacaze, R.; Chen, J.M.; Roujean, J.L.; Leblanc, S.G. Retrieval of vegetation clumping index using hot spot signatures measured by POLDER instrument. *Remote Sens. Environ.* **2002**, *79*, 84–95.

40. Epiphanio, J.C.N.; Huete, A.R. Dependence of NDVI and SAVI on sun/sensor geometry and its effect on fAPAR relationships in Alfalfa. *Remote Sens. Environ.* **1995**, *51*, 351–560.

41. Haltrin, V.I. Absorption and scattering of light in natural waters. In *Light Scattering Reviews: Single and Multiple Light Scattering*; Kokhanovsky, A.A., Ed.; Springer-Praxis: Berlin, Germany, 2006; pp. 445–486.

42. Fensholt, R.; Sandholt, I. Evaluation of MODIS and NOAA AVHRR vegetation indices with *in situ* measurements in a semi-arid environment. *Int. J. Remote Sens.* **2005**, *26*, 2561–2594.

43. Huber, S.; Tagesson, T.; Fensholt, R. An automated field spectrometer system for studying VIS, NIR and SWIR anisotropy for semi-arid savanna. *Remote Sens. Environ.* **2014**, *152*, 547–556.

44. Strachan, I.B.; Pattey, E.; Boisvert, J.B. Impact of nitrogen and environmental conditions on corn as detected by hyperspectral reflectance. *Remote Sens. Environ.* **2002**, *80*, 213–224.

45. Inoue, Y.; Peñuelas, J.; Miyata, A.; Mano, M. Normalized difference spectral indices for estimating photosynthetic efficiency and capacity at a canopy scale derived from hyperspectral and CO_2 flux measurements in rice. *Remote Sens. Environ.* **2008**, *112*, 156–172.

Monitoring Spatio-Temporal Distribution of Rice Planting Area in the Yangtze River Delta Region Using MODIS Images

Jingjing Shi and Jingfeng Huang

Abstract: A large-area map of the spatial distribution of rice is important for grain yield estimations, water management and an understanding of the biogeochemical cycling of carbon and nitrogen. In this paper, we developed the Normalized Weighted Difference Water Index (NWDWI) for identifying the unique characteristics of rice during the flooding and transplanting period. With the aid of the ASTER Global Digital Elevation Model and the phenological data observed at agrometeorological stations, the spatial distributions of single cropping rice and double cropping early and late rice in the Yangtze River Delta region were generated using the NWDWI and time-series Enhanced Vegetation Index data derived from MODIS/Terra data during the 2000–2010 period. The accuracy of the MODIS-derived rice planting area was validated against agricultural census data at the county level. The spatial accuracy was also tested based on a land use map and Landsat ETM+ data. The decision coefficients for county-level early and late rice were 0.560 and 0.619, respectively. The MODIS-derived area of late rice exhibited higher consistency with the census data during the 2000–2010 period. The algorithm could detect and monitor rice fields with different cropping patterns at the same site and is useful for generating spatial datasets of rice on a regional scale.

Reprinted from *Remote Sens.* Cite as: Shi, J.; Huang, J. Monitoring Spatio-Temporal Distribution of Rice Planting Area in the Yangtze River Delta Region Using MODIS Images. *Remote Sens.* **2015**, *7*, 8883–8905.

1. Introduction

Paddy rice fields provide essential food for more than half of the population of the entire world [1]. Rice is widely cultivated in Asian countries, especially China. Recent FAO (Food and Agriculture Organization) estimates indicate that to satisfy the projected demand of the year 2050, global agricultural production must increase 60 percent above the level of 2005–2007 [2]. Large-area assessments of potential food production regions and their impact on biogeochemical cycling require the acquisition of the best possible information on the distribution of paddy rice fields [3].

The official statistical data on rice sowing areas have been generated based on ground sample surveys and extrapolated to the provincial and national scales.

Large-scale census data cannot provide accurate spatial distributions of paddy rice, and a time lag is present in the datasets. Huke developed Asian rice datasets using agricultural statistical datasets collected at the sub-country level [4]. Leff *et al.* generated a global rice map at a spatial resolution of five arcminutes as part of a global cropland product using satellite-derived land cover data and agricultural census data [5]. Frolking *et al.* generated 0.5°-resolution maps of the distribution of rice agriculture in mainland China using a combination of county-scale agricultural census data and land cover maps derived from Landsat images collected during the 1995–1996 period [3]. A thematic land use map of China at a scale of 1:100,000 was generated via the visual interpretation of Landsat TM (Thematic Mapper) data [6]. A classification system of 25 land use categories, including paddy rice, was used in this work. The land use map was converted to 1-km gridded data. However, more updated datasets of annual rice distribution with finer resolution are needed at the regional scale.

Approximately half of the cropland in China is multi-cropped each year, and this land has a significant influence on the biogeochemical cycling of carbon and nitrogen. To date, many studies have been conducted to map paddy rice using fine-resolution satellite images, such as Landsat MSS, TM, ETM+ and NOAA/AVHRR images, by applying image classification procedures, but few of these studies have provided detailed information regarding the locations of multi-cropping [7–10]. Furthermore, because of the fine resolution of these images, it is difficult to obtain more comprehensive images covering an entire region simultaneously over a large area. Rice distribution maps have also been produced via multi-temporal analysis of NOAA/AVHRR and SPOT4/VEGETATION data with a resolution of ~1 km, which is rather coarse for rice mapping [11–13].

The Moderate-Resolution Imaging Spectroradiometer (MODIS) aboard the Terra and Aqua satellites, with its advantages of a high revisit period, moderate spatial resolution, wide field of view (FOV) and free access, has been applied for paddy rice mapping. Decision tree algorithms and spectral matching techniques were used to map rice-growing areas using temporal MODIS data [14,15]. A MODIS time-series analysis of spectral indices was found to be more useful for monitoring the phenological variations of paddy rice over a long period [16,17]. A paddy rice field is typically prepared by flooding a few days before the rice seedlings are transplanted. The wet growing season is regarded as a unique and significant characteristic of rice compared with other crops [18]. Thus, the flooding period is recognized as the best phase for rice identification. Spectral indices and bands that are sensitive to water and green vegetation are needed for monitoring the flooding and transplanting period of rice crops. These spectral indices are always calculated using two or more spectral bands to enhance the contrast between target and background and to reduce the effects of the atmosphere and solar illumination geometry. The Normalized

Difference Vegetation Index (NDVI), developed using the red and near-infrared bands, is correlated with the Leaf Area Index and chlorophyll content and has been widely used for crop yield estimations and the detection of changes in land use/cover [19,20]. The Enhanced Vegetation Index (EVI) was proposed because of the saturation of the NDVI in high-biomass regions to adjust for residual atmospheric contamination and background reflectance [21,22]. The infrared range is useful for estimating the water content of vegetation and in discriminating water from land. The Normalized Difference Water Index (NDWI) was developed using the reflectance in the near-infrared and green bands to enhance the detection of water features while eliminating soil and terrestrial vegetation features [23]. The modified NDWI (MNDWI) substitutes the near-infrared band with a middle-infrared band, such as Band 5 of Landsat TM, to efficiently enhance open-water signals and suppress or remove the signals from built-up land, as well as vegetation and soil [24]. However, because the reflectance of rice pixels during the flooding and transplanting stage is a mixture of water and vegetation, the sensitivity of the spectral index to flooding features should be further improved for rice mapping. The Land Surface Water Index (LSWI), which was formulated by combining the red and shortwave infrared channels of MODIS, has been used for the identification of rice pixels [25]. However, the threshold between the LSWI and the EVI was determined by considering local practices and rice cropping systems. Qiu *et al.* proposed a method for mapping rice planting areas by considering the vegetation phenology and surface water variations. The ratios of the changes in amplitude of the LSWI to the two-band Enhanced Vegetation Index 2 (EVI2) during the period from the tillering to the heading stage were used as one indicator to discriminate rice from non-rice fields [26]. Mosleh and Hassan developed a method for mapping "Boro" rice in Bangladesh using the MODIS-derived 16-day composite NDVI at a spatial resolution of 250 m [27]. The ISODATA clustering and the formulation of the mathematical model were the key procedures of this algorithm.

The objectives of the present study are to: (1) develop a Normalized Weighted Difference Water Index for identifying the flooding period of paddy rice fields; (2) map the early, single cropping and double cropping late rice distributions of the Yangtze River Delta region in the 2000–2010 period; and (3) validate the results using land use maps, Landsat ETM+ data and agricultural statistical data.

2. Study Area and Data

2.1. Study Area

The study area is the Yangtze River Delta region, which is one of the major rice-producing areas in China and spans three provinces (Figure 1). This region extends from 118°50′5″E to 134°46′26″E in longitude and from 38°43′15″N to

$53°33'39''N$ in latitude, with a territory of 2.1×10^5 km^2. The climate of the Yangtze River Delta is humid subtropical and is largely controlled by the East Asian monsoon [28]. Rice is the major food crop in the study area, with a high level of production and a wide distribution. The cropping system in Jiangsu Province and Shanghai City consists essentially of one crop of rice and another crop of winter wheat or oil rape, whereas single and double rice cropping systems are the two major planting patterns in Zhejiang Province.

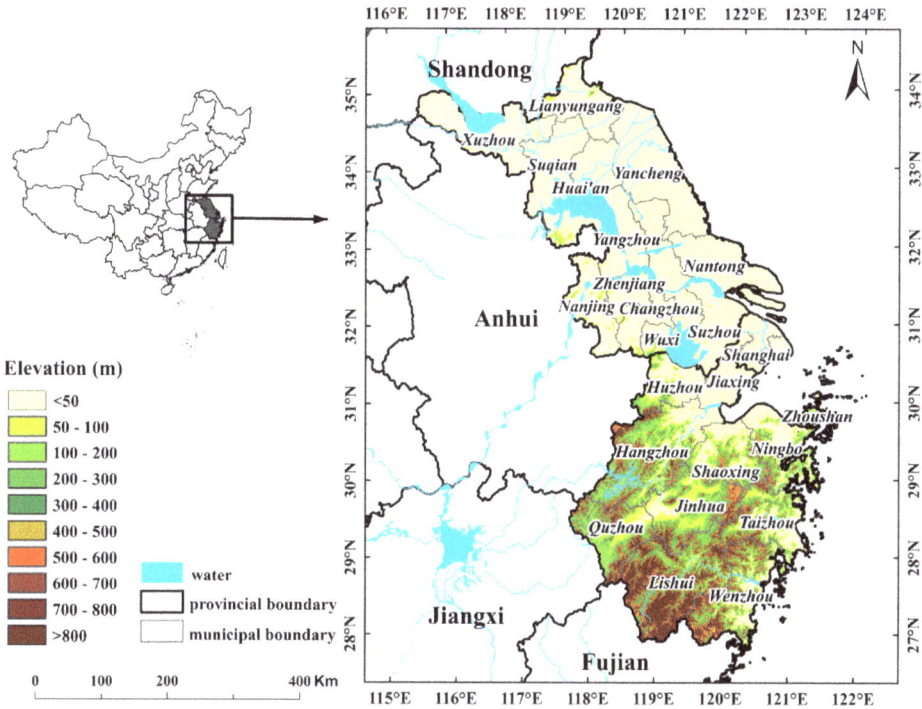

Figure 1. Location and ASTER Global Digital Elevation Model (GDEM) of the study area.

2.2. Data Acquisition

2.2.1. Field Data

Field experiments can yield accurate data under controlled conditions. A field experiment was conducted at the Experimental Farm of Zhejiang University, Hangzhou, Zhejiang Province, from June to October in 2004. Two rice cultivars (*i.e.*, Xieyou 9308 and Xiushui 110) were planted in 18 plots with three different nitrogen fertilization treatments: 0, 140 and 240 kg/ha. Each treatment was repeated three times. Rice seedlings were transplanted into the field on 8 July 2004, and the

185

canopy reached full closure in August. The rice canopy reflectance of each plot was acquired using an Analytical Spectral Devices (ASD) Field Spec Pro Full Range (350–2500 nm) spectroradiometer on 20 July, 8 August, 28 August, 22 September, 5 October and 27 October. At each plot, 10 reflectance measurements were acquired with a nadir view of $25°$ from a height of 1.0 m above the rice. The spectrum of each plot was recorded as the average of the 10 measurements.

2.2.2. Satellite Data

The MODIS sensor records data in 36 spectral bands and products at spatial resolutions of 250 m, 500 m and 1000 m. In this study, MODIS/Terra eight-day composite surface reflectance products (MOD09A1) were chosen for the mapping of rice planting regions.

The eight-day composite surface reflectance products were routinely processed for atmospheric and radiometric correction for the effects of aerosols and cirrus clouds, as well as to select the best observation and the lowest value in the blue band for each pixel over the eight-day period [29]. Three tiles (h27v05, h27v06 and h28v06) for the 2000–2010 period were acquired from the project website (https://lpdaac.usgs.gov/). The downloaded MODIS data were then mosaicked and reprojected to Albers equal-area conic projection using the MODIS Reprojection Tool (MRT).

A Landsat ETM+ image acquired on 13 May 2000 (path/row: 118/41) was downloaded from the International Scientific Data Service Platform (http://datamirror.csdb.cn/). The region spanned by the image covers the main rice-producing zones in Wenzhou City. According to the phenological data recorded at the local agricultural meteorological station, rice seedlings were generally transplanted into the fields in early May. Radiometric calibration was applied to the Landsat ETM+ image. The fast line-of-sight atmospheric analysis of spectral hypercubes (FLAASH) model was selected for atmospheric correction. The Landsat ETM+ image was resized to a 90 km × 90 km subset and reprojected to the Albers equal-area conic projection.

In addition, ASTER Global Digital Elevation Model (GDEM) data covering the study area were freely obtained from the Earth Remote Sensing Data Analysis Center of Japan (http://gdem.ersdac.jspacesystems.or.jp/).

2.2.3. Ancillary Data

A digital administrative map of China was obtained from the National Fundamental Geographic Information System. A land use map of Wenzhou City in 2005 was obtained from the Land and Resources Bureau of Wenzhou City. The annual sowing areas of paddy rice for each county in the study area during the 2000–2010 period were provided by the bureau of statistics.

3. Methodology

3.1. Spectral Characteristics of Rice during the Flooding and Transplanting Period

Canopy reflectance data collected in the field at a spectral resolution of 1 nm were used to simulate the reflectance in the first seven bands of the MODIS sensor (ρ_{MOD}) based on its spectral response function (Figure 2). The reflectance in the near-infrared and shortwave infrared wavelength bands was very low, whereas the reflectance in the visible bands (Bands 1, 3 and 4) was greater than in other growth periods. During the transplanting period, the water in the rice field was found to absorb most of the incident radiant flux, especially in the shortwave infrared region. It was also observed that the reflectance in Band 6 was lower than that in Band 4 (green band). With an increase in the tiller number and leaf area index, the reflectance in the visible bands decreased on 8 August 2004; however, the reflectance in the near-infrared and shortwave infrared bands increased significantly. The reflectance in Band 6 became higher than that in Band 4. In the previous literature, many water indices have been developed using the visible and infrared bands [23–25]. The green spectral range is highly sensitive to the Chl-a concentration over a wide range of variation and, thus, is helpful for the remote sensing of vegetation [30]. The near-infrared and shortwave infrared regions are the best wavelength regions for discriminating land from water. Because 1 of the 20 detectors in Terra MODIS Band 5 is noisy, there are stripes in the image. Band 6 was selected as the band sensitive to water, and Band 4 was used as the band sensitive to the presence of green seedlings.

Xu proposed the MNDWI (Equation (1)) to enhance the features of open water in remotely-sensed imagery [24]. Water pixels will have positive values of this index; however, pixels corresponding to flooded rice fields, built-up land and vegetation will have negative values. To enhance the features of rice pixels in the transplanting stage, we introduced a weight in the green band. Thus, we developed a Normalized Weighted Difference Water Index (NWDWI) based on the MNDWI. The ρ_{band4} and ρ_{band6} values described in Equation (2) denote the reflectances in MODIS Bands 4 and 6, respectively. A threshold of zero was applied to the NWDWI to separate flooded rice pixels from vegetation pixels. As shown in Equation (3), the values of the Ratio Vegetation Index (RVI) during different rice growth periods were calculated using the ρ_{MOD} values of Bands 6 and 4. Figure 3 shows that the RVI was the lowest during the transplanting stage and reached its peak at the heading stage, after which the RVI slowly deceased.

$$MNDWI = \frac{\rho_{green} - \rho_{MIR}}{\rho_{green} + \rho_{MIR}} \tag{1}$$

$$NWDWI = \frac{\rho_{band6} - a \cdot \rho_{band4}}{\rho_{band6} + a \cdot \rho_{band4}} \tag{2}$$

$$RVI = \frac{\rho_{band6}}{\rho_{band4}} \qquad\qquad (3)$$

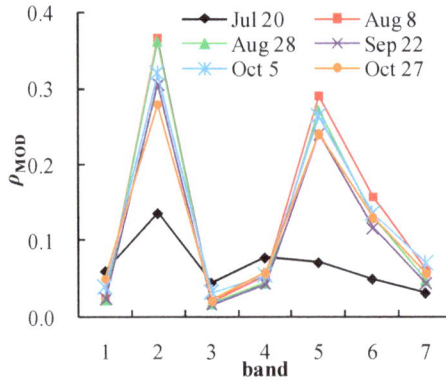

Figure 2. Simulated MODIS reflectances throughout the entire growth cycle of rice.

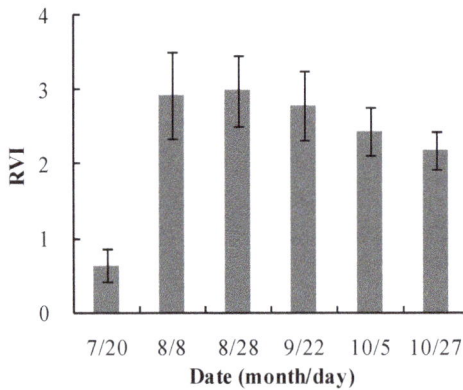

Figure 3. Statistical analysis of the Ratio Vegetation Index (RVI) throughout the entire growth cycle of a rice paddy in the field.

The acquisition date of the ETM+ image was consistent with the flooding and transplanting date for rice in Wenzhou, Zhejiang Province. Figure 4 is a false-color composite image of the ETM+ data. The paddy rice binary map was obtained from the ETM+ image using the maximum likelihood method and then degraded to the same resolution as the MODIS data using the pixel aggregate method. Because the overpass time of Landsat 7 is close to that of Terra, the aggregated rice map could be used as ground-truth data for validation. The MOD09A1 data (day of year: 2000129) were used to validate the performance of the NWDWI. When $a = 1$, flooded rice field and vegetation pixels both had positive NWDWI values. When $a = 1.5$, 54.8% of the rice pixels had negative values, and 21.2% of the negative pixels in the

188

MODIS-derived NWDWI image were labeled as rice pixels in the reference map. When a = 2, approximately 91.4% of the rice pixels had negative NWDWI values, and 24.7% of the negative pixels in the MODIS-derived NWDWI image were labeled as rice pixels in the reference map. The results of a simple density slice classification for the NWDWI (a = 2) demonstrated its ability to discriminate water pixels (Figure 5). In the ocean areas, the pixels had the lowest NWDWI values, whereas in the non-water areas, the pixels had positive values. In a comparison with the rice map derived from the ETM+ image, the omitted rice pixels were mainly distributed at the edges of the rice fields because of the mixed-pixel phenomenon and the uncertainty of edge pixels near large tracts of rice fields. When a = 2.5, approximately 98.5% of the rice pixels had negative NWDWI values. However, there were only 18.1% of pixels with negative values in the NWDWI image labeled as rice pixels in the reference map. Many of the forest, shrub and bare land pixels had negative value, as well as flooded rice pixels. Figure 3 also shows that the RVI was greater than 2 from the tillering stage to the harvest stage. Therefore, when a = 2, NWDWI $\leqslant$0 can be used to identify possible flooded rice pixels. Figure 5 also indicates that built-up pixels and pixels corresponding to natural water bodies also had low NWDWI values, which should allow them to be distinguished from rice pixels.

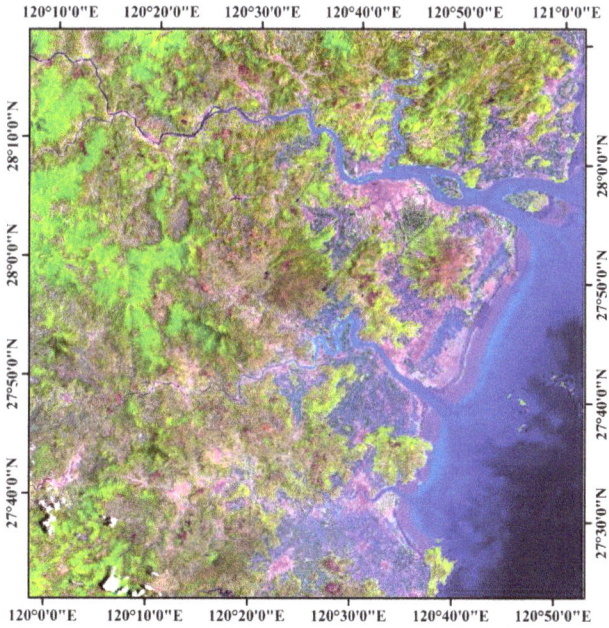

Figure 4. Color composite image of ETM+ data at the test site (R: Band 5; G: Band 4; B: Band 3).

189

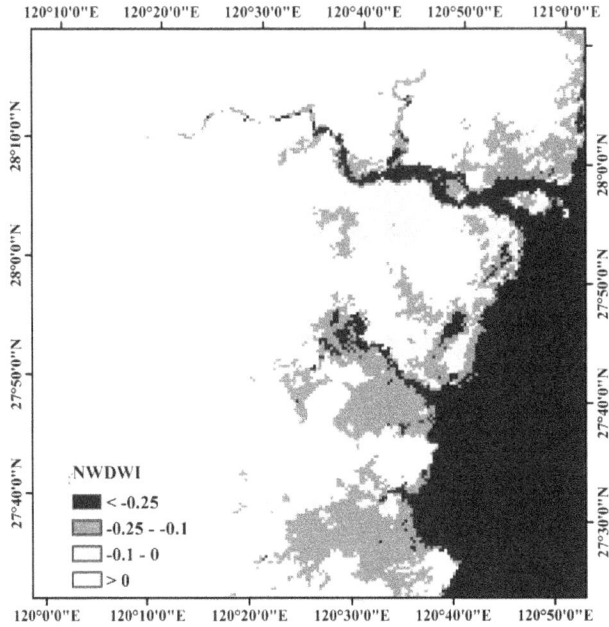

Figure 5. Spatial Normalized Weighted Difference Water Index (NWDWI) distribution of the MODIS image.

3.2. Reconstruction of the Spectral Index Profile

Although the 8-day composite surface reflectance data were routinely processed, some pixels were still affected by clouds. The pixels with Band 3 reflectances of greater than 10% were labeled as cloud pixels and removed as abnormal data [16]. Cloud masks for each MOD09A1 image were generated individually. Cloud-free data are important requirements for the operational monitoring of rice distributions using optical sensors [31]. To fill in the gaps in the EVI and NWDWI time series caused by clouds, the conditional temporal interpolation method was used in this study [32]. Compared with wavelet analysis and the Savitzky–Golay filter, the advantage of this method is its ability to retain the values of good pixels and repair the bad ones using valid pixels in the previous and subsequent images. If a pixel was contaminated in all three adjacent images, it was removed for further analysis.

3.3. Algorithm for Mapping Rice Planting Areas Using Time Series MODIS Data

The Yangtze River Delta region can be separated into two zones: one is the single rice planting area, including Jiangsu and Shanghai, and the other is Zhejiang Province [33]. In hilly regions, the elevation and slope are considered to be two important geographical factors for improving the accuracy and stability of classification in rice mapping [34]. In the study area, rice is grown in regions with elevations of less than 800 m and slopes of less than $10°$; thus, the elevation and slope were used to exclude non-paddy rice regions.

According to the rice growth calendar for the period of 2000–2010 collected from agrometeorological stations, the transplanting stages for early rice, single cropping rice and double cropping late rice occurred in late April–early May, mid-June–early July and mid-to-late July, respectively. The MODIS data corresponding to the transplanting periods were used to identify flooded rice pixels using the NWDWI to reduce the interference of other wetland plants or crops with short-term precipitation.

The time series of spectral indices are essential for analyzing the annual variability of vegetation activityFigure 6 shows the seasonal EVI and NWDWI profiles of various types of land cover in 2005. Figure 6a–c shows the seasonal spectral index profiles of three major cropping systems in the Yangtze River Delta region. It is obvious that the NWDWI decreased significantly when a pixel was labeled as a cloud pixel. The time series of the EVI and NWDWI revealed the growth stages of the crops. When the rice seedlings were transplanted into the field, the NWDWI was less than zero, because the reflectance of the rice pixels was dominated by water. However, the seasonal NWDWIs of rain-fed crops maintained consistently positive values. Forests in the study area exhibited high EVI and NWDWI values throughout the entire growth period (Figure 6e). Natural water and built-up pixels could be distinguished by their long periods of consistently low NWDWI and EVI values. In the study area, 40 single cropping rice samples and 35 double cropping rice samples were selected to perform a decision tree classification algorithm for mapping the planting areas of early, single cropping and double cropping late rice. For early rice, a pixel with a negative NWDWI and EVI <0.26 was labeled as a potential rice pixel during the transplanting stage. According to the characteristics of the growth period of early rice, the maximum EVI throughout the entire growth period was greater than 0.35, and the EVI decreased below 0.35 in the eleventh 8-day period after that identified as the transplanting stage. Because of the longer growth period of single cropping rice, the EVI decreased to less than 0.35 in the fifteenth 8-day period after the rice seedlings were transplanted. Late rice was transplanted to the same field after the early rice was harvested. If a pixel had NWDWI <0.05 and EVI <0.35, it was recognized as a transplanted rice pixel. The maximum EVI throughout the entire growth period was higher than 0.35, and in the twelfth 8-day period after the transplanting stage, the EVI decreased below 0.35. Natural water bodies and built-up pixels could be excluded by removing pixels with an EVI that was less than 0.3 for fourteen consecutive 8-day periods between April and September. Figure 7 shows the decision tree for the mapping of single cropping, early and double cropping late rice planting areas. Finally, the spatial distribution maps of early, single cropping and double cropping late rice were generated.

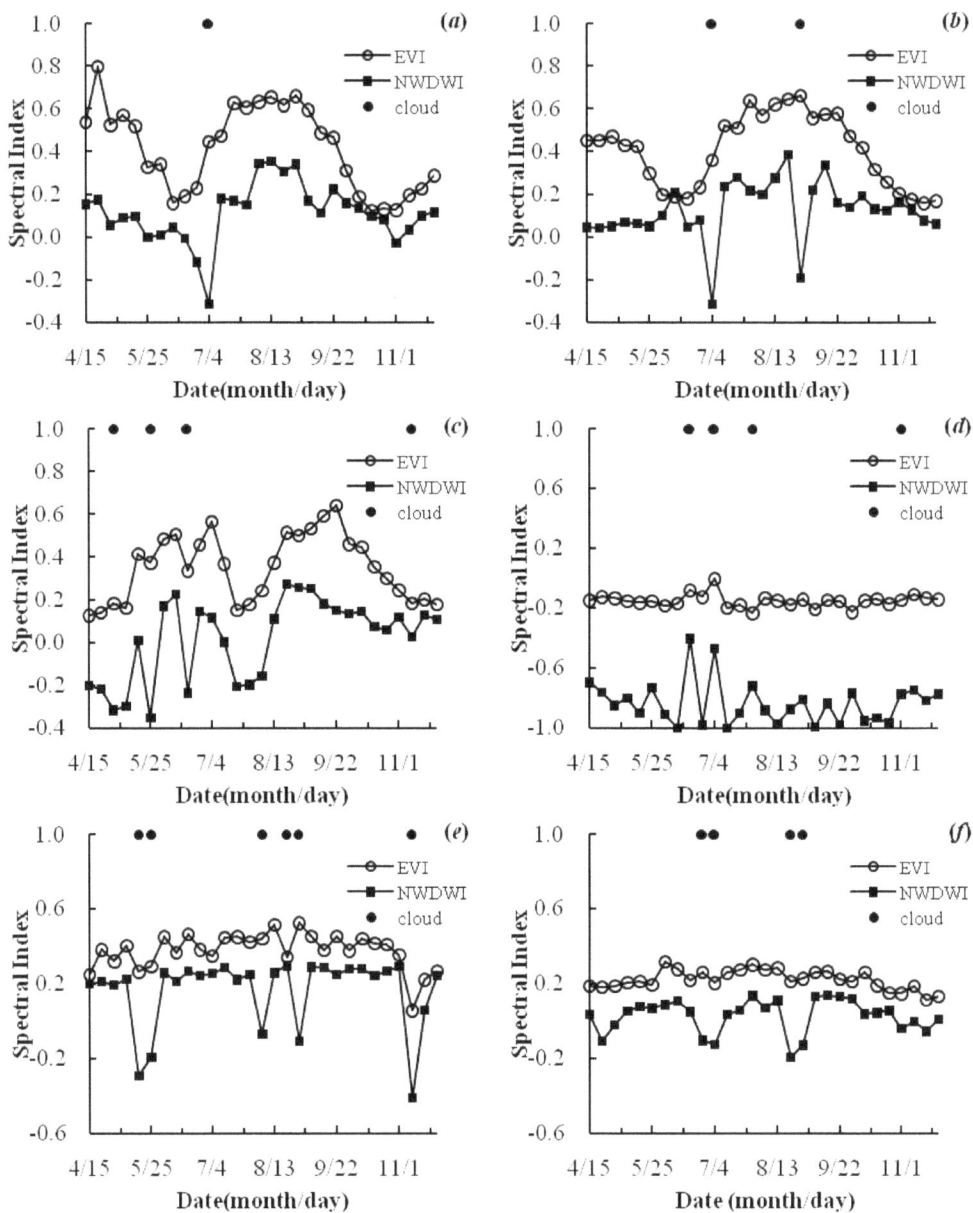

Figure 6. Spectral index time series corresponding to various types of land cover at the test sites in 2005: (**a**) winter wheat and single cropping rice, (**b**) winter wheat and rain-fed crops, (**c**) early and late rice, (**d**) lakes, (**e**) forests and (**f**) built-up regions.

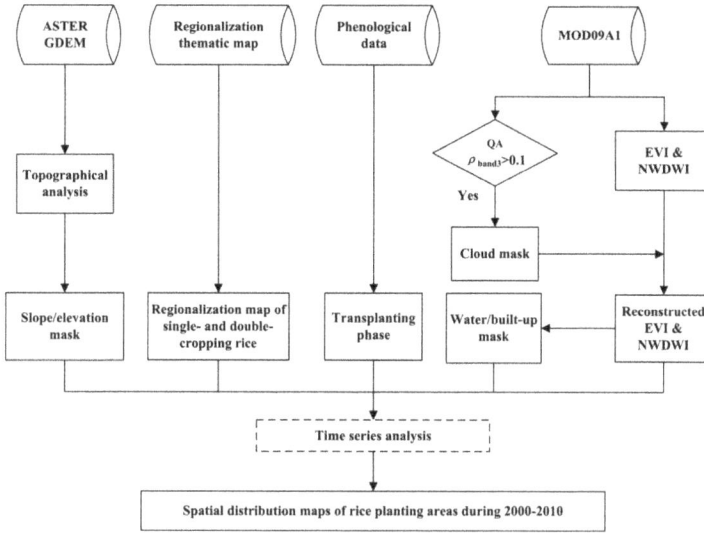

Figure 7. Flow chart for the extraction of rice planting regions.

3.4. Accuracy Assessment

The classification results for the Landsat ETM+ imagery at the studied site were used as the reference rice map for validation. An error confusion matrix was applied to evaluate the agreement of the MODIS-derived rice map with the reference rice map. The commission error, omission error, user's accuracy and producer's accuracy were calculated as follows:

$$\text{Commission error (\%)} = \frac{N_{\text{commit}}}{N_{\text{MODIS}}} \times 100\% \tag{4}$$

$$\text{User's accuracy (\%)} = 100 - \text{commission error} \tag{5}$$

$$\text{Omission error (\%)} = \frac{N_{\text{omit}}}{N_{\text{ETM}}} \times 100\% \tag{6}$$

$$\text{Producer's accuracy (\%)} = 100 - \text{omission error} \tag{7}$$

Here, N_{commit} and N_{omit} represent the numbers of committed and omitted rice pixels, respectively, in the MODIS-derived result, and N_{MODIS} and N_{ETM} represent the numbers of rice pixels in the MODIS-derived map and the aggregated reference rice map, respectively.

The error matrix was analyzed at the pixel level. Furthermore, because of the edge effects originating from the spatial aggregation of the ETM+ data and the geometric mismatch between the ETM+ and MODIS data, the error matrix was calculated using a 3×3 moving window [35].

4. Results and Discussion

4.1. Spatial and Temporal Distribution of Rice Planting Areas in the Yangtze River Delta Region

The spatial distributions of early, single cropping and double cropping late rice planting areas in the Yangtze River Delta region from 2000 to 2010 were generated using the presented algorithm, and the results are shown in Figures 8–10. As shown in Figure 8, single cropping rice was mainly distributed in Jiangsu, Shanghai, Hangzhou-Jiaxing-Huzhou plain, Jinhua-Quzhou basin, Ningbo-Shaoxing plain and the coastal plain of southeastern Zhejiang Province. Double cropping rice was mainly distributed in the Jinhua-Quzhou basin, Ningbo-Shaoxing plain and coastal plain of southeastern Zhejiang Province (Figures 9 and 10). The complexity of the terrain posed a considerable challenge in the extraction of scattered rice fields using MODIS 500-m data because of the mixed-pixel phenomenon. The rice fields were scattered throughout hilly regions, and most of them were distributed along rivers or in terraced planting regions.

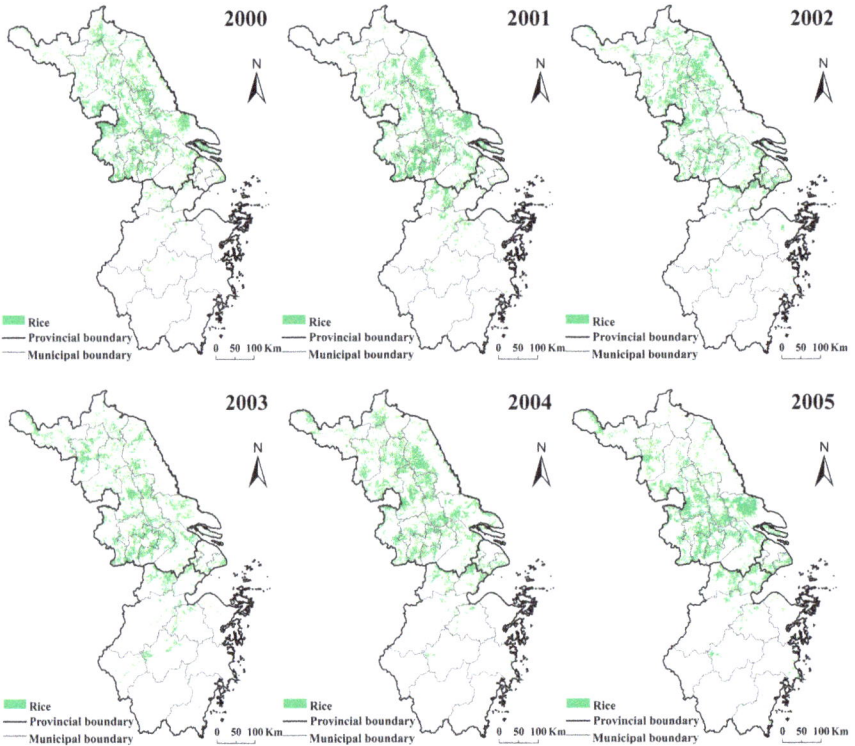

Figure 8. *Cont.*

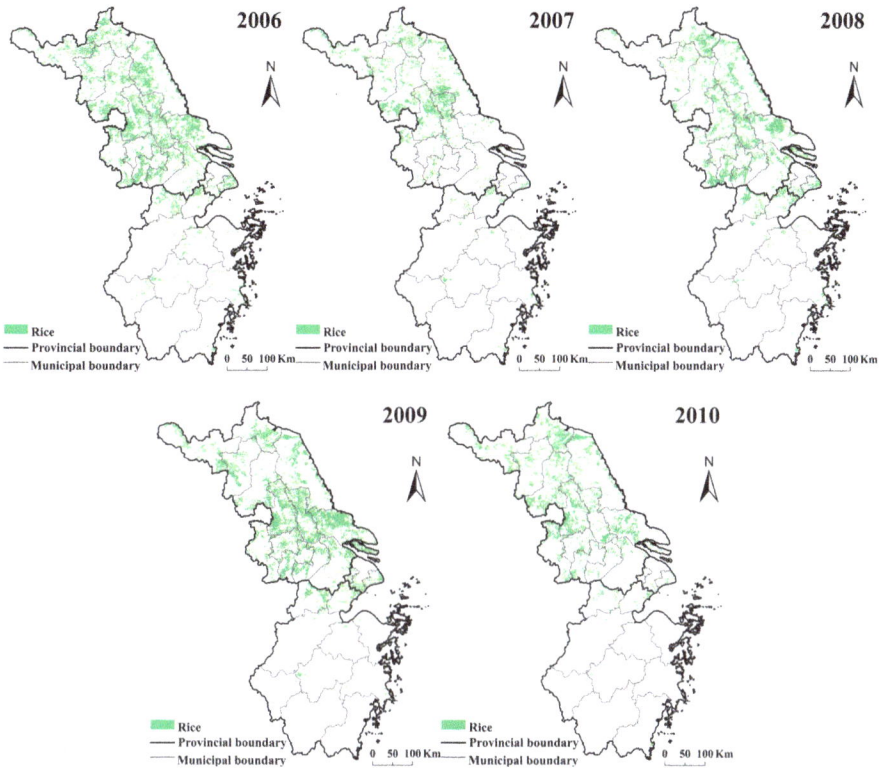

Figure 8. Spatio-temporal distribution of single cropping rice in the Yangtze River Delta region during the period of 2000–2010.

The total planting area of early rice decreased from one year to the next from 2000 to 2003 and then remained stable afterward. Single cropping rice began to be planted instead of double cropping rice in some areas of the Yangtze River Delta region. Figure 10 shows that the early rice area decreased significantly in Ningbo-Shaoxing plain and Jinhua-Quzhou basin. The results reveal the change in the cropping systems used in the Yangtze River Delta region over the studied decade.

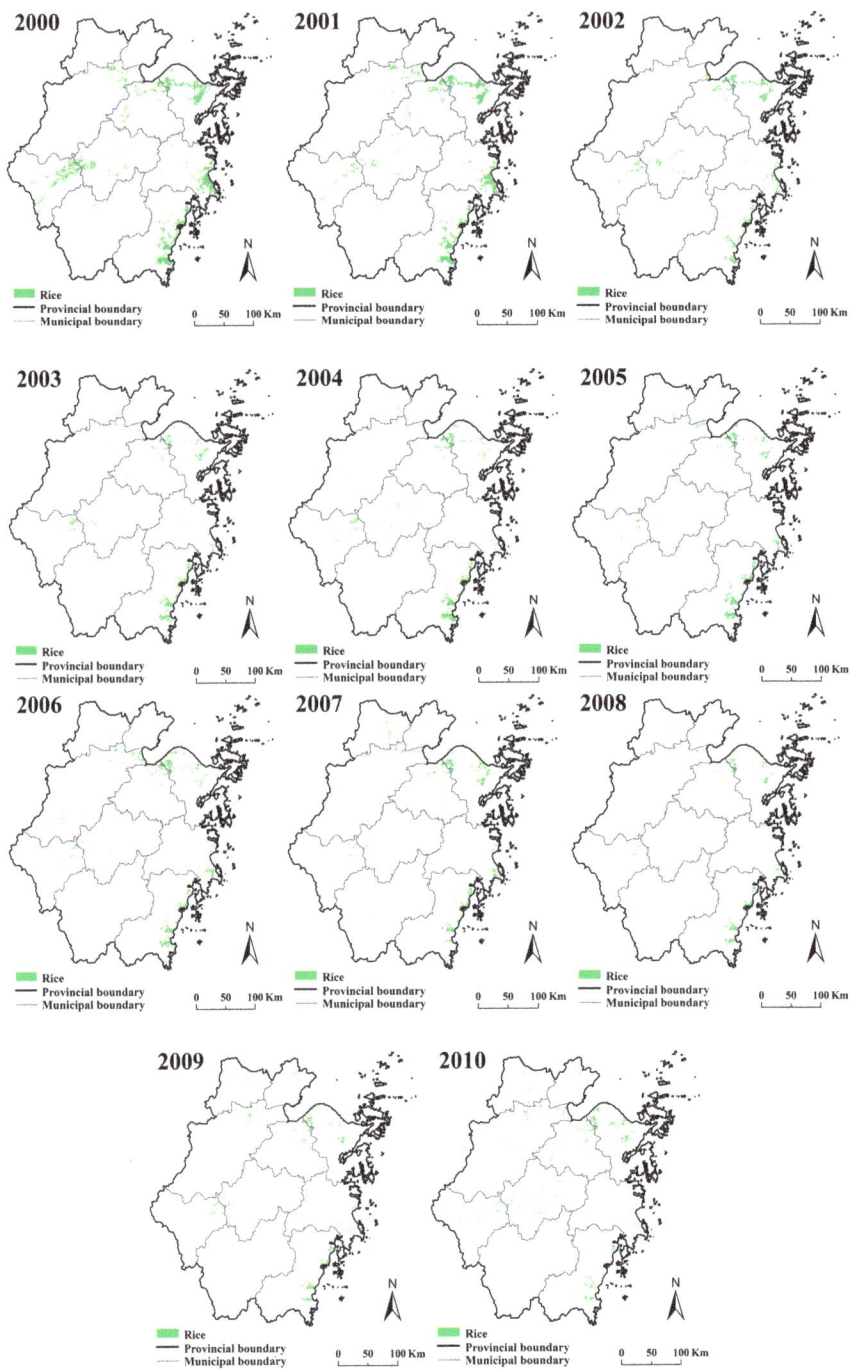

Figure 9. Spatio-temporal distribution of early rice in the Yangtze River Delta region during the period of 2000–2010.

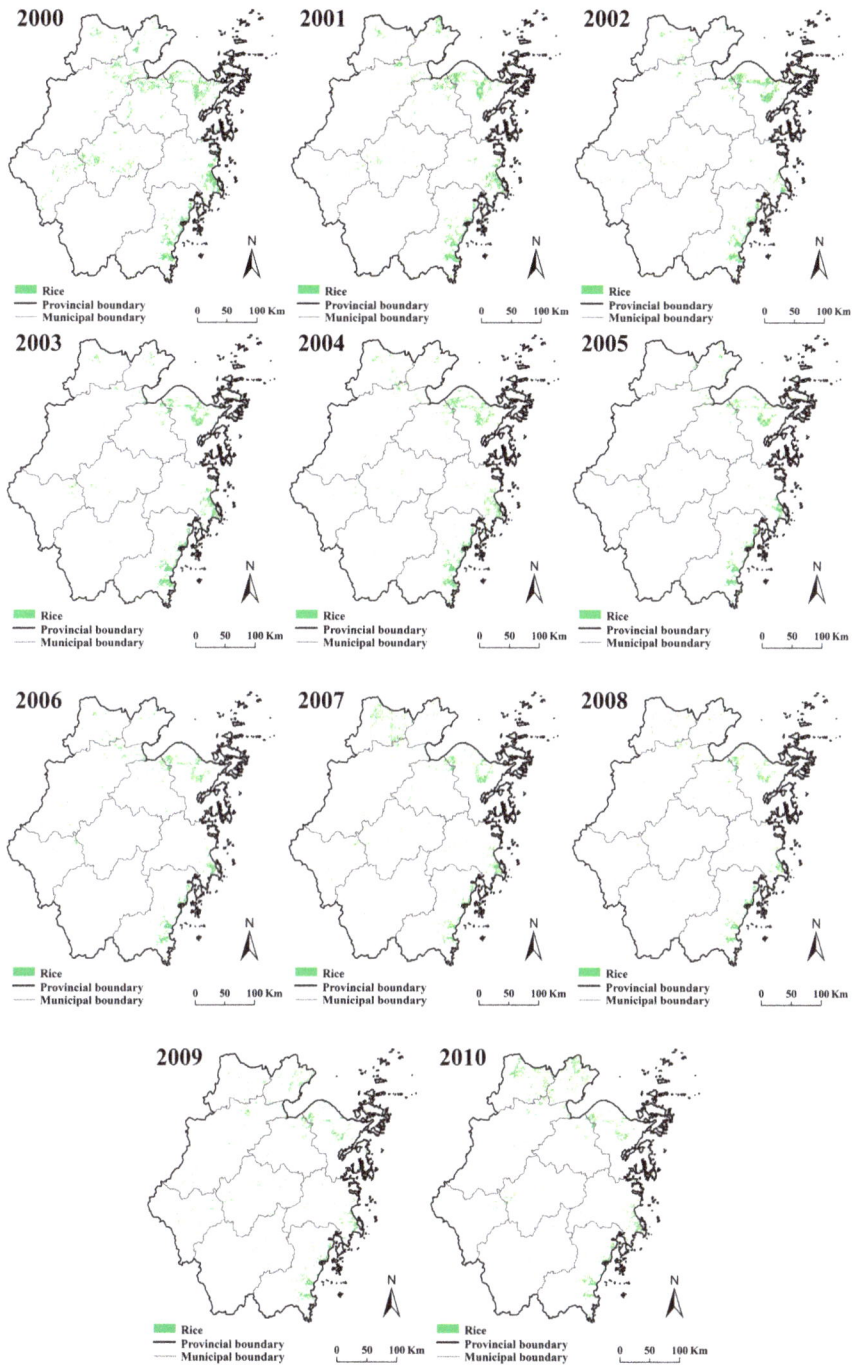

Figure 10. Spatio-temporal distribution of double cropping late rice in the Yangtze River Delta region during the period of 2000–2010.

4.2. Comparison of the Estimated Paddy Rice Planting Areas with Agricultural Census Data

It is a time-consuming and labor-intensive task to implement a large-scale regional survey of rice planting region and to obtain an annual spatial map of the study region. Agricultural census data were used as reference data to test the accuracy and stability of our algorithm. Table 1 presents a comparison of the annual total paddy planting areas derived from MODIS and agricultural census data in the Yangtze River Delta region. The absolute errors of the extracted annual total rice areas were less than 15%, except for 2007 and 2010. The MODIS-derived single cropping rice areas were underestimated in 2007 and 2010. The relative error was highest in 2007. The rice planting areas in this year were severely underestimated, especially those in the south of Jiangsu Province, Shanghai City and the north of Zhejiang Province. In 2010, underestimation mainly occurred in northern Nantong City, Taizhou, western Wuxi, northeastern Suzhou and central-southern Zhejiang. In these regions, the cloud occurrence frequency was greater than 60% during the transplanting stage of single cropping rice. Cloud cover during the rainy season may obscure optical observations. The existence of clouds and cloud shadows can result in abnormal changes in the spectral index. Continuous cloud contamination during the transplanting period was the major cause of the underestimation. Although eight-day composite surface reflectance products were generated by selecting the date within the eight-day window with the clearest atmospheric conditions for each pixel, the effects of cloud contamination cannot be neglected. In this study, the conditional temporal interpolation method was applied to reconstruct invalid pixels contaminated by clouds, but if three consecutive eight-day composite data points were all invalid during the flooding and transplanting period of the rice crop, that pixel was eliminated from further analysis. Radar images are a potential alternative means of rice mapping in these regions, especially during the rainy season, because they are independent on the time of day and unimpaired by weather conditions.

Furthermore, county-level validation of the rice planting area extraction results was performed. Because of the different standards for agricultural census data collected at the county level, single cropping rice and double cropping late rice were combined and treated simply as late rice for the comparison. The comparison results for early and late rice are shown in Figure 11. The solid line in the plot is the 1:1 line. The points in the plot are clustered near the 1:1 line, indicating that the MODIS-derived area of early rice is well correlated with the agricultural census data at the county level. The decision coefficients (R^2) for early rice and late rice are 0.560 and 0.619, respectively. The MODIS-derived area of late rice demonstrates a higher consistency with the census data during the 2000–2010 period, and the extracted early-rice area exhibits greater bias than that of late rice.

Table 1. Comparison of rice planting areas derived using the MODIS algorithm and from the agricultural census data.

Year	Census Data (kha)	* Rice$_{MOD}$ (kha)	Relative Error (%)
2000	3928.50	3882.06	−11.8
2001	3476.30	3753.93	7.99
2002	3239.83	3472.60	7.18
2003	2842.23	3161.56	11.24
2004	3285.70	3576.88	8.86
2005	3409.26	3609.66	5.88
2006	3439.45	3638.51	5.79
2007	3338.49	2140.38	−35.89
2008	3359.37	2893.90	−13.86
2009	3335.40	3546.10	6.32
2010	3309.10	2636.50	−20.33

* Rice$_{MOD}$ denotes the rice planting area derived from MODIS.

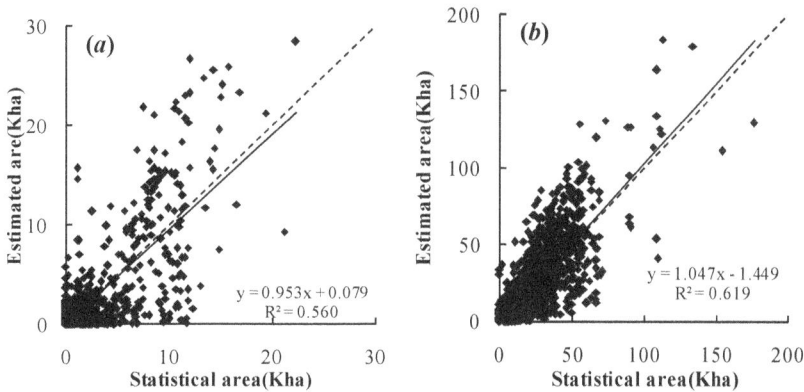

Figure 11. Correlation between areas of (**a**) early rice and (**b**) late rice derived using the MODIS algorithm and from the census data at the county level for the 2000–2010 period.

The topography in Zhejiang Province is very complicated, including plains, hills and mountains. Xu and Wang studied regionalization for rice yield estimation in Zhejiang Province by considering the local rice cropping systems, agroclimates, landforms, surface feature structures and rice yield levels. The county borders were treated as the region boundaries in the regionalization [36]. According to the regionalization map, the MODIS-derived early-rice area was close to that indicated by the census data in counties dominated by plains, but a large error was still observed in counties that grew less rice (Figure 12). The MODIS-derived late-rice area in counties dominated by plains was very close to that indicated by the census data. The Jinhua-Quzhou basin, located in central Zhejiang Province, is the major rice-producing region in the study area. The planting areas of early and late rice derived from the MODIS data were underestimated in the counties located in the

Jinhua-Quzhou basin region (Figure 13). The results were unsatisfactory because of the influence of the terrain on the land surface reflectance. In counties located in mountainous and hilly regions, the MODIS-derived areas of early and late rice were underestimated to different extents (Figure 14). The rice planting areas derived using the MODIS algorithm were severely underestimated in counties located in mountainous and hilly regions, where the rice fields were typically fragmentary and smaller than a MODIS pixel. Because the spatial resolution of the MODIS data used in the study was 500 m, it was unfeasible to recognize a pixel with a low abundance of rice as a rice pixel. The rice fields were not successfully identified in regions with complicated topographies. However, the MODIS-derived results are still useful for developing large-scale, timely and relatively accurate spatial datasets of paddy rice fields, especially in plain regions, and for providing vital information for yield estimation, growth monitoring, water management and greenhouse gas emission estimation.

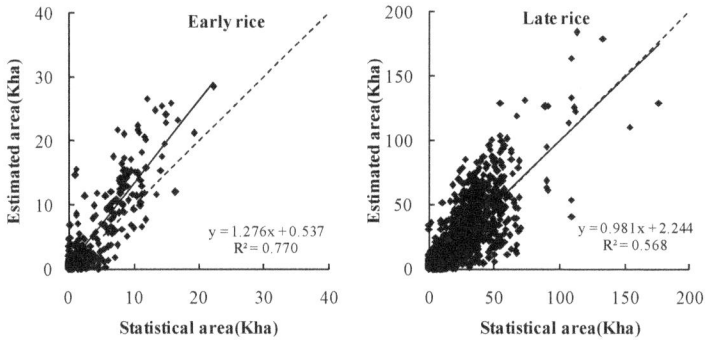

Figure 12. Comparison of MODIS-derived areas with census data in counties dominated by plains in the 2000–2010 period.

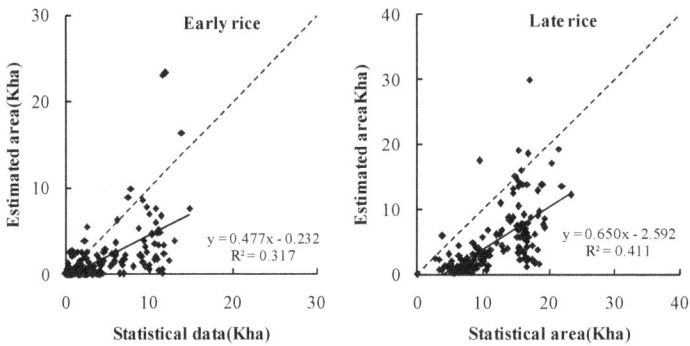

Figure 13. Comparison of MODIS-derived areas with census data in counties located in the Jinhua-Quzhou basin in the 2000–2010 period.

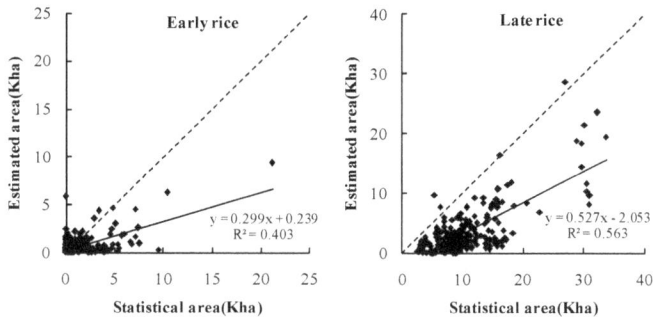

Figure 14. Comparison of MODIS-derived areas with census data in counties located in mountainous and hilly regions in the 2000–2010 period.

4.3. Spatial Comparison of Extracted Paddy Rice Planting Areas

In addition to the accuracy of the total estimated area, the spatial matching of the MODIS-derived results is also very important for practical applications. Table 2 summarizes the accuracy assessment of the MODIS-derived rice results. At the pixel level, the commission and omission errors were 26.30% and 22.67%, respectively. The user's and producer's accuracies were 73.70% and 77.33%, respectively. It is difficult to co-register Landsat ETM+ imagery with MODIS data because of the extremely large difference in spatial resolution between the two datasets. The pixels at the edges of discriminated rice pixels can give rise to considerable error when the accuracy validation is conducted for individual pixels.

Table 2. Accuracy assessment of MODIS-derived results at the studied sites.

Level	Commission Error (%)	User's Accuracy (%)	Omission Error (%)	Producer's Accuracy (%)
Pixel level	26.30	73.70	22.67	77.33
3 × 3 window	3.23	96.77	0.04	99.96

Therefore, the commission and omission errors for moving windows of 3 × 3 pixels were also calculated. If a pixel were identified as a rice pixel in the MODIS-derived result, but the eight pixels surrounding it were all labeled as non-rice pixels in the aggregated reference rice map, we considered it to be a committed pixel. If a pixel were labeled as a rice pixel in the aggregated rice map, but no pixels in the moving 3 × 3 pixel window surrounding it were identified as rice in the MODIS-derived result, it was considered to be an omitted pixel. In this analysis, the user's and producer's accuracies were found to be 96.77% and 99.96%, respectively.

Although an irrigated paddy indicated in the land use map may be used not only for planting paddy rice, but also for planting aquatic plants, such as reeds and lotus roots, any MODIS-derived rice region should be located in an irrigated paddy.

201

Therefore, a pixel was considered to be misclassified if it was a MODIS-derived rice pixel located in a non-irrigated paddy region. The irrigated paddy regions in the land use map of Wenzhou City for 2005 were extracted for the validation of the spatial matching of the results derived using the MODIS algorithm. In Figure 15a, it is seen that the majority of irrigated paddies were concentrated in the eastern coastal region, especially in Yueqing, Rui'an, Pingyang and Cangnan, and that the total area of irrigated paddies in Wenzhou City in 2005 was 109.7 kha. The two maps shown in Figure 15 were overlaid to examine the agreement between them. The number of pixels extracted using the MODIS algorithm was 4309, 88.95% of which were located in irrigated paddy regions. As indicated by the spatial matching analysis at the county level, the accuracy was lowest in counties with less than 5 kha (Table 3).

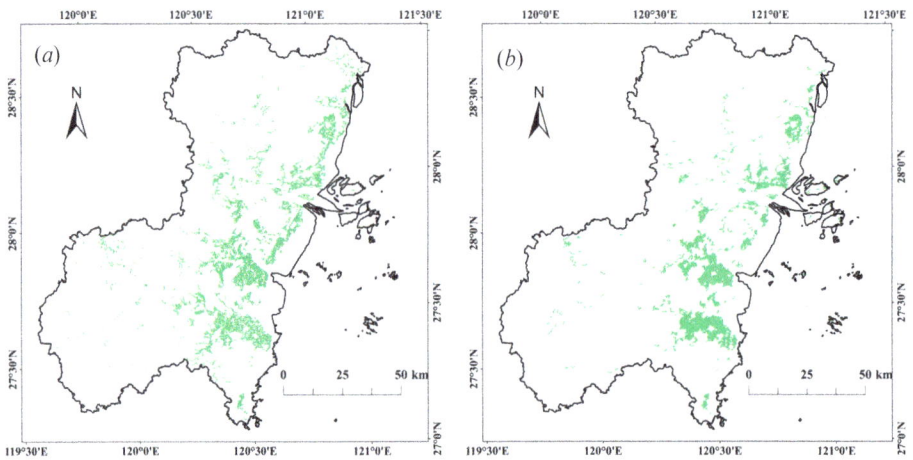

Figure 15. Irrigated paddies in the land use map (**a**) and the rice distribution map derived from MODIS (**b**) for Wenzhou City in 2005.

Table 3. Consistency analysis of MODIS-derived rice pixels with irrigated paddies in Wenzhou City, 2005.

City	Number of Rice Pixels Located in Irrigated Paddies	Number of Rice$_{MOD}$ *	Overlap Proportion (%)	Area of Irrigated Paddies (kha)	Area of Rice$_{MOD}$ (km^2)
Yueqing	832	888	93.69	21.38	19.06
Pingyang	586	625	93.76	16.50	13.42
Wencheng	55	93	59.14	3.51	2.00
Yongjia	178	238	74.79	7.99	5.11
Taishun	5	53	9.43	0.66	1.14
Dongtou	3	52	5.77	0.06	1.12
Wenzhou	371	440	84.32	12.44	9.44
Rui'an	810	878	92.26	25.08	18.85
Cangnan	993	1042	95.30	22.10	22.37
Total	3833	4309	88.95	109.70	92.50

* Rice$_{MOD}$ denotes the rice pixels derived from MODIS.

202

5. Conclusions

In this study, the NWDWI was proposed to enhance the signal of flooding regions in remotely-sensed images. The algorithm for the identification of flooded pixels was evaluated based on spectral data measured in the field, as well as ETM+ and MODIS data. Built-up regions and natural watersheds could be readily separated using the NWDWI and EVI. The spatial distribution maps of rice with different cropping patterns (*i.e.*, early rice, single cropping rice and double cropping late rice) in the period of 2000–2010 were generated using a decision tree classification algorithm. The accuracy of the extracted annual total rice area was greater than 85%, except for 2007 and 2010, for which it was poorer, because of the large areas of cloud masking during the transplanting period. The identified rice areas were also validated at the county level. The MODIS-derived area of late rice demonstrated a higher consistency with the census data during the period of 2000–2010. The user's and producer's accuracies for moving windows of 3×3 pixels were both greater than 95%. The algorithm also revealed the interannual variations in single and double cropping rice in the Yangtze River Delta region.

However, there were several factors that may have affected the accuracy of the results, such as cloud contamination, spatial resolution and topography. The accuracy was not satisfactory in counties with complex terrain. The value of a in the formula for the NWDWI used in this study was determined by field experimental data and the ETM+ image and was appropriate for discriminating rice pixels at a spatial resolution of 500 m during the transplanting stage. The higher value of a in NWDWI could lead to misclassifying numbers of non-rice pixels as rice pixels. Otherwise, rice pixels could be omitted by using a lower value of a in NWDWI. The developed algorithm was found to be unsuitable for use in regions with continuous rainy weather.

Despite the uncertainties in this algorithm, MODIS data are a suitable choice for generating rice distribution maps at large scales, which are useful for long-term grain yield estimations and the detection of changes in land use/cover change. The application of Aqua/MODIS data in combination with Terra/MODIS data could improve the accuracy of our algorithm.

Acknowledgments: This study was supported by the Agricultural Project of Scientific and Technological Research of Shanghai, China (2011-2-11), National High Technology Research and Development Program of China (Grant No. 2012AA12A30703) and Scientific Research Foundation of Ningbo University of Technology. We are grateful for the data support provided by the Land Processes Distributed Active Archive Center (LPDAAC, https://lpdaac.usgs.gov/), the International Scientific Data Service Platform (http://datamirror.csdb.cn/) and the National Meteorological Bureau of China. We would like to thank the anonymous reviewers for their valuable suggestions and comments.

Author Contributions: Jingjing Shi and Jingfeng Huang conceived of and designed the research. Jingjing Shi analyzed the data and wrote the manuscript.

Conflicts of Interest: The authors declare no conflict of interest.

References

1. Khush, G.S. What it will take to feed 5.0 billion rice consumers in 2030? *Plant Mol. Biol.* **2005**, *59*, 1–6.
2. Food and Agriculture Organziation of the United Nations. *Statistical Yearbook 2013: World Food and Agriculture*; FAO: Rome, Italy, 2013.
3. Frolking, S.; Qiu, J.J.; Boles, S.; Xiao, X.M.; Liu, J.Y.; Zhuang, Y.H.; Li, C.S.; Qin, X.G. Combining remote sensing and ground census data to develop new maps of the distribution of rice agriculture in China. *Glob. Biogeochem. Cycles* **2002**, *16*, 31–38.
4. Huke, R.E.; Huke, E.H. *Rice Area by Type of Culture: South, Southeast, and East Asia. A Review and Updated Data Base*; IRRI: Los Baños, Philippines, 1997.
5. Leff, B.; Ramankutty, N.; Foley, J.A. Geographic distribution of major crops across the world. *Glob. Biogeochem. Cycles* **2004**, *18*, GB1009.
6. Liu, J.Y.; Liu, M.L.; Zhuang, D.F.; Zhang, Z.X.; Deng, X.Z. Study on spatial pattern of land-use change in China during 1995–2000. *Sci. China Ser. D: Earth Sci.* **2003**, *46*, 373–384.
7. McCloy, K.R.; Smith, F.R.; Robinson, M.R. Monitoring rice areas using Landsat MSS data. *Int. J. Remote Sens.* **1987**, *8*, 741–749.
8. Oguro, Y.; Suga, Y.; Takeuchi, S.; Ogawa, H.; Tsuchiya, K. Monitoring of a rice field using Landsat-5 TM and Landsat-7 ETM+ data. *Adv. Space Res.* **2003**, *32*, 2223–2228.
9. Okamoto, K.; Fukuhara, M. Estimation of paddy field area using the area ratio of categories in each mixel of Landsat TM. *Int. J. Remote Sens.* **1996**, *17*, 1735–1749.
10. Li, Y.Z.; Zeng, Y. Study on methods of rice planting area estimation at regional scale using NOAA/AVHRR data. *J. Remote Sens.* **1998**, *2*, 125–130. (In Chinese)
11. Andres, L.; Salas, W.A.; Skole, D. Fourier analysis of multi-temporal AVHRR data applied to a land cover classification. *Int. J. Remote Sens.* **1994**, *15*, 1115–1121.
12. Xiao, X.; Boles, S.; Frolking, S.; Salas, W.; Moore, B.; Li, C.; He, L.; Zhao, R. Observation of flooding and rice transplanting of paddy rice fields at the site to landscape scales in China using VEGETATION sensor data. *Int. J. Remote Sens.* **2002**, *23*, 3009–3022.
13. Quarmby, N.A. Towards continental scale crop area estimation. *Int. J. Remote Sens.* **1992**, *13*, 981–989.
14. Gumma, M.K.; Gauchan, D.; Nelson, A.; Pandey, S.; Rala, A. Temporal changes in rice-growing area and their impact on livelihood over a decade: A case study of Nepal. *Agric. Ecosyst. Environ.* **2011**, *142*, 382–392.
15. Gumma, M.K.; Nelson, A.; Thenkabail, P.S.; Singh, A.N. Mapping rice areas of south Asia using MODIS multitemporal data. *J. Appl. Remote Sens.* **2011**, *5*.
16. Sakamoto, T.; Yokozawa, M.; Toritani, H.; Shibayama, M.; Ishitsuka, N.; Ohno, H. A crop phenology detection method using time-series MODIS data. *Remote Sens. Environ.* **2005**, *96*, 366–374.
17. Zhang, X.Y.; Friedl, M.A.; Schaaf, C.B.; Strahler, A.H.; Hodges, J.C.F.; Gao, F.; Reed, B.C.; Huete, A. Monitoring vegetation phenology using MODIS. *Remote Sens. Environ.* **2003**, *84*, 471–475.

18. Neue, H.U. Methane emission from rice fields. *Bioscience* **1993**, *43*, 466–474.
19. Quarmby, N.A.; Milnes, M.; Hindle, T.L.; Silleos, N. The use of multi-temporal NDVI measurements from AVHRR data for crop yield estimation and prediction. *Int. J. Remote Sens.* **1993**, *14*, 199–210.
20. Lunetta, R.S.; Knight, J.F.; Ediriwickrema, J.; Lyon, J.G.; Worthy, L.D. Land-cover change detection using multi-temporal MODIS NDVI data. *Remote Sens. Environ.* **2006**, *105*, 142–154.
21. Huete, A.R.; Liu, H.Q.; Batchily, K.; van Leeuwen, W. A comparison of vegetation indices over a global set of TM images for EOS-MODIS. *Remote Sens. Environ.* **1997**, *59*, 440–451.
22. Huete, A.; Didan, K.; Miura, T.; Rodriguez, E.P.; Gao, X.; Ferreira, L.G. Overview of the radiometric and biophysical performance of the MODIS vegetation indices. *Remote Sens. Environ.* **2002**, *83*, 195–213.
23. McFeeters, S.K. The use of the Normalized Difference Water Index (NDWI) in the delineation of open water features. *Int. J. Remote Sens.* **1996**, *17*, 1425–1432.
24. Xu, H.Q. Modification of Normalised Difference Water Index (NDWI) to enhance open water features in remotely sensed imagery. *Int. J. Remote Sens.* **2006**, *27*, 3025–3033.
25. Xiao, X.M.; Boles, S.; Liu, J.Y.; Zhuang, D.F.; Frolking, S.; Li, C.S.; Salas, W.; Moore, B. Mapping paddy rice agriculture in southern China using multi-temporal MODIS images. *Remote Sens. Environ.* **2005**, *95*, 480–492.
26. Qiu, B.W.; Li, W.J.; Tang, Z.H.; Chen, C.C.; Qi, W. Mapping paddy rice areas based on vegetation phenology and surface moisture conditions. *Ecol. Indic.* **2015**, *56*, 79–86.
27. Mosleh, M.K.; Hassan, Q.K. Development of a remote sensing-based "Boro" rice mapping system. *Remote Sens.* **2014**, *6*, 1938–1953.
28. Yi, S.; Saito, Y.; Zhao, Q.H.; Wang, P.X. Vegetation and climate changes in the Changjiang (Yangtze river) delta, China, during the past 13,000 years inferred from pollen records. *Quat. Sci. Rev.* **2003**, *22*, 1501–1519.
29. Vermote, E.F.; Vermeulen, A. *Atmospheric Correction Algorithm: Spectral Reflectance (MOD09), MODIS Algorithm Technical Background Document, version 4.0*; University of Maryland: Maryland, MD, USA, 1999.
30. Gitelson, A.A.; Kaufman, Y.J.; Merzlyak, M.N. Use of a green channel in remote sensing of global vegetation from EOS-MODIS. *Remote Sens. Environ.* **1996**, *58*, 289–298.
31. Spruce, J.P.; Sader, S.; Ryan, R.E.; Smoot, J.; Kuper, P.; Ross, K.; Prados, D.; Russell, J.; Gasser, G.; McKellip, R.; *et al.* Assessment of MODIS NDVI time series data products for detecting forest defoliation by gypsy moth outbreaks. *Remote Sens. Environ.* **2011**, *115*, 427–437.
32. Groten, S. NDVI—Crop monitoring and early yield assessment of Burkina Faso. *Int. J. Remote Sens.* **1993**, *14*, 1495–1515.
33. Sun, H.S.; Huang, J.F.; Li, B.; Wang, H.S. Study on the regionalization of paddy rice information acquirement through remote sensing technology in China. *Sci. Agric. Sin.* **2008**, *41*, 4039–4047.

34. Cheng, Q.; Wang, R.C. Estimation of the rice planting area using digital elevation model and multitemporal moderate resolution imaging spectroradiometer. *Trans. Chin. Soc. Agric. Eng.* **2005**, *21*, 89–92. (In Chinese)

35. Zhan, X.; Defries, R.; Townshend, J.R.G.; Dimiceli, C.; Hansen, M.; Huang, C.; Sohlberg, R. The 250 m global land cover change product from the Moderate Resolution Imaging Spectroradiometer of NASA's Earth Observing System. *Int. J. Remote Sens.* **2000**, *21*, 1433–1460.

36. Xu, H.W.; Wang, K. Regionalization for rice yield estimation by remote sensing in Zhejiang Province. *Pedosphere* **2001**, *11*, 175–184.

Rice Fields Mapping in Fragmented Area Using Multi-Temporal HJ-1A/B CCD Images

Jing Wang, Jingfeng Huang, Kangyu Zhang, Xinxing Li, Bao She, Chuanwen Wei, Jian Gao and Xiaodong Song

Abstract: Rice is one of the most important crops in the world; meanwhile, the rice field is also an important contributor to greenhouse gas methane emission. Therefore, it is important to get an accurate estimation of rice acreage for both food production and climate change related studies. The eastern plain region is one of the major single-cropped rice (SCR) growing areas in China. Subjected to the topography and intensified human activities, the rice fields are generally fragmented and irregular. How remote sensing can meet this challenge to accurately estimate the acreage of the rice in this region using medium-resolution imagery is the topic of this study. In this study, the applicability of the Chinese HJ-1A/B satellites and a two-band enhanced vegetation index (EVI2) was investigated. Field campaigns were carried out during the rice growing season and ground-truth data were collected for classification accuracy assessments in 2012. A stepwise classification strategy utilizing the EVI2 signatures during key phenology stages, *i.e.*, the transplanting and the vegetative to reproductive transition phases, of the SCR was proposed, and the overall classification accuracy was 91.7%. The influence of the mixed pixel and boundary effects to classification accuracy was also investigated. This work demonstrates that the Chinese HJ-1A/B data are suitable data source to estimating SCR cropping area under complex land cover composition.

Reprinted from *Remote Sens.* Cite as: Wang, J.; Huang, J.; Zhang, K.; Li, X.; She, B.; Wei, C.; Gao, J.; Song, X. Rice Fields Mapping in Fragmented Area Using Multi-Temporal HJ-1A/B CCD Images. *Remote Sens.* **2015**, *7*, 3467–3488.

1. Introduction

Rice is one of the most important crops in the world and provides the main source of energy for more than half of the world population [1]. Additionally, the seasonally flooded rice fields contribute about 5%–19% of total global methane emission, an important greenhouse gas source, to the atmosphere [2,3]. China produced about one third of the world's rice on about one fifth of the world's paddy rice land [4]. During the past two decades, the arable land in China declined at a speed of 0.25 million hectares per year [5]. The trend was more obvious in the eastern plain region of China, where intensified human activities have changed the land use and land cover (LULC) patterns dramatically in the last decades. This region has long been one of the major rice growing areas in China, and the cultivar is dominated

207

by the single-cropped rice (SCR). From the food safety, ecological and policy making points of view, a timely and efficient monitoring and mapping of rice cropping area is critical [6,7]. Conventionally, the local government usually estimates the cropping area of rice by field survey; however, it is time-consuming and costly. As a powerful alternative, remote sensing has proved its effectiveness in estimating rice cropping areas from regional to global scales [8–10].

In the literature, many different kinds of optical remote sensing data, e.g., the Advanced Very High Resolution Radiometer (AVHRR), Moderate Resolution Imaging Spectroradiometer (MODIS), SPOT VEGETATION and Landsat-MSS, and techniques have been applied in rice cropping area estimating practices [7,8,11–13]. The data mentioned above have demonstrated advantages in rice monitoring at regional to global scales due to wide range of coverage and relative long data archiving. However, coarse resolution satellite data is not suitable for precise rice crop mapping in the eastern plain region of China because the rice fields in this region are relatively small, irregular, and fragmented by well-developed roads and dense water networks, and generally mixed with other land cover types. As a consequence, the mixed-pixel problem is prominent and induces temporal uncertainty in discriminating the spectral signatures of rice and the other land cover types [6].

Middle to high spatial resolution satellite data, e.g., Landsat TM/ETM+/OLI, SPOT and China Brazil Earth Resources Satellite (CBERS), are promising in capturing small patches of crop fields [14,15]. However, the cost and relatively long revisit cycles partially offset their advantages in spatial resolution. Specifically, the cloud cover during monsoon season, which is partially overlapped with the major growing season of the SCR, makes it more difficult to obtain qualified remote sensing imageries [16,17]. For applications where the rice phenology information is critically needed, the satellite data with acceptable spatial resolution and more frequent revisit cycle should be more desirable.

The small sun-synchronous satellites for environment and disaster monitoring and forecasting (HJ-1A/B) of China were launched in 2008. HJ-1A/B satellites have a spatial resolution of 30 m and a revisit cycle of four days (the revisit cycle of the constellation is 2 days), with imaging swath of 700 km. The CCD camera onboard HJ-1A/B includes four bands, *i.e.*, blue, green, red and near-infrared, and the spectral range is 0.43–0.90 μm. HJ-1A/B CCD data have been applied in rice area estimation [18,19] and yield prediction [20]. In this study, however, it is our interest to explore the potential of using HJ-1A/B data to extract small, irregular SCR growing area in the eastern plain region of China, where the mixed-pixel problem is serious as mentioned above. Specifically, it is our interest to take advantage of its high revisit feature of the HJ-1A/B data to capture the key phenology spectral signatures of the SCR to facilitate the classification.

The unique physical feature of rice fields and the phenology of the SCR may provide valuable information for remote sensing classification. The rice grows on flooded soils, and the rice fields are a mixture of rice plant and open water during the transplanting and early period of the growing season [21]. As new leaves and tillers emerged, there is an accelerated increase in canopy height and leaf area of the rice. About 50 to 60 days after transplanting, the rice canopy would cover most of the surface area [22], but the leaf area is still increased till the heading stage. After that, the leaf area of rice starts to decease and the leaf color turns to yellow until the ripening and harvest stages. By using time series remote sensing images, the combined field and phenology features of rice, which differentiate the rice field from the other land cover types, may increase the classification accuracy.

To minimize the interference of external environmental factors, various vegetation indices (VIs) are commonly used in practice [23–25]. For example, the well-recognized normalized difference vegetation index (NDVI) [26] has been testified to be closely correlated with leaf area, biomass, percent ground cover and crop productivity [27–30]. Due to the saturation effect, however, NDVI may fail to capture the difference in well-vegetated areas, compared with the enhanced vegetation index (EVI) [31]. In practice, the time series signatures of NDVI and EVI derived from the MODIS and SPOT data had been used to map the area, species (single, early, and late), and key phenologies of rice [13,18,32]. Recently, a novel VI, *i.e.*, the 2-band EVI (EVI2), has been proposed and testified to be comparable with the traditional EVI, and more importantly, it may achieve greater consistencies across sensors because only 2 bands are involved, as compared with 3 bands in EVI [33,34].

In addition to the data used, it is of critical importance to select appropriate classification method to properly mapping the rice fields. It is our interest to compare the classification efficiencies of the commonly used parametric and nonparametric classification algorithms, *i.e.*, the maximum likelihood classifier (MLC) and support vector machines (SVM), with a two-step classification method proposed in this study and specifically designed to classify the rice fields from the other land cover types. The MLC is one of the most commonly used classification techniques [35–37]. It is a parametric classification algorithm with the assumption that the class signatures are normally distributed. The SVM is a nonparametric classifier, which projects the training data in the input space into a high dimensional space using a kernel function where the classes are linearly separable [38]. The SVMs have no limitation about the probability distribution forms of the class signature, but its performance largely depends on the kernel used, the parameter choice for the specific kernel, and the method used to generate the SVM [39–41].

Our study aimed to investigate the capability of EVI2 in SCR growth monitoring, and to test the feasibility of using HJ-1A/B CCD data to estimate the SCR growing area in the eastern plain region of China. For this purpose, we proposed a simple but effective classification method, which makes use of the time series HJ-1A/B imageries and the specific signatures of EVI2 (including its 1st derivative) at key phenology stages of the SCR. An extensive field campaign was carried out for verification simultaneously. We compared the effectiveness of this method with the parametric and nonparametric classification algorithms, namely MLC and SVM. We also discussed the influence of the mixed-pixel which was typical in the study area and may affect the classification accuracy.

2. Data and Methods

2.1. Study Area

Deqing County lies in the west of Hangjiahu Plain, with mean annual temperature ranging between 13 °C and 16 °C and annual precipitation of 1379 mm (Figure 1). The plain areas mainly distribute at the eastern Deqing, with the altitudes ranging from 4 m along the Beijing-Hangzhou Grand Canal to 721 m on the Tianmu Mountains. Deqing County is part of the SCR growing region in the water network area of north Zhejiang [42], where countless lakes, ponds and winding rivers scattered throughout this region, with the addition of well-developed road networks, leading to fragmented patches of irregular crop land plots. Deqing has a total area of 936 km^2, and the SCR area in Deqing accounts for more than 91% of the major crop areas according to the statistical data of local agriculture department. The SCR fields mainly concentrate in the eastern regions of Deqing with average elevation less than 20 m.

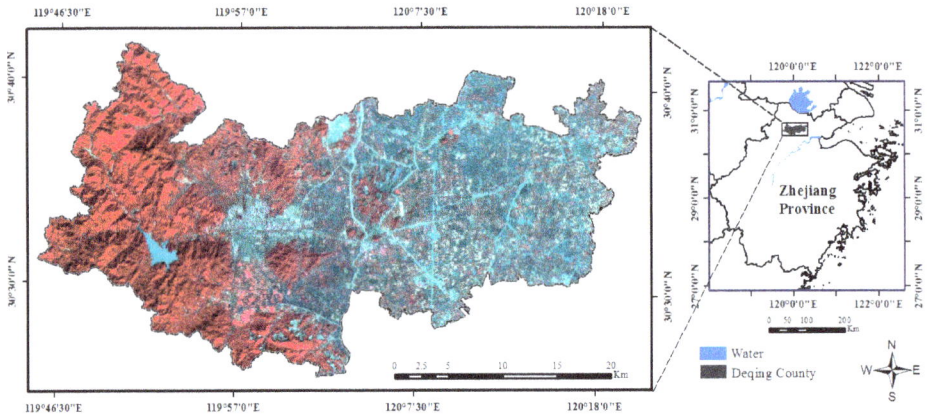

Figure 1. HJ-1A CCD false-color composite image of the study area on 19 November 2012.

2.2. Field Campaigns

To facilitate the remote sensing classification and verification, a continuous field campaign was carried out to record the phenologies of the SCR. Additionally, five field sites named *A* to *E* were also selected at the east of Deqing County. All the sites were larger than 1 km², and were surveyed using a handheld GPS receiver (Trimble Juno-SB). For each land cover patch, the boundary and the corresponding land cover type were recorded. The land cover types were classified as rice, trees, water bodies, economic crops and other nonvegetated areas. The vector format maps of the five field sites in 2012 were shown in Figure 2. These maps were then reclassified into SCR and non-rice area and converted into raster format at 30 m resolution as ground-truth data for accuracy assessment.

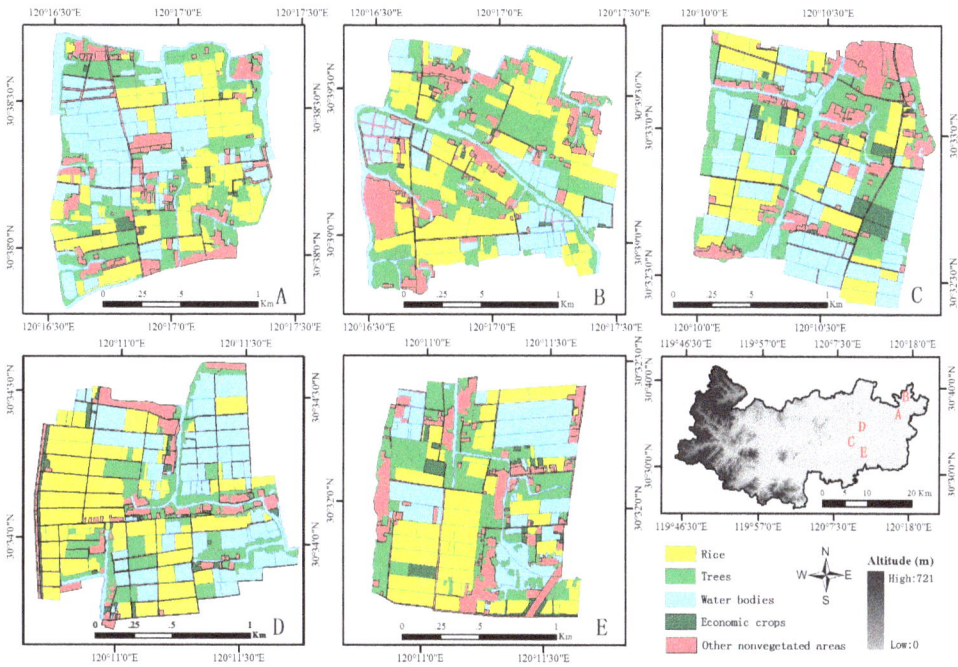

Figure 2. Vector maps and geo-locations of the five field sites.

2.3. Remote Sensing Data

HJ-1A/B data from 17 May to 5 December 2012 over the study area were collected for time-series VI analysis and downloaded from the China Center for Resources Satellite Data and Application. The sensor characteristics are presented in Table 1. Total 14 HJ-1A/B images with cloud cover less than 10% during the key phenology periods of SCR were selected for the following classification procedures (Table 2).

Table 1. Technical specification of HJ-1-A/B CCD and ZY1-02C P/MS sensors.

Satellite	Payload	Band No.	Spectral Range (μm)	Nadir Spatial Resolution (m)	Swath Width (km)	Repetition Cycle (day)
HJ-1A/B	Multispectral CCD camera	1	0.43–0.52	30	360 (700 for two)	4
		2	0.52–0.60	30		
		3	0.63–0.69	30		
		4	0.76–0.90	30		
ZY1-02C	P/MS camera	1	0.51–0.85	5	60	3–5
		2	0.52–0.59	10		
		3	0.63–0.69	10		
		4	0.77–0.89	10		

Table 2. Dates of the selected HJ-1A/B CCD images, field campaigns, and the corresponding SCR phenology stages.

NO.	Satellite	Date	Field Campaign Date	Phenology Stage
1	HJ-1B	2012/05/17	/	Fallow
2	HJ-1B	2012/05/28	/	Site preparation
3	HJ-1A	2012/06/29	2012/06/29	Sowing-transplanting
4	HJ-1B	2012/07/05	/	Vegetative stage
5	HJ-1B	2012/07/19	2012/07/20	Vegetative stage (tillering)
6	HJ-1A	2012/07/29	2012/07/30	Vegetative stage (maximum tiller number)
7	HJ-1A	2012/08/17	2012/08/15	Reproductive stage (ear differentiation)
8	HJ-1B	2012/09/02	2012/08/31	Reproductive stage (heading)
9	HJ-1B	2012/09/18	2012/09/16	Reproductive stage (panicle initiation and flowering)
10	HJ-1B	2012/09/29	2012/09/25	Reproductive stage
11	HJ-1B	2012/10/10	2012/10/13	Ripening stage (grain filling)
12	HJ-1A	2012/10/23	2012/10/27	Ripening stage (milk)
13	HJ-1B	2012/11/06	/	Ripening stage
14	HJ-1A	2012/11/19	2012/11/18	Harvest/fallow

To assist the selection of training samples for classification, the Chinese Resource-1 02C satellite (ZY1-02C), which provides multispectral and panchromatic images at 10 m and 5 m spatial resolutions, respectively, was used as an auxiliary data source (Table 1). The multispectral and panchromatic images of ZY1-02C were fused to facilitate location identity in field campaigns and visual interpretation.

All the HJ 1-A/B and ZY1-02C images were geometrically corrected using the Second National Soil Survey Vector Map (scale 1:10,000), and the Root Mean Square Error (RMS error) was less than one pixel (30 m). Additionally, the radiometric calibration and atmospheric correction of the HJ 1-A/B CCD data were performed, respectively. Figure 3 showed the images of HJ-1A/B CCD and ZY1-02C of field site *B* at different phenology stages of the SCR.

The remote sensing classification system (five land cover types) was same as the one used in the field campaign. The training samples were randomly located and visually interpreted from the ZY1-02C fused image (5 m in spatial resolution). The class separability of the training data set was analyzed using the Jeffries-Matusita

(J-M) distance metric between classes [43,44]. A larger J-M distance indicates more distinct distributions between two classes. The training data were modified if the J-M distance was close to 2 between rice and the other land cover types [45,46]. The final set of training samples were 800 pixels in total. There were 354 training pixels for rice, 92 training pixels for trees, 195 training pixels for water bodies, 76 training pixels for economic crops and 83 training pixels for other nonvegetated areas. The vector data of the 5 field sites were rasterized into 30 m resolution as ground-truth data for accuracy assessment.

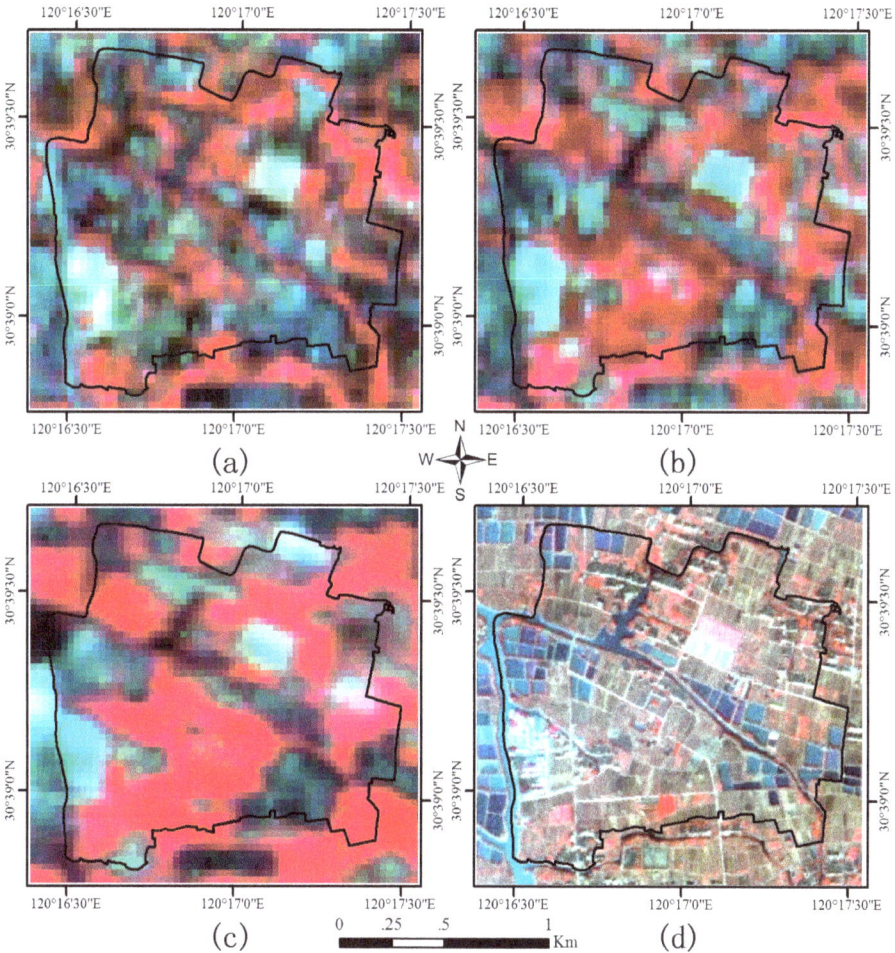

Figure 3. HJ-1A/B CCD and ZY1-02C false color images of the field site B at different phenology stages of the SCR: (**a**) to (**c**) were acquired from HJ-1A/B on 29 June (sowing-transplanting stage), 29 July (vegetative stage), and 2 September (reproductive stage) 2012, respectively; and (**d**) ZY1-02C on 19 February 2012.

213

2.4. Classification Methods

2.4.1. Characteristics of EVI2 Time-Series Data during SCR Growing Periods

EVI2 may achieve greater consistencies across sensors because only 2 bands are involved. EVI2 is defined as follows [33]:

$$EVI2 = 2.5 \times \frac{\rho_{nir} - \rho_{red}}{\rho_{nir} + 2.4 \times \rho_{red} + 1} \tag{1}$$

where ρ_{nir} and ρ_{red} are estimated surface reflectance values for near-infrared and visible red bands (HJ-1A/B CCD bands 4 and 3, respectively).

Since there are nearly always disturbances in optical remote sensing applications caused by unfavorable atmospheric conditions and sun zenith angle changes in year around and show up as undesirable noise [47,48], noise reduction is necessary before further analysis. In this study, we used the Savitzky-Golay (S-G) filters to smooth the EVI2 time-series data. The S-G filters are suitable to smooth the irregular spacing data points, e.g., the time-series HJ-1A/B CCD data used in this study [49]. The S-G filters apply an iterative weighted moving average filter to time series data, with weighting given as a polynomial of a particular degree [50,51], which can be summarized as:

$$g_i = \frac{\sum_{n=-nL}^{nR} c_n f_{i+n}}{n} \tag{2}$$

where f_i represents original value at data point i; g_i is the smoothed value; n is the width of the moving window to perform filtering; nL and nR corresponding to the left and right edges of the signal component. For a specific uneven time-series data in a moving window, c_n is not a constant but a polynomial fitting function, depending on the user's preference. The fitting function can be defined as quadratic polynomial for a specific f_i:

$$c_n(t) = c_1 + c_2 t + c_3 t^2 \tag{3}$$

where t corresponds to the day of year in EVI2 time-series.

The S-G filter was implemented using IDL 8.0 programming language to perform an image-based EVI2 time-series filtering for the HJ-1A/B CCD data from 17 May to 5 December 2012 in the study area. In this way, the time-series EVI2 curves of the five land cover types could be used to identify the most critical stages to distinguish between different land cover types.

2.4.2. Single-Cropped Rice Classification Method

In this study, we proposed a classification method which is based on the assumption that the probability distribution functions (PDFs) of the land cover types follow normal distributions [52,53]. For this purpose, we tested the samples' probability distributions of the five land cover types using the Quantile-Quantile Plot (Q-Q Plot) [54], and it showed that normal distribution assumption was acceptable.

Using the training data set, the mean (μ) and standard deviation (σ) of each land cover type can be obtained, and then we can define the normal distribution function for each land cover type using these two parameters. To properly differentiate one specific land cover type from the others, it is crucial to minimize the overlaps between the target and the neighboring normal PDFs. For two land cover types L_1 and L_2, assuming $L_1 \sim N(\mu_1, \sigma_1^2)$ and $L_2 \sim N(\mu_2, \sigma_2^2)$, then the intersection between L_1 and L_2 should be as follows [55]:

$$x = \frac{\sigma_1 \mu_2 + \sigma_2 \mu_1}{\sigma_1 + \sigma_2} \tag{4}$$

where if x is out of $[\mu - 2\sigma, \mu + 2\sigma]$ (hereafter $\mu(\sigma)$ can be $\mu_1(\sigma_1)$ or $\mu_2(\sigma_2)$ either), the two classes can be assumed distinguishable; if x is out of $[\mu - \sigma, \mu + \sigma]$ but within $[\mu - 2\sigma, \mu + 2\sigma]$, the two classes are mildly overlapped; if x is within $[\mu - \sigma, \mu + \sigma]$, then the two classes are seriously overlapped. Generally, the two classes can be thought separable if x is out of $[\mu - \sigma, \mu + \sigma]$.

Instead of using the whole growing period dataset, only key phenology stages images (during which the SCR are most differentiable from the other land cover types as explained later) were investigated in SCR field extraction. We used both of EVI2 and its 1st derivative, calculated by three consecutive images, to minimize the probability of mis-classification. For EVI2, we selected the image on 29 June 2012 (transplanting stage), whilst the spectral characteristic of the SCR was similar to water but not to the other land cover types, especially the trees. In addition, we made use of the quick change rate of EVI2, *i.e.*, the 1st derivative of SCR during the vegetative stages (here we used the image on 29 July 2012) to gather further information to refine the classification results [56,57].

2.4.3. Parametric and Nonparametric Classification Algorithms

The MLC assumes that the class signatures are normally distributed and calculates the probabilities of a given pixel belonging to each class. The pixel is assigned to the class with the highest probability [58]. The SVM classifier is a kernel-based machine learning technique; it separates the classes with a decision surface which maximizes the margin between the classes. The success of the SVM depends on how well the process is trained. In this study, a well-known radial basis function (RBF) kernel was used in the SVM [19,38,41].

We applied the MLC and SVM using the same training samples and parameters for each classifier. The multi-temporal HJ-1 CCD data, *i.e.*, six scenes from 2012/06/29 to 2012/09/02 during SCR transplanting to early reproductive stages, were used. The six reflectance/EVI2 imageries were composited and classified using the MLC and SVM separately, and the results were compared with the method proposed in Section 2.4.2.

2.4.4. Classification Accuracy Assessment

To assess classification accuracy, the ground-truth data (30 m resolution) were used. The ground-truth pixel numbers for field site A to E were 2842, 3078, 3186, 3248, and 2350, respectively. The proposed classification results were compared with the local agricultural statistic data in 2012. The user's and producer's accuracies, overall accuracy and Kappa statistic were also used to evaluate the SCR classification accuracy among the proposed method and the traditional methods, *i.e.*, MLC and SVM.

2.5. Influence of the Mixed-Pixel

To analyze the relationship between the land cover structure (or fragmentation) and classification accuracy, we calculated the landscape metrics, *i.e.*, class area (CA), percent of landscape (%LAND), patch density (PD), mean patch size (MPS), area-weighted mean shape index (AWMSI), and mean nearest-neighbor distance (MNN), for each ground-truth site at class level using FRAGSTATS to quantify its structure property [59]. Among the landscape metrics used here, CA is a measure of how much of the landscape is composed of a particular land cover type; %LAND is the percent of each land cover type; PD is the number of patches on a per unit area; MPS is the average area of patches for a certain class; AWMSI measures the area-weighted average patch shape; and MNN measures the mean average nearest distance among patches in a class. The landscape indices, e.g., PD, MPS, AWMSI and MNN, can be used to represent the fragmentation of land cover for a specific field site. The larger the values of PD, AWMSI and MNN, the more fragmented the site was; and vice versa for MPS.

To evaluate the influence of the land cover composition of a specific pixel on the classification accuracy, the vector maps of the five ground-truth sites were further divided into cells of size 30 m × 30 m using the gridlines derived from the HJ CCD images (only the SCR fields were kept and the other land cover types were taken as background). We calculated the area proportion of SCR in each cell, and divided the cells, in which the SCR area were greater than zero, into three grades, *i.e.*, 75%–100%, 50%–75%, and <50%, according to the SCR area proportion. For each grade in per site, the proportion of cells, which were classified as rice field in the HJ CCD images, to the total cell number in that specific grade was calculated. We further calculated the number of misclassified pixels, *i.e.*, the commission and omission errors, and analyzed the corresponding spatial distribution of the misclassified pixels. For a specific pixel, the commission error means that the pixel's SCR area proportion is less than 50% but is classified as rice field; while the omission error means that the pixel contains more than 50% SCR area but is classified as the other land cover types.

3. Results

3.1. Time-Series EVI2 Characteristics

The temporal dynamics of time-series EVI2 of the SCR processed by S-G filters and the other land cover types calculated from HJ-1A/B images were shown in Figures 4 and 5. The EVI2 of water bodies varied slightly over the growing season of SCR in the range of 0.07–0.14. During the transplanting and early part of the SCR growing period, rice fields were flooded and its spectral signature was similar to that of the water bodies. Not surprisingly, the EVI2 value of rice fields was very close to water bodies but obviously lower than that of trees and economic crops on June 29 (DOY = 181, about 10 days after transplanting in 2012). After transplanting, the EVI2 of SCR increased rapidly and maximized at about 0.6 between the ear differentiation and early heading stages, about 75 days after transplanting. Caused by the etiolation and senescence of the SCR leaves, the EVI2 started to decrease after the heading period till harvest.

The EVI2 values of the other nonvegetated areas, including residential areas, roads and bare land, were similar with less fluctuation but relatively higher compared with water bodies. The trees class had relatively high EVI2 values around 0.30 to 0.47. The economic crops were generally planted during a similar period as the SCR were transplanted, but usually have a longer life cycle and relativly small changing rate of EVI2 compared with SCR, especially during the vegetative stages of SCR. During the transplanting period, the water like spectral characteristic of the SCR made its EVI2 signature a little lower than the economic crops.

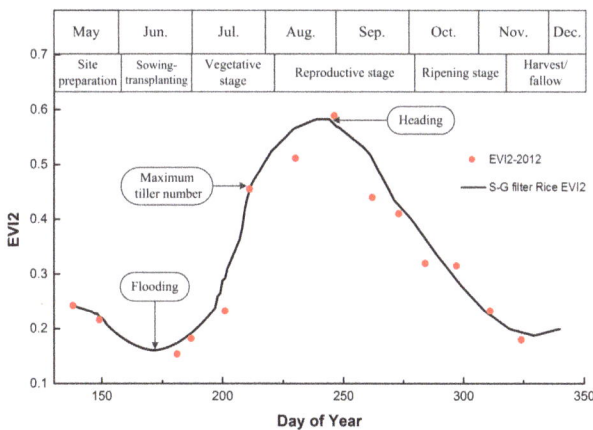

Figure 4. Time-series EVI2 data fitted by the S-G filters.

217

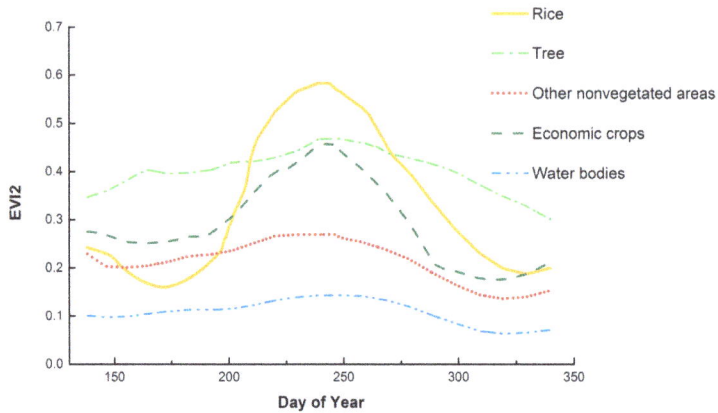

Figure 5. Time-series EVI2 of the five land use types processed by the S-G filters.

3.2. Classification Thresholds

The normal distributions of EVI2 on 29 June 2012 and its first derivative on 29 July 2012 of the five land cover types were shown in Figure 6. During the transplanting stage, the rice fields were flooded and the PDF of EVI2 of the SCR was close to that of the water bodies, and it also mixed with the nonvegetated areas and economic crops classes. Therefore, except the trees, which were mildly separable with respect to the economic crops but highly distinguishable from the other classes, the EVI2 signature of SCR was seriously overlapped with the other three land cover types (Figure 6a). Obviously, it is insufficient to identify the SCR just using images in the transplanting period.

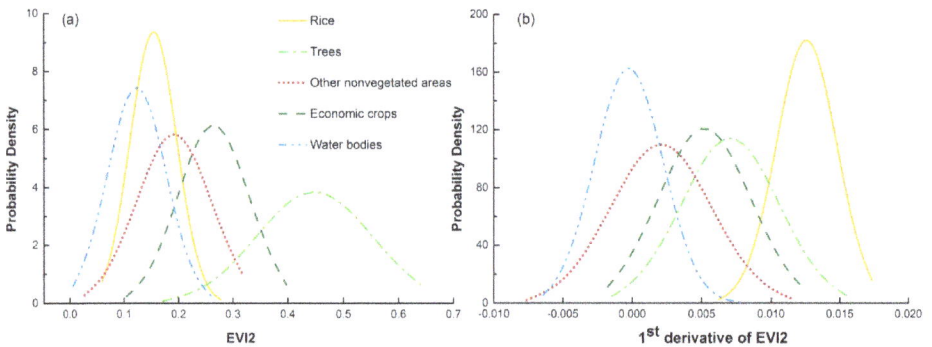

Figure 6. Normal distributions of the five land cover types: (**a**) EVI2 on 29 June 2012; and (**b**) the 1st derivative of EVI2 on 29 July 2012.

218

During the vegetative stages of the SCR, there was a quick increase of EVI2 due to the formation of additional tillers (Figure 4), while the increase rates of the economic crops and trees were not as steep as SCR. The other two land cover types didn't show obvious changes during this period. In this situation, the 1st derivative of EVI2 based on the image on 29 July 2012, when the maximum tiller number had arrived, demonstrated that the rice fields could be confidently distinguished from the water bodies and the other nonvegetated areas (Figure 6b). As shown in Figure 6a, the rice fields class was distinguishable from the trees using EVI2 signatures during the transplanting stage, and it is mildly separable from the economic crops class. By using Equation (4), one pixel could be classified as rice field if its EVI2 value on 29 June 2012 equal or less than 0.24, whilst its 1st derivative of EVI2 on 29 July 2012 was equal or greater than 0.010.

3.3. Classification Accuracy Assessment

Based on the coupling thresholds of EVI2 and its 1st derivative of the SCR, during the key phenology stages, i.e., the transplanting and the vegetative to reproductive transition phases, the rice fields in Deqing County was classified using HJ-1A/B data in 2012 (Figure 7). The classified rice fields, with an area of about 94.0 km^2, mainly concentrated in the eastern plain region of the study area with altitudes around 4 m. According to the statistical data of the local agriculture department in 2012, the total acreage of the SCR was 86.4 km^2, so the relative classification accuracy was about 91.2%.

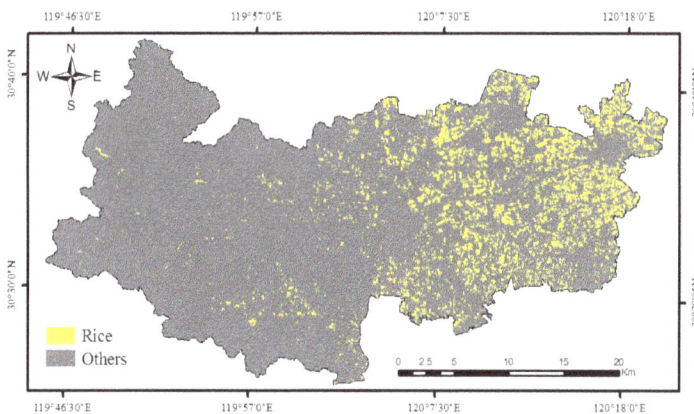

Figure 7. Classification result of the SCR fields using HJ-1A/B data in 2012.

We compared the classification accuracies of the 5 ground-truth sites (Table 3). The overall classification accuracy and Kappa coefficient for all the sites were 91.68% and 0.79, respectively. For each site, the producer's and user's accuracies of rice,

219

overall classification accuracy and Kappa coefficient are listed in Table 3, in which the site D had the highest user's accuracy and overall classification accuracy (86.31% and 94.21%, respectively), followed by site E, with user's accuracy of 83.42% and overall accuracy of 93.40%. All the ground-truth sites had producer's accuracies higher than 90.72% (site C). The accuracy assessment demonstrated a satisfactory result of the proposed classification method.

Table 3. Classification accuracies and Kappa coefficients of the five ground-truth sites for SCR. The last column corresponded to that if all sites were treated as a whole.

Classification Accuracy	A	B	C	D	E	All
Producer's accuracy (%)	95.43	94.12	90.72	94.98	95.67	94.35
User's accuracy (%)	75.1	77.29	69.17	86.31	83.42	78.70
Overall accuracy (%)	92.43	90.68	91.93	94.21	93.40	91.68
Kappa coefficient	0.76	0.75	0.71	0.85	0.83	0.79

3.4. Comparison of Classification Methods

The classification accuracy of the proposed method used in this study outperformed the MLCs and SVMs (Tables 4 and 5). By using EVI2 instead of the reflectance data, the MLC-EVI2 and SVM-EVI2 improved the classification accuracies to certain extent, but not significant, compared with their counterparts, respectively. The MLCs and SVMs also showed better classification accuracies in site D and site E. While site C had a lower classification result compared with the other four field sites.

Table 4. Classification accuracies and Kappa coefficients of the five ground-truth sites for SCR using MLC and SVM methods. The last column is the corresponding results if all sites were treated as a whole.

Classification Methods	Classification Accuracy	A	B	C	D	E	All
MLC-EVI2	Producer's accuracy (%)	67.77	70.77	60.68	77.01	70.69	62.41
	User's accuracy (%)	67.08	63.83	61.43	73.25	73.68	60.58
	Overall accuracy (%)	88.85	85.10	84.03	89.17	87.81	84.88
	Kappa coefficient	0.71	0.68	0.68	0.60	0.74	0.65
MLC-Reflectance	Producer's accuracy (%)	68.60	66.48	52.97	79.93	70.27	62.55
	User's accuracy (%)	65.23	61.21	56.15	69.49	70.27	57.52
	Overall accuracy (%)	88.42	83.74	81.54	88.58	86.72	83.83
	Kappa coefficient	0.70	0.63	0.64	0.70	0.72	0.65
SVM-EVI2	Producer's accuracy (%)	82.23	83.38	60.05	83.76	81.07	73.97
	User's accuracy (%)	50.90	52.43	50.90	59.14	61.16	54.33
	Overall accuracy (%)	83.46	80.17	76.96	85.37	85.34	81.09
	Kappa coefficient	0.63	0.62	0.57	0.55	0.69	0.61
SVM-Reflectance	Producer's accuracy (%)	84.30	85.82	60.96	85.40	79.10	74.80
	User's accuracy (%)	50.00	51.63	51.51	58.40	63.87	52.87
	Overall accuracy (%)	82.97	78.23	76.35	84.78	84.27	80.00
	Kappa coefficient	0.63	0.59	0.56	0.55	0.61	0.57

Table 5. Landscape indices of the five land cover types in the ground-truth sites.

Site	Class	CA (ha)	%LAND	PD	MPS (ha)	AWMSI	MNN (m)
	Rice	50.08	25.78	1.56	0.64	1.73	27.41
	Trees	44.99	23.16	2.20	0.45	2.83	19.45
A	Water bodies	69.12	35.58	1.22	0.82	2.54	18.03
	Economic crops	2.32	1.19	6.89	0.15	1.40	126.17
	Other nonvegetated areas	27.77	14.29	4.83	0.21	7.76	30.91
	Rice	60.85	30.25	1.25	0.80	2.11	21.90
	Trees	56.73	28.20	1.73	0.58	3.23	21.38
B	Water bodies	48.12	23.92	1.41	0.71	3.86	20.32
	Economic crops	3.75	1.86	5.33	0.19	5.67	80.80
	Other nonvegetated areas	31.72	15.77	4.48	0.22	4.55	17.64
	Rice	40.50	19.66	1.31	0.76	1.57	22.14
	Trees	45.47	22.07	1.83	0.55	2.56	15.98
C	Water bodies	74.08	35.95	0.90	1.11	2.10	20.15
	Economic crops	7.55	3.67	2.25	0.44	1.34	92.04
	Other nonvegetated areas	38.44	18.66	4.63	0.22	7.34	21.30
	Rice	70.31	36.03	0.95	1.05	1.25	13.85
	Trees	32.12	16.46	1.96	0.51	2.20	13.24
D	Water bodies	62.61	32.08	1.15	0.87	1.94	7.00
	Economic crops	5.95	3.05	15.12	0.07	4.38	21.55
	Other nonvegetated areas	24.15	12.38	8.49	0.12	18.53	13.50
	Rice	55.62	31.15	1.11	0.90	1.40	13.64
	Trees	41.63	23.32	2.45	0.41	1.97	14.60
E	Water bodies	43.74	24.50	0.78	1.29	2.60	22.36
	Economic crops	5.88	3.29	5.44	0.18	1.58	65.06
	Other nonvegetated areas	31.67	17.74	5.27	0.19	8.29	19.73

3.5. Influence of the Mixed-Pixel

In the five ground-truth sites (Table 5), the average area percentage of water bodies was the highest among the 5 land cover types (larger than 0.71 in five field sites), followed by rice and trees (larger than 0.64 and 0.41, respectively). About 36.03% area of site D was rice, compared with the smallest proportion of 19.66% in site C. The site D had the smallest trees area proportion of 16.46%. The economic crops had the smallest area, and its average patch size in site D was only 0.07 ha, smaller than the area of one pixel of HJ-1 CCD image (30 m × 30 m); the high values of MNN also indicated the highly scattered status of the economic crops (see also Figure 2), and economic crops had the largest value of MNN of the five categories for five field sites. The other nonvegetated areas had the highest AWMSI (except site B), reflecting the complex shape of the road system. The fragmentation statuses of rice, indicated by PD, AWMSI and MNN, of sites D and E were the lowest compared with sites A–C; while the site D and E had the highest MPS. The sites D and E had less fragmented degrees compared with site C, while sites A and B had intermediate level of fragmentation statuses of rice.

Figure 8 showed the ratios of the pixels which were classified as rice field in the HJ CCD images to the total pixel number (ground-truth data, pixels in which the area proportion of rice field is greater than 50%) in each grade for sites A–E. It

221

is obvious that the recognition ratio increased as the area proportion of rice field in pixel ascending, *i.e.*, the grade 75%–100% had the highest classification accuracy. The site *D* and *E* had the highest recognition ratio in each grade, while site *C* the lowest. This result also demonstrated the difficulties in classification in fragmented areas where the mixed-pixel problems were more serious.

More than 69.03% commission pixels concentrated at the boundaries, while at least 63.89% omission pixels lay on the boundaries; that is, most of the misclassified pixels concentrated at the boundaries of the rice fields (Table 6 and Figure 9). The omission pixel numbers of sites *D*–*E* were 40 and 36 respectively, obviously less than sites *A*–*C* (65, 70 and 80 respectively). The misclassification error was largely determined by the commission error. As shown in Figure 9, the commission error pixels (red color) were more than the omission ones (blue color).

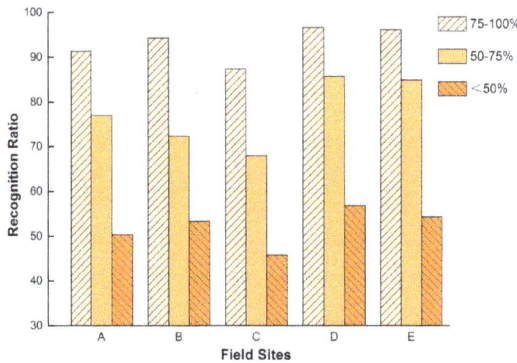

Figure 8. Recognition ratios for the ground-truth sites A–E, compared with the classification results of HJ CCD images.

Table 6. Statistics of the pixel numbers in classification for sites *A*–*E*. The boundary pixels are the pixels which contain certain area of rice field and interface with the other land cover types; the commission pixels represent pixels in which the rice field area proportion is less than <50%, but is misclassified as rice field; and the omission pixels are pixels in which the rice field area proportion is great than 50% but is wrongly classified as the other land cover types.

Statistics of Classification	A	B	C	D	E
Rice pixel number	779	995	678	942	742
Boundary pixel number	476	594	433	542	369
Commission pixel number	150	217	155	144	119
Commission pixels on boundary	122	169	107	101	91
Omission pixel number	65	70	80	44	36
Omission pixels on boundary	47	57	53	40	23
Classification error(%) [1]	27.60	28.84	34.66	19.96	20.89

Note: [1] The classification error is calculated as (omission number + commission number)/rice pixel number.

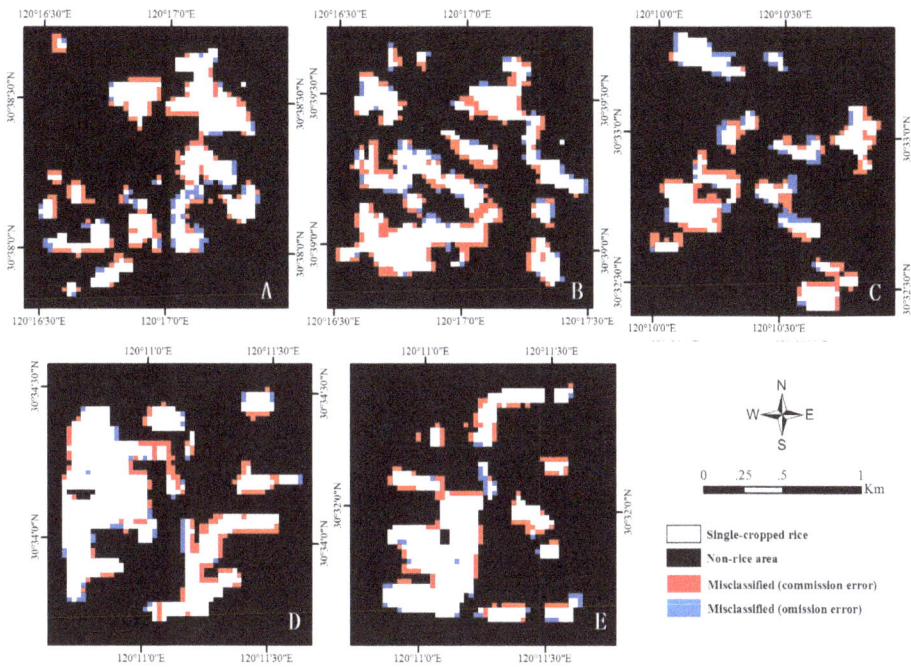

Figure 9. Spatial distribution of the misclassified pixels in sites (**A–E**).

4. Discussion

It is generally acknowledged that using a single-temporal image to well discriminate a specific kind of crop at various phenology stages from the other vegetation (or land cover types) is an enormous challenge [19,60,61]. However, using the spectral characteristics (or vegetation indices) determined by the key phenologies of a specific crop species, *i.e.*, multi-temporal remote sensing imageries, is a promising way to improve the classification accuracy [62,63]. To effectively discriminate the rice field in eastern plain region of China, where the rice field is generally fragmented and irregular due to the topography and widely distributed water bodies and road networks, a specifically designed stepwise remote sensing classification strategy was applied in this study.

The time-series EVI2 data for the major land cover types in the study area were built from the HJ-1 A/B CCD imageries and the S-G filters was applied to smooth the EVI2 time-series. With the reference field campaign data, the EVI2 showed efficient discriminating capability in capturing the spectral differences between SCR and the other land cover types during the key SCR phenology stages (Figure 5). It is prominent that the EVI2 of SCR increased rapidly during the transplanting and ear differentiation (including early heading) stages, and the temporal resolution of HJ-1 A/B CCD data was testified to be suitable to capture these features. The stepwise classification algorithm proposed in this study can be seen as an exemplar of the decision tree classification category, and it outperformed the parametric (MLC) and nonparametric (SVM) classification algorithms, respectively (Tables 3 and 4). By using EVI2 instead of the reflectance data, the classification accuracies improved to certain extents for both of MLC and SVM. The results also implied that by treating the satellite-derived vegetation classification information hierarchically, the mixtures among spectral feature spaces can be effectively alleviated. For MLC and SVM, total six scenes during SCR transplanting to early reproductive stages (from 2012/06/29 to 2012/09/02) were used, including the transplanting, vegetative to reproductive transition phases. However, it is noteworthy that time-series EVI2 of SCR during this period increased rapidly and intersected with the EVI2s of all the other land cover types, except water bodies (Figure 5). Therefore, the classification accuracies of MLC and SVM should unavoidably be decreased, because both of the methods treated the spectral signatures contained in the six scenes collectively.

The influence of the mixed-pixel is a primary concern in remote sensing classification practices. We used five ground-truth sites as an example and analyzed the relationship between the purity of pixels (measured as the area proportion of rice field in a specific cell) and the corresponding recognition ratios. The mixed-pixel analysis showed that the recognition ratio was positively correlated with the rice field area proportion at each ground-truth site (Table 5 and Figure 8). The sites D and E showed the best recognition ratio of SCR among the five ground-truth sites, and it is in accordance with the fragmentation statuses indicated by the landscape indices (Table 5). It is not unexpectedly that as the area proportion of rice field increased in each cell, the possibility of misclassification decreased consequently, especially for the grade of 75%–100% (refer in particular to rice field).

As large part of the classification error can be attributed to the influence of mixed-pixels where the area proportion of rice field was less than 75%, and most of the mixed-pixels concentrated at the boundaries of the rice fields (Table 6 and Figure 9). We further analyzed the classification error caused by the commission and omission errors due to the mixed-pixels and boundary effects, respectively. The results showed that the ratio of the edge pixels to the total rice pixel number correlated with the fragmentation states of each site, $i.e.$, the number of the edge pixel was positively correlated with land fragmentation states of each site due to the increased rice field perimeter. As a consequence, the classification errors of sites D–E were less than sites A–C as shown in Table 6. For rice fields, the misclassification caused by the commission errors was more common, compared with the omission errors ($i.e.$, cells in which rice field area was less than 50% but was classified as rice field, see Figure 8).

However, it should be noted that due to the existence of spatial autocorrelation, the classification accuracy reported in this study may be overestimated [64]. Spatial autocorrelation might be present due to large pixel size [65] or points sampled in close proximity [66]. To avoid the artificially increased classification accuracy caused by the random cross-validation using autocorrelated dataset, more than one permanent training/test dataset should be utilized in accuracy assessment [64]. In this study, five ground-truth sites with different land cover percentages were selected for classification accuracy assessment, however the authors acknowledged that the autocorrelation may still unavoidable and quantitative evaluation of its influence is still a challenge. Further studies should be focused on field data collection, with subsampling and cross-validation like k-fold method [64] to improve the classification accuracy assessment.

The extrapolation of the findings in this study must be cautious due to various changes in the environmental factors (e.g., dry or wet) and vegetation status in different regions and years. The aim of this study was to provide a general methodology in the classification of single-cropped rice. However, when applying it to another region or year, the VI thresholds, which are used to distinguish different land cover types, must be decided according to the specific time series satellite images, *i.e.*, the VI thresholds and the timestamps (according to the key phenologies) are variable.

5. Conclusions

In this study, we applied a simple but robust stepwise algorithm to estimate the single cropped rice (SCR) growing area in irregular and fragmented regions. The multi-temporal HJ-1A/B images and specific signatures of EVI2 at the key phenology stages, *i.e.*, the transplanting and the vegetative to reproductive transition phases, of the SCR were used to classify the rice fields from the other land use types with satisfactory results, compared with the traditional MLC and SVM methods. Due to the fragmented land use composition in the study area, we also assessed the influence of mixed-pixel quantified by using the landscape indices and it showed that the classification accuracy ratio of rice field was positively correlated with its compactness. We showed that by making full use of the key phenological information, and under the support of high-temporal resolution remote sensing data, e.g., HJ-1A/B, the SCR can be mapped at a relative high confidence. The crucial point in the proposed method was the construction of high-quality time-series VI curves, which were then used to identify the key phenology stages to differentiate different land cover types. However, due to the variation of environmental factors and the corresponding changes of vegetation status, due care should be taken when extrapolating the results to other regions or periods. Additionally, we noted that the influence of spatial autocorrelation should also be taken into consideration in classification accuracy evaluation in further study.

Acknowledgments: This research was funded by The National High Technology Research and Development Program of China (Grant No. 2012AA12A30703) and Agricultural Project of Scientific and Technological Research of Shanghai, China (2011-2-11). We are grateful to the lab members Qiaoli Ge, Sujuan Wang, Jingjing Shi, Liwen Zhang, Qiaoying Guo, Zhuokun Pan, Zhewen Zhao, Yao Zhang, Weijiao Huang, Chen Wei, Dilong Gan, Zhen Zhou, Yuanyuan Chen, Bing Han, Mengting Jin, Xiaoqiang Zhang, Jingbo Zhu, and Ran Huang for their great assistance during the field campaigns and data processing. We would also like to express our sincere thanks to the anonymous reviewers for their constructive comments.

Author Contributions: Jing Wang had the original idea for the study and wrote the original manuscript. Jingfeng Huang supervised the process of field campaign and data analysis. Kangyu Zhang provided partial source codes for image analysis. Xinxing Li was responsible for image data collection. Bao She, Chuanwen Wei and Jian Gao processed field campaign data. Xiaodong Song offered valuable comments to the manuscript and was responsible to manuscript revisions. All authors read and approved the final manuscript.

Conflicts of Interest: The authors declare no conflict of interest.

References

1. Gnanamanickam, S.S. Rice and its importance to human life. In *Rice and Its Importance to Human Life*; Springer: Berlin, Germany, 2009; pp. 1–11.

2. Gon, H.D. Changes in CH_4 emission from rice fields from 1960 to 1990s: 1. Impacts of modern rice technology. *Glob. Biogeochem. Cycles* **2000**, *14*, 61–72.

3. Zhang, W.; Yu, Y.Q.; Huang, Y.; Li, T.T.; Wang, P. Modeling methane emissions from irrigated rice cultivation in China from 1960 to 2050. *Glob. Change Biol.* **2011**, *17*, 3511–3523.

4. FAOSTAT. Statistical Database of the Food and Agricultural Organization of the United Nations. Available online: http://faostat.fao.org/default.aspx?lang=en (accessed on 30 December 2014).

5. Zhai, H. Prospects for grain demand and supply in the 21st century. In Proceedings of the 12th Toyota Conference: Challenge of Plant and Agricultural Sciences to the Crisis of the Biosphere on the Earth in the 21st Century, Shizuoka, Japan, 25–28 November 1998; pp. 29–37.

6. Chen, C.F.; Huang, S.W.; Son, N.T.; Chang, L.Y. Mapping double-cropped irrigated rice fields in Taiwan using time-series Satellite Pour l'Observation de la Terre data. *J. Appl. Remote Sens.* **2011**, *5*.

7. Xiao, X.M.; Boles, S.; Frolking, S.; Li, C.S.; Babu, J.Y.; Salas, W.; Moore, B., III. Mapping paddy rice agriculture in South and Southeast Asia using multi-temporal MODIS images. *Remote Sens. Environ.* **2006**, *100*, 95–113.

8. Fang, H.L.; Wu, B.F.; Liu, H.Y.; Huang, X. Using NOAA AVHRR and Landsat TM to estimate rice area year-by-year. *Int. J. Remote Sens.* **1998**, *19*, 521–525.

9. Gumma, M.K.; Thenkabail, P.S.; Hideto, F.; Nelson, A.; Dheeravath, V.; Busia, D.; Rala, A. Mapping irrigated areas of Ghana using fusion of 30 m and 250 m resolution remote-sensing data. *Remote Sens.* **2011**, *3*, 816–835.

10. Yu, L.; Wang, J.; Clinton, N.; Xin, Q.C.; Zhong, L.H.; Chen, Y.L.; Gong, P. FROM-GC: 30 m global cropland extent derived through multisource data integration. *Int. J. Digit. Earth* **2013**, *6*, 521–533.

11. Kamthonkiat, D.; Honda, K.; Turral, H.; Tripathi, N.K.; Wuwongse, V. Discrimination of irrigated and rainfed rice in a tropical agricultural system using SPOT VEGETATION NDVI and rainfall data. *Int. J. Remote Sens.* **2005**, *26*, 2527–2547.

12. Sakamoto, T.; Van Nguyen, N.; Ohno, H.; Ishitsuka, N.; Yokozawa, M. Spatio–temporal distribution of rice phenology and cropping systems in the Mekong Delta with special reference to the seasonal water flow of the Mekong and Bassac Rivers. *Remote Sens. Environ.* **2006**, *100*, 1–16.

13. Sun, H.S.; Huang, J.F.; Huete, A.R.; Peng, D.L.; Zhang, F. Mapping paddy rice with multi-date moderate-resolution imaging spectroradiometer (MODIS) data in China. *J. Zhejiang Univ-Sc. A.* **2009**, *10*, 1509–1522.

14. Carfagna, E.; Gallego, F.J. Using remote sensing for agricultural statistics. *Int. Stat. Rev.* **2005**, *73*, 389–404.

15. Wu, B.F.; Li, Q.Z. Crop planting and type proportion method for crop acreage estimation of complex agricultural landscapes. *Int. J. Appl. Earth Obs.* **2012**, *16*, 101–112.

16. Asner, G.P. Cloud cover in Landsat observations of the Brazilian Amazon. *Int. J. Remote Sens.* **2001**, *22*, 3855–3862.

17. Pyongsop, R.I.; Zhangbao, M.A.; Qingwen, Q.I.; Gaohuan, L. Cloud and shadow removal from Landsat TM data. *J. Remote Sens.* **2010**, *14*, 534–545.

18. Chen, J.S.; Huang, J.X.; Hu, J.X. Mapping rice planting areas in southern China using the China Environment Satellite data. *Math. Comput. Model.* **2011**, *54*, 1037–1043.

19. Jia, K.; Wu, B.F.; Li, Q.Z. Crop classification using HJ satellite multispectral data in the North China Plain. *J. Appl. Remote Sens.* **2013**, *7*.

20. Li, W.G.; Li, H.; Zhao, L.H. Estimating Rice Yield by HJ-1A Satellite Images. *Rice Sci.* **2011**, *18*, 142–147.

21. Xiao, X.M.; Boles, S.; Frolking, S.; Salas, W.; Moore, B., III; Li, C.; He, L.; Zhao, R. Observation of flooding and rice transplanting of paddy rice fields at the site to landscape scales in China using VEGETATION sensor data. *Int. J. Remote Sens.* **2002**, *23*, 3009–3022.

22. Le Toan, T.; Ribbes, F.; Wang, L.F.; Floury, N.; Ding, K.H.; Kong, J.A.; Fujita, M.; Kurosu, T. Rice crop mapping and monitoring using ERS-1 data based on experiment and modeling results. *IEEE Trans. Geosci. Remote Sens.* **1997**, *35*, 41–56.

23. Baret, F.; Guyot, G. Potentials and limits of vegetation indices for LAI and APAR assessment. *Remote Sens. Environ.* **1991**, *35*, 161–173.

24. Gitelson, A.A. Remote estimation of crop fractional vegetation cover: the use of noise equivalent as an indicator of performance of vegetation indices. *Int. J. Remote Sens.* **2013**, *34*, 6054–6066.

25. Meng, J.H.; Wu, B.F.; Chen, X.Y.; Du, X.; Niu, L.M.; Zhang, F.F. Validation of HJ-1 B charge-coupled device vegetation index products with spectral reflectance of Hyperion. *Int. J. Remote Sens.* **2011**, *32*, 9051–9070.

26. Rouse, J.W.; Haas, R.H.; Schell, J.A.; Deering, D.W.; Harlan, J.C. *Monitoring the Vernal Advancement and Retrogradation of Natural Vegetation*; NASA/GSFC, Type III, Final Report; Texas AM University: College Station, TX, USA, 1974.

27. De Rosnay, P.; Calvet, J.-C.; Kerr, Y.; Wigneron, J.-P.; Lemaître, F.; Escorihuela, M.J.; Sabater, J.M.; Saleh, K.; Barrié, J.; Bouhours, G. SMOSREX: A long term field campaign experiment for soil moisture and land surface processes remote sensing. *Remote Sens. Environ.* **2006**, *102*, 377–389.

28. Gao, S.; Niu, Z.; Huang, N.; Hou, X. Estimating the Leaf Area Index, height and biomass of maize using HJ-1 and RADARSAT-2. *Int. J. Appl. Earth Obs.* **2013**, *24*, 1–8.

29. Hansen, P.M.; Schjoerring, J.K. Reflectance measurement of canopy biomass and nitrogen status in wheat crops using normalized difference vegetation indices and partial least squares regression. *Remote Sens. Environ.* **2003**, *86*, 542–553.

30. Lu, L.; Li, X.; Huang, C.L.; Ma, M.G.; Che, T.; Bogaert, J.; Veroustraete, F.; Dong, Q.H.; Ceulemans, R. Investigating the relationship between ground-measured LAI and vegetation indices in an alpine meadow, north-west China. *Int. J. Remote Sens.* **2005**, *26*, 4471–4484.

31. Huete, A.; Didan, K.; Miura, T.; Rodriguez, E.P.; Gao, X.; Ferreira, L.G. Overview of the radiometric and biophysical performance of the MODIS vegetation indices. *Remote Sens. Environ.* **2002**, *83*, 195–213.

32. Xiao, X.M.; Boles, S.; Liu, J.Y.; Zhuang, D.F.; Frolking, S.; Li, C.S.; Salas, W.; Moore, B., III. Mapping paddy rice agriculture in southern China using multi-temporal MODIS images. *Remote Sens. Environ.* **2005**, *95*, 480–492.

33. Jiang, Z.Y.; Huete, A.R.; Didan, K.; Miura, T. Development of a two-band enhanced vegetation index without a blue band. *Remote Sens. Environ.* **2008**, *112*, 3833–3845.

34. Kim, Y.; Miura, T.; Jiang, Z.; Huete, A.R. Spectral compatibility of vegetation indices across sensors: band decomposition analysis with Hyperion data. *J. Appl. Remote Sens.* **2010**, *4*.

35. DeFries, R.S.; Townshend, J. NDVI-derived land cover classifications on a global scale. *Int. J. Remote Sens.* **1994**, *15*, 3567–3586.

36. Hubert-Moy, L.; Cotonnec, A.; Le Du, L.; Chardin, A.; Perez, P. A comparison of parametric classification procedures of remotely sensed data applied on different landscape units. *Remote Sens. Environ.* **2001**, *75*, 174–187.

37. Yang, C.H.; Everitt, J.H.; Murden, D. Evaluating high resolution SPOT 5 satellite imagery for crop identification. *Comput. Electron. Agr.* **2011**, *75*, 347–354.

38. Chen, C.F.; Son, N.T.; Chen, C.R.; Chang, L.Y. Wavelet filtering of time-series moderate resolution imaging spectroradiometer data for rice crop mapping using support vector machines and maximum likelihood classifier. *J. Appl. Remote Sens.* **2011**, *5*.

39. Foody, G.M.; Mathur, A. Toward intelligent training of supervised image classifications: directing training data acquisition for SVM classification. *Remote Sens. Environ.* **2004**, *93*, 107–117.

40. Mountrakis, G.; Im, J.; Ogole, C. Support vector machines in remote sensing: A review. *ISPRS J. Photogramm. Remote Sens.* **2011**, *66*, 247–259.

41. Rao, T.V. Supervised classification of remote sensed data using support vector machine. *Glob. J. Comput. Sci. Technol.* **2014**, *14*, 71–76.

42. Xu, H.W.; Wang, K. Regionalization for rice yield estimation by remote sensing in Zhejiang Province. *Pedosphere* **2001**, *11*, 175–184.

43. Lobo, A.; Chic, O.; Casterad, A. Classification of Mediterranean crops with multisensor data: per-pixel *versus* per-object statistics and image segmentation. *Int. J. Remote Sens.* **1996**, *17*, 2385–2400.

44. Panigrahy, S.; Parihar, J.S.; Patel, N.K. Kharif rice acreage estimation in Orissa using NOAA-AVHRR data. *J. Indian Soc. Remote Sens.* **1992**, *20*, 35–42.

45. Congalton, R.G. A Review of Assessing the Accuracy of Classifications of Remotely Sensed Data. *Remote Sens. Environ.* **1991**, *37*, 35–46.

46. Wardlow, B.D.; Egbert, S.L.; Kastens, J.H. Analysis of time-series MODIS 250 m vegetation index data for crop classification in the US Central Great Plains. *Remote Sens. Environ.* **2007**, *108*, 290–310.

47. Pan, Z.; Huang, J.; Zhou, Q.; Wang, L.; Cheng, Y.; Zhang, H.; Blackburn, G.A.; Yan, J.; Liu, J. Mapping crop phenology using NDVI time-series derived from HJ-1 A/B data. *Int. J. Appl. Earth Obs.* **2015**, *34*, 188–197.

48. Sakamoto, T.; Wardlow, B.D.; Gitelson, A.A.; Verma, S.B.; Suyker, A.E.; Arkebauer, T.J. A Two-Step Filtering approach for detecting maize and soybean phenology with time-series MODIS data. *Remote Sens. Environ.* **2010**, *114*, 2146–2159.

49. Cong, N.; Piao, S.; Chen, A.; Wang, X.; Lin, X.; Chen, S.; Han, S.; Zhou, G.; Zhang, X. Spring vegetation green-up date in China inferred from SPOT NDVI data: A multiple model analysis. *Agr. Forest Meteorol.* **2012**, *165*, 104–113.

50. Hird, J.N.; McDermid, G.J. Noise reduction of NDVI time series: An empirical comparison of selected techniques. *Remote Sens. Environ.* **2009**, *113*, 248–258.

51. Jönsson, P.; Eklundh, L. TIMESAT—A program for analyzing time-series of satellite sensor data. *Comput. Geosci.* **2004**, *30*, 833–845.

52. Li, A.N.; Jiang, J.G.; Bian, J.H.; Deng, W. Combining the matter element model with the associated function of probability transformation for multi-source remote sensing data classification in mountainous regions. *ISPRS J. Photogramm. Remote Sens.* **2012**, *67*, 80–92.

53. Voisin, A.; Krylov, V.A.; Moser, G.; Serpico, S.B.; Zerubia, J. Supervised classification of multisensor and multiresolution remote sensing images with a hierarchical copula-based approach. *IEEE Trans. Geosci. Remote Sens.* **2013**, *52*, 3346–3358.

54. Zhang, D.X.; Zhang, C.R.; Li, W.D.; Cromley, R.; Hanink, D.; Civco, D.; Travis, D. Restoration of the missing pixel information caused by contrails in multispectral remotely sensed imagery. *J. Appl. Remote Sens.* **2014**, *8*.

55. Hu, L.H.; Yu, Z.F.; Liu, Y.F. An algorithm of decision-tree generating automatically based on classification. In Proceedings of the First International Workshop on Education Technology and Computer Science, Wuhan, China, 7–8 March 2009; Vol. I, pp. 823–827.

56. Jonsson, P.; Eklundh, L. Seasonality extraction by function fitting to time-series of satellite sensor data. *IEEE Trans. Geosci. Remote Sens.* **2002**, *40*, 1824–1832.

57. Sakamoto, T.; Yokozawa, M.; Toritani, H.; Shibayama, M.; Ishitsuka, N.; Ohno, H. A crop phenology detection method using time-series MODIS data. *Remote Sens. Environ.* **2005**, *96*, 366–374.

58. Otukei, J.R.; Blaschke, T. Land cover change assessment using decision trees, support vector machines and maximum likelihood classification algorithms. *Int. J. Appl. Earth Obs.* **2010**, *12S*, S27–S31.

59. Li, X.; Lu, L.; Cheng, G.D.; Xiao, H.L. Quantifying landscape structure of the Heihe River Basin, north-west China using FRAGSTATS. *J. Arid Environ.* **2001**, *48*, 521–535.

60. Murthy, C.S.; Raju, P.V.; Badrinath, K. Classification of wheat crop with multi-temporal images: performance of maximum likelihood and artificial neural networks. *Int. J. Remote Sens.* **2003**, *24*, 4871–4890.

61. Shao, Y.; Fan, X.T.; Liu, H.; Xiao, J.H.; Ross, S.; Brisco, B.; Brown, R.; Staples, G. Rice monitoring and production estimation using multitemporal RADARSAT. *Remote Sens. Environ.* **2001**, *76*, 310–325.

62. Han, H.B.; Ma, M.B.; Wang, X.F.; Ma, S.C. Classifying cropping area of middle Heihe River Basin in China using multitemporal Normalized Difference Vegetation Index data. *J. Appl. Remote Sens.* **2014**, *8*.

63. Nuarsa, I.W.; Nishio, F.; Hongo, C.; Mahardika, I.G. Using variance analysis of multitemporal MODIS images for rice field mapping in Bali Province, Indonesia. *Int. J. Remote Sens.* **2012**, *33*, 5402–5417.

64. Mannel, S.; Price, M.; Hua, D. Impact of reference datasets and autocorrelation on classification accuracy. *Int. J. Remote Sens.* **2011**, *32*, 5321–5330.

65. Friedl, M.A.; Woodcock, C.; Gopal, S.; Muchoney, D.; Strahler, A.H.; Barker-Schaaf, C. A note on procedures used for accuracy assessment in land cover maps derived from AVHRR data. *Int. J. Remote Sens.* **2000**, *21*, 1073–1077.

66. Mannel, S.; Price, M.; Hua, D. A method to obtain large quantities of reference data. *Int. J. Remote Sens.* **2006**, *27*, 623–627.

A Hidden Markov Models Approach for Crop Classification: Linking Crop Phenology to Time Series of Multi-Sensor Remote Sensing Data

Sofia Siachalou, Giorgos Mallinis and Maria Tsakiri-Strati

Abstract: Vegetation monitoring and mapping based on multi-temporal imagery has recently received much attention due to the plethora of medium-high spatial resolution satellites and the improved classification accuracies attained compared to uni-temporal approaches. Efficient image processing strategies are needed to exploit the phenological information present in temporal image sequences and to limit data redundancy and computational complexity. Within this framework, we implement the theory of Hidden Markov Models in crop classification, based on the time-series analysis of phenological states, inferred by a sequence of remote sensing observations. More specifically, we model the dynamics of vegetation over an agricultural area of Greece, characterized by spatio-temporal heterogeneity and small-sized fields, using RapidEye and Landsat ETM+ imagery. In addition, the classification performance of image sequences with variable spatial and temporal characteristics is evaluated and compared. The classification model considering one RapidEye and four pan-sharpened Landsat ETM+ images was found superior, resulting in a conditional kappa from 0.77 to 0.94 per class and an overall accuracy of 89.7%. The results highlight the potential of the method for operational crop mapping in Euro-Mediterranean areas and provide some hints for optimal image acquisition windows regarding major crop types in Greece.

Reprinted from *Remote Sens.* Cite as: Siachalou, S.; Mallinis, G.; Tsakiri-Strati, M. A Hidden Markov Models Approach for Crop Classification: Linking Crop Phenology to Time Series of Multi-Sensor Remote Sensing Data. *Remote Sens.* **2015**, *7*, 3633–3650.

1. Introduction

The problem of ensuring food security for an increasing population is currently one of the main concerns globally. To solve economic and social issues resulting from current and predicted food shortage, one billion hectares of new cropland would be required in order to meet the demand for food by 2050 [1]. However, taking into consideration environmental restrictions, the potential to expand cropland at the expense of other lands, such as forests or rangelands, is limited [2]. The challenge for agronomists, farmers and their allied partners is to produce humanity's food in an ecologically sustainable manner, through socially accepted production systems [3]. These trends suggest an increasing demand for dependable and accurate agricultural monitoring to ensure sustainable crop production and investigation of land management practices [4].

Within this framework, spatially explicit cropland information, such as cropland extent and crop type, is crucial to sustain agriculture and preserve natural resources. A basic prerequisite for the implementation of a land-management strategy is the development of up-to-date Land Use/Land Cover (LULC) databases over agricultural landscapes. Indeed, LULC data, regarding the spatial distribution of crop types, is considered as key information from a geostrategic point of view. The research community is moving towards providing and timely agricultural maps at national or global level of detail [5].

Over the past decade satellite images offer a valuable source of information concerning the monitoring of the Earth's surface in fine spectral and spatial scales. Satellite based earth observation has been used to map crop types under a variety of environmental conditions, providing synoptic coverage of fields in several spectral regions, smooth integration with existing geographical databases under a cost-effective and time-saving approach than traditional statistical surveys [6]. Through these studies, remote sensing techniques have proven to be cost-effective in widespread agricultural lands in Africa, America, Europe and Australia.

Monitoring and mapping vegetation involves investigating vegetation dynamics, such as phenological states and the seasonal growth of crop types. Spectral behavior of agricultural parcels is constantly changing; different crop types may be at a certain instant in the same phenological state, depicting similar spectral attributes, but diverge remarkably in another instant. Classification of parcels based on single-date images, even if they are acquired at critical growth states, cannot offer reliable results in the case of crops with similar growing cycle. As a result, the significance of classification based on multi-temporal images has been well-recognized and especially regarding vegetation mapping, the usage of seasonal imagery is vital [5,7–12].

Low resolution images with high revisit frequency have been processed on the continental and global scale, providing consistent information at high temporal resolution while covering large areas at low costs [13]. However, because of sub-pixel heterogeneity, the spatial resolution of the imagery may result in significant errors in the estimated crop areas [14,15].

At the regional level, crop area estimations have been significantly improved since the introduction of the MODIS sensor with 250 meter ground resolution [13]. MODIS offered unprecedented capabilities for large-area LULC mapping by providing global coverage, half-day revisit capacity and intermediate spatial resolution. Several studies have already demonstrated successfully the potential of these data for detailed LULC mapping in an agricultural setting [15], especially for areas where the typical field size is large [5,16].

Multi-spectral, medium-high resolution images, acquired mainly by Landsat Thematic Mapper/Enhanced Thematic Mapper Plus (TM/ETM+), SPOT and RapidEye [17–20], have been used for regional to local scale crop mapping, either on mono-temporal basis or under a multi-temporal perspective, accounting for small sized fields or heterogeneous crop patterns. To solve the issue of high within-class variability originating from various agronomic practices, the synergy of multi-temporal optical and Synthetic Aperture Radar (SAR) data has also been proposed; the increased set of multi-temporal imagery enables the continuous monitoring of all stages of vegetation development [21]. While multi-temporal data of medium-to-high resolution offers high potential for crop discrimination in fine-structured agricultural landscapes, integration of the temporal information in the classification process is not trivial. Precise annual mapping of crops, through an approach that could be used routinely over large areas, remains challenging [22,23].

Although a variety of algorithms has been employed in crop mapping studies, including among others, the minimum distance, Mahalanobis distance, maximum likelihood, spectral angle mapper and support vector machines, an approach integrating phenological models in the classification process has been given little attention so far. Through phenology, remote sensing observations and biophysical changes during vegetation's growth can be linked statistically in order to discriminate crop types. A possible way of incorporating knowledge of phenology into the classification process lies on the adoption of the stochastic Hidden Markov Models (HMMs). HMMs allow the simulation of crop dynamics, exploiting the spectral information of their phenological states and their relations. In this regard, a common assumption is made that the vegetation signal and the different phenological states are considered random variables [24]. The correlation of phenological states is described by different transition state probabilities in each crop model. As far as the different cultivation practices, the algorithm reckons the possibility of temporal variation in the phenological cycle. Different growing states of the same crop type per image can be introduced instead of using a generalized crop model. These states correspond to different spectral attributes and are used jointly to define each model. This is the basic advantage of HMMs compared to other techniques that produce simulations of average seasonal phenology [25] and may fail to account for a restrained or accelerated phenological progress.

Previous work has tested the application of HMMs in Landsat time series to classify mountain vegetation in Norway [26] and arable land in Brazil [27], in MODIS-NDVI time series covering cultivated areas of the United States [28] and NDVI data derived from the Advanced Very High Resolution Radiometer (AVHRR) over the West African savanna [24]. In all the aforementioned studies, the low and medium resolution images have been reported to be adequate for the classification of large-sized agricultural holdings. However, Mediterranean regions that are characterized by distinct environmental and climate settings, high spatio-temporal ecological heterogeneity [29,30], variety of crop types and high fragmentation of farming lands [23,31], require a different approach.

The main aim of this study is the development of a robust crop mapping technique adopting the theory of Markov chains and phenological models, over a Euro-Mediterranean agricultural area. Specifically, this work proposes a crop classification approach that integrates high and medium resolution remote sensing images in order to monitor constant variations in the ecological process of the cropping systems. A pixel-based methodology was selected, instead of using the segment-based approach proposed by [27], to avoid errors produced by segmentation algorithms. The per-segment approach applied to small-sized crop parcels, found over Euro-Mediterranean areas, may have the disadvantage of falsely including within field-crop objects small non-vegetated classes (*i.e.*, roads, canals) leading to an overestimation of the total vegetated area [32].

In particular, the objectives of the study are: (1) the identification of different crop types using a sequence of four seasonal multispectral Landsat ETM+ and a RapidEye image, processed simultaneously through Hidden Markov Models, (2) the assessment of the impact on the accuracy of a pan-sharpening procedure applied to the lower resolution Landsat ETM+ imagery and (3) the investigation of the role of the temporal resolution and extent of the image sequence used, in relation to the phenological cycle of each crop type. The multi-sensor and multi-temporal approach is motivated by the acknowledgment of the potential of coarser spatial resolution data to cover large geographic extends, the demands of complex territories and the growing interest in exploiting multi-scale data synergistically [4]. Furthermore, the definition of optimal temporal acquisition windows is considered vital by several crop mapping studies [8,11,18,33] while it can improve classification accuracy significantly.

2. Study Area

The research site is an irrigated agricultural area, near the city of Thessaloniki, Greece. The study area is dominated by rice and cotton while maize, sugar beet, wheat and alfalfa are planted to a smaller extent. The cropping calendar (planting and harvesting dates) of the area's crop types is presented in Figure 1.

Rice, cotton, maize and sugar beet are summer crops and those fields are characterized by dense vegetation during summer months. Wheat is harvested before June and is a spring crop. Rice, cotton, maize, sugar beet and wheat are considered annual crops and have a 12-months cycle. Alfalfa on the other hand can have 3–4 cuttings and flowerings per year, usually between May and September. Thus, the cropping pattern of the study area can be considered heterogeneous regarding the dates of planting, emergence, and harvesting. The majority of the parcels are rectangular but small sized. Despite the applied land consolidation the size of the parcels ranges from 0.006 to 10 ha. The terrain across the study area is relatively flat. The average annual temperature of the study area is 15.8 °C. The area is characterized by modest annual rainfall, averaging 441 mm/y.

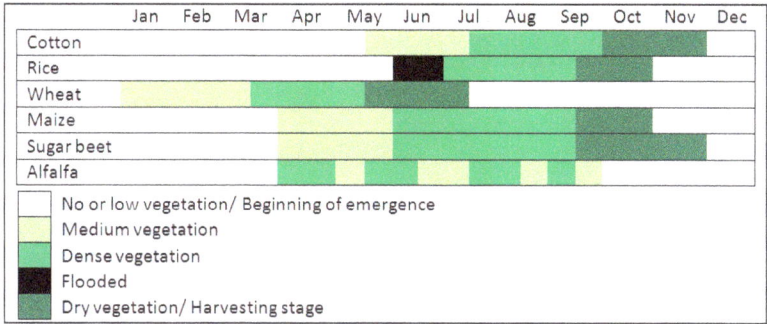

Figure 1. Idealized cropping calendar of the main crop types grown in the study area.

3. Materials and Methods

3.1. Outline of the Methodology

Originally, the Landsat ETM+ images were registered to the higher resolution RapidEye image, which has been georeferenced using ground control points (GCPs) identified over VHR orthophotographs (Figure 2) in the same geodetic system with the vector dataset representing field entities of the area (Land Parcel Identification System-LPIS). LPIS was visually corrected for small inconsistencies. The multispectral ETM+ images were pan-sharpened using the panchromatic band and the High Pass Filter (HPF) algorithm. Four synthetic images were produced with a spatial resolution of 15 meters. Nine different classifications experiments were applied on image sequences with variable spatial and temporal characteristics. A common set of training data, derived from the LPIS, was used to estimate the parameters of the crop models. For each HMM, we calculated the probability that the specific set of temporal observations corresponded to a class. Each pixel is assigned to the class whose crop-model emits the maximum probability. The results of the classification tests were evaluated in terms of overall accuracy and kappa coefficients.

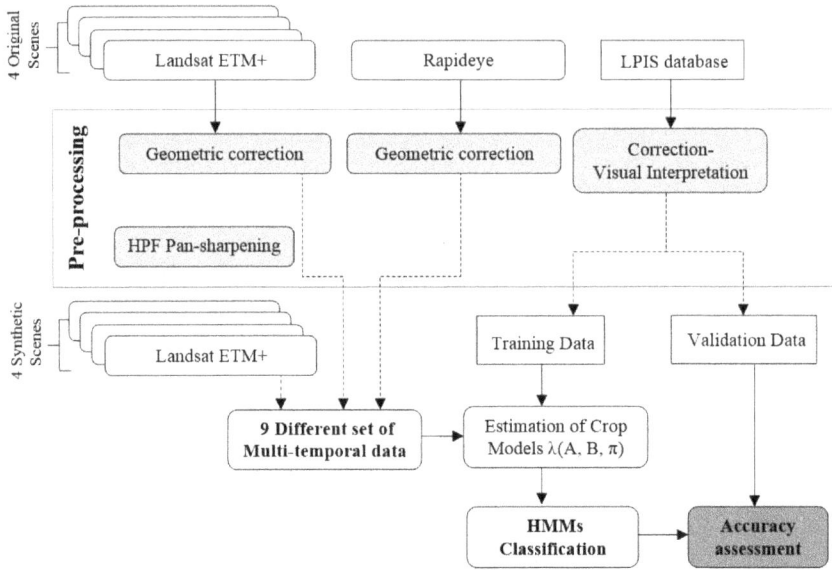

Figure 2. Overall process diagram of this research study.

3.2. Satellite Data and Preprocessing

Imagery acquired from sensors with different spectral, spatial and radiometric characteristics was used in the analysis. More specifically, the multispectral (excluding the thermal bands) and panchromatic components of four Landsat ETM+ images (184/32) and one multispectral RapidEye image, all acquired on different dates of 2010 (Figures 3 and 4), were employed in this study (Table 1).

Table 1. Description of the satellite data used in the study. *: blue; G: green; R: red; NIR: near-infrared; SWIR: shortwave-infrared bands.

Time Step	Sensor	Date of Acquisition	Spatial Resolution Multi/Pan	Radiometric Resolution	Spectral Bands *
t_1	Landsat ETM+	07/05/2010	30/15 m	8-bit	B, G, R, NIR, SWIR1, SWIR2
t_2	Landsat ETM+	08/06/2010	30/15 m	8-bit	B, G, R, NIR, SWIR1, SWIR2
t_3	RapidEye	05/08/2010	5 m	16-bit	R, G, B, Red edge, NIR
t_4	Landsat ETM+	27/08/2010	30/15 m	8-bit	B, G, R, NIR, SWIR1, SWIR2
t_5	Landsat ETM+	30/10/2010	30/15 m	8-bit	B, G, R, NIR, SWIR1, SWIR2

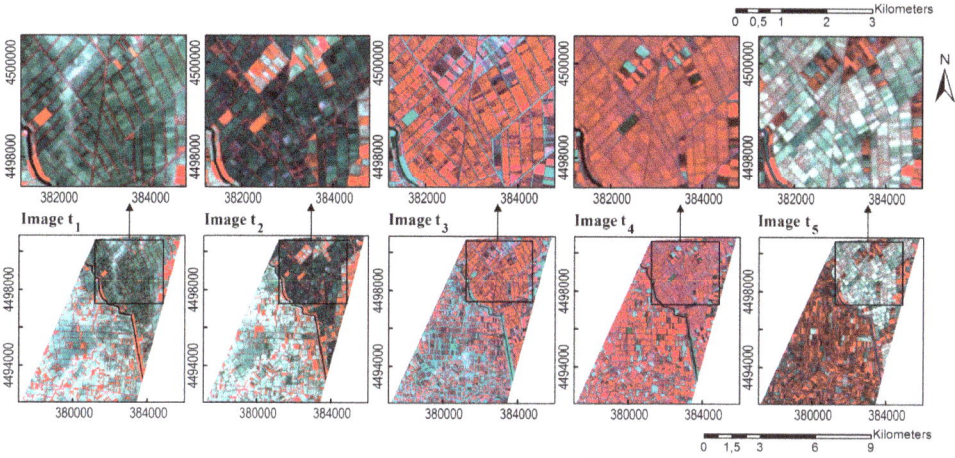

Figure 3. The set of images used included four Landsat ETM+ images (image t_1, image t_2, image t_4, image t_5) illustrated in false-color composite (R: NIR, G:Red, B:Green) and one Rapideye image (image t_3) in false-color composite (R: NIR, G:Red, B:Green).

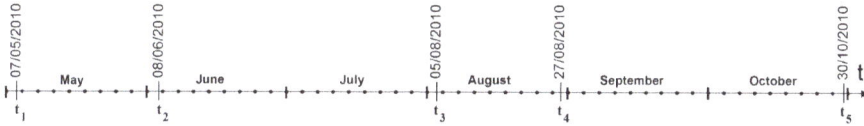

Figure 4. The acquisition dates of the images used in the study.

Landsat ETM + sensor launched in April 1999 has a spatial resolution of 30 meters for the six reflective bands, 60 meters for the thermal band, and 15 meters for the panchromatic (pan) band. On 31 May 2003, the ETM+ Scan Line Corrector (SLC) failed causing the scanning pattern to exhibit wedge-shaped, scan-to-scan gaps, which are most pronounced along the edge of the scene. The scans give near-contiguous coverage of the surface scanned below the satellite in the center of the image (approximately 22 km wide).

RapidEye is a constellation of 5 multispectral satellite sensors launched in August 2008 with a primary focus on agricultural applications. The RapidEye sensor has a multispectral push broom imager with a spatial resolution of 6.25 meters. It captures data in five spectral bands covering visible–infrared part of the electromagnetic spectrum: blue (440–550 nm), green (520–590 nm), red (630–685 nm), red edge (690–730 nm), and near infrared (760–850 nm). In our study, we used the RapidEye (Level 3A) in which radiometric, sensor, and geometric correction have been applied and resampled to a 5 meters spatial resolution. We georeferenced the RapidEye imagery to the Greek Geodetic Reference System 1987, using ground control points, identified on natural color orthoimages with 50 cm spatial resolution acquired on 2007. Between 20 and 30 ground control points were used to co-register the Landsat scenes to the RapidEye imagery, with a Root Mean Square error (RMS) of less than a pixel. We did not apply any atmospheric correction or radiometric normalization since the adopted classification approach does not employ any direct comparison of pixel DN values between the temporal sequences of images. Instead, the classification scheme assigns pixels to crop classes, according to their similarity of states within each image separately and in a subsequent step the temporal images are linked using statistical relationships. In this context, atmospheric correction was not considered a prerequisite [32].

The same geographic subset was identified on every Landsat imagery, with no effective clouds or sensor defects, such as the "SLC-off problem", covering approximately 7500 ha of cultivated area (1760 by 1710 RapidEye pixels). Additionally, regarding the Landsat images, the panchromatic images were merged with the multispectral ones, using the High Pass Filter (HPF) algorithm and four synthetic images were produced with a spatial resolution of 15 meters [21]. Finally, in order to achieve spatial correspondence for each pixel, all Landsat ETM+ images (original and pan-sharpened) were re-sampled to 5 meters, using a nearest-neighbor algorithm to match the spatial resolution the RapidEye image (Figure 3).

3.3. Reference Data

The Land Parcel Identification System (LPIS) is a fundamental part of the Integrated Administration and Control system that has been developed and adopted in 1992 by the EU as the spatial component for the implementation and supporting of the Common Agricultural Policy (CAP) and land management across Europe [34]. The main functions of the LPIS are localization, identification and quantification of agricultural land via very detailed geospatial data, in order to spatially represent the activities of farmers on their land and facilitate the geographical identification of the agricultural parcels declared annually to receive funding [35].

Although the regulatory requirements for the LPIS are uniform across the EU sector, the particular implementations are subject to member states. The Greek GIS-based LPIS integrates information about the crop type, the acreage of a parcel, the identity of the farmer and relates it to a vector layer comprised of the declared parcels. Since the information of this database is gathered from the declaration of the farmers it cannot be considered flawless. In this respect, an expert from "Greek Payment Authority of Common Agricultural Policy Aid Schemes of the Ministry of Rural Development and Food" visually examined and corrected the parcels' crop type and boundaries manually, taking into account the cropping calendar of the study area and the spectral- temporal profile of each parcel. The detailed delineation of the boundaries was guided mainly by the high resolution RapidEye image. In total, 3319 declared parcels were found in the study area. A set of 55 parcels was used as training data and the rest was used during the accuracy assessment of the classification.

3.4. Description of the Proposed HMMs Classification Algorithm

The temporal evolution of vegetation can be described effectively by the state-oriented approach of Hidden Markov Models (HMMs). Each cultivated parcel has a dynamic behavior that depends on cropping phenology, climatic conditions, drought, water irrigation and chemical nutrients (Figure 5).

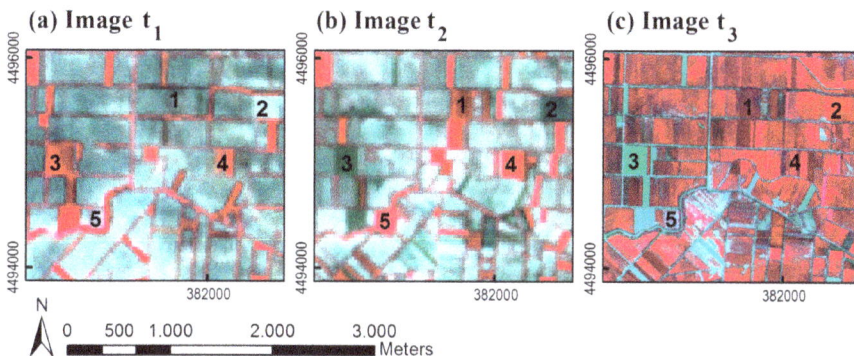

Figure 5. Temporal sequence of images t_1, t_2 and t_3 covering the same area containing parcels cultivated by various crop types (1 = maize, 2 = rice, 3 = wheat, 4 = alfalfa and 5 = sugarcane). It is indicated that the different phenological states of crops at each time step impose significant variance in the between-class separability.

Furthermore, neighboring parcels of the same crop type, over the same time step, may be at different states of growth due to varying agronomic practices; different planting or harvesting dates and fertilizers can accelerate or restrain the phenological progress (Figure 6).

Given that each parcel changes constantly from state to state (Figure 5) and that each state cannot be directly linked to a remote sensing measurement but to a probability distribution of observations, an HMM can be used to simulate the cycle of vegetation based on statistical relations. In this case, an HMM is a doubly embedded stochastic process comprised by two chains: the external chain of the remote sensing observations and the internal chain of states, which are unknown [24] (Figure 7).

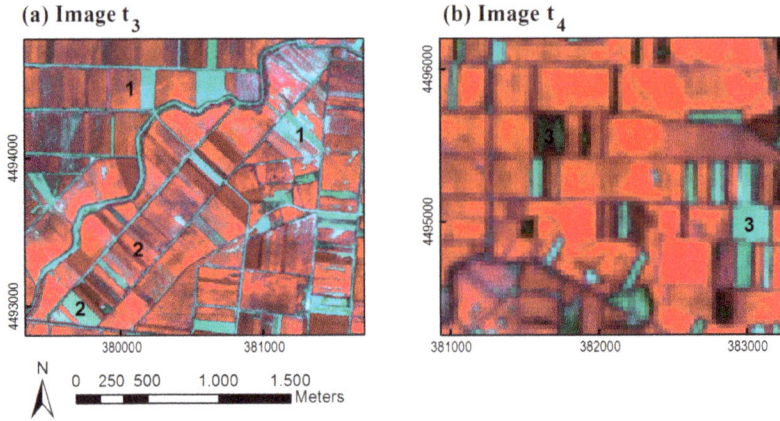

(a) Image t$_3$

(b) Image t$_4$

Figure 6. Different subsets of satellite images t$_3$ and t$_4$ containing parcels with various crop types (1 = cotton, 2 = alfalfa and 3 = maize). In the left subset it can be observed that certain parcels of cotton and alfalfa may resemble according to their phenological state, while other parcels of the same crops can have distinct spectral properties. This can be also observed in the right subset referring to maize fields with different spectral characteristics.

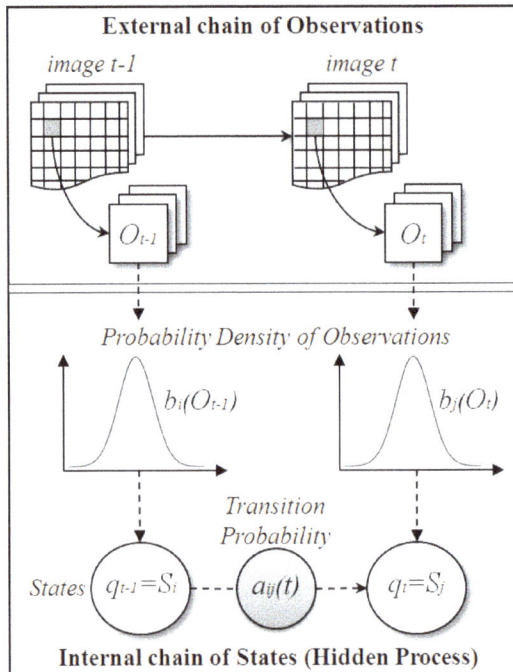

External chain of Observations

image t-1   *image t*

O_{t-1} O_t

Probability Density of Observations

$b_i(O_{t-1})$ $b_j(O_t)$

Transition Probability

States $q_{t-1}=S_i$ --- $a_{ij}(t)$ --→ $q_t=S_j$

Internal chain of States (Hidden Process)

Figure 7. Schematic description of the basic elements of the proposed HMMs.

240

In this study, a crop classification algorithm, based on Markov-chain analysis, was designed and implemented using Matlab®. Accordingly, a model was built describing the phenological states during the dates of the study for each crop type. The sequence of observations consisted of a set of remote sensing measurements $O = \{O_{t1}, ..., O_{tn}\}$, where $t = \{t_1, ..., t_n\}$ are the acquisition dates of the images. The hidden states correspond to the different phenological states $S = \{S_1, ..., S_m\}$ of each crop type and $Q = \{q_{t1}, ..., q_{tn}\}$ is the fixed sequence of hidden states. During the acquisition of the satellite data, the identified states of the parcels in the study area were: S_1, no vegetation or beginning of emergence, S_2, medium vegetation or growth state, S_3, dense vegetation or flowering and S_4, dry vegetation or harvesting state.

An HMM is characterized by the following elements:

I   The state transition probability matrix A, where $a_{i,j}(t) = P[q_t = S_j \mid q_{t-1} = S_i]$, denotes the transition probability from state S_i to state S_j at time t. The emission probability matrix B, defines the probability that O_t is emitted by state S_j, i.e., $b_j(O_t) = P[O_t \mid q_t = S_j]$. In order to estimate the symbol probability distributions B, a multivariate Gaussian distribution is assumed for the observed spectral data. The mean vector μ_i and the covariance matrix Σ_i were calculated by the training data for each crop type, for each state and for every image by the equation,

$$b_i(O_t) = \frac{1}{\sqrt{(2\pi)^d \ |\Sigma_i|}} \exp\left(-\frac{(O_t - \mu_i)^T \Sigma_i^{-1} (O_t - \mu_i)}{2} \right) \tag{1}$$

II   The initial probability π_i is the probability of being in state Si at time t1, i.e., $\pi_i = P[q_{t1} = S_i]$. The parameters A, B and π_i were estimated by the set of training data. The set of training samples was selected and defined according to our knowledge of local agronomic practices and the cropping calendar of the district. To ensure classification success, all classes need to be described by representative training samples. The samples define the different states of crop types according to their different spectral attributes. The set of the training data was evenly distributed in the study area, located in homogenous fields and not in boundary mixed pixels. It should also be noted that in each image, fields of the same crop type may be in different states; usually a state before or a state after (Figure 7). Judging by the cropping calendar and the dates of the used images, not all transitions between states are possible in this case study. Once we have estimated the parameters A, B, π_i of each model λ

and given the sequence of observations O for each pixel, the probability that the sequence O was generated by these models λ is defined by:

$$P\left(q_1 = S_i, \ldots, q_t = S_j, O_1, \ldots, O_t \mid \lambda\right) = P\left(O_1, \ldots, O_t \mid q_1 = S_i, \ldots, q_t = S_j\right) \times$$

$$P\left(q_1 = S_i, \ldots, q_t = S_j \mid \lambda\right) = \pi_{q_1} \times \prod_{l=2}^{t} a_{q_{t-1},\, q_t}(l) \times \prod_{l=1}^{t} b_{q_t}(O_l) \tag{2}$$

Detailed mathematical explanation of HMMs has been reported in previous studies [24,36] which propose the implementation of the Forward algorithm to simplify computations of Equation (2).

In this study, five models were built (one for each crop class), and five measurements of probability were estimated for each pixel (Equation (2)). Let us consider a temporal spectral sequence of pixel x belonging to l_m, the most probable crop, where $L = \{l_1, \ldots, l_k\}$ is the set of possible crop classes. For each pixel x, the most likely crop-class is determined by the following rule:

$$x \in \lambda_m \text{ when } \lambda_m = \text{argmax} \left(P\left(q_1 = S_i, \ldots, q_t = S_j, O_1, \ldots, O_t \mid \lambda_m\right)\right) \tag{3}$$

Finally, nine different HMM models were developed in order to evaluate the role of the temporal resolution and extent of the image sequence used, in relation to the phenological cycle of each crop type, as well as to assess the utility of the pan-sharpening procedure in terms of classification accuracy (Table 2).

Table 2. Selections of different images included in the evaluation of our proposed methodology.

	Original ETM+				Pan-Sharpened ETM+				RapidEye	Rationale of the Classification Experiment
	t1	t2	t4	t5	t1	t2	t4	t5	t3	
HMM-1	X	X	X	X					X	Assessment of the HMMs approach
HMM-2					X	X	X	X	X	Influence of pan-sharpening
HMM-3					X	X	X	X		Influence of spatial resolution-RapidEye
HMM-4					X				X	Influence of temporal extent and resolution
HMM-5						X		X	X	Influence of temporal extent and resolution
HMM-6						X	X		X	Influence of temporal extent and resolution
HMM-7	X	X							X	Influence of temporal extent and resolution
HMM-8	X		X							Influence of temporal extent and resolution
HMM-9		X	X							Influence of temporal extent and resolution

3.5. Accuracy Assessment

Each classification test was evaluated in terms of overall accuracy (OA) and Kappa coefficient, by comparing the reference data with the classified images, pixel by pixel. Overall accuracy represents the proportion of the correctly classified pixels relative to the total number of validation pixels. Kappa coefficient takes into account all the elements of the error matrix and is a measure of the proportional improvement by the classifier over a purely random assignment to classes [37]. Despite the fact that both the OA and Kappa coefficient measure the agreement between the classified map and the reference data, Kappa is often considered a better indicator of classification performance because it excludes chance agreement [38]. Finally, the conditional Kappa coefficient was used for assessing the agreement for the individual crop categories of the maps.

4. Results and Discussion

The results of the accuracy assessment of the classification tests are presented in Table 3 including overall accuracy, overall kappa coefficient and kappa coefficient of each class and are further discussed in the following sections.

Table 3. Overall accuracy, overall kappa coefficient and the kappa coefficient of each class for all experiments according to the selected images.

		HMM-1	HMM-2	HMM-3	HMM-4	HMM-5	HMM-6	HMM-7	HMM-8	HMM-9
Overall Accuracy		84.7%	89.7%	88.5%	87.5%	90.5%	92.5%	75.9%	76.0%	91.1%
Overall Kappa coefficient		0.774	0.843	0.825	0.811	0.852	0.881	0.658	0.655	0.859
Conditional Kappa Coefficient	Cotton	0.662	0.770	0.743	0.722	0.823	0.894	0.504	0.570	0.871
	Rice	0.911	0.882	0.875	0.848	0.924	0.907	0.730	0.640	0.898
	Sugar beet	0.954	0.936	0.934	0.930	0.922	0.906	0.875	0.854	0.900
	Alfalfa	0.789	0.852	0.897	0.795	0.606	0.788	0.694	0.858	0.883
	Maize	0.742	0.861	0.831	0.859	0.867	0.805	0.834	0.641	0.702
	Wheat	0.930	0.916	0.869	0.953	0.659	0.690	0.895	0.705	0.681

4.1. Multitemporal Classification Using the Original ETM+ and RapidEye Imagery

It has been observed [6] that the spatial resolution of the imagery should be at or below the size of the fields. Nevertheless, detailed information provided by high resolution images does not meet the requirements for temporal availability and cost-effective processing framework. When lower resolution sensors are selected, the accuracy of the classification is affected by the mixed pixel problem. For this reason, we explored the processing of time series of high and medium resolution images simultaneously. In the first classification experiment (HMM-1), considering one RapidEye and the original multispectral Landsat ETM+ images (Table 2), the classification model achieved an overall accuracy of 84.7% and an overall Kappa coefficient of 0.774. The conditional kappa coefficients ranging from 0.662 to 0.954, suggest that the spectral information was adequate for discriminating the majority of the crop types.

4.2. Multitemporal Classification Using the Pan-Sharpened ETM+ and RapidEye Imagery

In order to evaluate the contribution of the pan-sharpening of the Landsat ETM+ images in the improvement of the classification process, the developed HMM was tested using the synthetic ETM+ bands along with the multispectral RapidEye imagery (HMM-2). As the spatial resolution of the Landsat ETM+ images increased, the performance of the classification model improved, reaching 89.7% in overall accuracy and 0.843 in overall kappa coefficient (which corresponds to an increase of 5% and 0.069 respectively).

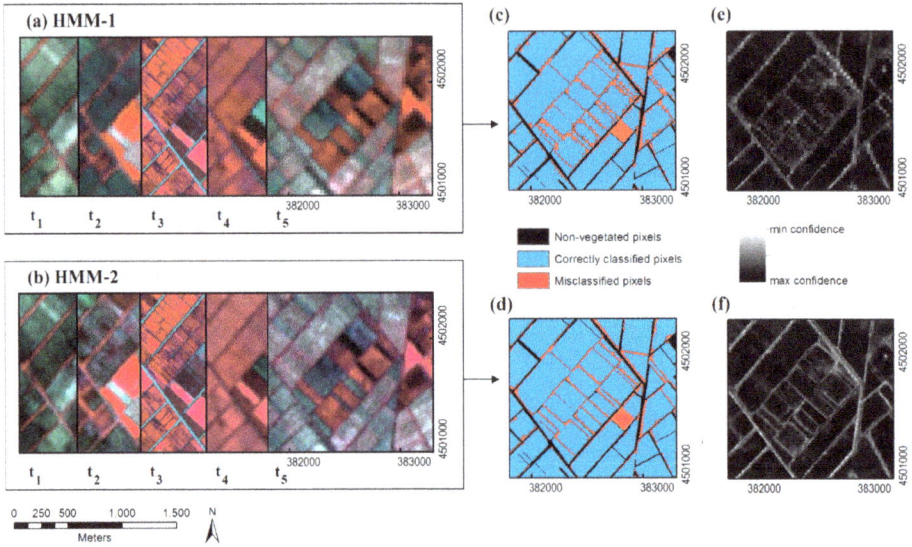

Figure 8. Visual assessment of the classification results obtained from experiment HMM-1 considering the original ETM+ and RapidEye images (**a**) and from experiment HMM-2 considering the pan-sharpened ETM+ and RapidEye images (**b**). HMM-1 experiment resulted to more extended classification errors along the parcel's borderline (**c**) and the respective classification errors (**d**) compared to lower classification confidence (**e**) and confidence score (**f**) obtained from experiment HMM-2.

Regarding the conditional kappa coefficients of individual crops, the highest increase of ~0.110 is observed for "cotton" and "maize" classes, which presented the lowest discrimination ability in the previous classification experiment (HMM-1). This sharp increase can be attributed to the shape of the respective crop fields, being more elongated and narrow compared to the other crop fields of the area. The spatial explicit assessment of the classification errors distribution resulting from the HMM-1 and HMM-2 maps (Figure 8), verifies the contribution of the pan-sharpening procedure in the improvement of the classification result. Differences are observed along the boundaries of the different crop fields with a larger proportion of misclassified area found along the borderline of the fields in the case of the original dataset (HMM-1). In addition, the confidence score derived by the computed probabilities of the HMMs and the corresponding confidence images (Figure 8) prove that a larger proportion of each agriculture field is classified with higher confidence in the case of the HMM-2 model.

4.3. Multitemporal Classification Using the Pan-Sharpened ETM+ Imagery

To quantify the contribution of the RapidEye image in the classification scheme the third classification experiment (HMM-3) included only the pan-sharpened Landsat ETM+ images. The overall accuracy of this experiment was 88.5% while the overall kappa coefficient was 0.825. Compared to the results obtained from the previous experiment, integrating the RapidEye imagery, a decrease in both accuracy metrics is evident (1.2% in OA and 0.018 in Kappa values); this can be attributed to the information content inherent to the higher spatial resolution RapidEye imagery.

Parcels not being wide enough to be mapped in a 15 meter resolution may be distinguished in a RapidEye image of 5 meters pixel size. Thus, the RapidEye image actually adds information involving narrow parcels and boundary pixels that cannot be viewed in a Landsat ETM+ scene. Comparison of the individual class results obtained by classification experiments HMM-2 and HMM-3 indicates that the lowest differences exist for the conditional kappa coefficient of class "sugar beet". This relates to the fact that these crop fields within the study area, have a mean size of 1.28 ha. Their relatively large size allows satisfactory discrimination despite the coarse spatial resolution of the pan-sharpened ETM+ bands.

4.4. Multitemporal Classification Considering Different Temporal Extents

One of our objectives was to assess classification accuracy obtained by decreasing the number of images used in our classification model. Usually five images per year are used to perform multi- temporal classification in agriculture applications [12]. During classification experiments of HMM-4 to HMM-9, the number of images employed in the model decreases, but the resulting overall accuracy does not decrease proportionally. The selection of different temporal images affects the discrimination of certain crop types. By using three images (HMM-4, HMM-5 and HMM-6) instead of five, the conditional kappa coefficient of each summer crop remains relatively high (above 0.722). As a matter of fact, in HMM-6, where the three selected images were all acquired during summer, the corresponding summer crops attain the highest accuracy values ranging between 0.805 and 0.907. When using just three or two images neither the overall accuracy nor the overall kappa coefficient necessarily decrease, but certain individual crop accuracies fall below average. In the cases of the classification experiments HMM-5, HMM-6 and HMM-9, which do not incorporate the ETM+ image acquired on May (t_1), the overall accuracy increases, ranging between 90.5% and 92.5% with an overall kappa coefficient between 0.852 and 0.881. However, these classification experiments perform poorly for class "wheat" because it is an "early crop". This implies that only in the early May image (t_1) "wheat" fields appear vegetated. In fact, the kappa coefficient of "wheat" drops at 0.659–0.690, while it reaches values of 0.869–0.953 in all other tests. As for class "alfalfa", it has been already stated that it has a different phenological cycle, depending on the varied dates of the cuttings. This crop type does not have a seasonal pattern like the other crops because the state of "emergence" can be repeated 3–4 times per year and even neighboring fields can differ spectrally. For this crop type at least 4 images are required to reach a stable classification result, as indicated by the classification results.

Comparison of all the classification experiments, suggests that HMM-2 involving four pan-sharpened ETM+ and one RapidEye images, provided the best results, as far as crop-specific accuracies are concerned. It ranges between 0.770 and 0.936, whereas in other tests it falls below 0.60. The visual assessment of the classification errors and the confidence images, verifies the findings of the accuracy assessment, and highlights the importance of the spatial information inserted into this model through the pan-sharpening procedure for improving the classification of the borderline pixels. The pan-sharpening procedure also improved classification of pixels located within narrow fields by better preserving their spectral attributes. Even though, the computed measures of 89.7% in the overall accuracy and 0.843 in the overall kappa coefficient are not the highest, all accuracies per class reach satisfactory levels. Similar performance metrics were reached in [27] and [18], where five images were sufficient to reach an overall accuracy of 85%–89%. The further decrease in the number of images had a significant impact deteriorating the average accuracy per class. As far as the kappa coefficients are concerned, the highest values are observed on class "sugar beet" and class "wheat" (0.936 and 0.916 respectively). This is due to the size of the sugar beet parcels which can be monitored by the Landsat ETM+ scenes. The high accuracy of "wheat" is justified by its different phenology compared to the rest of the crop types. The poorest results were obtained for class "cotton" (0.77 in the kappa index), which was confused with the other "summer crops" due to its high internal spectral variability, resulting from the different farming practices applied. Apart from errors related to sub-pixel heterogeneity, classification errors related to whole field misallocation might arise (Figure 8). This kind of errors stem from various agronomic practices (*i.e.*, different dates of planting and harvesting, usage of herbicides and fertilizers, *etc.*). This is a common problem in crop mapping studies [21] and could be resolved with the insertion of additional images representing more adequately high within-class variability.

In conclusion, the results suggest optimal dates of scenes according to which crop types are to be examined. It should be noted that deciding on the optimal number and dates of scenes depends on the study area, the number of classes and the variety of cropping systems. In [11] it was proposed that four scenes were adequate for the classification of six crop classes in Japan, while in [32] 2–3 multi-date images were used to discriminate seven classes including "grassland" in the Netherlands. In this paper, by selecting 3–5 scenes the overall accuracy ranges between 87.5% and 92.5%, the overall kappa coefficient between 0.811 and 0.881 and the least conditional kappa coefficient is about 0.60. However, reducing the number of images to two, led to an overall accuracy and overall kappa coefficient of about 75% and 0.65 respectively and a conditional kappa index of 0.50 which are considered moderate. Yet, if we need to constrain to using only two multi-date images we should choose a combination of May and August to account for both "winter" and "summer" cropping systems.

5. Conclusions

In this study, the classification approach was directed to meet the needs of a Mediterranean agricultural area. The approach integrated the following ideas: (a) the theory of HMMs to describe the dynamics of vegetation, (b) the combination of multi-sensor data and (c) the implementation of image enhancement techniques.

The challenge of crop mapping stems from the variety of cropping systems, distinct climate settings and cultivation practices. By using Hidden Markov Models we were able to set a dynamic model per crop type to represent the biophysical processes of agricultural land. Due to the high fragmentation of the land a multi-resolution approach was introduced. Experimental results demonstrated that the integration of even one very high resolution image and pan-sharpening of the set of Landsat images improved the overall classification accuracy by 1.2% and 5% respectively. Spatial explicit classification errors demonstrated that our methodology succeeded in mapping even small-sized fields enhancing classification even on the boundaries of different crop fields. It is worth mentioning that our approach can incorporate and process different satellite data without the implementation of atmospheric correction because the covariance matrix and the mean vector are computed by the samples for each image separately.

However, an evident shortcoming is that models cannot be directly transferred to another year or a different region. This can be attributed to two factors: the inter-annual variation of climate conditions and the heterogeneity of cropping patterns of distinct territories. Even within similar agricultural areas, crops may grow in a different rate depending on the soil or irrigation system. To this end, the algorithm can be extended to integrate weather information and improve the estimation of the transition probabilities. Another drawback is that ground reference data are required each year and over the specific study area to train the classifier. In this paper, to avoid time consuming and costly on-the-field visits, we proposed the use of the ancillary crop maps.

Depending on the application and the investigated crop types, different sets of temporal images may be used. If we are interested in mapping only "summer" or "winter" crops a set of only two images may provide adequate classification results.

However, it has been shown that for more complex agricultural landscapes a robust classification scheme requires at least 4–5 images to achieve a kappa index per class above 0.70.

The increased availability of imagery by recent and up-coming satellite missions (*i.e.*, Landsat-8 and Sentinel-2) will offer a more dense set of observations and will broaden the mapping capabilities of the proposed technique. Additionally, further research may explore the integration of vegetation indices (VIs) to examine the impact on the classification performance. Considering the computational complexity of the algorithm, a reduction in the dimension of the input data by using VIs will improve the efficiency of the processing. In the future, classification errors could be eliminated by extending the model to incorporate contextual knowledge and accounting for possible interactions of neighboring pixels.

Acknowledgments: Landsat-7 ETM+ data used were available at no-cost from the U.S. Geological Survey. The USGS home page is http://www.usgs.gov. The authors would like to thank the Greek Payment Authority of Common Agricultural Policy Aid Schemes of the Ministry of Rural Development and Food for providing reference data from the Land Parcel Identification System and especially the agronomist Katerina Ioannidou for correcting the errors of the database.

Author Contributions: Sofia Siachalou conceived the idea, designed the HMMs algorithm, performed the experiments and prepared the manuscript. Giorgos Mallinis contributed to image-pre-processing and manuscript writing and gave conceptual advice on designing the experiments. Maria Tsakiri-Strati supervised the research and commented on the manuscript at all stages. All authors contributed to the interpretation of the results and manuscript revisions.

Conflicts of Interest: The authors declare no conflict of interest.

References

1. Gong, P.; Wang, J.; Yu, L.; Zhao, Y.; Zhao, Y.; Liang, L.; Niu, Z.; Huang, X.; Fu, H.; Liu, S.; *et al.* Finer resolution observation and monitoring of global land cover: First mapping results with Landsat TM and ETM+ data. *Int. J. Remote Sens.* **2012**, *34*, 2607–2654.

2. Thenkabail, P.S. Global croplands and their importance for water and food security in the twenty-first century: Towards an ever green revolution that combines a second green revolution with a blue revolution. *Remote Sens.* **2010**, *2*, 2305–2312.

3. Miller, F.P. After 10,000 years of agriculture, whither agronomy? *Agron. J.* **2008**, *100*, 22–34.

4. Löw, F.; Duveiller, G. Defining the spatial resolution requirements for crop identification using optical remote sensing. *Remote Sens.* **2014**, *6*, 9034–9063.

5. Wardlow, B.D.; Egbert, S.L.; Kastens, J.H. Analysis of time-series MODIS 250 m vegetation index data for crop classification in the U.S. Central great plains. *Remote Sens. Environ.* **2007**, *108*, 290–310.

6. Ozdogan, M.; Yang, Y.; Allez, G.; Cervantes, C. Remote sensing of irrigated agriculture: Opportunities and challenges. *Remote Sens.* **2010**, *2*, 2274–2304.

7.   Arvor, D.; Jonathan, M.; Meirelles, M.S.P.; Dubreuil, V.; Durieux, L. Classification of MODIS EVI time series for crop mapping in the state of Mato Grosso, Brazil. *Int. J. Remote Sens.* **2011**, *32*, 7847–7871.

8.   Carrão, H.; Gonçalves, P.; Caetano, M. Contribution of multispectral and multitemporal information from MODIS images to land cover classification. *Remote Sens. Environ.* **2008**, *112*, 986–997.

9.   Foerster, S.; Kaden, K.; Foerster, M.; Itzerott, S. Crop type mapping using spectral-temporal profiles and phenological information. *Comput. Electron. Agr.* **2012**, *89*, 30–40.

10. Jia, K.; Liang, S.; Wei, X.; Yao, Y.; Su, Y.; Jiang, B.; Wang, X. Land cover classification of Landsat data with phenological features extracted from time series MODIS NDVI data. *Remote Sens.* **2014**, *6*, 11518–11532.

11. Murakami, T.; Ogawa, S.; Ishitsuka, N.; Kumagai, K.; Saito, G. Crop discrimination with multitemporal SPOT/HRV data in the Saga plains, Japan. *Int. J. Remote Sens.* **2001**, *22*, 1335–1348.

12. Zhong, L.; Gong, P.; Biging, G.S. Efficient corn and soybean mapping with temporal extendability: A multi-year experiment using Landsat imagery. *Remote Sens. Environ.* **2014**, *140*, 1–13.

13. Lunetta, R.S.; Shao, Y.; Ediriwickrema, J.; Lyon, J.G. Monitoring agricultural cropping patterns across the Laurentian Great Lakes Basin using MODIS-NDVI data. *Int. J. Appl. Earth Obs. Geoinf.* **2010**, *12*, 81–88.

14. Atzberger, C.; Rembold, F. Mapping the spatial distribution of winter crops at sub-pixel level using AVHRR NDVI time series and neural nets. *Remote Sens.* **2013**, *5*, 1335–1354.

15. Chang, J.; Hansen, M.C.; Pittman, K.; Carroll, M.; DiMiceli, C. Corn and soybean mapping in the United States using MODIS time-series data sets. *Agron. J.* **2007**, *99*, 1654–1664.

16. Yao, F.; Feng, L.; Zhang, J. Corn area extraction by the integration of MODIS-EVI time series data and China's environment satellite (HJ-1) data. *J. Indian Soc. Remote Sens.* **2014**, *42*, 859–867.

17. Oetter, D.R.; Cohen, W.B.; Berterretche, M.; Maiersperger, T.K.; Kennedy, R.E. Land cover mapping in an agricultural setting using multiseasonal thematic mapper data. *Remote Sens. Environ.* **2001**, *76*, 139–155.

18. Conrad, C.; Dech, S.; Dubovyk, O.; Fritsch, S.; Klein, D.; Löw, F.; Schorcht, G.; Zeidler, J. Derivation of temporal windows for accurate crop discrimination in heterogeneous croplands of Uzbekistan using multitemporal RapidEye images. *Comput. Electron. Agr.* **2014**, *103*, 63–74.

19. Yang, C.; Everitt, J.H.; Murden, D. Evaluating high resolution SPOT 5 satellite imagery for crop identification. *Comput. Electron. Agr.* **2011**, *75*, 347–354.

20. Turker, M.; Ozdarici, A. Field-based crop classification using SPOT4, SPOT5, IKONOS and Quickbird imagery for agricultural areas: A comparison study. *Int. J. Remote Sens.* **2011**, *32*, 9735–9768.

21. Forkuor, G.; Conrad, C.; Thiel, M.; Ullmann, T.; Zoungrana, E. Integration of optical and Synthetic Aperture Radar imagery for improving crop mapping in Northwestern Benin, West Africa. *Remote Sens.* **2014**, *6*, 6472–6499.

22. Gumma, M.; Pyla, K.; Thenkabail, P.; Reddi, V.; Naresh, G.; Mohammed, I.; Rafi, I. Crop dominance mapping with IRS-P6 and MODIS 250-m time series data. *Agriculture* **2014**, *4*, 113–131.

23. Zhong, L.; Hawkins, T.; Biging, G.; Gong, P. A phenology-based approach to map crop types in the San Joaquin Valley, California. *Int. J. Remote Sens.* **2011**, *32*, 7777–7804.

24. Viovy, N.; Saint, G. Hidden markov models applied to vegetation dynamics analysis using satellite remote sensing. *IEEE Trans. Geosci. Remote Sens.* **1994**, *32*, 906–917.

25. Bradley, B.A.; Jacob, R.W.; Hermance, J.F.; Mustard, J.F. A curve fitting procedure to derive inter-annual phenologies from time series of noisy satellite NDVI data. *Remote Sens. Environ.* **2007**, *106*, 137–145.

26. Aurdal, L.; Huseby, R.B.; Eikvil, L.; Solberg, R.; Vikhamar, D.; Solberg, A. Use of Hidden Markov Models and Phenology for Multitemporal Satellite Image Classification: Applications to Mountain Vegetation Classification. Available online: http://publications.nr.no/directdownload/rask/old/multitemp05.pdf (accessed on 14 January 2015).

27. Leite, P.B.C.; Feitosa, R.Q.; Formaggio, A.R.; Da Costa, G.A.O.P.; Pakzad, K.; Sanches, I.D. Hidden markov models for crop recognition in remote sensing image sequences. *Pattern Recogn. Lett.* **2011**, *32*, 19–26.

28. Shen, Y.; Wu, L.; Di, L.; Yu, G.; Tang, H.; Yu, G.; Shao, Y. Hidden Markov models for real-time estimation of corn progress stages using MODIS and meteorological data. *Remote Sens.* **2013**, *5*, 1734–1753.

29. Berberoglu, S.; Lloyd, C.D.; Atkinson, P.M.; Curran, P.J. The integration of spectral and textural information using neural networks for land cover mapping in the Mediterranean. *Comput. Geosci.* **2000**, *26*, 385–396.

30. Mallinis, G.; Koutsias, N. Spectral and spatial-based classification for broad-scale land cover mapping based on logistic regression. *Sensors* **2008**, *8*, 8067–8085.

31. Cohen, Y.; Shoshany, M. A national knowledge-based crop recognition in Mediterranean environment. *Int. J. Appl. Earth Obs. Geoinf.* **2002**, *4*, 75–87.

32. De Wit, A.J.W.; Clevers, J.G.P.W. Efficiency and accuracy of per-field classification for operational crop mapping. *Int. J. Remote Sens.* **2004**, *25*, 4091–4112.

33. Van Niel, T.G.; McVicar, T.R. Determining temporal windows for crop discrimination with remote sensing: A case study in south-eastern Australia. *Comput. Electron. Agr.* **2004**, *45*, 91–108.

34. Kay, S. *Monitoring and Evaluation of IACS Implementation for the Identification of Agricultural Parcels in Member States of the EU*; Agriculture and Fisheries Unit: Ispra, Italy, 2002.

35. Sagris, V.; Wojda, P.; Milenov, P.; Devos, W. The harmonised data model for assessing land parcel identification systems compliance with requirements of direct aid and agri-environmental schemes of the cap. *J. Environ. Manag.* **2013**, *118*, 40–48.

36. Rabiner, L.R. Tutorial on hidden markov models and selected applications in speech recognition. *Proc. IEEE* **1989**, *77*, 257–286.
37. Cohen, J. A coefficient of agreement for nominal scales. *Educ. Psychol. Meas.* **1960**, *20*, 37–46.
38. Congalton, R.G. A review of assessing the accuracy of classifications of remotely sensed data. *Remote Sens. Environ.* **1991**, *37*, 35–46.

MODIS-Based Fractional Crop Mapping in the U.S. Midwest with Spatially Constrained Phenological Mixture Analysis

Cheng Zhong, Cuizhen Wang and Changshan Wu

Abstract: Since the 2000s, bioenergy land use has been rapidly expanded in U.S. agricultural lands. Monitoring this change with limited acquisition of remote sensing imagery is difficult because of the similar spectral properties of crops. While phenology-assisted crop mapping is promising, relying on frequently observed images, the accuracies are often low, with mixed pixels in coarse-resolution imagery. In this paper, we used the eight-day, 500 m MODIS products (MOD09A1) to test the feasibility of crop unmixing in the U.S. Midwest, an important bioenergy land use region. With all MODIS images acquired in 2007, the 46-point Normalized Difference Vegetation Index (NDVI) time series was extracted in the study region. Assuming the phenological pattern at a pixel is a linear mixture of all crops in this pixel, a spatially constrained phenological mixture analysis (SPMA) was performed to extract crop percent covers with endmembers selected in a dynamic local neighborhood. The SPMA results matched well with the USDA crop data layers (CDL) at pixel level and the Crop Census records at county level. This study revealed more spatial details of energy crops that could better assist bioenergy decision-making in the Midwest.

Reprinted from *Remote Sens.* Cite as: Zhong, C.; Wang, C.; Wu, C. MODIS-Based Fractional Crop Mapping in the U.S. Midwest with Spatially Constrained Phenological Mixture Analysis. *Remote Sens.* **2015**, *7*, 512–529.

1. Introduction

The Midwest is one of the major agricultural regions in the United States. In 2007, the Midwestern states had a market value of over $76 billion for crops (corn, soybean, wheat, and forage grass) and livestock [1]. Currently corn grain is still the most commonly utilized feedstock for ethanol [2]. With increased biofuel demand, food security and environmental contamination from intensified corn cropping become major concerns in this region [3]. Perennial native prairie grasses are recognized as promising alternative energy crops for cellulosic feedstock [4]. The USDA National Agricultural Statistics Service (NASS) publishes annual cropland data layer (CDL) products, in which annual crops are classified from satellite images at 30–56 m resolutions [5]. Perennial crops, however, are not examined in these products. To assist with sustainable bioenergy land use, there is a need for accurate mapping of energy crops in this important agricultural region.

Remote sensing has been proven to be an effective tool of regional and global agricultural monitoring. Aside from the U.S. CDL products from medium-resolution imagery, global cropland extents and irrigated areas have been extracted from frequently observed, coarse-resolution data [6,7]. A comprehensive review was presented in [8] about studies of global croplands and their water use with remote sensing and non-remote sensing approaches by the world's leading researchers. Yet, even with high spatial-resolution imagery, mapping of individual crop types is often limited to large fields with homogeneous structures [9]. In complex agricultural areas with a diversity of crops, different crops often have similar spectra during the growing season. It is thus difficult to delineate crops using regular binary classifiers [10]. For satellite images at coarse resolutions, the mixed-pixel problem also results in uncertainties in crop delineation.

In the temporal domain, seasonal variations of the normalized difference vegetation index (NDVI) are closely related to phenological features such as the onset dates of green-up, peak growth, and senescence along vegetation development [11]. Phenology of annual crops is associated with their planting dates and development cycles in growing seasons. For example, corn is usually planted 1–2 weeks earlier than soybeans, but both have similar growth cycles along the season. Winter wheat is planted in winter and primarily grows in early spring. Spring wheat has a narrow growth cycle due to cold weather in the north. For perennial crops, cool-season grass (CSG) starts its growth in early spring and reaches peak growth in May, while warm-season grass (WSG) starts in later spring and has delayed Peak dates [12]. In addition, CSG turns to dormancy in hot, dry summers and has a second growth peak in the fall, while WSG remains green in summer. These phenological differences provide important information for crop mapping with repeated satellite observations [13–15]. Relying on these unique phenological features, multi-temporal, medium-resolution satellite imagery has been used for national mapping products such as the annual CDL maps [4] and the National Land Cover Databases (NLCD) [16] in the conterminous United States. Due to the tremendous amount of such satellite scenes needed in these products, the classification processes are time-consuming and labor-/cost-intensive. Moreover, limited by their coarse revisit cycles (e.g., 16-day interval for Landsat) and frequent contamination from cloud cover in the growing season, it is often difficult to extract stable phenological features from these data series for regional mapping processes.

The Moderate Resolution Imaging Spectroradiometer (MODIS) satellite product has been available since 2000. Its capabilities for daily observation and global coverage allow for efficient monitoring of seasonal crop development in large regions [17] and operational cropland estimation [18]. Algorithms using MODIS time series to derive phenological parameters have also been developed for crop mapping and monitoring its interannual dynamics [12,15]. At 250–1000 m resolution, a MODIS pixel often covers multiple crop fields on the ground. It reduces the accuracies of crop mapping, and smaller crop clusters are often lost with conventional binary classification approaches [15]. This may severely affect regional crop analysis that relies on the accuracies of crop delineation [19].

Spectral mixture analysis (SMA) has been widely used to extract sub-pixel information of land covers based on their spectral differences [20–22]. Recently, some efforts were made to extract sub-pixel land covers with crop phenology from MODIS time series [23,24], the so-called phenological mixture analysis (PMA). Both methods share the same principle, *i.e.*, the SMA approaches improve the spatial resolutions with spectral signature of crops, while the PMA approaches perform the same process with their temporal signatures. However, two major challenges remain: (1) appropriate selection of endmembers [20,25]; and (2) identifying the correct signatures that characterize crops. Differences in spectral or temporal signatures of the same class (endmember variability) may significantly affect the accuracies of sub-pixel land cover fractions [21,25]. To reduce these in-class differences, Deng and Wu [22] developed a spatially adaptive spectral mixture analysis, in which spatial patterns were used to overcome the endmember variability in extracting sub-pixel impervious surfaces in urban lands. The idea of spatial adjustment could also be employed in crop unmixing and regional crop mapping.

This study aims to combine the phenology-based mapping and unmixing approaches to extract in-pixel fractional crop covers in the Midwest using MODIS time series in 2007. A spatially constrained PMA (SPMA) approach was developed to overcome the challenges in regular unmixing processes. The results were validated with the 56-m CDL products in the region. The extracted crop planting acreages were also compared with the county-level Crop Census records in the same year. Findings in this study provide spatially detailed information about bioenergy land use in the Midwest.

2. Materials and Methods

2.1. Study Area and Data Sets

The U.S. Midwest is composed of 12 states across the central United States (Figure 1). Topography gradually changes from the arid highlands in the west to gently rolling hills and semi-humid flat lands along the west-east gradient. Prairies cover most of the Great Plains in the western states, while cultivated lands dominate other states across the region—the so-called Corn Belt. Agriculture is the largest driver of local economies in the Midwest, accounting for billions of dollars' worth of exports of grain and livestock production [1]. Corn/soybean shift planting is the common cropping pattern. The extended area grows winter wheat in southern states and spring wheat in the north. The warm-season prairie grasses grow natively across the tallgrass prairie, covering two-thirds of the Midwest [2]. Although more than 90% of tallgrass prairie lands have been cultivated since the European settlement in the 1830s, WSG remains in prairie remnants and is often mixed with introduced, highly productive CSG species in pasturelands.

Accompanying the increased corn ethanol production, expansion of corn planting areas has been recorded in national Crop Census records surveyed by USDA NASS. In 2007, corn acreage reached its historical record (after 1944) of 93.6 million acres, as high as 19% up from 2006 [1]. Ten of the 12 Midwestern states (except Missouri and Minnesota) are marked among the nation's top 10 states for ethanol production capacity [26]. Although cellulosic technology is still in its early stage, crop residuals and prairie native grasses are treated as a promising alternative biofuel feedstock in the Midwest [2,5].

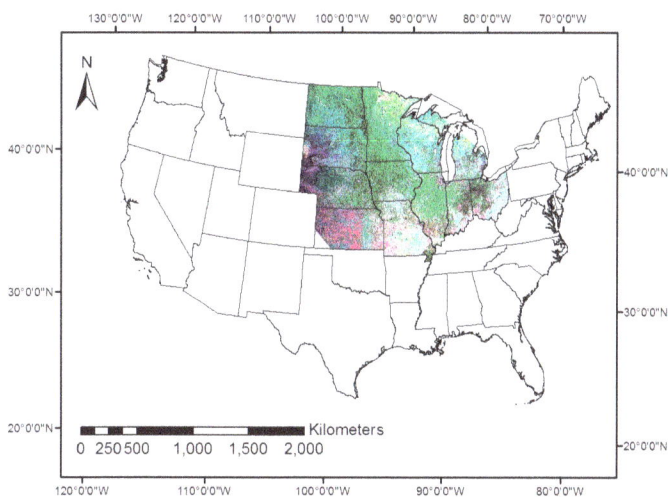

Figure 1. The study area of the Midwest and an example MODIS NDVI composite in 2007 (Day-of-Year 137/177/105 as B/G/R).

The 8-day, 500-m MODIS surface reflectance products (MOD09A1) in 2007 are the primary satellite data in this study. The Midwest can be almost fully covered by four MODIS tiles. For each MODIS tile, all MOD09A1 scenes in 2007 were downloaded from the Land Processes Distributed Active Archive Center and re-projected to the North America Datum 1983. NDVI was then extracted from the red and near infrared spectral bands of each MODIS scene, mosaicked and clipped to the Midwest region, and stacked to 46-scene NDVI time series. The time series was finally smoothed with a five-point median filter followed by the Savitzky-Golay filtering method [2,27]. The example NDVI composite in Figure 1 demonstrates the clustering patters of crops in the Midwest. The 250-m MODIS NDVI products such as

MOD13Q1 [14] were not used in this study because of their longer temporal intervals (16-day). As revealed in previous studies [2], at such temporal resolution the subtle biophysical differences in critical growing stages are often lost.

The CDL products were downloaded from USDA NASS [28]. In 2007 the CDL maps were classified from the Indian RESOURCESAT-1 Advanced Wide Field Sensor (AWiFS) imagery acquired in the growing season, with generally 85%–95% accuracies for major crop-specific land cover categories [4]. At 56-m resolution, they could serve as the reference for crop classification in this study. To simplify the process, only corn, soybean, winter wheat, spring wheat fields, and perennial lands were considered in this study. Other crops and non-crop covers were blacked out from the CDL map. Additionally, the NASS county-level Crop Census records were downloaded and used as a secondary data source for crop planting acreages.

Training data of annual crops (corn, soybean, winter wheat, and spring wheat) were selected from the 2007 CDL map. A layer covering the four annual crops was extracted from the raw CDL map and aggregated to 500 m cell size with the following process. In a given 500-m cell, the percentage of each crop was calculated as the ratio of the number of CDL pixels of this crop to the total CDL pixels in the 500×500 m^2 square. Therefore, the re-processed 500 m CDL data actually contained the percent cover layers of the four annual crops. In the percent cover layer of each crop, subsets of 5×5 pure cells (100%) were extracted and their central locations were collected as ground truth sample points. To reduce errors induced from CDL classification, NDVI time series of these samples were individually examined based on a crop's specific phenological patterns in accordance with crop calendars. A total of 172 sample points for corn, 96 for soybean, 103 for spring wheat, and 100 for winter wheat were randomly collected as their training samples (Figure 2).

For perennial crops, the WSG and CSG distributions were not available in any public agricultural databases. In this study, their ground truth samples were extracted from a previously published grass abundance map of plant function types (PFT) in the Great Plains [15], in which a phenology-assisted decision tree was developed to identify different PFT (C3 and C4) grasses in two floristic regions (shortgrass and tallgrass) with yearly MODIS NDVI time series in 2000–2009. The overall classification accuracies reached about 22%. At a given pixel with 10 years' classification results, abundance of a grass type was approximated by the frequency of its occurrence of a grass type in 10 years (in percent). The C3 grass was actually the CSG while C4 was WSG of our study here. Considering geographic differences in both PFT and floristic regions of the Great Plains, we randomly extracted pure (>80%) samples of four grass types: shortgrass CSG (132), shortgrass WSG (143), tallgrass CSG (182), and tallgrass WSG (188). Only four Midwestern states (ND, SD, NE, and KS) are covered in the Great Plains and, therefore, sample points of these grass types were clustered in the west of the study region (Figure 2).

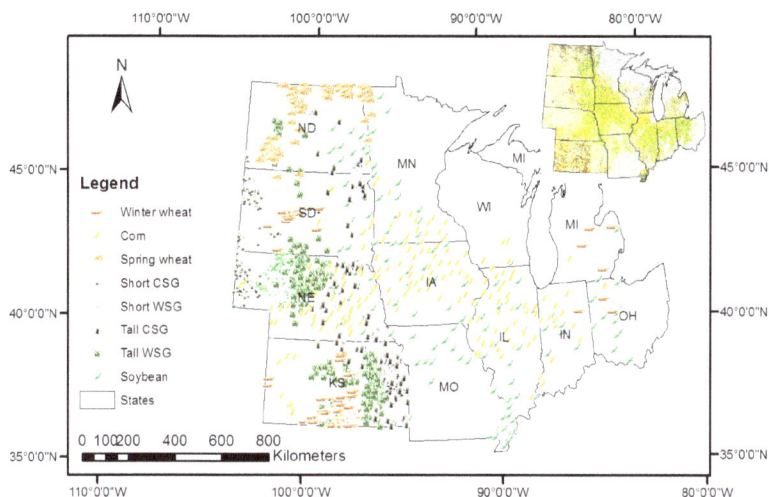

Figure 2. Training data sets of the eight crops in the study area. CSG and WSG represent cool-season and warm-season grasses, while Tall and Short indicate tallgrass and shortgrass floristic regions. The inset is the 2007 CDL map of the region.

2.2. Methodology

2.2.1. NDVI Time Series and Crop Phenology

For each crop, the NDVI values of all training samples were averaged at each time interval. The averaged NDVI time series in Figure 3a reflects the crop's general growth cycle and phenological characteristics. Corn was planted slightly earlier than soybeans and had similar growth cycles throughout the season. Winter wheat had the earliest growth in spring. Spring wheat had the narrowest growth cycle. Perennial grasses had much longer growing season than annual crops. Tall grasses always had higher NDVI values than short grasses. These unique phenological features were useful in delineating these crop types.

The phenological curves in Figure 3a may vary at different geographical locations. It is commonly recognized that crop phenology relies on local environmental conditions. As an apparent phenomenon, for example, a crop's planting or greenness onset dates in the north could be a few weeks later than those in the south. The Midwest covers 15° in latitude across the region. The shift of crop phenology in such a huge region could be dramatic, which leads to considerable phenological variability in the crop. Figure 3b demonstrates the phenological variability of corn by displaying the 95% envelopes of its NDVI variation at each interval. Variations in critical growing stages such as start of growth and senescence (the NDVI variation along the slopes) are much larger than other stages. The average variation for

257

each crop could be simply calculated as the mean of the 46 standard deviations of NDVI along the year. In Table 1, the average variation reaches 5–10 (in the unit of NDVI × 100) for the eight crops. Tallgrass crops had higher average variation because of their heterogeneous growth conditions. The average variation of winter wheat was even higher than 10 due to the dramatic difference of winter conditions in such a large region. The "stddev" in the table is the standard deviation of the 46 standard deviations of NDVI along the year. Its large values indicated large dynamics of the variability in different growing periods along the year. Therefore, it is not suitable to assign the globally averaged phenological curves in Figure 3a as crop endmembers in the unmixing process. Here we established a spatially constrained rule to reduce these uncertainties.

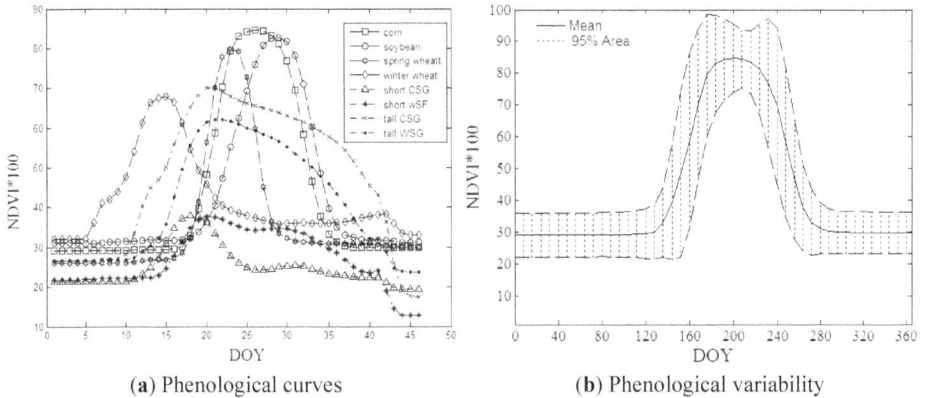

(a) Phenological curves   (b) Phenological variability

Figure 3. The smoothed NDVI time series of the eight crops (a) and phenological variability using corn as an example (b). The X-axis represents the Day-of-Year (DOY) of data acquisition dates in 2007.

Table 1. Descriptive statistics of the 46 standard deviations of the eight crops.

Crops	Corn	Soybean	Spring Wheat	Winter Wheat	Short CSG	Short WSG	Tall CSG	Tall WSG
mean	5.47	6.56	6.17	10.4	4.70	3.79	7.31	6.61
stddev *	3.13	2.64	2.14	2.94	1.28	1.79	3.40	2.49

* stddev denotes the standard deviation.

2.2.2. Spatially Constrained Phenological Mixture Analysis (SPMA)

The phenological curves in Figure 3a are not always distinctive in all MODIS pixels. At coarse resolution, a pixel often covers multiple crops on the ground. Its NDVI time series is thus a combination of multiple phenological curves. Following the logic of linear spectral mixture analysis, we assumed that the NDVI of a mixed pixel was a linearly weighted contribution from multiple crops in the pixel. It can be written as:

$$R_i = \sum_{i=1}^{N} r_{ij}x_j + \varepsilon_i$$

$$\text{with} \sum_{j=1}^{N} x_j = 1 \text{ and } 0 \leqslant x_j \leqslant 1$$

(1)

where R_i is the NDVI value at time i of the mixed pixel ($1 \leqslant i \leqslant 46$), r_{ij} is the NDVI of endmember j at time i. x_j is the fractional cover of endmember j, and N is the number of endmembers ($N \leqslant 8$ in this study). ε_j is the error term at each time.

Endmember selection is a challenge to all unmixing methods. In a mixed pixel, non-representative endmembers result in dramatic classification errors. In this study, phenological variation of crops (as shown in Figure 3b) at different geographic locations cannot be ignored in such a large region. Here we developed a spatially constrained rule to optimize the endmember selection process. The basic hypothesis of the approach is that crop growth and distribution are spatially constrained to environmentally similar conditions. For the same crop type, plants growing nearby share similar water, soil, weather conditions, and management activities, and therefore have more similar phenological curves than those planted farther apart. Past studies have shown that crops have their own clustering patterns across the Midwest. For example, winter wheat is most common in southern states such as Kansas while spring wheat only grows in northern states such as North Dakota and South Dakota, and short grasses are only available in the arid and cold uplands in western states. Therefore, it is reasonable to assume that the phenological curves of a crop growing in a local cluster do not vary much.

Under this hypothesis, pure pixels of a crop from a spatial neighborhood are selected and their averaged phenological curve serves as its endmember in Equation (1). Scanning method is commonly employed to find objects in neighborhoods [21]. When dealing with huge area and relatively less pure pixels in this study, however, it becomes time-consuming and unreliable. Here the spatial distance between any given pixel and the training data set is employed to find the nearest pure pixels. To be computationally efficient in this large region, we used the absolute distance to substitute the Euclidean distance:

$$D_{xy} = |x_i - y_i| + |x_j - y_j|$$

(2)

To initiate the process, at a given pixel the nearest 10 pure pixels (from training data) were set as an empirical threshold of spatial constraints. If a crop had multiple pure pixels in this constraint, its phenological curve was averaged and used in Equation (1). The threshold 10 was determined here based on our repeated tests to balance the accuracy and computation efficiency. When the threshold was too small, the phenological curves of crops were less representative and resulted in large confusion in mixed pixels. When the threshold was set larger (>10), the process became time-consuming and small clusters of a crop were omitted due to its limited pure pixels in the neighborhood. Also, the CDL map in Figure 2 (inset) demonstrates the clustering patterns of annual crops. To better identify these small crop clusters, it was reasonable to assume that at least three crops grow in a local area (e.g., corn, soybean, and CSG grass in the Corn Belt). However, in Figure 2 pure pixels of crops at 500×500 m^2 cell size are not randomly distributed all over the Midwest. In cases where pure pixels of fewer than three crops can be found in this constraint, the search moves to the nearest 20 pure pixels, and so on. It should be noted that the method heavily relied on the distribution of pure pixels. Pure pixels of grasses were clustered in the west of the region (Figure 2) because their training data were only available in four Midwestern states (ND, SD, NE, and KS). Of the CDL products, tallgrass still commonly grows in pasturelands of other states. Therefore, we always used the nearest five pure pixels of tallgrass (WSG or CSG) in the search, although the distance could be larger than 20 pixels.

2.2.3. Accuracy Assessment

Validation samples of annual crops were re-processed from the 2007 CDL product in the study region. For a given crop, a number of sample points were randomly selected on the aggregated CDL percent cover layers (500 m cell size). At any point, percent cover of a crop was the average of a 3×3 local window centered at this point. One hundred fifty random samples were selected for corn and soybeans and 50 samples for spring wheat and winter wheat, respectively. As shown in Figure 2 (inset), the growth of wheat is clustered in the Midwest, with spring wheat in the northern states (ND and SD) and winter wheat in the southern states (especially KS). Their distributions are much lower than for corn and soybeans. Therefore, we selected a smaller size of validation samples for each crop. Validation samples of perennial grasses were randomly extracted from the ranked (at a 10% interval) grass abundance map published by [15]. One hundred fifty random samples were selected for tall grasses, and 50 samples for short grasses, similarly. To be comparable, we grouped the SPMA-extracted results of each perennial crop into the corresponding ranks. Finally, the SPMA-extracted results were also averaged in a 3×3 local window centered at each sample point.

With these validation samples, two common indicators were used to assess the accuracies of the SPMA-extracted crop percent covers by assuming the CDL outputs as references. The root mean square error (RMSE) quantifies the relative errors of the SPMA abundance at the pixel level, while the systematic error (SE) indicates an overall tendency of upward or downward estimation bias [29]. These accuracy metrics can be calculated as:

$$RMSE = \sqrt{\frac{\sum\limits_{i=1}^{N} (r_i - x_i)^2}{N}} \tag{3}$$

$$SE = \frac{\sum\limits_{i=1}^{N} (r_i - x_i)}{N} \tag{4}$$

where x_i is the SPMA-extracted crop percent cover at sample i, r_i is the corresponding percent cover from the CDL-summarized reference, and N is the total number of samples.

Crop planting acreages can be summarized at the county level from the SPMA-extracted percent covers. To assess their accuracies by assuming the Crop Census records as references, we used a mean relative area error (MRAE) to reduce the size effects of different counties:

$$MRAE = \sum\limits_{i=1}^{N} \frac{(r_i - x_i)}{r_i} / N \tag{5}$$

where x_i is the SPMA-extracted crop area (in acres) in county i, r_i is the surveyed crop area from the NASS Crop Census records, and N is the total number of counties in the Midwest.

3. Results and Discussion

3.1. SPMA-Extracted Crop Percent Covers

The SPMA approach calculated the percent covers of all crops. Assigning the pixel to a crop that has the largest cover, crop distributions in the Midwest were extracted from MODIS time series (Figure 4). The general patterns agreed with the CDL product, showing the apparent corn and soybean domination in the Corn Belt, spring wheat in the north, and winter wheat in the south. Grasses dominated the western states. The native prairie remnants identified in [15], such as the Flint Hills (tallgrass prairie) in KS and Sand Hills (tallgrass/shortgrass mixed-grass prairie) in NE, were also extracted in the figure.

In the percent cover maps, corn (Figure 5a) was the primary crop of the region and was mainly distributed in Iowa, Illinois, Indiana, southern Minnesota, and eastern Nebraska. The predominant distribution of corn in 2007 was consistent with the Crop Census records that corn planting areas reached a historical record in this year. Soybeans (Figure 5b) grew in the same area but had less coverage than corn. Most corn and soybean pixels were a mixture of each other because of their shift planting patterns, commonly observed in the Midwest. Spring wheat (Figure 5c) was clustered in North Dakota, South Dakota, and Minnesota, adapting to the cold weather in this area. Winter wheat (Figure 5d) was primarily clustered in Kansas, with expanded planting areas in Nebraska and South Dakota. Their spatially different clustering patterns revealed the climatic impacts on crop planting, and supported the spatial constraint strategy in this study.

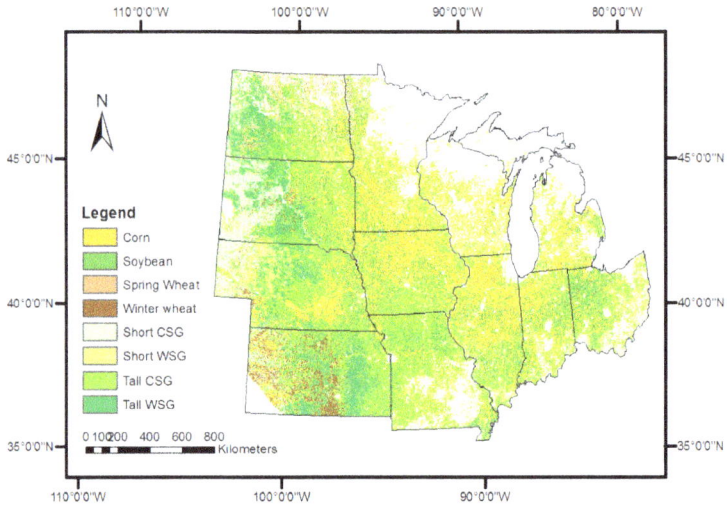

Figure 4. The SPMA-extracted dominant crop distributions in the Midwest.

Shortgrass in the west of the region extended from the shortgrass prairie in the Great Plains. Cool-season shortgrass (Figure 5e) dominated in the Black Hills along the west edge of South Dakota, adapting to the cold, upland climates. Warm-season shortgrass (Figure 5f) mainly grew in the Sand Hills, Nebraska, a typical mixed-grass prairie and the largest unplowed prairie remnant in the United States. Cool-season tallgrass (Figure 5g) distributes all over the Midwest because it has been commonly planted as a productive forage species in pasturelands across the region. Their percent covers varied, in mixed conditions with grasses in western states and annual crop fields (especially corn and soybeans) in eastern states. Warm-season tallgrass (Figure 5g) was native to the tallgrass prairie in the central United States. Today it is mostly observed in prairie remnants in the south (e.g., the Flint Hills in KS and the Sand Hills in NE) and upland mixed-grass grazing prairies in the north. Compared with binary classifications, the percent cover maps in Figure 5 present more quantitative information about crop distribution, and improve the level of detail in coarse-resolution images.

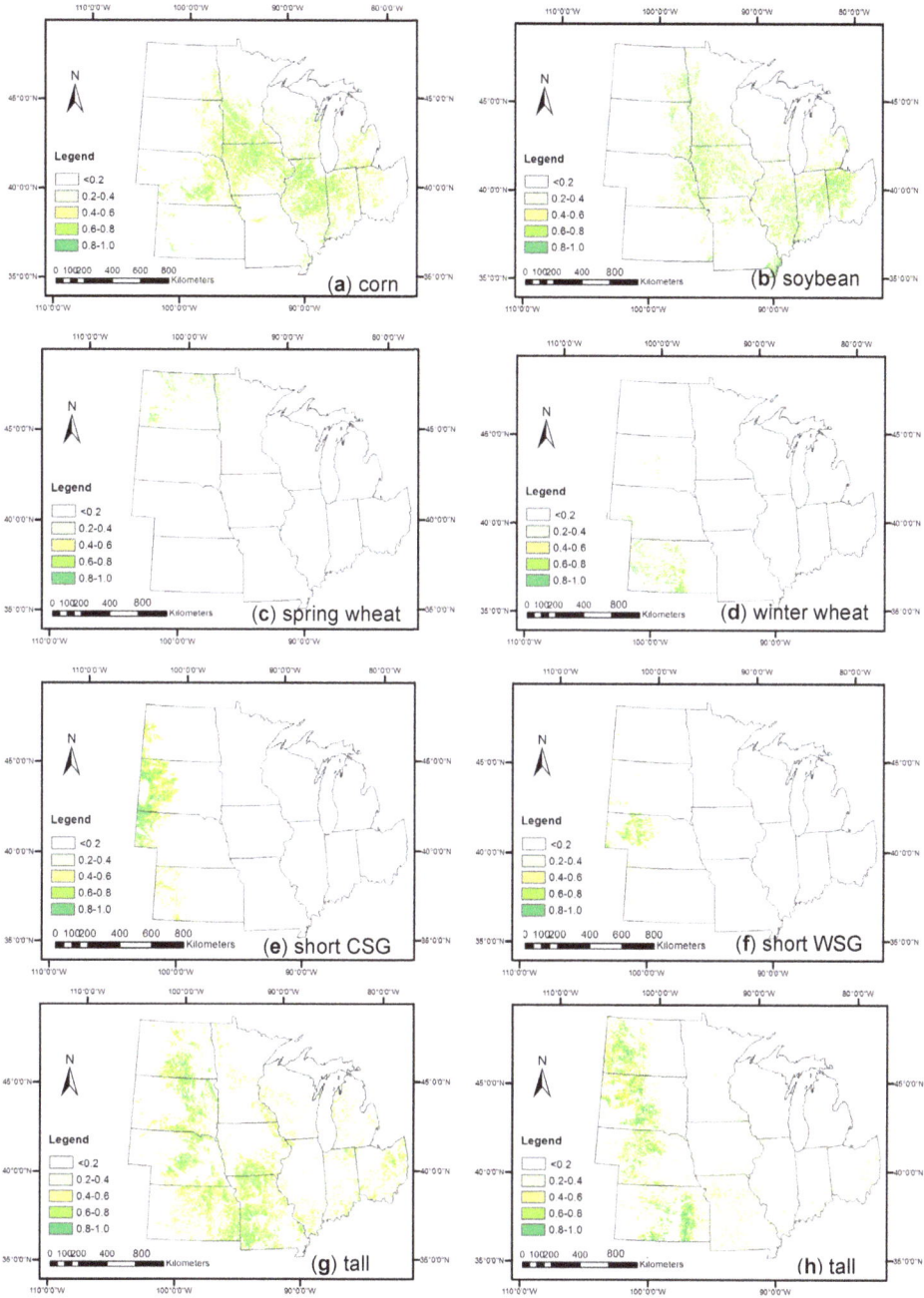

Figure 5. The SPMA-extracted percent cover distributions of the eight crops: corn (**a**), soybean (**b**), spring wheat (**c**), winter wheat (**d**), cool-season shortgrass (**e**), warm-season shortgrass (**f**), cool-season tallgrass (**g**), and warm-season tallgrass (**h**).

3.2. Comparison with References at Pixel Level

The SPMA-extracted crop percent covers were compared with the references from the CDL-summarized crop covers and published grass covers at sampled validation points. Table 2 lists the two accuracy measures for each crop. The RMSE values of all crops reached 14%–20%. When absolute errors were considered, all crops except tallgrass had negative SE values, indicating that the SPMA-extracted percent covers were generally higher than the CDL-summarized references. Tallgrass may be misclassified as corn and soybeans in the Corn Belt if there is a lack of training samples (pure pixels).

Table 2. Accuracy assessment of validation samples.

Types	Corn	Soybean	Spring Wheat	Winter Wheat	Short CSG	Short WSG	Tall CSG	Tall WSG
RMSE	0.163	0.162	0.195	0.160	0.187	0.165	0.136	0.156
SE	−0.096	−0.099	−0.152	−0.081	−0.124	−0.016	0.049	0.059
Student's t	21.84	21.30	15.08	11.52	8.97	6.54	8.37	7.52

When percent covers at all validation samples were considered (Figure 6), the Pearson's correlation coefficients (r) in the scatterplots ranged from 0.706 to 0.860 for these crops, suggesting fair agreement between the SPMA-extracted and CDL-summarized percent covers. The relationships were statistically significant ($p < 0.001$) under the Student's t test (degree of freedom = 149 for corn, soybeans, and tall grasses, and 49 for other crops). Table 2 also lists the Student's t values of the test for all crops. Although all tests were significant, perennial crops had smaller t values than annual crops, which may be attributed to the uncertainties when their samples were extracted from the coarse-resolution grass abundance maps [15]. For annual crops (Figure 6(a–d)), the SPMA could effectively extract their percent covers in the lower end. In the higher end, however, the SPMA results turned out to be overestimated, because other crops (e.g., alfalfa, sorghum) and small plots of non-crop land covers were not considered in this study. For this reason, the correlation line was lower than the diagonal line in each plot. For perennial crops (Figure 6(e–h)), the references were from the ranked grass cover maps extracted from the MODIS time series. The points were more scattered in their scatterplots, but the RMSE reached similar values to those of annual crops.

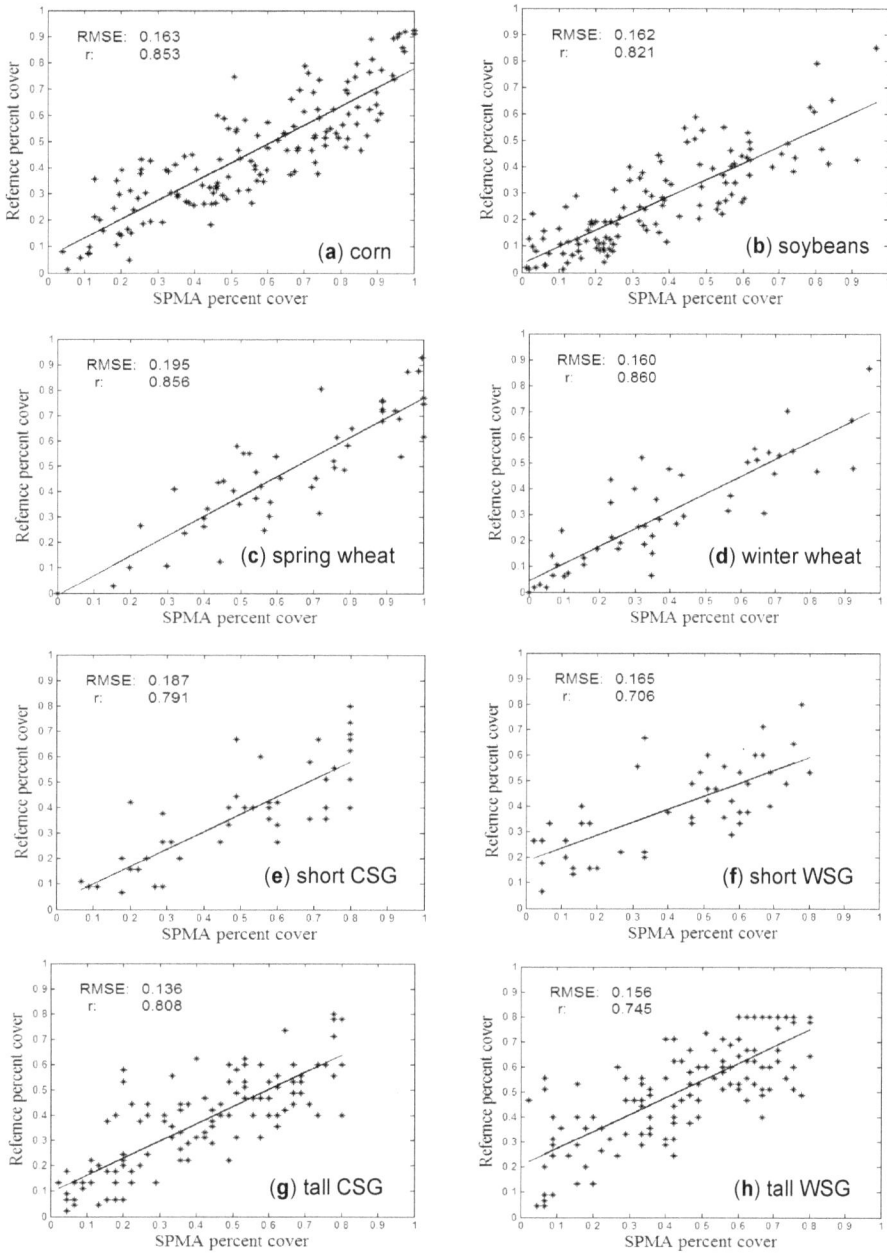

Figure 6. The scatterplots between the SPMA-extracted percent covers (X-axis) and references (Y-axis) for the eight crops: corn (**a**), soybean (**b**), spring wheat (**c**), winter wheat (**d**), cool-season shortgrass (**e**), warm-season shortgrass (**f**), cool-season tallgrass (**g**), and warm-season tallgrass (**h**).

3.3. Comparison with Crop Census Records at County Level

As the primary crops in the Midwest, corn and soybeans maintained good records of planting areas from the county-level NASS Crop Census data in 2007. To be comparable with these records, the pixel-level SPMA results were summarized into county-level crop planting acreages. The census records in a total of 1046 counties were extracted in the Midwest. Only corn and soybeans were compared. Other crops were not examined because spring wheat and winter wheat only grew in limited counties in the region. The warm-season and cool-season grasses were not specifically recorded in the census data.

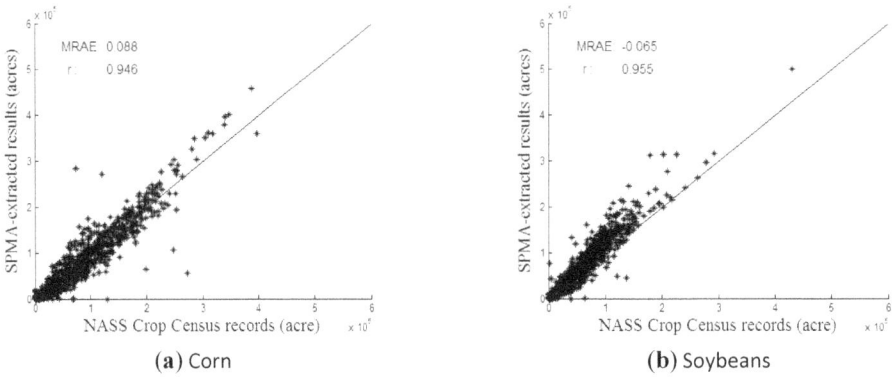

(a) Corn   (b) Soybeans

Figure 7. The county-level comparison between the Crop Census records (X-axis) and the SPMA-extracted planting areas of corn (**a**) and soybeans (**b**).

In Figure 7, the county-level comparison had a Pearson's r of 0.946 for corn and 0.955 for soybeans, indicating high agreement between the SPMA results and the census records. The mean relative area errors (MRAE) reduced the uncertainties from county sizes and were quite low for both crops. For counties with corn planting areas higher than 300 k acres (Figure 7a), the SPMA estimations became much higher than those in the census records. For soybeans (Figure 7b), a majority of counties had their SPMA estimations higher than the census records, which resulted in a negative MRAE value of -0.065. Figure 7 agrees with the Crop Census and CDL products that soybean planting areas were lower than corn in most counties. When compared all over the region, the SPMA approach in this study extracted a similar range of the total corn planting area (93.99%) as the census records (Table 3). Soybean planting area was overestimated, and reached 116.91% of the census records.

Table 3. Accuracy assessment at county level.

Types	County Level		Region-Level (Midwest)		
	MRAE	r	SPMA Results (Million Acres)	NASS Census (Million Acres)	Ratio (SPMA/Census)
Corn	0.088	0.946	72.297	76.921	93.99%
Soybean	−0.065	0.955	61.616	52.702	116.91%

Regional agricultural monitoring requires large-area coverage of remotely sensed data and standardized image analysis. While the USDA CDL products are published annually and provide high-quality crop maps in the conterminous United States, the process involves a huge amount of medium-resolution satellite images and repeated classification and validation algorithms for individual states [4]. It is time-consuming and labor-intensive, and thus cannot meet the requirement of quick responses for regional or global studies. Coarse-resolution satellite sensors such as MODIS have the ability to make daily observations all over the globe. Although not suitable for crop mapping with individual scenes, time series of these data effectively reveal crop development throughout a growing season. In our recent study [30], the same set of MODIS time series was input to a Support Vector Machine (SVM) classifier to extract major crop types, especially bioenergy crops including corn and perennial native grass. The classification, however, was binary without considering the in-pixel mixed growth of crops. Although relatively high accuracies (~90%) were achieved using large CDL clusters as validation source, dominant crops were apparently overestimated at an expense of omission of small crop clusters in local areas. In comparison, this study addressed the mixed pixel problem in crop delineation. Our study extracted percent covers of major crops using an unmixing algorithm based on their unique phenological curves. While crop phenology may vary at different geographical locations, the spatially constrained endmember selection reduces these uncertainties by only considering endmembers in the local neighborhood. The randomly selected small-size CDL clusters (3 × 3 of the aggregated 500 m cells) provided a more explicit spatial representation of the validation source. An overall RMSE range of 14%–20% for all crops in the Midwest indicates the feasibility of our SPMA approach for fractional crop mapping with MODIS time series in major agricultural regions.

One advantage of our approach is the computation efficiency for regional mapping. With only four MODIS tiles covering the Midwest region, computation time of our study is significantly improved in comparison with hundreds of Landsat tiles in the same region that were processed in national products. The recent publications about data fusion between these multi-resolution images [31] may bring in further opportunities to improve the classification accuracies with multi-source satellite time series. Also, the 8-day, 250-m MODIS products (MOD09Q1 and

267

MYD09Q1) have become available in recent years. At finer resolution than the 500-m imagery that we used in this study, these new datasets could have better SPMA results for regional crop mapping. One concern about the phenology-based analysis in this study rises from the uncertainties when only one-year satellite data were applied. The NDVI time series fairly reflects crop development cycles along the growing season. Its temporal variation, however, is highly influenced by seasonal weather conditions that vary year to year. It is less of a concern for annual crops in this study because our ground truth samples were extracted from CDL products in the same year. In areas where real-time truth data were not available (e.g., perennial grasses in this study), large uncertainties could be introduced during the endmember selection process of our SPMA approach. As our further research, multi-year time series will be explored to investigate these challenges, and to examine land use change in the long run.

Agricultural land use patterns are changing all over the world. With advanced biofuel demand, for example, corn planting expansion, environmental contamination, corn grain price increase, and food shortages are of increasing concern to sustainable agriculture in the U.S. Midwest. In our recent research [30], we performed phenology-assisted binary classifications of the MODIS time series in 2006–2008 to explore the corn boom in 2007, the year with the historical record for corn expansion. Multi-year satellite time series effectively identified this type of bioenergy land use change. By taking advantage of multi-crops in mixed pixels, fractional mapping conducted in this study provides more quantitative information for rapid assessment of crop planting areas, crop production, and land use conversion.

4. Conclusions

This study developed a spatially constrained phenology-assisted unmixing (SPMA) approach to extracting crop percent covers in the U.S. Midwest using MODIS time series in 2007. The NDVI time series of a pixel was assumed to be a linear combination of phenological curves from multiple crops in the pixel, and endmembers in a dynamic local neighborhood were selected in the unmixing process. The resulted spatial distributions of major crops agreed with the CDL products at randomly selected validation points, reaching an overall RMSE range of 14%–20% for all crops. The planting areas of major crops (corn and soybeans) also fit well with county-level Crop Census records in the region, although soybeans were slightly overestimated and corn underestimated in comparison with census data. In corn- and soybean-dominated counties (>300 k acres), the extracted acreages of both crops were higher than census data. With crop percent covers, the SPMA results verified the hypothesis of the spatial constraint and presented more spatial details of crop distributions from coarse-resolution satellite imagery. The SPMA approach developed in this study shows great potential for regional crop monitoring in U.S.

agricultural lands. With multi-year satellite time series involved, it could provide spatially explicit information about the rapid growth of bioenergy land use for decision-making at a regional scale.

Acknowledgments: This research is supported by Agriculture and Food Research Initiative Competitive Grant (# 2012-67009-22137) from the USDA National Institute of Food and Agriculture. We thank the USDA NASS for publishing the CDL products and Crop Census records that serve as an excellent reference in this project.

Author Contributions: Cheng Zhong did the experiments and wrote the original manuscript. Cuizhen Wang supervised the process of data analysis and was responsible to manuscript revisions. Changshan Wu provided partial source codes for image analysis and offered valuable comments to the manuscript.

Conflicts of Interest: The authors declare no conflict of interest.

References

1. 2007 Corn Crop A Record Breaker. Available online: http://www.nass.usda.gov/Newsroom/2008/01_11_2008.asp (accessed on 15 May 2014).
2. Wang, C.; Fritschi, F.B.; Stacey, G.; Yang, Z. Phenology-based assessment of perennial energy crops in North American Tallgrass Prairie. *Ann. Assoc. Am. Geogr.* **2011**, *101*, 741–751.
3. Smith, W.K.; Cleveland, C.C.; Reed, S.C.; Miller, N.L.; Running, S.W. Bioenergy potential of the United States constrained by satellite observations of existing productivity. *Environ. Sci. Tech.* **2012**, *46*.
4. Boryan, C.; Yang, Z.; Mueller, R.; Craig, M. Monitoring US agriculture: The US department of agriculture, national agricultural statistics service, Cropland Data Layer Program. *Geocarto Int.* **2011**, *26*, 341–358.
5. Solomon, B.D.; Barnes, J.R.; Halvorsen, K.E. Grain and cellulosic ethanol: History, economics, and energy policy. *Biomass Bioenerg.* **2007**, *31*, 416–425.
6. Pittman, K.; Hansen, M.C.; Becker-Reshef, I.; Potapov, P.V.; Christopher, O.J. Estimating global cropland extent with multi-year MODIS data. *Remote Sens.* **2010**, *2*, 1844–1863.
7. Thenkabail, P.S.; Biradar, C.M.; Noojipady, P.; Dheeravath, V.; Li, Y.J.; Velpuri, M.; Gumma, M.; Reddy, G.P.O.; Turral, H.; Cai, X.L.; *et al.* Global Irrigated Area Map (GIAM), derived from remote sensing, for the end of the last millennium. *Int. J. Remote Sens.* **2009**, *30*, 3679–3733.
8. Thenkabail, P.S.; Hanjra, M.A.; Dheeravath, V.; Gumma, M.A. A holistic view of global croplands and their water use for ensuring global food security in the 21st century through advanced remote sensing and non-remote sensing approaches. *Remote Sens.* **2010**, *2*, 211–261.
9. Santos, U.L. Spectral identification of native and non-native plant species. In Proceedings of ASD and IEEE GRS; Art, Science and Applications of Reflectance Spectroscopy Symposium, Boulder, CO, USA, 23–25 February 2010.
10. Xie, Y.; Sha, Z.; Yu, M. Remote sensing imagery in vegetation mapping: A review. *J. Plant Ecol.* **2008**, *1*, 9–23.

11. McCloy, K.R.; Lucht, W. Comparative evaluation of seasonal patterns in long time series of satellite image data and simulations of a global vegetation model. *IEEE Trans. Geosci. Remote Sens.* **2004**, *42*, 140–153.

12. Wang, C.; Jamison, B.; Spicci, A. Trajectory-based warm season grass mapping in Missouri prairies with multi-temporal ASTER imagery. *Remote Sens. Environ.* **2010**, *114*, 531–439.

13. Peterson, D.L.; Price, K.P.; Martinko, E.A. Discriminating between cool season and warm season grassland cover types in Northeaster Kansas. *Int. J. Remote Sens.* **2002**, *23*, 5015–5030.

14. Wardlow, B.D.; Egbert, S.L.; Kastens, J.H. Analysis of time-series MODIS 250 m vegetation index data for crop classification in the U.S. Central Great Plains. *Remote Sens. Environ.* **2007**, *108*, 290–310.

15. Wang, C.; Hunt, E.R.; Zhang, L.; Guo, H. Phenology-assisted classification of C3 and C4 grasses in the U.S. Great Plains and their climate dependency with MODIS time series. *Remote Sens. Environ.* **2013**, *138*, 90–101.

16. Homer, C.; Huang, C.; Yang, L.; Wylie, B.; Coan, M. Development of a 2001 national landcover database for the United States. *Photogramm. Eng. Remote Sens.* **2004**, *70*, 829–840.

17. Geerken, R.A. An algorithm to classify and monitor seasonal variations in vegetation phenologies and their inter-annual change. *ISPRS J. Photogramm. Remote Sens.* **2009**, *64*, 422–431.

18. Wardlow, B.D.; Egbert, S.L. Large-area crop mapping using time-series MODIS 250 m NDVI data: An assessment for the US Central Great Plains. *Remote Sens. Environ.* **2008**, *112*, 1096–1116.

19. Lobell, D.B.; Asner, G.P. Cropland distributions from temporal unmixing of MODIS data. *Remote Sens. Environ.* **2004**, *93*, 412–422.

20. Weng, Q.; Hu, X.; Lu, D. Extracting impervious surface from medium spatial resolution multispectral and hyperspectral imagery: A comparison. *Int. J. Remote Sens.* **2008**, *29*, 3209–3232.

21. Weng, Q. Remote sensing of impervious surfaces in the urban areas: Requirements, methods, and trends. *Remote Sens. Environ.* **2012**, *117*, 34–49.

22. Deng, C.; Wu, C. A spatially adaptive spectral mixture analysis for mapping subpixel urban impervious surface distribution. *Remote Sens. Environ.* **2013**, *133*, 62–70.

23. Busetto, L.; Meroni, M.; Colombo, R. Combining medium and coarse spatial resolution satellite data to improve the estimation of sub-pixel NDVI time series. *Remote Sens. Environ.* **2008**, *112*, 118–131.

24. Ozdogan, M. The spatial distribution of crop types from MODIS data: Temporal unmixing using independent component analysis. *Remote Sens. Environ.* **2010**, *114*, 1190–1204.

25. Somers, B.; Asner, G.P.; Tits, L.; Coppin, P. Endmember variability in spectral mixture analysis: A review. *Remote Sens. Environ.* **2011**, *115*, 1603–1616.

26. Ethanol Facilities' Capacity by State. Available online: http://www.neo.ne.gov/statshtml/121.htm. (accessed on 15 May 2014).

27. Savitzky, A.; Golay, M.J.E. Smoothing and differentiation of data by simplified least squares procedures. *Anal. Chem.* **1964**, *36*, 1627–1639.

28. Cropland Layer Data. Available online: http://nassgeodata.gmu.edu/CropScape/ (accessed on 15 May 2014).

29. Hu, X.; Weng, Q. Estimating impervious surfaces from medium spatial resolution imagery using the self-organizing map and multi-layer perceptron neural networks. *Remote Sens. Environ.* **2009**, *113*, 2089–2102.

30. Wang, C.; Zhong, C.; Yang, Z. Assessing bioenergy-driven agricultural land use change and biomass quantities in the U.S. Midwest with MODIS time series. *J. Appl. Remote Sens.* **2014**, *8*.

31. Zhu, X.; Cheng, J.; Gao, F.; Chen, X.; Masek, J.G. An enhanced spatial and temporal adaptive reflectance fusion model for complex heterogeneous regions. *Remote Sens. Environ.* **2010**, *114*, 2610–2623.

Feature Selection of Time Series MODIS Data for Early Crop Classification Using Random Forest: A Case Study in Kansas, USA

Pengyu Hao, Yulin Zhan, Li Wang, Zheng Niu and Muhammad Shakir

Abstract: Currently, accurate information on crop area coverage is vital for food security and industry, and there is strong demand for timely crop mapping. In this study, we used MODIS time series data to investigate the effect of the time series length on crop mapping. Eight time series with different lengths (ranging from one month to eight months) were tested. For each time series, we first used the Random Forest (RF) algorithm to calculate the importance score for all features (including multi-spectral data, Normalized Difference Vegetation Index (NDVI), Normalized Difference Water Index (NDWI), and phenological metrics). Subsequently, an extension of the Jeffries–Matusita (JM) distance was used to measure class separability for each time series. Finally, the RF algorithm was used to classify crop types, and the classification accuracy and certainty were used to analyze the influence of the time series length and the number of features on classification performance; the features were added one by one based on their importance scores. Results indicated that when the time series was longer than five months, the top ten features remained stable. These features were mainly in July and August. In addition, the NDVI features contributed the majority of the most significant features for crop mapping. The NDWI and data from multi-spectral bands also contributed to improving crop mapping. On the other hand, separability, classification accuracy, and certainty increased with the number of features used and the time series length, although these values quickly reached saturation. Five months was the optimal time series length, as longer time series provided no further improvement in the classification performance. This result shows that relatively short time series have the potential to identify crops accurately, which allows for early crop mapping over large areas.

Reprinted from *Remote Sens*. Cite as: Hao, P.; Zhan, Y.; Wang, L.; Niu, Z.; Shakir, M. Feature Selection of Time Series MODIS Data for Early Crop Classification Using Random Forest: A Case Study in Kansas, USA. *Remote Sens*. **2015**, *7*, 5347–5369.

1. Introduction

Crop-type information is important for food security, and the demand for accurate crop maps is increasing in society and in the plant industry [1–3]. In addition,

272

crop maps can be incorporated into a range of environmental models to improve understanding of the overall agricultural response to environmental issues [4,5]. Remote sensing data have shown potential for mapping crop distributions at both regional and local scales [4,6,7], and substantial efforts have been made toward monitoring agricultural land and accurately assessing crop acreage [8,9].

Multi-temporal remote sensing data can be used to describe the vegetation conditions over different periods, and have been widely employed to produce crop distribution maps [10–13]. Images of several key periods, such as the "initial spring green-up phase" and the "late senescence phase" are sufficient for accurate crop mapping [14–16]. In addition, Hao *et al.* [17] merged Landsat and Huan Jing (HJ) data, which have similar spatial resolution to Landsat and higher temporal resolution, to obtain an image time series with relatively high temporal resolution, and increase the possibility of acquiring images in the optimal periods for crop identification. The timeline is an important consideration for crop classification because obtaining an early classification result benefits both decision makers and the private sector [18]. Zhou *et al.* [19] found that reducing the time series length had little influence on the average accuracy of land cover classification, except for a slight increase in the classification variance when different training samples were used. However, few studies have determined the effect of time series length on crop-type mapping.

Apart from multi-spectral time series data, several vegetation indices (VIs) and phenological metrics derived from VI time series have been used to enrich the information available for vegetation mapping and monitoring [20–23]. However, using all these features involves a large volume of data, which may increase computation times with little improvement in accuracy [24]. To solve the problems associated with large volumes of data, various feature selection methods have been employed [4,24–26]. Most previous studies have focused on the effects of feature-space size reduction on classification accuracy and certainty, but the contributions of different features remain unclear.

The majority of the statistical measures used to assess land-cover classification accuracy are based on the confusion matrix [27]. The information contained in this matrix is a location-independent pattern of misclassification, which can only provide a generally accurate measurement for the user [26]. Classification certainty can be defined as a quantitative measure of doubt regarding a specific single class assignment [28]. Additionally, several newly proposed classifiers, such as Support Vector Machine (SVM), Random Forest (RF), and C5.0, provide a soft output (a probability for each class), and this information is used to derive the certainty for each pixel [4,29,30]. In contrast to accuracy measures, pixel-based certainty measures allow spatial representation of the map quality, and provide a better understanding of location error in classification [24,26].

The objectives of this research were to use the MODIS reflectance product on a regional scale to analyze (1) the effect of time series lengths on crop classification, (2) the importance of multi-spectral band data and indices (NDVI and NDWI) at different time series lengths, and (3) the influence of the number of features on crop identification. The cropland data layer (CDL) data at a spatial resolution of 30 m was used as ground reference data [31]. In addition, both classification accuracy and certainty were utilized to better understand the quality of the crop mapping.

2. Study Area and Datasets

2.1. Study Area

This study was conducted in the State of Kansas (37°N–40°N, 94°W–102°W) in the U.S. Central Great Plains (Figure 1). Kansas is a state dominated by agriculture with 46.9% (10.0 million ha) of its total area dedicated to crop production [32]. The major crop types are alfalfa, corn, sorghum, soybeans, and winter wheat [33]. Although each crop has a well-defined crop calendar and unique seasonal growth pattern, the growth situation varies throughout the state. On one hand, the state has a significant east-west precipitation gradient that has a strong influence on crop growth. For example, western Kansas receives on average 460–510 mm of precipitation per year, whereas eastern Kansas receives 890–1020 mm [16]. Therefore, eastern Kansas receives adequate precipitation, and corn and soybeans are the two primary crops grown in this part of the state. However, semiarid western Kansas commonly experiences drought events; as a result, drought-tolerant crops, such as winter wheat and sorghum, are widely planted. In addition, water-requiring crops, such as alfalfa, corn, and soybean require irrigation from aquifers. Another complicating factor is that planting times for many crops in Kansas differ by more than one month "along a general southeast (earliest) to northwest (latest)" trend [34]. For example, the recommended planting date for corn is 25 March to 25 April for southeast Kansas, but 20 April to 20 May for northwest Kansas. Moreover, the size of fields varies across the state. According to [32], western Kansas has "large individual fields (sizes commonly range from 65 to 245 ha)" while the fields in eastern Kansas are small (less than 65 ha) and the cropland areas are fragmented.

2.2. MODIS Data and Derived Phenological Variables

A 30-date time series of 8-day composite MODIS 500-m reflectance data (MOD09A1), spanning 7 April to 25 November, 2013 was created for Kansas. Data were required from four MODIS tiles (h09v04, h09v05, h10v04, and h10v05) for statewide coverage. The tiled MODIS data were acquired from the Land Processes Distributed Active Archive Center (LP-DAAC) [35], reprojected from the Sinusoidal to UTM projection (WGS 84 zone 14N), and subset over Kansas for each composite

274

period and then sequentially stacked to produce the time series dataset. Two additional indices, the Normalized Difference Vegetation Index (NDVI) and the Normalized Difference Water Index (NDWI), were derived from the time series of MODIS reflectance data using Equations (1) and (2) [36]:

$$NDVI = \frac{\rho\,(B2) - \rho\,(B1)}{\rho\,(B2) + \rho\,(B1)} \tag{1}$$

$$NDWI = \frac{\rho\,(B2) - \rho\,(B5)}{\rho\,(B2) + \rho\,(B5)} \tag{2}$$

where $\rho\,(B1)$, $\rho\,(B2)$, and $\rho\,(B5)$ are the reflectance values of MODIS bands 1, 2, and 5, respectively. Each variable has the same length as the MODIS reflectance time series.

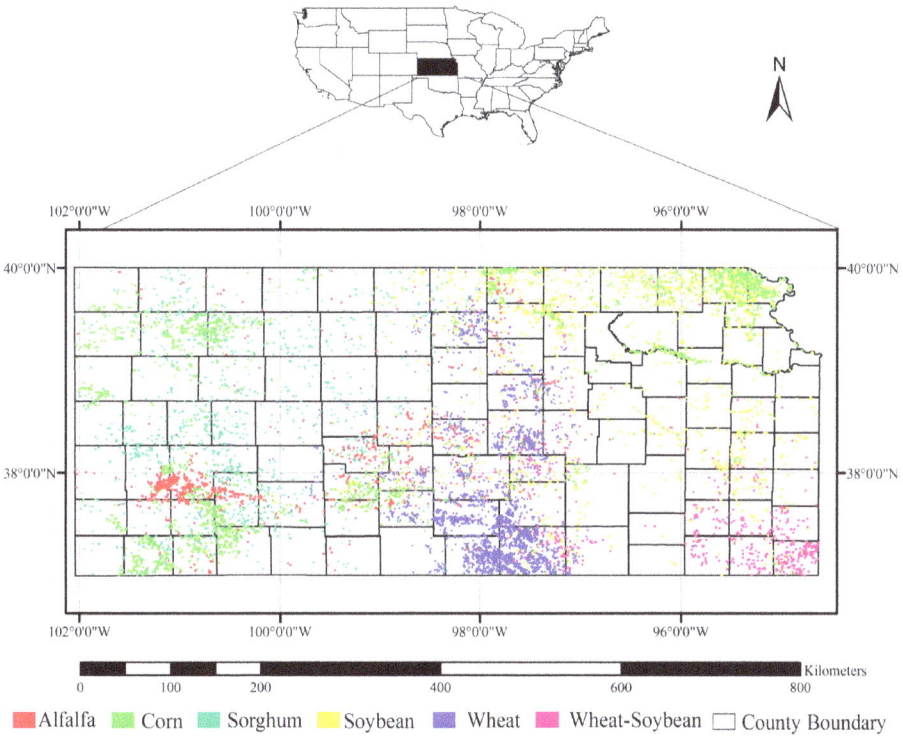

Figure 1. Study area and distribution of ground reference samples.

Land surface phenological metrics represent stages of plant growth or development that occurs during a growing season, and several phenological characters can be identified using multi-temporal remote sensing data [37]. In this research, nine unique annual phenology metrics were employed for crop

275

classification. The metrics included start-of-season time (SOST), start-of-season NDVI (SOSN), end-of-season time (EOST), end-of-season NDVI (EOSN), maximum NDVI (MaxN), maximum NDVI time (MAXT), duration of season (DOS), amplitude of NDVI (AON), and seasonal time-integrated NDVI (TIN) [38]. These metrics were derived from250-m weekly eMODIS NDVI using a curve derivative method [4]. This method employed a delayed moving average (DMA), in which predicted values were based on previous observations along a time-series NDVI curve. Smoothed NDVI data values were compared to a moving average of the previous observations to identify departures from an established trend. For example, if smoothed NDVI values became larger than those predicted by the DMA, this departure point was labeled as the start of the growing season (SOS) [38]. All available annual 250-m phenology metrics for 2013 were obtained from [38] and then resampled to a spatial resolution of 500-m [38]. The variables used in this research are shown in Table 1.

Table 1. Groups of input variables for classification. Notes: The wavelengths of the reflectance bands are *c.* 620–670 nm (B1), *c.* 841–876 nm (B2), *c.* 459–479 nm (B3), *c.* 545–565 nm (B4), *c.* 1230–1250 nm (B5), *c.* 1628–1652 nm (B6), and *c.* 2105–2155 nm (B7).

Group Name	Number of Variables in the Group	Denotation of Variables
Reflectance	210 (30 images ×7 bands)	BxDy, x = 1,2, … 7, y = 1,2, … 30
Indices (NDVI and NDWI)	60 (30 image × 2 bands)	NDVI_Dy, NDWI_Dy, y = 1,2, … 30
Phenological metrics	9	SOST, SOSN, EOST, EOSN, MaxN, MaxT, DOS, AON, TIN

2.3. Reference Dataset

The crop-type reference data used in this study were obtained from the National Agricultural Statistics Service (NASS) Cropland Data Layers (CDL) for 2013 [31]. Table 2 shows the classification accuracy of CDL in Kansas ordered by areal proportion. The bold-font crops were selected as reference crops in this study because both producer's and user's accuracies were higher than 85%, and the areal proportions of these crops were more than 1% [39]. To obtain pure pixels on 500-m spatial resolution MODIS images, we first obtained the MODIS pixel grid from the MODIS image and then calculated the fraction for every crop in each MODIS pixel using CDL data. If one crop filled more than 80% of a MODIS pixel, we defined that pixel as a "pure" pixel and used it as a reference pixel. In this procedure, we selected 80% as the threshold because it provided a balance between pixel quality and sample number. A higher threshold (such as 90%) might increase the purity of the pixel, but the sample number would drop substantially because the spatial resolution was 500-m in this study. Conversely, a lower threshold (such as 70%) could increase the pixel number, but would lead to more sample pixels of low purity. We then extracted the time series quality assessment of the MODIS 500-m reflectance product using the reference pixels. If there were more than 25 "corrected products produced at ideal quality in all bands" periods in a reference pixel [40], the pixel was retained. Otherwise, the pixel was removed from the reference dataset. For the remaining pixels, the reflectance for low-quality periods (if the quality of a pixel is not 'ideal quality', we define it as 'low-quality') was replaced by the average reflectance of the previous and following periods. Subsequently, the reference pixels were randomly divided into two parts: training samples and validation samples. The numbers of these samples are shown in Table 3, and their distributions are shown in Figure 1.

Table 2. 2013 Kansas cropland data layer statewide agricultural accuracy.

Crop Type	CDL Code	Producer's Accuracy	User's Accuracy	Areal Proportions
Winter Wheat	24	94.37%	94.45%	38.33%
Corn	1	93.21%	93.6%	16.99%
Soybeans	5	92.97%	92.97%	13.66%
Sorghum	4	89.32%	89.27%	11.25%
Fallow/Idle Cropland	61	87.47%	87.81%	11.02%
(Double Crop) Winter Wheat/Soybeans	26	85.9%	85.25%	3.00%
Other Hay/Non Alfalfa	37	56.07%	90.39%	2.85%
Alfalfa	36	85.95%	91.21%	1.95%
(Double Crop) Winter Wheat/Sorghum	236	36.64%	65.03%	0.37%
Canola	31	78.22%	90.75%	0.12%
Rye	27	37.55%	76.76%	0.11%
Oats	28	37.63%	72.27%	0.10%

Table 3. Number of training and validation samples.

Crop Type	Training	Validation
Alfalfa	562	561
Corn	1441	1441
Sorghum	847	847
Soybean	1005	1006
Wheat	1665	1664
Wheat-soybean	437	437
Total	5957	5956

3. Method

The overall methodology used in this study is presented in Figure 2. First, we extracted time series multi-spectral band data and indices (NDVI and NDWI) from the MOD09 product and phenological metrics from eMODIS phenological data using the ground reference data. We then exploited the Random Forest (RF) algorithm to calculate an importance score for all available features containing multi-spectral, NDVI, NDWI, and phenological metrics for each time series length using the training samples. To simplify the analysis, we used one month as the unit of the time-series period (Table A1 in Appendix 6). Therefore, the time series length varied from one month to eight months. Then, an extension of the Jeffries–Matusita (JM) distance was used to calculate the separability among all crops. Furthermore, the RF algorithm was used to classify crop types and obtain a probability output for each crop. The classification accuracy and certainty were then obtained to measure the classification performance. When calculating the extension of the JM distance and classifying crop types, features were added one by one based on the importance score acquired from the RF algorithm for each time series length.

3.1. Random Forest

The classification algorithm employed for this research was the Random Forest (RF) algorithm. The RF algorithm is an ensemble machine learning technique that combines multiple trees [30]. Each tree is constructed using two-thirds of the original cases. Then, the remaining one-third of cases is employed to generate a test classification, with an error referred to as the "out-of-bag error" (OOB error). Subsequently, the model output is determined by the majority vote of the classifier ensemble [26]. Two free parameters can be optimized in the RF algorithm: the number of trees (*ntree*) and the number of features to split the nodes (*mtry*). The advantages of the RF algorithm, such as the relatively high efficiency with large datasets, the probability output for each class, and the generated OOB error (an internal unbiased estimate of the generalization error) make it suitable for remote sensing applications [41]. In this research, both the feature importance score and crop classification were obtained using the RandomForest package for R [42]. The *ntree* parameter was set to a relatively high value of 1000 to allow convergence of the OOB error statistic, and *mtry* was set to the square root of the total number of input features [43]. Additionally, the decreased accuracy (*i.e.*, the difference in prediction accuracy before and after permutation of the variable of interest)was used to measure the importance of the features. As for classification, the RF algorithm allowed quantification of the prediction probability at the pixel level, together with the class label. The probability p(i) of a pixel being classified as class i was defined as

$$p\,(i) = \frac{k_i}{k} \tag{3}$$

where k was the total number of trees involved in the classification process, and k_i was the number of trees classifying the pixel as cover type i [26].

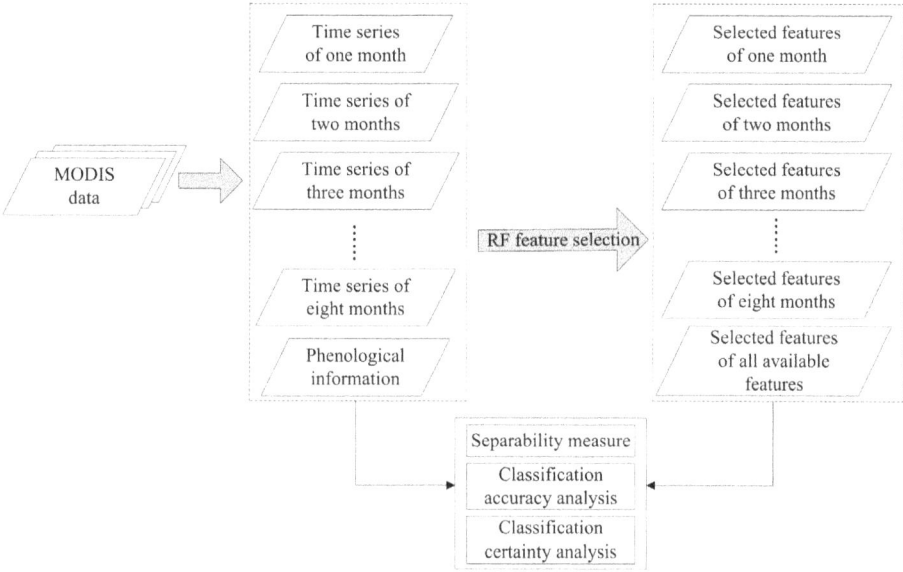

Figure 2. Methodology used in this study.

3.2. Extension of the Jeffries–Matusita Distance

In this study, we used the JM distance to measure the separability for each pair of crops, because previous research had shown that JM distance can provide a more accurate separability indicator than other distance measures, such as Euclidean distance or divergence [14,44]. The JM distance between a pair of class-specific functions was given by:

$$JM\left(c_i, c_j\right) = \int_x \left(\sqrt{p\left(x \mid c_i\right)} - \sqrt{p\left(x \mid c_j\right)}\right)^2 dx \qquad (4)$$

where x denoted a span of VI time series values, and c_i and c_j (lowercase c) denoted the two crop classes under consideration. Under normality assumptions, Equation (4) was reduced to $M = 2\left(1 - e^{-B}\right)$, where

$$B = \frac{1}{8}\left(\mu_i - \mu_j\right)^T \left(\frac{C_i + C_j}{2}\right)^{-1}\left(\mu_i - \mu_j\right) + \frac{1}{2}\ln\left(\left|\frac{\left|C_i + C_j\right|}{2\sqrt{\left|C_i\right| \times \left|C_j\right|}}\right|\right) \qquad (5)$$

and C_i and C_j (uppercase C) were the covariance matrices of classes i and j, respectively. Additionally, $\left|C_i\right|$ and $\left|C_j\right|$ were the determinants of C_i and C_j, respectively. The JM distance ranged from 0 to 2, with a high value indicating a high level of separability between the two classes [45].

279

When considering the separability of multiple classes, different classes were given different weights to account for the different sample sizes of each class. The extension of the JM distance (J_{Bh}) was used for this purpose. J_{Bh} was calculated from Equation (6) based on Bhattacharyya bounds, and it gave greater importance to classes with high *a priori* probabilities in the selection process [46]:

$$J_{Bh} = \sum_{i=1}^{N} \sum_{j>i}^{N} \sqrt{p(w_i) \times p(w_j) \times JM^2(i,j)} \tag{6}$$

where N was the number of classes, and $p(w_i)$ and $p(w_j)$ were the *a priori* probabilities of classes i and j, respectively, which were calculated using the combination of training samples in Table 3.

3.3. Accuracy and Certainty Measures

A series of accuracy metrics were employed to evaluate the classification accuracy. First, overall accuracy (OA), producer's accuracy (PA), and user's accuracy (UA) were used for the hard results (class labels) [27]. For the probability result, a soft answer was provided by the random forest algorithm in the form of a vector containing probability estimates belonging to each class:

$$p(x) = \{p_1(x), \cdots, p_k(x), \cdots, p_K(x), k = 1, 2, \cdots, K\} \tag{7}$$

where $p_k(x)$ was the probability that x belongs to class k, and K was the number of classes. In this study, the probability vector was first sorted in descending order. Then, we used the specificity measure to calculate the certainty, as in Equation (8) [47]:

$$C(x) = \sum_{k=1}^{K-1} (\hat{p}_k(x) - \hat{p}_{k+1}(x)) \tag{8}$$

The advantage of the specificity measure is that it applies all the information in the probability vector. The certainty of a pixel is equal to 1 if the maximum probability in its probability vector is 1. On the other hand, if the all the classes have the same probability ($\hat{p}_k(x) = 1/K$), the certainty of the pixel is 0.

4. Results

4.1. Importance of Features for Crop Mapping

An analysis of the ten most informative features for each time series length was shown in Figure 3. The selected features varied among the different time series lengths. During April and May (Figure 3a,b), the most important features were mainly multi-spectral bands data. However, the selection of these features may not have a phenological component because most summer crops were immature, and the standard deviations of the importance scores from 20 model runs were relatively high (higher than 1.5 in most cases). NDVI and NDWI were selected when the time series length was longer than three months (Figure 3c). In Figure 3d,e, the time series lengths were one month longer than in Figure 3c,d, and several features unique to the additional months (such as the features in July for Figure 3d and August for Figure 3e) were selected as the most important features. However, in Figure 3f–h, the time series were longer than five months, but no features from the extra months were found to be among the most important.

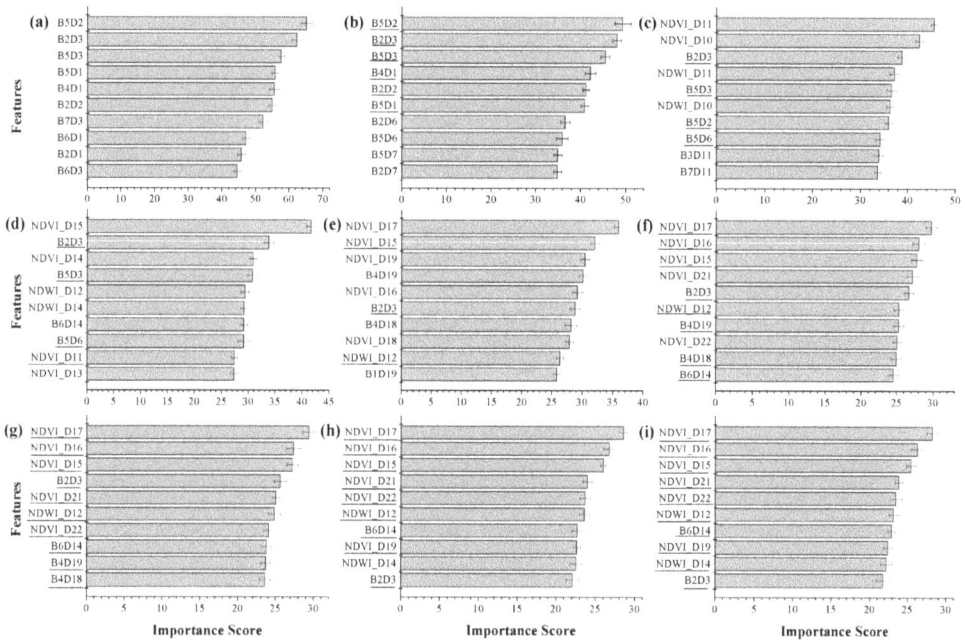

Figure 3. Mean relative feature importance of the ten most important features for each time series length. Error bars indicate the standard deviation of the variable score for 20 model runs. In addition, "B1–B7" in the feature names indicates the number of the MODIS band; "D1–D30'" indicates the acquisition date: (**a**) D1–D3 in April, (**b**) D1–D7 for April to May, (**c**) D1–D11 for April to June, (**d**) D1–D15 for April to July, (**e**) D1–D19 for April to August, (**f**) D1–D22 for April to September, (**g**) D1–D26 for April to October, (**h**) D1–D30 for April to November, and (**i**) D1–D30 and phenological metrics. The underlined features indicate that these features were among the ten most important features in the previous shorter time series.

When both NDVI and NDWI were selected in the top ten features, NDVI obtained a higher importance score than NDWI (Figure 3c); moreover, when the time series was longer than five months, more NDVI features were selected for the top ten important features. Furthermore, the phenological features were not among the features selected as most important when combining phenological information with multi-spectral metrics, vegetation indices, and water indices metrics (Figure 3i).

The importance score of the selected features decreased with the augmentation of the time series length. Figure 3a showed that the importance scores of the most important features were nearly 60, but the scores were around 20 for the most important features selected from the eight-month time series (Figure 3h). This was because the importance score was measured by the difference in prediction accuracy before and after permutations of the feature. Therefore, the importance score reduced when more features were employed to build the RF model.

4.2. Class Separability

Month-by-month JM distances in Kansas are shown in Figure 4, and the time series for different crops are shown in Figure 5. Winter wheat was highly separable from the summer crops (JM distance larger than 1.5). During the early growing season (April and May), wheat had a relatively high vegetation fraction but the summer crops had not yet been sown. Then, in summer (July and August), the wheat had been harvested and the summer crops had developed (Figure 5). In October and November, wheat was sown again and the summer crops were harvested. As for winter wheat and the double crop wheat-soybean, the JM distance was high (larger than 1.5) in August because the soybean had developed during this period. Among the three summer crops, sorghum and soybean had high separability in June (JM distance larger than 1.5) and September because of their different rates of emergence and senescence. Corn was highly separable from sorghum because of its earlier planting and emergence (corn is mostly sown in May and emerges in early June, whereas sorghum is planted in June). Additionally, the JM distance between corn and soybean was lower than 1.0 throughout the growing season because of the similar planting, reproduction, and harvest periods (Figure 5). Alfalfa had a relatively high JM distance (larger than 1.2) from the other crops throughout almost the entire growing season, except when compared with several summer crops, such as corn, sorghum, and soybean, during June and August (JM distance around 1). The relatively low separability in this period was mainly because of the fact that the summer crops had developed and the separability between alfalfa and these summer crops was relatively low (Figure 5). At the beginning of the growing season (during April and May), soybean was the least separable crop compared to alfalfa; this result is unexpected because alfalfa and wheat are well developed during this time, whereas soybean is not developed [16]. Thus, soybean should have a better separability than wheat. There are two possible explanations for this unexpected outcome: (1) the misclassification of CDL data in Kansas and (2) the use of mixed pixels as training samples. Firstly, the producer's accuracy for alfalfa was 85.95% (Table 1), which indicated that several other crops were mislabeled as alfalfa. Additionally, the average NDVI of alfalfa in this period was substantially lower than values found in [16]. On the other hand, the misclassification also led to the relatively high standard deviation of alfalfa NDVI profiles (Figure 5). As for the second reason, both alfalfa and soybean had relatively small field sizes, which resulted in more mixed pixels with higher NDVI profile variations (standard deviations larger than 0.1). Thus, the lower alfalfa NDVI and the higher standard deviation of the two crops contributed to the unexpected low separability between alfalfa and soybean during April and May.

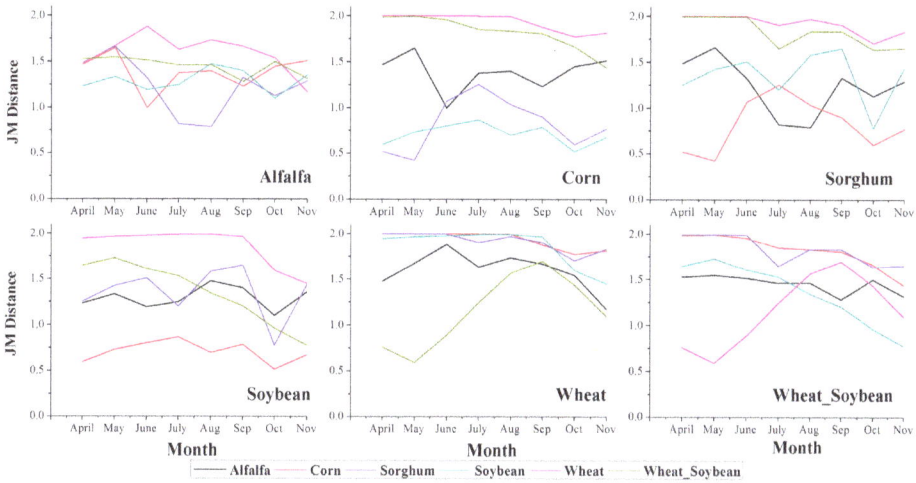

Figure 4. JM distance values for all crop pair comparisons in each month using training samples from Kansas.

Figure 6 showed the relationship between the JM distance of crop pair comparisons and the time series length. The figure showed that the JM distance increased with the time series length. For example, alfalfa had a relatively high JM distance when the time series was two months, and the JM distances between alfalfa and other crops then increased until the time series reached five months (when JM distances were 2). For winter wheat, when the time series length was three months, the JM distance between wheat and other summer crops was almost 2. In addition, the JM distance among the three summer crops increased substantial when the time series length was shorter than four months. Beyond four months, further increases in the time series did not meaningfully increase the JM distance. However, the JM distance between corn and soybean was still low (lower than 1). Furthermore, wheat and wheat-soybean demonstrated good separability when the time series length was longer than five months.

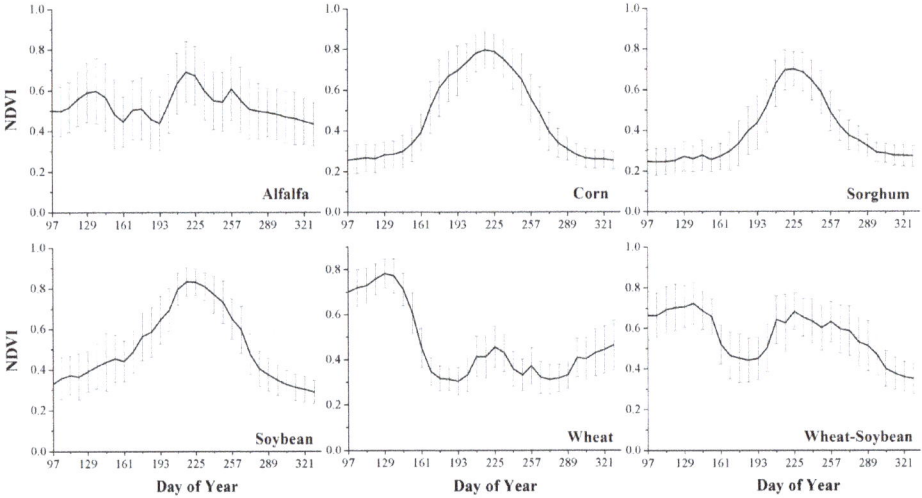

Figure 5. Average NDVI time series for the crops used in this study. Error bars indicate the standard deviation of the NDVI of the samples.

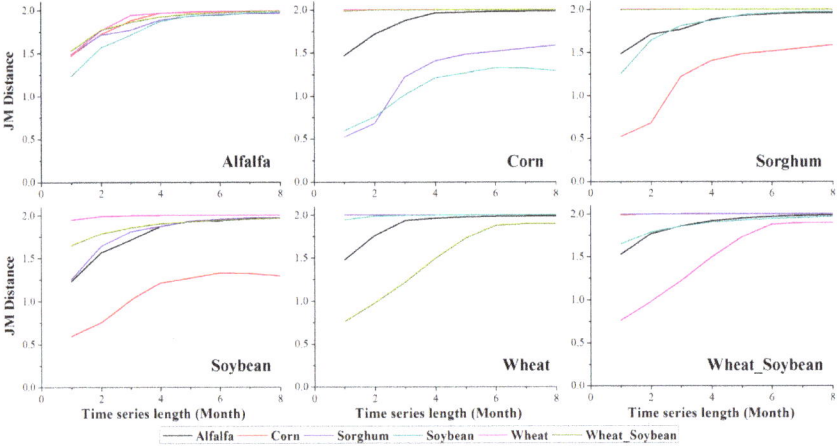

Figure 6. JM distance values for all crop pair comparisons with different time series lengths using training samples from Kansas.

Figure 7 showed the relationship between the number of features used and J_{BH} (extension of the JM) distance. The maximum J_{BH} increased substantially when the time series length was shorter than five months (April–August). For example, the maximum J_{BH} was 5.67 when only images from April were employed but increased to 6.61 when the time series length was two months. However, with the additional inclusion of images from October and November and the phenological metrics, the separability did not increase substantially. When images from April to September were used, the J_{BH} was 8.43, and when images from all eight months were used, J_{BH} increased only slightly to 8.6. In addition, the separability increased substantial when a few features were used for each time series, but did not increase substantial when more features were employed. For example, J_{BH} increased from 2.2 to 7.2 when the number of features increased from 1 to 24 for the April–August time series, but only increased to 8.4 when all 171 features in this period were used.

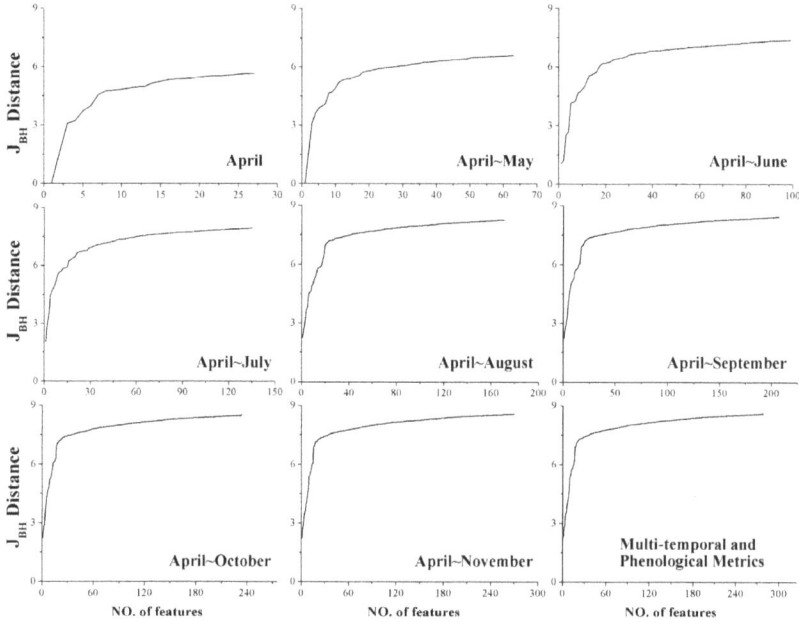

Figure 7. J_{BH} obtained from training samples using the different input features suggested by the RF importance scores for different time series lengths.

4.3. Classification Accuracy

The influence of the time series length and the number of features on the classification accuracy was shown in Figure 8. For each time series, the overall classification accuracy increased with the number of features used for classification until a saturation point was reached, after which the accuracy did not increase further. For example, the overall accuracy increased to a saturation point at 14 features (72.18%) and 23 features (88.56%) for the April and the April–November time series, respectively. Among the different time series lengths, combinations of only one or two months could not achieve a classification accuracy of more than 80%. For the April time series, the maximum overall accuracy was 72.77%; and when the time series length increased to two months (April–May), the maximum overall accuracy was 77.83%. When the time series was longer than five months, the overall accuracy also reached a saturation point. The maximum overall accuracy increased from 88.45% to just 88.81% when the time series length was increased from five months (April–August) to six months (April–September).

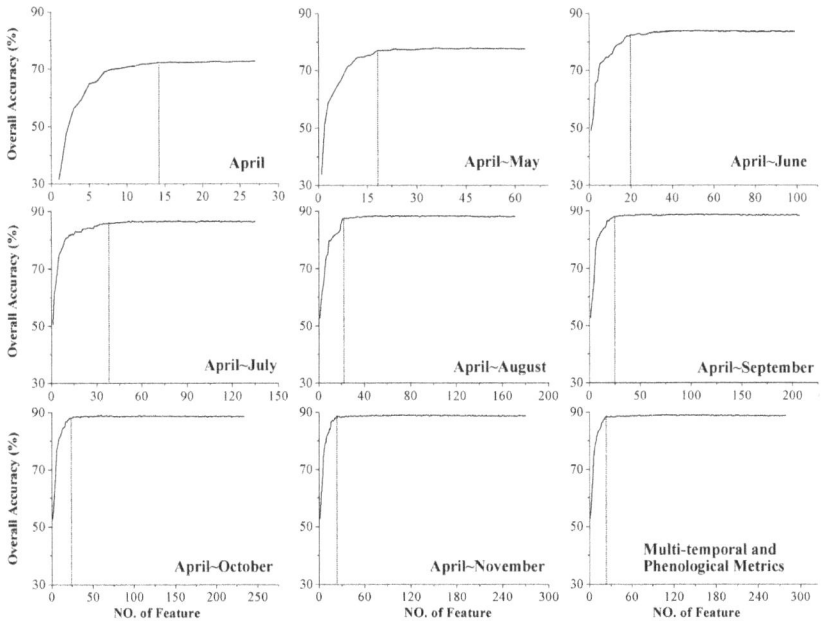

Figure 8. Overall accuracy (%) of validation samples found from the RF algorithm using the different input features suggested by the importance score for different time series lengths. Only the average overall accuracy is shown in this figure because the standard deviation of the 10–model run is approximately 0.1%. The dotted line shows the saturation point for each time series length.

The producer's and user's accuracies achieved from the different time series are shown in Table 4. When the time series was short (one month or two months), wheat had relatively high producer's and user's accuracies (PA = 96.75% and UA = 88.03%), while both producer's and user's accuracies for other crops were less than 70% in most cases. In addition, the accuracies increased with the time series length and remained stable when the time series was longer than five months, which was consistent with the trend of the overall accuracy. Among all the crop types, wheat had the highest classification accuracy (UA = 95.58% and PA = 95.39%) for the five-month time series. Alfalfa had high UA (92.13%) but relatively low PA (85.56%). Although the accuracies were relatively low compared with those for the other crops, the three summer crops were distinguishable because all PA and UA values were above 80%, which was consistent with the separability of these three crops discussed in Section 4.2.

Table 4. Producer's and user's accuracies obtained from the different time series for each crop.

	Alfalfa	Corn	Sorghum	Soybean	Wheat	Wheat-Soybean
	PA/UA	PA/UA	PA/UA	PA/UA	PA/UA	PA/UA
April	62.2%/75.9%	70.0%/63.1%	55.5%/67.2%	65.7%/61.3%	96.8%/88.0%	56.1%/84.2%
April ~ May	71.3%/87.3%	79.3%/66.4%	62.6%/74.5%	66.5%/70.9%	97.8%/88.9%	58.4%/87.3%
April ~ June	75.4%/92.4%	83.9%/78.5%	79.8%/81.8%	75.8%/74.8%	99.1%/91.3%	61.1%/87.5%
April ~ July	81.3%/92.5%	86.1%/82.0%	82.3%/84.0%	78.8%/78.7%	99.6%/93.6%	70.3%/89.8%
April ~ August	85.6%/92.1%	86.0%/83.2%	83.7%/87.1%	81.8%/80.2%	99.6%/95.4%	76.9%/91.1%
April ~ September	85.4%/92.3%	85.8%/83.4%	84.1%/87.3%	82.1%/79.7%	99.5%/96.6%	80.8%/90.8%
April ~ October	85.2%/92.1%	86.4%/83.6%	83.9%/87.5%	82.1%/79.9%	99.5%/96.7%	81.0%/90.5%
April ~ November	85.7%/93.0%	85.7%/84.5%	84.5%/87.1%	82.8%/79.8%	99.4%/96.7%	81.0%/91.0%
Add Phe	85.6%/93.3%	85.9%/84.2%	84.2%/87.5%	83.2%/79.6%	99.5%/96.5%	81.0%/91.1%

4.4. Classification Certainty

The influences of the time series length and the number of features on the average classification certainty are shown in Figure 9. When only one feature was used for classification, the classification certainty was relatively high (nearly 0.9 in most cases) with a low classification accuracy (less than 60%). Then, the certainty decreased substantially until reaching a minimum. After this point, the certainty began to increase until it reached a saturation point, and then remained generally stable at, for example, nine features (certainty = 0.68) and 21 features (certainty = 0.81) for the April and April–August time series, respectively. Similarly, the stable classification certainties increased with augmentation of the time series length. Additionally, when the time series was longer than five months, the certainty did not continue to increase. For example, from the three-month (April–June) to the four-month (April–July) time series, the stable certainty increased from 0.71 to 0.78; whereas from the five-month time series (April–August) to the six-month (April–September) series, stable certainty remained unchanged (0.81).

The distributions of certainty for correctly and wrongly classified validation samples are shown in Figure 10. For the correctly labeled samples, certainty was mainly in the range [0.8, 1]; for the wrongly labeled samples, certainty was mainly in the range [0.4, 0.8]. In addition, when the time series was relatively short, the certainty of several correctly classified samples was low. For example, in the April time series, the certainties of nearly 20% of the correctly labeled samples were in the range [0.4, 0.6], and the certainties of only about 30% of the correctly labeled samples were in the range [0.8, 1]. However, when the time series was longer than five months, more than 60% of the validation samples were correctly labeled with high certainty (between [0.8, 1]).

287

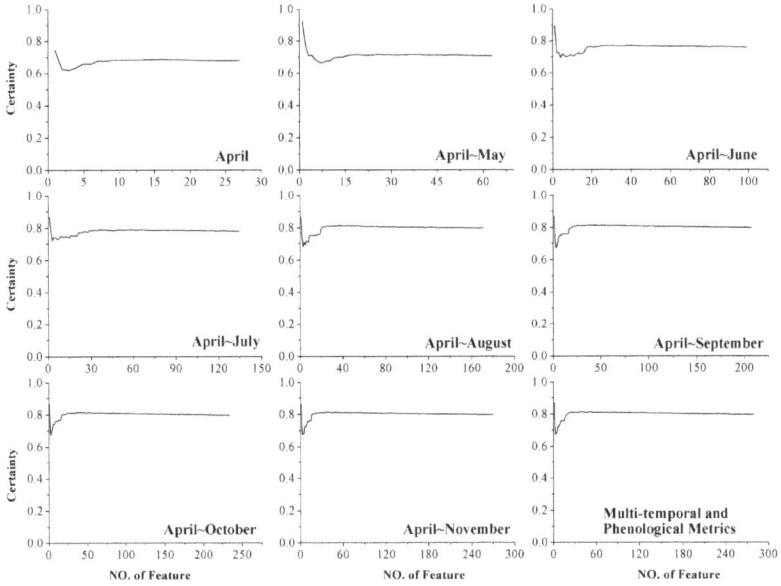

Figure 9. Classification certainty of validation samples derived from the RF algorithm using the different input features suggested by the importance score for different time series lengths.

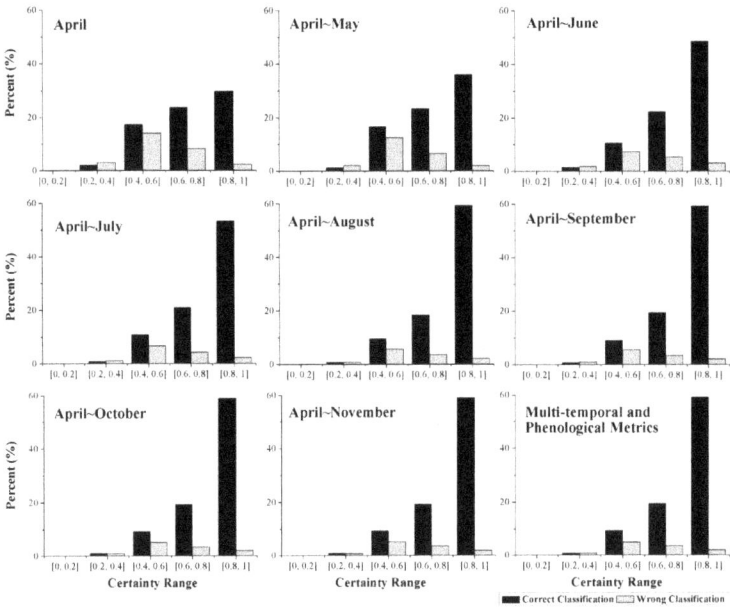

Figure 10. Frequency distributions of classification certainties for correctly and wrongly classified validation samples.

288

5. Discussion

In this study, image time series of different lengths were used to identify crop types, and all three measurements (crop separability, classification accuracy, and certainty) showed that the five-month time series has the potential to classify the crops in the study area accurately, and that longer time series cannot improve the classification result. However, the crop types determine the optimal time series length. In Kansas, the dominant crops are wheat, alfalfa, corn, sorghum, and soybean. The winter crop, wheat, is separable from all the other crops when the time series length is only one month; alfalfa also shows high separability during the early growing season, when the summer crops have not yet developed. For the three summer crops, corn, sorghum, and soybean, the time periods of different emergence rate contribute most substantially to the high separability.

Previous studies have shown that rather than using the entire growing season, images of several optimal time periods can achieve high classification accuracy [14,17]. Additionally, according to [16], the most separable time periods for the summer crops in Kansas are during the initial spring green-up phase and/or the late senescence phase in June and early October, respectively. Although the short time series in this research (such as the April–August time series) cover only a part of the optimal time periods for crop identification, they still have the potential to correctly classify crops. More importantly, the earlier classification using these short periods makes the crop map more valuable.

For each time series, separability, classification accuracy, and certainty can be achieved using a portion of the features similar to using all the features available for that time series. Low *et al* [24] sorted features by their RF importance score and used SVM to detect the relationship between classification accuracy and the number of features exploited. The accuracy reached a peak when a fraction of the available features was employed, and the accuracy declined substantially with the addition of other features. The same situation has been observed in SVM classification using hyper-spectral data [48]. In the presented research, the RF algorithm was employed to classify the crop types, and the classification accuracy and certainty remained stable when additional features were used, which is also consistent with the findings of [24] that the RF algorithm was less affected by the number of features than SVM.

In this study, we calculated all month-by-month crop pair comparisons using both multi-spectral data and indices (NDVI and NDWI). Compared with previous research using only NDVI and EVI time series [16], the separability of crop pairs in this research is relatively high. For example, the JM distance between corn and sorghum was more than 1 in July. This relatively high separability is because of several factors. First, the temporal unit in this research is one month, with three or four time periods in each month. However, the separability analysis of [16] is based on the 16-day NDVI and EVI, and the temporal unit is the single time period.

Second, in addition to NDVI, both multi-spectral data and NDWI were included in this research. Although NDVI features comprised the majority of the top ten features selected for time series longer than five months, several multi-spectral and NDWI features, such as NDWI_D12 and B6D19 (Figure 3), were also selected as key features, and these features increased the separability when identifying crops over short time series.

The phenological metrics features were not selected for the top ten features for crop identification (Figure 3), which indicates that phenological metrics may not classify crops as good as the other features. This is mainly because of phenological character variations in crop development schedules due to local weather conditions and farm management. For example, the recommended planting date for corn varies by nearly one month from southeast to northwest Kansas. Another complicating factor is that the phenological metrics are sensitive to the signal noise introduced by pre-crop vegetation. This pre-crop vegetation mainly consists of weeds and "volunteer crops" (in particular, winter wheat), and can lead to a misleading early estimation of the green-up onset of the crops. As a result, estimates for several phenological metrics, such as SOST, SOSN, and DOS, are prone to errors and inconsistencies [49]. Moreover, several other phenological features, such as EOST, EOSN, and DOS, can only be acquired after harvest, which may delay the completion of the crop map.

When using short time series to identify the crops, both the classification accuracy and certainty were low (Figure 10), and even several correctly classified samples had low certainty (between 0.4 and 0.6). This was because of the low separability among the different crops when the length of the time series was one or two months (Figure 5). The classification certainty increased with the time series length, but several correctly labeled validation samples still had low classification certainty (between 0.4 and 0.6). Figure 11 shows the average probability of validation samples for each crop type within different certainty ranges. For corn samples, the low certainty samples have high probability for soybean and sorghum. Similarly, the sorghum samples with low certainty also have relatively high probabilities for corn and soybean, which is consistent with the low JM distance among these three crops (Figure 4). Generally, low separability leads to low classification certainty.

CDL data were used as ground reference data in this research, which may introduce some uncertainty regarding our conclusions. Some misclassification of CDL data may lead to variations in a crop's features (such as the NDVI time series), and the underestimation of the separability between several crops (such as alfalfa and soybean during April and May). Another complication is that the use of CDL as ground reference data may lead to overly optimistic classification accuracies in this research. This is because both MODIS and Landsat (the basis for CDL) data are dependent on similar local atmospheric and ground conditions. The reflectance from the two sensors may therefore correlate to some extent during the crop-growing season. As a result, the reported accuracy of this research is likely overstated to some degree. Additionally, we defined 80% as the threshold for a 'pure' pixel to increase the number of reference samples, but the relatively heterogeneous pixels may be a limiting factor for this research.

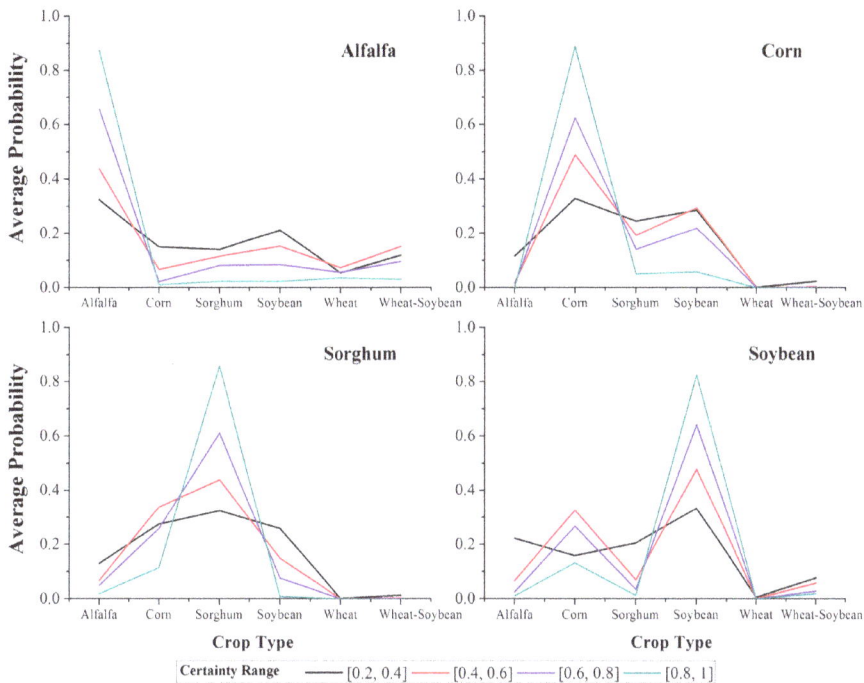

Figure 11. Average probability for the validation samples for each crop type. Different line types indicate different levels of uncertainty. All 270 features of the eight-month time series were used to calculate the probability in Figure 10.

6. Conclusion

In this study, we investigated the influence of the time series length on crop identification using 8-day composite MODIS 500-m reflectance data (MOD09A1) in Kansas, USA, with CDL data as ground reference data. The main conclusions are as follows.

1. The augmentation of the time series length can improve crop classification because the separability among different crops, the classification accuracy, and the certainty are increased. In addition, the five-month time series (April to August) was the optimal time series for identifying crops in Kansas because longer time series cannot improve the classification performance (accuracy and

certainty). The result also indicated that rather than the entire growing season, relatively short time series have the potential to accurately classify crops.

2. For each time series used in this research, additional features improved the classification, as measured by higher separability, classification accuracy, and certainty. Additionally, a portion of these features (such as the first 23 features during the April–November time series) was sufficient to classify the crops accurately, and adding more features after this point had no significant positive effect on crop identification.

3. Among the features used in this research, NDVI was the most important feature, as shown by the fact that NDVI features comprised the majority of the top ten features during the eight-month time series (April–November). In addition, the water content index (NDWI) and multi-spectral band data also contributed to distinguishing between the crop types. The phenological metrics features had a relatively low importance and were not selected as the most important features. Moreover, several phenological features, such as EOST and EOSN, can only be obtained after harvest and therefore, cannot contribute to early crop identification using short time series.

4. The RF algorithm was used in this research to calculate the importance score, classify the crops, and obtain the classification certainty. When the time series was longer than five months, little change was seen among the top ten features. In addition, the classification accuracy and certainty remained stable when additional features were employed. These results indicate that the RF algorithm is a suitable algorithm for selecting features and classifying crops using a large volume of data.

In this research, we investigated the potential of using multiple features, including NDVI, NDWI, and multi-spectral band data, to classify crops in short time series, which could contribute to early crop mapping over a large area. After all, crop separability and optimal crop discriminating periods are determined by the crop type. Therefore, more work is needed to evaluate the contributions of different features to identifying specific crops using relatively short time series in other study areas.

Appendix

Table A1. Relationship between dates and months in this research.

Month	Time Period in This Study	Date Flag	Corresponding Day of Year (DOY)	Date
April	1	097	097–104	7 April–14 April
	2	105	105–112	15 April–22 April
	3	113	113–120	23 April–30 April
May	4	121	121–128	1 May–8 May
	5	129	129–136	9 May–16 May
	6	137	137–144	17 May–24 May
	7	145	145–152	25 May–1 June
June	8	153	153–160	2 June–9 June
	9	161	161–168	10 June–17 June
	10	169	169–176	18 June–25 June
	11	177	177–184	26 June–3 July
July	12	185	185–192	4 July–11 July
	13	193	193–200	12 July–19 July
	14	201	201–208	20 July–27 July
	15	209	209–216	28 July–4 August
August	16	217	217–224	5 August–12 August
	17	225	225–232	13 August–20 August
	18	233	233–240	21 August–28 August
	19	241	241–248	29 August–5 September
September	20	249	249–256	6 September–13 September
	21	257	257–264	14 September–21 September
	22	265	265–272	22 September–29 September
October	23	273	273–280	30 September–7 October
	24	281	281–288	8 October–15 October
	25	289	289–296	16 October–23 October
	26	297	297–304	24 October–31 October
November	27	305	305–312	1 November–8 November
	28	313	313–320	9 November–16 November
	29	321	321–328	17 November–24 November
	30	329	329–337	25 November–2 December

Acknowledgments: This work was supported by the National Natural Science Foundation of China (41371416), (41371358) and National Science and Technology Major Project (14CNIC-032079-32-02). Moreover, we would thank the reviewers and editors for the constructive suggestions. In addition, the authors thank the NASS for providing the CDL data for free.

Author Contributions: Pengyu Hao designed the experiments, processed the data and wrote the paper. Yulin Zhan, Li Wang, Zheng Niu and Muhammad Shakir contributed important ideas and considerations

Conflicts of Interest: The authors declare no conflict of interest.

References

1. Vintrou, E.; Ienco, D.; Begue, A.; Teisseire, M. Data mining, a promising tool for large-area cropland mapping. *IEEE J. Sel. Top. Appl. Earth Obs. Remote Sens.* **2013**, *6*, 2132–2138.

2. Potgieter, A.B.; Lawson, K.; Huete, A.R. Determining crop acreage estimates for specific winter crops using shape attributes from sequential MODIS imagery. *Int. J. Appl. Earth Obs. Geoinf.* **2013**, *23*, 254–263.

3. Potgieter, A.B.; Apan, A.; Hammer, G.; Dunn, P. Early-season crop area estimates for winter crops in NE Australia using MODIS satellite imagery. *ISPRS J. Photogramm. Remote Sens.* **2010**, *65*, 380–387.

4. Howard, D.M.; Wylie, B.K. Annual crop type classification of the US Great Plains for 2000 to 2011. *Photogramm. Eng. Remote Sens.* **2014**, *80*, 537–549.

5. Atzberger, C. Advances in remote sensing of agriculture: Context description, existing operational monitoring systems and major information needs. *Remote Sens.* **2013**, *5*, 949–981.

6. Zhang, J.H.; Feng, L.L.; Yao, F.M. Improved maize cultivated area estimation over a large scale combining MODIS-EVI time series data and crop phenological information. *ISPRS J. Photogramm. Remote Sens.* **2014**, *94*, 102–113.

7. Senturk, S.; Sertel, E.; Kaya, S. Vineyards mapping using object based analysis. In Proceedings of 2013 Second International Conference on Agro-Geoinformatics (Agro-Geoinformatics), Fairfax, VA, USA, 12–16 August 2013.

8. Amoros-Lopez, J.; Gomez-Chova, L.; Alonso, L.; Guanter, L.; Zurita-Milla, R.; Moreno, J.; Camps-Valls, G. Multitemporal fusion of Landsat/TM and ENVISAT/MERIS for crop monitoring. *Int. J. Appl. Earth Obs. Geoinf.* **2013**, *23*, 132–141.

9. Kuenzer, C.; Knauer, K. Remote sensing of rice crop areas. *Int. J. Remote Sens.* **2013**, *34*, 2101–2139.

10. Cable, J.W.; Kovacs, J.M.; Shang, J.L.; Jiao, X.F. Multi-temporal polarimetric Radarsat-2 for land cover monitoring in northeastern Ontario, Canada. *Remote Sens.* **2014**, *6*, 2372–2392.

11. Wang, D.; Lin, H.; Chen, J.S.; Zhang, Y.Z.; Zeng, Q.W. Application of multi-temporal ENVISAT ASAR data to agricultural area mapping in the Pearl River Delta. *Int. J. Remote Sens.* **2010**, *31*, 1555–1572.

12. Jia, K.; Wu, B.; Li, Q. Crop classification using HJ satellite multispectral data in the North China Plain. *J. Appl. Remote Sens.* **2013**, *7*, 073576.

13. Brown, J.C.; Kastens, J.H.; Coutinho, A.C.; Victoria, D.D.; Bishop, C.R. Classifying multiyear agricultural land use data from Mato Grosso using time-series MODIS vegetation index data. *Remote Sens. Environ.* **2013**, *130*, 39–50.

14. Murakami, T.; Ogawa, S.; Ishitsuka, N.; Kumagai, K.; Saito, G. Crop discrimination with multitemporal SPOT/HRV data in the Saga Plains, Japan. *Int. J. Remote Sens.* **2001**, *22*, 1335–1348.

15. Van Niel, T.G.; McVicar, T.R. Determining temporal windows for crop discrimination with remote sensing: A case study in south-eastern Australia. *Comput. Electron. Agric.* **2004**, *45*, 91–108.

16. Wardlow, B.D.; Egbert, S.L.; Kastens, J.H. Analysis of time-series MODIS 250 m vegetation index data for crop classification in the US Central Great Plains. *Remote Sens. Environ.* **2007**, *108*, 290–310.

17. Hao, P.; Wang, L.; Niu, Z.; Aablikim, A.; Huang, N.; Xu, S.; Chen, F. The potential of time series merged from Landsat-5 tm and HJ-1 CCD for crop classification: A case study for Bole and Manas counties in Xinjiang, China. *Remote Sens.* **2014**, *6*, 7610–7631.

18. Gallego, J.; Craig, M.; Michaelsen, J.; Bossyns, B.; Fritz, S. *Best practices for crop area estimation with remote sensing*; European Commission Joint Research Centre: Ispra, Italy, 2010.

19. Zhou, F.Q.; Zhang, A.N.; Townley-Smith, L. A data mining approach for evaluation of optimal time-series of MODIS data for land cover mapping at a regional level. *ISPRS J. Photogramm. Remote Sens.* **2013**, *84*, 114–129.

20. Zhong, L.H.; Gong, P.; Biging, G.S. Phenology-based crop classification algorithm and its implications on agricultural water use assessments in California's Central Valley. *Photogramm. Eng. Remote Sens.* **2012**, *78*, 799–813.

21. Zhong, L.H.; Gong, P.; Biging, G.S. Efficient corn and soybean mapping with temporal extend ability: A multi-year experiment using Landsat imagery. *Remote Sens. Environ.* **2014**, *140*, 1–13.

22. Wardlow, B.D.; Egbert, S.L. A comparison of MODIS 250-m EVI and NDVI data for crop mapping: A case study for Southwest Kansas. *Int. J. Remote Sens.* **2010**, *31*, 805–830.

23. Da Silva, C.A.; Frank, T.; Rodrigues, T.C.S. Discrimination of soybean areas through images EVI/MODIS and analysis based on geo-object. *Rev. Bras. Eng. Agric. Ambient.* **2014**, *18*, 44–53.

24. Low, F.; Michel, U.; Dech, S.; Conrad, C. Impact of feature selection on the accuracy and spatial uncertainty of per-field crop classification using support vector machines. *ISPRS J. Photogramm. Remote Sens.* **2013**, *85*, 102–119.

25. Vieira, M.A.; Formaggio, A.R.; Renno, C.D.; Atzberger, C.; Aguiar, D.A.; Mello, M.P. Object based image analysis and data mining applied to a remotely sensed Landsat time-series to map sugarcane over large areas. *Remote Sens. Environ.* **2012**, *123*, 553–562.

26. Loosvelt, L.; Peters, J.; Skriver, H.; De Baets, B.; Verhoest, N.E.C. Impact of reducing polarimetric SAR input on the uncertainty of crop classifications based on the random forests algorithm. *IEEE Trans. Geosci. Remote Sens.* **2012**, *50*, 4185–4200.

27. Congalton, R.G. A review of assessing the accuracy of classifications of remotely sensed data. *Remote Sens. Environ.* **1991**, *37*, 35–46.

28. Bloch, I. Information combination operators for data fusion: A comparative review with classification. *IEEE Trans. Syst. Man Cybern. Part A: Syst. Humans* **1996**, *26*, 52–67.

29. Mountrakis, G.; Im, J.; Ogole, C. Support vector machines in remote sensing: A review. *ISPRS J. Photogramm. Remote Sens.* **2011**, *66*, 247–259.

30. Breiman, L. Random forests. *Mach. Learn.* **2001**, *45*, 5–32.

31. Cropscape-Cropland Data Layer. Available online: http://nassgeodata.gmu.edu/CropScape/ (accessed on 2 December 2014).

32. Wardlow, B.D.; Egbert, S.L. Large-area crop mapping using time-series MODIS 250-m NDVI data: An assessment for the U.S. Central Great Plains. *Remote Sens. Environ.* **2008**, *112*, 1096–1116.

33. Masialeti, I.; Egbert, S.; Wardlow, B.D. A comparative analysis of phenological curves for major crops in Kansas. *Gisci. Remote Sens.* **2010**, *47*, 241–259.

34. Kansas crop planting guide. Available online: http://www.ksre.k-state.edu/bookstore/pubs/l818.pdf (accessed on 23 April 2015).

35. Land Processes Distributed Active Archive Center. Available online: http://lpdaac.usgs.gov/ (accessed on 23 April 2015).

36. Gao, B.C. NDWI—A normalized difference water index for remote sensing of vegetation liquid water from space. *Remote Sens. Environ.* **1996**, *58*, 257–266.

37. Viña, A.; Tuanmu, M.-N.; Xu, W.; Li, Y.; Qi, J.; Ouyang, Z.; Liu, J. Relationship between floristic similarity and vegetated land surface phenology: Implications for the synoptic monitoring of species diversity at broad geographic regions. *Remote Sens. Environ.* **2012**, *121*, 488–496.

38. Remote Sensing Phenology. Available online: http://phenology.cr.usgs.gov/ (accessed on 11 December 2014).

39. USDA National agricultural statistics service, 2013 Kansas cropland data layer. Available online: http://www.nass.usda.gov/research/Cropland/metadata/metadata_ks13.htm (accessed on 11 December 2014).

40. Surface Reflectance 8-Day L3 Global 500m. Available online: https://lpdaac.usgs.gov/products/MODIS_products_table/mod09a1 (accessed on 12 December 2014).

41. Rodriguez-Galiano, V.F.; Ghimire, B.; Rogan, J.; Chica-Olmo, M.; Rigol-Sanchez, J.P. An assessment of the effectiveness of a random forest classifier for land-cover classification. *ISPRS J. Photogramm. Remote Sens.* **2012**, *67*, 93–104.

42. Liaw, A.; Wiener, M. Randomforest: Breiman and Cutler's Random Forests for Classification and Regression. Available online: http://cran.r-project.org/web/packages/randomForest/index.html (accessed on 15 December 2014).

43. Loosvelt, L.; Peters, J.; Skriver, H.; Lievens, H.; Van Coillie, F.M.B.; De Baets, B.; Verhoest, N.E.C. Random forests as a tool for estimating uncertainty at pixel-level in SAR image classification. *Int. J. Applied Earth Obs. Geoinf.* **2012**, *19*, 173–184.

44. Van Niel, T.G.; McVicar, T.R.; Datt, B. On the relationship between training sample size and data dimensionality: Monte Carlo analysis of broadband multi-temporal classification. *Remote Sens. Environ.* **2005**, *98*, 468–480.

45. Adam, E.; Mutanga, O. Spectral discrimination of papyrus vegetation (*Cyperus Papyrus* L.) in swamp wetlands using field spectrometry. *ISPRS J. Photogramm. Remote Sens.* **2009**, *64*, 612–620.

46. Bruzzone, L.; Roli, F.; Serpico, S.B. An extension of the Jeffreys-Matusita distance to multiclass cases for feature selection. *IEEE Trans. Geosci. Remote Sens.* **1995**, *33*, 1318–1321.

47. Huang, X.; Zhang, L. An SVM ensemble approach combining spectral, structural, and semantic features for the classification of high-resolution remotely sensed imagery. *IEEE Trans. Geosci. Remote Sens.* **2013**, *51*, 257–272.

48. Pal, M.; Foody, G.M. Feature selection for classification of hyperspectral data by SVM. *IEEE Trans. Geosci. Remote Sens.* **2010**, *48*, 2297–2307.

49. Wardlow, B.D.; Kastens, J.H.; Egbert, S.L. Using USDA crop progress data for the evaluation of greenup onset date calculated from MODIS 250-meter data. *Photogramm. Eng. Remote Sens.* **2006**, *72*, 1225–1234.

In-Season Mapping of Crop Type with Optical and X-Band SAR Data: A Classification Tree Approach Using Synoptic Seasonal Features

Paolo Villa, Daniela Stroppiana, Giacomo Fontanelli, Ramin Azar and Pietro Alessandro Brivio

Abstract: The work focuses on developing a classification tree approach for in-season crop mapping during early summer, by integrating optical (Landsat 8 OLI) and X-band SAR (COSMO-SkyMed) data acquired over a test site in Northern Italy. The approach is based on a classification tree scheme fed with a set of synoptic seasonal features (minimum, maximum and average, computed over the multi-temporal datasets) derived from vegetation and soil condition proxies for optical (three spectral indices) and X-band SAR (backscatter) data. Best performing input features were selected based on crop type separability and preliminary classification tests. The final outputs are crop maps identifying seven crop types, delivered during the early growing season (mid-July). Validation was carried out for two seasons (2013 and 2014), achieving overall accuracy greater than 86%. Results highlighted the contribution of the X-band backscatter ($\sigma°$) in improving mapping accuracy and promoting the transferability of the algorithm over a different year, when compared to using only optical features.

Reprinted from *Remote Sens.*. Cite as: Villa, P.; Stroppiana, D.; Fontanelli, G.; Azar, R.; Brivio, P.A. In-Season Mapping of Crop Type with Optical and X-Band SAR Data: A Classification Tree Approach Using Synoptic Seasonal Features. *Remote Sens.* **2015**, 7, 12859–12886.

1. Introduction

The increasing demand for information on crop acreage for agricultural monitoring in support of private and public decision makers requires the production of reliable crop maps [1,2]. Up-to-date information on agricultural land use is necessary for crop planning and management: e.g., for estimating biomass and yield, analyzing agronomic practices, assessing soil productivity, monitoring crop phenology and stress. Earth Observation (EO) techniques have been widely exploited in agriculture and agronomy for the advantages offered when compared to *in situ* and statistical surveys: frequency of acquisitions, synoptic view, and multi-dimensional content. Satellite remote sensing also constitutes the only source of consistent historical data for long-term analysis over large areas, e.g., for the identification of anomalous conditions in vegetation development driven by climatic variability [3]. Moreover, EO data are available already during the growing season, whereas official statistics on crop acreages are often provided at the end of the season or later, thus being not useful for supporting in-season crop management. Since crop productivity quickly responds to unfavorable growing conditions, timeliness in delivering information on crop status is an important operational requirement [4,5], e.g., for mitigating the impact of crop stress conditions, especially for summer crops, which are prone to water stress in the dry summer months [6,7].

EO satellite data have been used for agricultural monitoring since the launch of the Landsat-1 system in the early 1970s and their potential for distinguishing different crops has been shown across various environmental conditions, and with many different data sources and methodologies, e.g., in [8,9]. The Landsat archive constitutes the longest record of multi-spectral data available at medium spatial resolution, and has been used for crop mapping purposes at regional scale [10–12], using either spectral response and/or vegetation indices [13–16]. The opening of the Landsat archives in 2008 has pushed forward the implementation of data analysis and image classification techniques based on multi-temporal features and time series analysis [17]. Multi-temporal analysis techniques have been applied as well to coarser resolution data such as NOAA-AVHRR [18] and NASA-MODIS data [1,19], taking advantage of high revisit time for these sensors [20]. Other satellite data too, with spectral and spatial features similar to Landsat, have been used for crop mapping achieving satisfactory results, e.g., IRS LISS data [21,22].

Data acquired by Synthetic Aperture Radar (SAR) active sensors have also been exploited for crop mapping and monitoring, especially during the last two decades. C-band data have been used for mapping rice [23–26], wheat [27], and multiple crops [28–32]. L-band data have been used too, although with generally poorer performance [33,34]. More recently, with the launch of the TerraSAR-X and COSMO-SkyMed satellites, the use of X-band SAR data has largely expanded, mainly thanks to the higher spatial and temporal resolutions and theoretical flexibility of these platforms [35,36]. Concerning X-band SAR data, different polarimetric configurations have been tested for crop mapping, from vertical-based, e.g., in [37], to horizontal-based polarization, e.g., in [38]; comparative studies using multiple polarizations have been carried out as well, e.g., in [39]. However, to our knowledge no agreement has been reached so far on the best polarization configuration for crop mapping.

The integration of SAR and optical sensors for agricultural applications is a recent topic, aiming at reducing the impact of optical and SAR specific limitations (*i.e.*, dependence on solar and clear sky conditions for the former, and on signal noise and stability for the latter). SAR and optical data integration takes advantage of their complementarity in terms of sensitivity to vegetation and soil characteristics [40–42]:

plant biomass, soil moisture and surface texture for SAR, spectral response of canopy-background system and photosynthetic features for optical sensors [31,32,43]. Successful integration examples are the work of Michelson *et al.* [44], and more recently some large scale studies using multi-source data from RADARSAT-1, Envisat ASAR, SPOT and Landsat sensors [29,37,45,46]. The algorithms used for land cover mapping with both optical and SAR data range from maximum likelihood [46] and neural network ensembles [46,47] for crop classification, to maximum likelihood with iterated conditional modes [48] and Random Forest [49] for regional and urban land cover targets, up to fuzzy scores aggregation for burned area mapping [50]. With the advent of new generation satellites, e.g., Landsat 8, Sentinel-2, WorldView-3, as well as COSMO-SkyMed, TerraSAR-X, RADARSAT-2, and Sentinel-1, crop mapping applications can be more timely and reliable, in particular during the early growth stages, and the operational use of such techniques will be promoted.

This paper describes a classification tree approach for in-season crop mapping, which exploits features derived from multi-temporal optical, Landsat 8 Operational Land Imager (OLI), and X-band SAR, COSMO-SkyMed, data for producing reliable in-season crop maps over temperate areas. The proposed approach builds on the analysis of separability between different crops to identify the best performing proxy combinations and synoptic seasonal features as crop mapping inputs. Classification tree approaches can handle input features of different types and derived from different sources, and are directly interpretable and adaptable, being structured as a set of simple rules [51–53]. These characteristics make classification tree approaches both efficient and effective, especially for delivering mapping algorithms which are to be used in operational contexts.

The main objective is to define a classification tree approach for producing a crop map early in the summer season, *i.e.*, around mid-July [54], to support agricultural management in Northern Italy. Spectral features for the winter and summer crop seasons (named synoptic seasonal features) are extracted from the temporal profiles of a set of proxies derived from optical and SAR data. Different proxies were used, sensitive to vegetation and soil conditions and able to characterize the dynamics of different crop types throughout the growing season: Spectral Indices (SIs) from optical data and/or backscatter and interferometric coherence information from X-band SAR data. Most of the literature using multi-temporal information for crop mapping focus on the use of temporal profiles of spectral indices derived from optical data [55–58], and only recently some authors successfully included SAR backscatter profiles for rice mapping [59,60]. The novelty of our work are the use of synoptic seasonal features integrating optical and SAR data, and the delivery of crop type mapping already during the early stage of growth; eventually, the transferability to a growing season different from the one used for developing the approach is tested.

2. Materials

2.1. Study Area

The study area is located in Lombardy region, northern Italy (Figure 1); it lies south of Milan and it is bordered by the Po river. The area, covering around 1100 Km^2, is mostly flat and intensively cultivated. Climate is continental, with annual temperature changes between January and July up to 20 °C and average precipitations of 850 mm/year. The most economically valuable crop types cultivated in Lombardy are (in per cent of total cropland area): maize (38.5%), temporary and permanent meadows used for forages (34.1%), rice (10.1%), winter cereals (wheat and barley, 7.5%), soybean and other legumes (2.7%), and vegetables (1.4%) [61]. The two major crop seasons run from October to June and from April to October for winter and summer crops, respectively. Barley is the prevailing winter crop, typically flowering in April-May and harvested in May-June. The main summer crops, covering most of the cropland area and consuming most of irrigation resources, are maize and rice. Maize, sown between April and early May, reaches the peak of the vegetative phase in July and is harvested from the end of August through September. Maize is often also sown in double cropping practices for fresh forages or silage, after meadows or winter cereals, in integrated crop-livestock systems. Rice is usually sown later than maize, from the second half of April to late May, reaching the flowering stage in late July or early August, and it is harvested from late September onwards.

2.2. Satellite Data

The remotely sensed dataset is composed of 13 COSMO-SkyMed (CSK) and 14 Landsat 8 OLI images covering the spring/summer seasons of the years 2013 (18 April–23 July) and 2014 (05 April–19 July), as shown in Table 1. CSK data have been consistently acquired by the same satellite of the COSMO SkyMed constellation (CSK 1) in single polarization (HH) and interferometric mode. CSK dataset of both years is acquired from the same flight track and with constant orbital configuration. This configuration allowed us to exploit the dataset not only for X-band intensity calculation, but also for extracting repeat-pass interferometric coherence, with absolute values of perpendicular baselines ranging from 190 to 479 m for 2013 CSK dataset, and from 363 to 1020 m for 2014 CSK dataset.

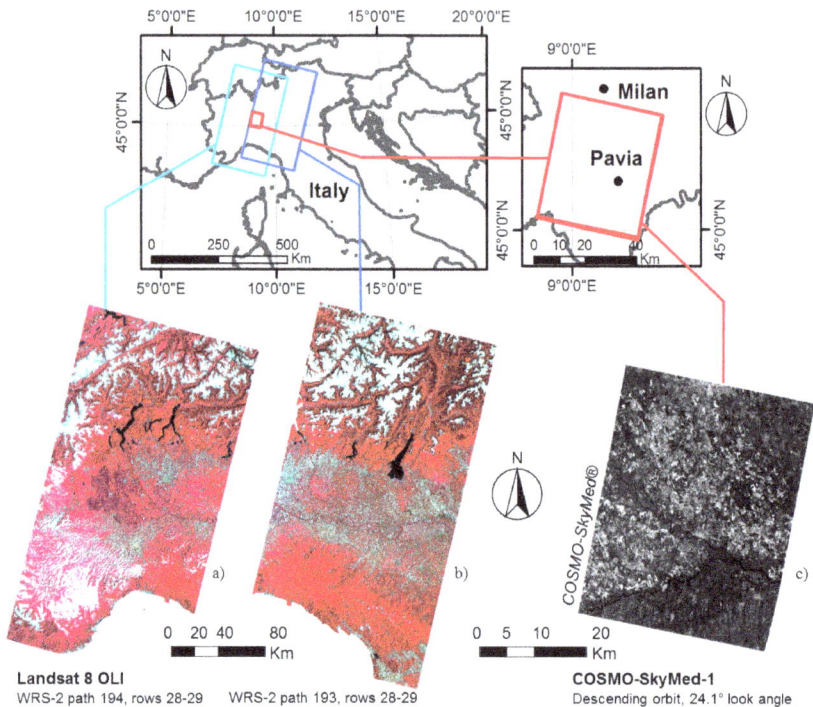

Figure 1. The study area in Northern Italy and an overview of the optical and SAR data coverage: (**a**) Landsat 8 OLI, path 194-rows 28-29 (03 July 2014, RGB = 543), (**b**) Landsat 8 OLI, path 193-rows 28-29 (10 June 2014, RGB = 543), (**c**) COSMO-SkyMed-1 (10 July 2014, Product processed under a license of the Italian Space Agency (ASI); Original COSMO-SkyMed Product - ©ASI - (2013)).

Table 1. The X-band SAR and optical satellite acquisitions divided into the development (2013) and the transferability dataset (2014).

Dataset	SAR Data (COSMO SkyMed-1)		Optical Data (Landsat 8 OLI)		
	Date	DOY	Date	DOY	WRS-2 (path/rows)
	18-04	108	13-05	133	194/28-29
	04-05	124	07-06	158	193/28-29
	20-05	140	14-06	165	194/28-29
2013	05-06	156	23-06	174	193/28-29
	21-06	172	30-06	181	194/28-29
	07-07	188	16-07	197	194/28-29
	23-07	204			
	05-04	95	14-04	104	194/28-29
	21-04	111	23-04	113	193/28-29
	07-05	127	09-05	129	193/28-29
2014	23-05	143	25-05	145	193/28-29
	08-06	159	01-06	152	194/28-29
	10-07	191	10-06	161	193/28-29
			03-07	184	194/28-29
			19-07	200	194/28-29

302

The CSK product was acquired as StripMap HIMAGE in Single Look Complex (SLC) format, HH polarization, descending pass, with look angle of 24.1 degrees. The OLI dataset was collected by taking advantage of two overlapping WRS-2 paths (193 and 194, rows 28-29) that guarantees a theoretical revisiting frequency of 7–9 days over the study area. Landsat 8 OLI scenes were retained only when overall cloud cover was less than 10%.

2.3. Reference Data

Three thematic levels of crop types were considered for covering the variability of crops cultivated in the study area: a detailed level (level 2—Lev2), an intermediate level (level 1—Lev1), and a generic level (level 0—Lev0), as summarized in Table 2. Lev2 is composed of 12 classes: *early maize* (Ma1), *medium maize* (Ma2), *late maize* (Ma3), *early rice* (R1), *late rice* (R2), *dry seeded rice* (R3), *early soybean* (Sb1), *late soybean* (Sb2), *winter crop* (WC), *double crop* (*i.e.*, winter crop followed by a summer crop; DC), *forages* (*i.e.*, permanent and temporary meadows used as fodder; Fo), and *forestry-woodland* (either natural or man-made; F-W). Lev1 groups the subclasses of maize, rice and soybean crops into three mono-type classes: *maize* (Ma), *rice* (R), and *soybean* (Sb), thus delivering a total of 7 crop cover classes (Ma, R, Sb, WC, DC, Fo, F-W). Indeed, Lev1 classes represent the target crop types for operational use of the early crop map, but since some Lev1 classes showed multimodal SIs temporal profiles due to different sowing dates of various cultivars (e.g., long and short cycle maize), we split some summer crop classes into different sub-classes in order to effectively calculating class separability, thus composing Lev2 classes. Finally, Lev0 was derived by further aggregating summer crop types (Ma, R, Sb) into a unique *summer crop* (SC) class, leading to a total of 5 land cover classes (SC, WC, DC, Fo, F-W).

For building the reference dataset to be used for crop type classification implementation, a set of 570 crop fields (almost 9000 pixels), belonging to 12 different crop classes, have been identified for the 2013 spring-summer season based on the Annual Agricultural Land Use Map of Lombardy region (Carta Uso Agricolo Annuale, CUAA); this map is produced and distributed by the "*Ente Regionale per i Servizi all'Agricoltura e alle Foreste*" (*i.e.*, the regional agency for agriculture and forest services of the Lombardy regional government) on an annual basis and relies on farmers' declarations as the primary source of information [62].

Table 2. Composition of the reference sample for 2013 development set, including training and validation sub-samples.

	Lev2				Lev1				Lev0		
	Crop Class	Training px/fields	Validation px/fields		Crop Class	Training px/fields	Validation px/fields		Crop Class	Training px/fields	Validation px/fields
Ma1	maize (early seeding)	881/34	520/17	Ma	maize	2233/96	1173/48				
Ma2	maize (medium seeding)	771/36	378/18					SC	summer crop (generic)	5439/260	2794/130
Ma3	maize (late seeding)	581/26	275/13								
R1	rice (early seeding)	628/34	335/17	R	rice	2296/112	1066/56				
R2	rice (late seeding)	962/44	431/22								
R3	rice (dry seeding)	706/34	300/17								
Sb1	soybean (early seeding)	462/26	319/13	Sb	soybean	910/52	561/26				
Sb2	soybean (late seeding)	448/26	242/13								
WC	winter crop	782/30	344/15	WC	winter crop	782/30	344/15	WC	winter crop	782/30	344/15
DC	double crop	1007/34	443/17	DC	double crop	1007/34	443/17	DC	double crop	1007/34	443/17
Fo	forages (artificial grassland)	447/22	212/11	Fo	forages (artificial grassland)	447/22	212/11	Fo	forages (artificial grassland)	447/22	212/11
F-W	forestry-woodland	1118/34	530/17	F-W	forestry-woodland	1118/34	530/17	F-W	forestry-woodland	1118/34	530/17

The crop categories of the CUAA legend are not consistent with crop classes defined on the basis of spectral response from remotely sensed data. For example, the CUAA crop category "maize" includes both single and double crop cultivations, which are characterized by different temporal profiles of the Spectral Indices derived from OLI data, thus leading to two distinct classes. Furthermore, the CUAA category "forages" includes all crops cultivated for animal consumption, *i.e.*, some winter cereals, fodder and managed grasslands (alfalfa and similar), which are grown and mowed several times per season. Finally, no official figures are provided for the accuracy of the CUAA product. Therefore, the CUAA 2013 map was not used as direct reference data source, but it was used as base information for extracting sample fields, which have been confirmed by visual assessment of high resolution satellite photos covering part of the study area (acquired on 22 March and 10 August 2013, from Google Earth), *in situ* observations for a limited number of fields (survey along main roads using camera and GPS) and interpretation of multi-temporal profiles of optical scenes (*i.e.*, for extracting Lev2 subclasses from Lev1 by assessing season timing of EVI peak, and for delineating double cropped fields).The reference dataset for 2013 growing season is described in Table 2, including the number of fields sampled for each class and level. The sampling was done on the basis of random selection of spatially distributed points from CUAA 2013 map within each crop class, followed by checking for correct class labelling consistently with semantic crop classes included into our target legend. Finally, a subdivision of the reference set was made on a per-field basis, with 2/3 of the fields used for training and 1/3 used for validation (see Table 2). This procedure allowed the attribution of crop type labels consistent with target crop classes in 2013 season, and it is used together with the satellite development set. For 2014 season, a validation set was constructed starting from CUAA 2014 map and using the same checking procedure described for 2013 reference set. 2014 validation set is composed of a total of 3759 pixels, with class cardinality ranging from 289 to 753 pixels, and it is consistently used for assessing mapping accuracy derived from satellite transferability set.

The satellite and reference data have been divided into two separate datasets: (i) a development set, used for best input feature selection, training of the classification algorithm and accuracy assessment (*i.e.*, for developing the crop mapping approach), made of the satellite data from the year 2013 (7 CSK, 6 OLI) and the training and validation samples extracted and checked from CUAA 2013 (Table 2) and (ii) a transferability set, used for validation of the crop mapping approach implemented for a different growing season (*i.e.*, testing the transferability of the approach), composed of the satellite data acquired in 2014 (6 CSK, 8 OLI) and the validation sample extracted and checked from CUAA 2014.

3. Methods

3.1. Satellite Data Pre-Processing

Landsat 8 OLI data [63] were converted to surface reflectance through atmospheric correction, performed with Atmospheric/Topographic CORrection for Satellite Imagery (ATCOR) [64]. Multi-temporal SIs have already demonstrated their efficacy in capturing cropland characteristics [54,65]. For our approach, three SIs were derived as proxies of crop conditions from optical data: Enhanced Vegetation Index (EVI, Equation (1)), Normalized Difference Flood Index (NDFI, Equation (2)), and Red Green Ratio Index (RGRI, Equation (3)).

$$\text{EVI} = 2\frac{\rho_{NIR(b5OLI)} - \rho_{RED(b4OLI)}}{\rho_{NIR(b5OLI)} + 6\rho_{RED(b4OLI)} - 7.5\rho_{BLUE(b2OLI)} + 1} \tag{1}$$

$$\text{NDFI} = \frac{\rho_{RED(b4OLI)} - \rho_{SWIR(b7OLI)}}{\rho_{RED(b4OLI)} + \rho_{SWIR(b7OLI)}} \tag{2}$$

$$\text{RGRI} = \frac{\rho_{GREEN(b3OLI)}}{\rho_{RED(b4OLI)}} \tag{3}$$

EVI was developed as an enhanced version of NDVI, including correction for background and atmospheric disturbances; the spectral bands of near infrared (NIR) and visible red (RED) are supplemented by information from the visible blue (BLUE) band, by using optimal weighting [66]. EVI provides information about vegetation vigor, linked to biomass and fractional cover. As a complement to EVI, we included

NDFI, which is an index developed for the detection of surface water in flooded rice areas; NDFI, originally introduced as $NDFI_2$ or $NDSI_{B2B7}$ [67], is the normalized difference of the RED and the short wave (SWIR, 2.1–2.2 µm) spectral bands. NDFI provides information about soil moisture and flooding conditions, especially relevant for paddy rice fields. The ratio of the RED and GREEN reflectance values (RGRI) was included due to its sensitivity to photosynthetic efficiency and leaf pigments [68].

Figure 2 shows EVI (a) and NDFI (b) multi-temporal profiles extracted from the 2014 OLI dataset (covering the whole growing season: 16 March–23 October), together with acquisition dates of CSK and OLI for the development dataset (2013); the grey bar highlights the temporal range adopted as early crop map production deadline (mid-July). The profiles qualitatively well describe the cycles of the major winter and summer crops of the study area, thus promoting the SIs selected as suitable candidates for monitoring crop dynamics during the season.

CSK images were pre-processed with MAPscape-RICE software [59] for (i) mosaicking single frames into slant range continuous strips and (ii) co-registration of images using orbital information and automatic spatial matching based on cross-correlation.

Two different proxies of crop conditions were derived from CSK data: X-band backscattering coefficient sigma nought (σ°), related to plant biomass and morphology and soil conditions (moisture and roughness), already used for crop mapping by Fontanelli *et al.* [69], and repeat-pass interferometric coherence (γ), related to the evolution of surface scattering properties of canopy/background system (plant height and density) during the season [70,71]. The σ° was derived through three processing steps: (a) multi-temporal speckle filtering according to the approach developed by De Grandi *et al.* [59,60,72], to balance differences in reflectivity between images at different times, (b) geocoding and radiometric calibration, using a Digital Elevation Model (SRTM DEM, at 90m equivalent ground resolution) and the radar equation, in which scattering area, antenna gain patterns and range spread loss were considered, and finally (c) normalization on local incidence angle, according to the cosine law. The interferometric coherence γ maps were produced using the complex data of image pairs of consecutive acquisitions [73], with a temporal baseline of 16 days (32 days for the 08 June–10 July 2014 pair). The multi-temporal σ° and γ maps were finally geocoded to UTM 32N WGS84 and spatially resampled to the same spatial resolution of L8 OLI images (30 m), by aggregating the average value of 10×10 pixels at the original resolution of 3 m.

3.2. Multi-Temporal Proxies Test

In order to capture the distinct seasonal patterns of different crops, we divided the multi-temporal dataset into two periods: (i) from April to the beginning of June, when winter crops are harvested; and (ii) from May-June, when summer crops emergence starts, to mid-July. Synoptic seasonal features over the two periods were computed for EVI, RGRI, $\sigma°$ and γ. For NDFI, only pre-summer features were derived, being it related to flooding in rice cultivation. Seasonal proxies used for the development set (2013) are summarized in Table 3. Each seasonal proxy is made a series of values corresponding to dates falling into the Day Of Year (DOY) range representative of winter crop and early summer crop seasons.

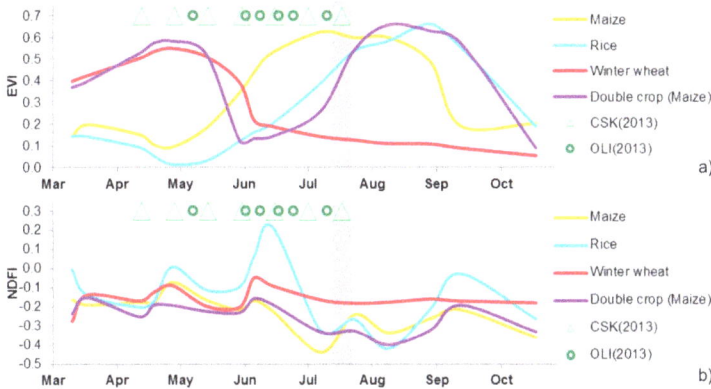

Figure 2. EVI (**a**) and NDFI (**b**) multi-temporal profiles extracted from the 2014 OLI dataset for major crops in Lombardy study site (covering the whole growing season from winter crop stem elongation phase to summer crops harvesting: March to October). Satellite acquisition dates of OLI and CSK images used for the development set (2013) are superimposed on the graphs. Grey color box represent the target temporal range for producing the early in-season crop map (*i.e.*, mid-July).

Table 3. Optical and X-band SAR seasonal proxies for the 2013 development set. EVI_W = EVI in winter crop growing season; EVI_S = EVI in early summer crop growing season; $RGRI_W$ = RGRI in winter crop growing season; $RGRI_S$= RGRI in early summer crop growing season; NDFI = NDFI before summer crop peak; $\sigma°_W$ = X-band HH backscatter in winter crop growing season; $\sigma°_S$ = X-band HH backscatter in early summer crop growing season; γ_W = interferometric coherence in winter crop growing season; γ_S = interferometric coherence in early summer crop growing season.

Seasonal Proxy	DOY Range (2013 Dataset)	Crop Vegetation and Soil Characteristics Connected
EVI_W	133–165	EVI during winter crop peak season (May–June)
EVI_S	174–197	EVI during early summer crop growth season (June–July)
$RGRI_W$	133–165	RGRI during winter crop peak season (May–June)
$RGRI_S$	174–197	RGRI during early summer crop growth season (June–July)
NDFI	133–197	NDFI before summer crop peak (May–July)
$\sigma°_W$	108–156	X-band backscattering coefficient during winter crop peak season (May–June)
$\sigma°_S$	172–204	X-band backscattering coefficient during early summer crop growth season (June–July)
γ_W	108–156	repeat-pass interferometric coherence during winter crop peak season (May–June)
γ_S	172–204	repeat-pass interferometric coherence during early summer crop growth season (June–July)

For each proxy, the following synoptic seasonal features were extracted for the winter and summer periods: minimum value over the seasonal range (min), maximum value over the seasonal range (max), mean value over the seasonal range (ave), standard deviation over the seasonal range (std) (Figure 3), and the asymmetry index of the seasonal proxy scores histogram, or skewness (ske), not included in Figure 3.

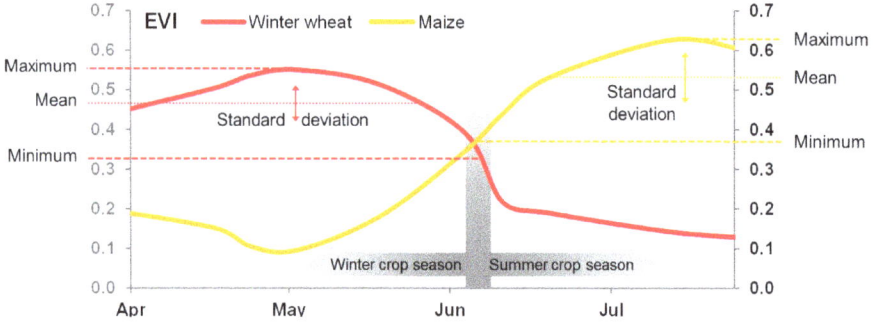

Figure 3. An example of synoptic seasonal features (min, max, ave, std) extracted for the winter and summer periods from EVI 2014 multitemporal profiles (April to mid-July) over winter wheat (red) and maize (yellow) sample fields.

3.3. In-Season Crop Type Classification

The classification scheme and the input features were selected to satisfy as much as possible crop mapping pre-operational requirements, thus (i) providing a product with high thematic mapping accuracy; (ii) being transferable to different years and (iii) building on rules both simple and interpretable, even to those who are non-experts in pattern recognition and remote sensing. Given these constraints, we implemented a rule-based classification tree, which grants both flexibility and robustness, and support the use of multi-source data [51–53]. The scheme is implemented using the Classification and regression Tree (CT) algorithm, in the extension of J48 java routine, programmed in WEKA 3.6 [74]. J48 CT routine is shaped on C4.5 [75] and consists of a recursive algorithm, that generates a classification tree through iterative partitioning of the feature space by using the information gain (computed from the entropy function) of each attribute for a set of cases [76]. Each node in a tree is associated to a set of two or more cases. The attribute with the highest information gain is selected for each node, and the optimal threshold for continuous attributes is computed. For avoiding too complicated tree structures and over-fitting, embedded pruning capabilities are implemented into J48 CT algorithm, according to a given confidence level. For our J48 CT implementation we allowed only binary splits for each node, and used online pruning, with confidence factor of 0.25 and sub-tree raising option. For dealing with possible over-fitting and minimizing the tree size, we set the minimum number of classified instances per each final node equal to 200, which is approximately half of the size of the smallest Lev1 crop class in the training dataset (*forages*, 447 pixels, see Table 2). CT outputs a set of hierarchical rules with optimized decision boundaries in form of thresholds, which can be straightforwardly implemented for image classification. An additional output of CT is the assessment of class attribution error for each tree node, which is a useful metric for ex-post tree re-structuring, in case of high accumulation of misclassified instances in some branches.

The performance of the CT schemes implemented was tested with different combination of proxies and features, by computing the confusion matrix and derived accuracy metrics [77]. During this phase, results achieved with the CT approach were compared to the ones achieved by Random Forest (RF) classification [78], which is currently acknowledged as the upper limit reachable using state of the art classification tree algorithms and multi-source data [79–81].

3.4. Selection of Input Features

Descriptive statistics were extracted for each Lev2 class and different synoptic seasonal features combinations: (i) σ° features; (ii) σ° and γ features; (iii) EVI features; (iv) EVI and NDFI features; (v) EVI, NDFI and RGRI features; (vi) EVI, NDFI, RGRI, and σ° features; (vii) EVI, NDFI, RGRI, σ° and γ features. Lev2 class-by-class separability was computed for these combinations using the Jeffries-Matusita Distance (J-M_{DIST}) [82], and aggregated as per-class separability by averaging all possible pairings comprising a specific class. As a rule of the thumb, good separability is generally set at J-M_{DIST} higher than 1.9 [83]. Lev2 classes used for extracting class-by-class separability were further summarized into average class separability scores at Lev1. The best performing combinations of seasonal proxies were then selected by maximizing separability, with J-M_{DIST}~2. Following the selection of best proxy combinations, the feasibility of reducing the synoptic seasonal features input set was assessed based on the overall performance of preliminary crop classification tests using CT and RF, with three different sets of features: (i) minimum, maximum, mean, standard deviation, and skewness (min-max-ave-std-ske); (ii) minimum, maximum, and mean (min-max-ave); and(iii) minimum and maximum (min-max). The rationale behind this choice is to assess the effect in terms of mapping accuracy, when discarding features mostly affected by year-to-year variability of satellite image frequency: *i.e.*, we first excluded standard deviation and skewness, and then the average, which in case of few scenes available can be more biased than extreme values (min-max).

At the end of the test phase, we retained the input set (*i.e.*, the combinations of proxies and synoptic seasonal features) that granted as much reduction as possible in the number of features, granting the lowest difference of overall accuracy between CT and RF results, at Lev1.

3.5. Validation

The CT scheme derived for the selected input set was applied and validated over 2013 and 2014 seasonal data, development and transferability sets, by calculating confusion matrices and derived metrics: Overall accuracy (OA), Kappa coefficient of agreement (κ) and per-class Commission (CE) and Omission (OE) Errors [77]. Validation was carried out on samples independent from the training set, with per-class cardinality either proportional to the one of the training set (case t), or of actual cropland cover calculated from CUAA reference information (case r). Crop mapping performance was assessed at two thematic levels: Lev1, and Lev0.

An overview of the whole methodological approach described in Section 3 is given in Figure 4.

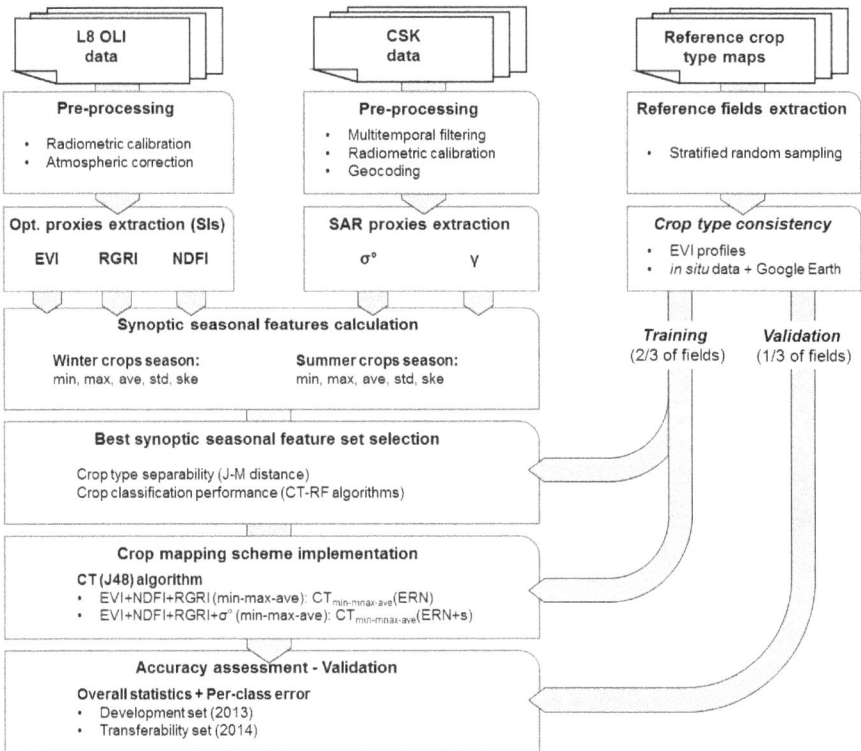

Figure 4. Flow chart of the overall methodological approach.

4. Results and Discussion

4.1. Selection of Synoptic Seasonal Features

Since the separability scores for Lev2 classes were generally not high enough for expecting good performance in early mapping, Lev1 (*i.e.*, 7 crop classes) was chosen as the target level for our crop mapping approach. Class separability scores aggregated at Lev1 (Table 4) show some interesting response to the different input proxy combinations (derived from optical and/or X-band SAR data).

Using only X-band SAR information, J-M_{DIST} increases consistently for all classes when adding γ to $\sigma°$: from +0.041 for better separated classes (*rice* and *forages*), to +0.150 and up to +0.204 for crop classes difficult to separate using only $\sigma°$ (*maize*, *soybean*, and *double crop*). Using optical proxies, a slighter but consistent increment in J-M_{DIST} (0.037 to 0.053) is observed by adding NDFI to EVI for summer crops (*maize*, *rice*, and *soybean*). Further addition of RGRI brings a small, yet consistent, increment in J-M_{DIST} (0.009 to 0.023) over summer crops. As regards separability achieved by using optical and SAR proxies together, an increment is observed adding $\sigma°$ to the full optical feature set; since separability is already close to the maximum value (J-M_{DIST}~2) the increment is lower for *maize* and *soybean* classes (+0.007 to +0.010). No significant increment (+0.000 to +0.002) is granted by further adding γ.

Based on separability scores shown in Table 4, we kept only the best performing combinations in terms of overall separability (minimum JM_{DIST} > 1.98): EVI+NDFI+RGRI (ERN, J-M_{DIST} > 1.983) and EVI+NDFI+RGRI+$\sigma°$ (ERN+s, J-M_{DIST} > 1.997). Since the scores achieved using EVI+NDFI+RGRI+$\sigma°$ and EVI+NDFI+RGRI+$\sigma°$+γ synoptic features are not significantly different, we decided to discard the combination including γ to keep the feature set as simple as possible. These two combinations were used as input for the classification approach development.

Table 4. Mean J-M_{DIST} for each crop type class at Lev1, as a function of the combination of OLI and SAR proxies. Maximum separability corresponds to J-M_{DIST} = 2.

Crop Type (Lev1)	Combination of Seasonal Proxies Used						
	$\sigma°$	$\sigma°$+γ	EVI	EVI+NDFI	EVI+NDFI+RGRI	EVI+NDFI+RGRI+$\sigma°$	EVI+NDFI+RGRI+$\sigma°$+γ
Maize	1.781	1.944	1.928	1.973	1.988	1.998	1.999
Rice	1.936	1.977	1.951	1.988	1.997	2.000	2.000
Soybean	1.784	1.934	1.907	1.960	1.983	1.997	1.999
Winter crop	1.921	1.988	1.998	1.999	1.999	2.000	2.000
Double crop	1.755	1.959	1.998	1.999	1.999	2.000	2.000
Forages	1.949	1.990	1.999	2.000	2.000	2.000	2.000
Forestry-woodland	1.927	1.990	1.999	2.000	2.000	2.000	2.000

Table 5 shows OA of preliminary crop classification tests using synoptic seasonal features extracted from ERN and ERN+s combinations at Lev1 and Lev0 as input for CT, with RF scores as reference. Results achieved with the complete features set (min-max-ave-std-ske) show that: (i) at Lev1, OA achieved with CT increases

311

from 85.3% (ERN) to 89.0% (ERN+s), with a gap towards RF of 5.8%–8.9%; (ii) at Lev0, very high OA is scored by CT (96.7–97.8%), reducing the gap towards RF to 1.3%–2.5%. Reducing the input features to min-max-ave set did not produce a sensible decrement in OA, with maximum decrement in OA of −0.8% across different proxy combinations and thematic levels, while increments in OA for CT are observed using ERN: +1.2% (at Lev0) and +1.7% (at Lev1). Further reducing input features to min-max did not significantly change OA at Lev0, but a decrement up to 2.1% was observed for Lev1.

Figure 5 shows per-class omission and commission errors (OE and CE) at Lev1 for CT and RF. When the min-max-ave set is used, no significant increase of per-class errors is observed, compared to the use of a full set of features (min-max-ave-std-ske; Figure 5b–e). Instead, the use of min-max-ave set and CT fed with optical only features (ERN) contributes to a reduction of OE for *forages* (25%, Figure 5b), and of CE for *forestry-woodland* (9%, Figure 5e). When the input feature set is further reduced to min-max (Figure 5c–f), higher errors for *double crop* (+11% OE, +5% CE), and some overestimation of *winter crop* (+8% CE) are observed. The best performing synoptic seasonal feature set was therefore identified as min-max-ave, which was therefore selected as best option input for early crop mapping.

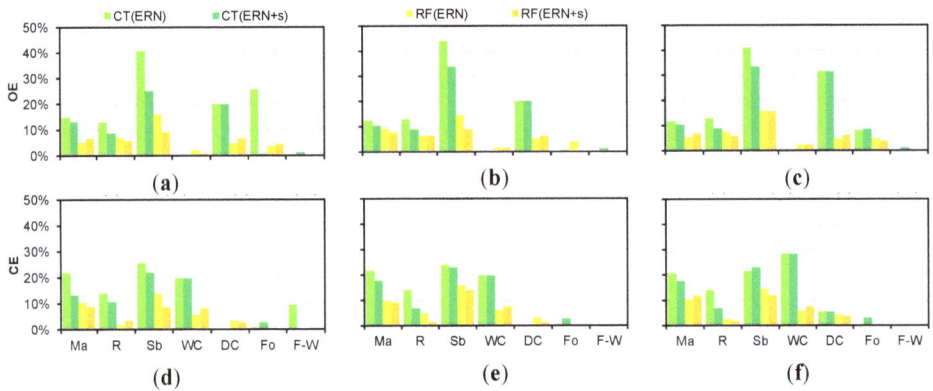

Figure 5. Per-class omission (OE) and commission (CE) errors for Lev1 classes achieved by CT with ERN and ERN+s proxies and different synoptic seasonal feature sets: (**a**) OE with min-max-ave-std-ske; (**b**) OE with min-max-ave; (**c**) OE with min-max; (**d**) CE with min-max-ave-std-ske; (**e**) CE with min-max-ave; (**f**) CE with min-max. Ma = maize; R = rice; Sb = soybean; WC = winter crop; DC = double crop; Fo = forages; F-W = forestry-woodland; ERN = EVI+NDFI+RGRI; ERN = EVI+NDFI+RGRI+$\sigma°$ = ERN+s. Results achievable using RF are shown as reference.

Table 5. Overall Accuracy (OA) for CT at Lev1 and Lev0, using ERN and ERN+s proxy combinations and min-max-ave-std-ske, min-max-ave and min-max seasonal features. ERN = EVI+NDFI+RGRI; ERN = EVI+NDFI+RGRI+σ° = ERN+s. Results achievable using RF are shown as reference.

OA		ERN			ERN+s		
Method	Level	min-max-ave-std-ske	min-max-ave	min-max	min-max-ave-std-ske	min-max-ave	min-max
CT	Lev1	85.3%	87.0%	84.8%	89.0%	88.7%	86.9%
	Lev0	96.7%	97.9%	96.4%	97.8%	97.8%	96.2%
RF	Lev1	94.2%	93.4%	93.9%	94.8%	94.6%	93.8%
(reference)	Lev0	99.2%	99.2%	99.1%	99.1%	99.3%	99.0%

4.2. In-Season Crop Type Classification

Figure 6 shows the two CT schemes implemented using the J48 algorithm for to the 2013 training dataset with the min-max-ave synoptic seasonal feature set, applied to the ERN (CT$_{min-max-ave}$(ERN), Figure 6a) and ERN+s (CT$_{min-max-ave}$(ERN+s), Figure 6b).

The CT scheme developed using only optical features (CT$_{min-max-ave}$(ERN)) shows the first split of tree nodes based on winter season RGRI maximum (RGRI$_W^{max}$ ≥ 1.361), separating pixels which are vegetated in May-June (right side) from non-vegetated ones (left side), thus distinguishing summer crops from winter crops, double crops and other agricultural land cover. The tree branches on the right side of Figure 6a further split winter and double crops from non-sown land covers (*forages, forestry-woodland*) based on two combinations of EVI mean in early summer (EVI$_S^{ave}$), since the latter classes show high green fractional cover already during spring, when summer crops are not yet sown. All these nodes show crop type attribution errors lower than 2.4%, except for *winter crop* class (9.6% cumulated node error). Left branches in Figure 6a are populated by summer crop classes, characterized by low RGRI$_W^{max}$. Below these branches, the main splits are based on NDFI maximum (>0.089) to separate flooded *rice* fields (very accurate, with node error of 0.1%), and EVI maximum in winter-spring season (EVI$_W^{max}$ > 0.298) to identify early-cycle crops (mostly *maize*). Crop type detection in lower level branches are due to a combination of EVI and RGRI features from both winter-spring and early summer features, and are meant to separate a mixture of *rice, maize* and *soybean*; these branches are characterized by cumulative node error above 20% (high misclassification rate).

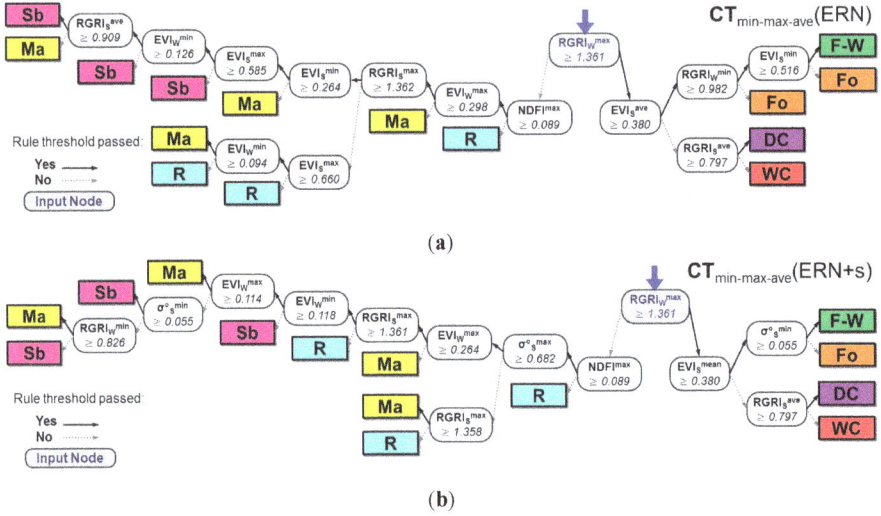

Figure 6. Classification tree schemes implemented using: (a) OLI data only, min-max-ave(ERN) input set, and (b) integrated OLI and CSK data, min-max-ave(ERN+s) input set. Ma = maize; R = rice; Sb = soybean; WC = winter crop; DC = double crop; Fo = forages; F-W = forestry-woodland.

The classification scheme developed using integrated optical and X-band SAR features (CTmin-max-ave(ERN+s)) shows a main split for $RGRI_W^{max} \geqslant 1.361$, consistently with $CT_{min-max-ave}(ERN)$. The branches on the right side highlight the contribution of X-band minimum backscatter in early summer for the discrimination of *forages* from *forestry-woodland* ($\sigma^{\circ}{}_S^{min} > 0.055$). In the left branches, the major *rice* class (flooded rice) is first identified based on maximum NDFI (as in $CT_{min-max-ave}(ERN)$) while *maize* or *rice* pixels showing very high X-band backscatter in early summer ($\sigma^{\circ}{}_S^{max} > 0.682$) are separated from a mixture of *rice, maize* and *soybean*. This mixture of summer crops is further untangled by a combination of winter season EVI, mean summer backscatter and RGRI peak scores. As previously noted for Figure 6a scheme, these leftmost branches are characterized by high class attribution errors (>20%). Figure 7 shows the in-season early crop maps produced using $CT_{min-max-ave}(ERN+s)$ scheme applied to the 2013 and 2014 datasets.

4.3. Validation

Table 6 summarizes accuracy metrics (OA and κ) for the crop maps shown in Figure 7 and computed for two different validation sets, with crop class cardinality proportional to either the training set (case *t*), or the actual cropland coverage of the area calculated from CUAA (case *r*). For 2013, the accuracy scores retrieved using the two sets are highly consistent, with case *r* giving slightly higher scores. OA achieved

314

at Lev1 with 2013 ERN+s dataset, peaking at 91.8% over case r set ($\kappa = 0.897$), are 1.7% higher than when using only optical data (ERN). At Lev0, the performance of ERN and ERN+s data are nearly the same, with OA (κ) around 98% (0.960) for both case r and case t validation sets.

Table 6. OA and κ computed for crop maps derived with the CT schemes of Figure 6 applied to the development (2013) and transferability (2014) sets, with ERN and ERN+s input features. case r = class distribution proportional to real case crop acreage (%); case t = same class distribution of training set; ERN = EVI+NDFI+RGRI; ERN+s = EVI+NDFI+RGRI+$\sigma°$.

	Level	Input Features	2013 Dataset		2014 Dataset	
		Validation Set	ERN	ERN+s	ERN	ERN+s
OA	Lev1	case t	87.0%	88.7%	-	-
		case r	90.1%	91.8%	66.9%	86.6%
	Lev0	case t	97.9%	97.8%	-	-
		case r	98.2%	98.2%	85.6%	92.4%
κ	Lev1	case t	0.839	0.860	-	-
		case r	0.875	0.897	0.572	0.826
	Lev0	case t	0.960	0.959	-	-
		case r	0.968	0.968	0.740	0.861

Classification performance over the transferability set (2014) shows very good accuracy scores at Lev0, yet lower than for 2013: OA = 85.6% ($\kappa = 0.740$) using optical features and OA = 92.4% ($\kappa = 0.861$) using optical and $\sigma°$ features. An increment of 6.8% in OA is achieved at Lev0 by integrating $\sigma°$ for 2014, while for 2013 no enhancement was observed. At Lev1, less consistent results are observed: using ERN input an OA = 66.9% was reached, 23.2% lower than 2013 results, while when ERN+s input are used, a rebound of +19.7% in OA is achieved, reaching 86.6% ($\kappa = 0.826$). This result highlights the significant contribution of X-band SAR backscattering in terms of transferability of the approach, *i.e.*, when the classification scheme is applied to a seasonal dataset different from the one used for algorithm development. The additional information brought by X-band SAR increases the robustness of the mapping approach at Lev1.

Per-class errors were analyzed at Lev1, providing some insights into the disaggregation of global accuracy results (Figure 8 and Table 7). Using the 2013 validation dataset (case r), depicted in light and dark green bars of Figure 8 and in the two upper matrices of Table 7, OE and CE are consistently lower than 25% for all classes, with the exception of *soybean* (OE = 25.3%–40.7%, CE = 50.7%–55.0%, with either ERN or ERN+s input), which is mainly misclassified as *maize.* For the *soybean* class, the use of X-band $\sigma°$ results in a OE reduction of 15.4%, due to less confusion with other summer crops, and in a CE reduction of 4.3%, due to less confusion with *rice*; a reduction of 4.9% CE for *rice* is also registered using ERN+s input, while errors over other crop types are stable. The best performances, with class errors not exceeding 10% and balanced between omission and commission, are achieved for *maize* (OE = 9.3%, CE = 10.6%), *rice* (OE = 9.3%, CE = 1.6%), *forages* (OE = 0.1%, CE = 0.5%), and *forestry-woodland* (OE = 0.6%, CE = 0.1%). Tendencies to overestimation for *winter crop* (CE = 16.8%) and to underestimation for *double crop* (OE = 23.7%) are observed, due to mutual confusion between these two classes.

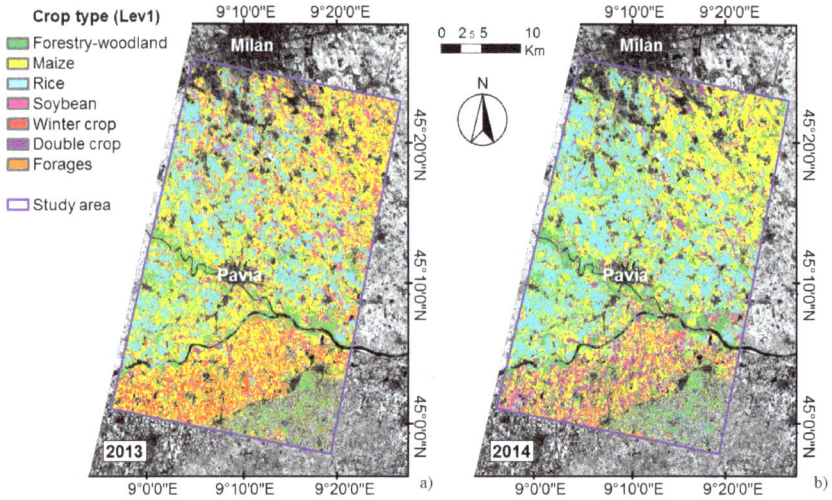

Figure 7. Early in-season crop maps produced using the $CT_{min-max-ave}$(ERN+s) scheme over the study area, for the year 2013 (**a**) and 2014 (**b**). Crop maps are overlaid on mid-July EVI (16 July 2013 for panel a, 19 July 2014 for panel b), represented in grey tones.

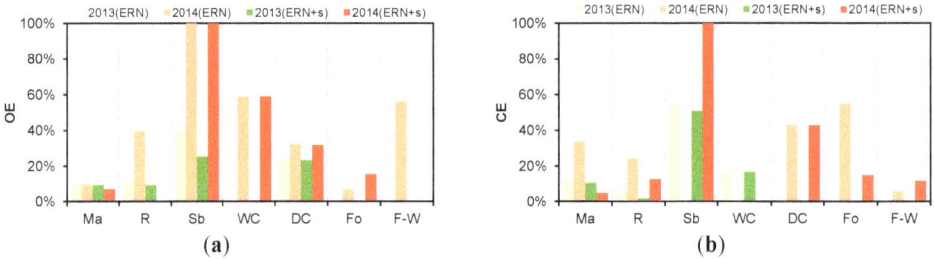

Figure 8. Class omission (OE) (**a**) and commission (CE) (**b**) errors at Lev1 for $CT_{min-max-ave}$ approach implemented over development (year 2013) and transferability (year 2014) sets using different input features (ERN and ERN+s). Ma = maize; R = rice; Sb = soybean; WC = winter crop; DC = double crop; Fo = forages; F-W = forestry-woodland; ERN = EVI+NDFI+RGRI; ERN = EVI+NDFI+RGRI+$\sigma°$ = ERN+s.

Table 7. Confusion matrices of early in-season crop maps produced using the CT scheme implemented over development (2013) and transferability (2014) sets, with ERN and ERN+s input features. Figures are expressed in hectares [ha]. ERN = EVI+NDFI+RGRI; ERN = EVI+NDFI+RGRI+σ° = ERN+s.

				Reference Dataset						
				Maize	Rice	Soybean	Winter crop	Double Crop	Forages	Forestry-woodland
Development Set	$CT_{min\text{-}max\text{-}ave}$ (ERN)	2013	Maize	95.6	8.6	14.6	0.0	0.5	0.0	0.1
			Rice	4.3	85.1	6.2	0.0	0.0	0.0	0.0
			Soybean	5.7	2.3	29.7	0.0	0.0	0.0	0.0
			Winter crop	0.0	0.0	0.0	29.1	7.7	0.0	0.0
			Double Crop	0.0	0.0	0.0	0.0	28.3	0.0	0.0
			Forages	0.0	0.0	0.0	0.0	0.0	19.1	0.0
			Forestry-woodland	0.0	0.0	0.0	0.0	0.0	0.0	47.4
	$CT_{min\text{-}max\text{-}ave}$ (ERN+s)	2013	Maize	95.8	8.6	12.4	0.0	0.0	0.0	0.1
			Rice	0.4	87.4	4.2	0.0	0.0	0.0	0.0
			Soybean	9.5	0.0	33.8	0.0	0.4	0.0	0.0
			Winter crop	0.0	0.0	0.0	29.1	7.7	0.0	0.0
			Double Crop	0.0	0.0	0.0	0.0	28.4	0.0	0.0
			Forages	0.0	0.0	0.0	0.0	0.0	19.1	0.3
			Forestry-woodland	0.0	0.0	0.0	0.0	0.0	0.0	47.2
Transferability Set	$CT_{min\text{-}max\text{-}ave}$ (ERN)	2014	Maize	61.0	26.8	8.6	0.0	2.5	0.0	0.1
			Rice	6.6	41.0	30.2	0.0	0.0	0.0	0.0
			Soybean	0.0	0.0	0.0	0.0	0.0	0.0	0.0
			Winter crop	0.0	0.0	0.0	14.6	0.0	0.0	0.0
			Double Crop	0.0	0.0	0.0	19.9	27.6	0.5	0.0
			Forages	0.0	0.0	0.0	0.0	8.9	24.2	34.7
			Forestry-woodland	0.0	0.0	0.0	0.0	1.4	0.0	29.8
	$CT_{min\text{-}max\text{-}ave}$ (ERN+s)	2014	Maize	62.8	0.0	13.3	0.0	0.0	0.0	0.0
			Rice	4.8	67.8	25.2	0.0	0.0	0.0	0.1
			Soybean	0.0	0.0	0.3	0.0	2.3	0.0	0.0
			Winter crop	0.0	0.0	0.0	14.6	0.0	0.0	0.0
			Double Crop	0.0	0.0	0.0	19.9	28.1	0.5	0.0
			Forages	0.0	0.0	0.0	0.0	5.3	22.4	0.1
			Forestry-woodland	0.0	0.0	0.0	3.4	3.2	0.0	64.4

When the $CT_{min\text{-}max\text{-}ave}$ scheme is applied to 2014 data (transferability set) in case *r* (Figure 8, light and dark orange bars, Table 7, two lower confusion matrices), some different patterns emerge: both CE and OE are greater than 2013 over most of the classes. A remarkable case is represented by *soybean*, with OE = 100.0%, meaning that this class is not represented in the classified pixels belonging to 2014 validation set. These pixels are mistakenly classified as *rice* (78%) and *maize* (22%). Other major discrepancies between 2013 and 2014 per-class accuracy occur for *winter crop* and *double crop* classes: for the former, the 2014 crop map is strongly underestimating (OE = 58%), while for the latter the rather conservative 2013 performance (CE = 0%) is not repeated for 2014 (CE > 40%, OE > 30%). The overestimation of *winter crop* is due to confusion with *double crop*, while the errors observed for *double crop* come from misclassification not only with *winter crop*, but also with *forages*. This could be due to the timing of 2014 acquisition dates, which less effectively capture the single and double crop dynamics compared to 2013 data. The high OE for *forestry-woodland* (56%) is instead due to class confusion with *forages*.

The integration of σ° for the 2014 dataset (ERN+s) results in very small changes of OE for *soybean*, *winter crop* and *double crop*, while it brings an improvement for *maize* (OE = 7.0%, CE = 4.9%), *rice* (OE = 0.2%, CE = 12.4%), *forages* (OE = 14.9%, CE = 20.0%), and *forestry-woodland* (OE = 0.3%, CE = 11.7%); this is mainly due to the reduced confusion between *rice* and *maize* and between *forages* and *forestry-woodland*. Still, per-class performances using ERN+s over 2014 are generally worse than for 2013, especially over crop classes more sensitive to the seasonal climatic conditions (e.g., *double crop* and *soybean*, usually sown late in summer season in Lombardy and thus prone to meteorological fluctuations). A different behavior is shown by *rice*, with a slight overestimation of class extent for 2014 dataset (<OE, >CE), compared to 2013 results. In summary, $CT_{min-max-ave}$ performance using ERN+s input is better than with ERN for most of the crop types at Lev1 and outputs acceptable class accuracies ranging from 62.8% for *double crop*, to >90% for *forestry-woodland* (94.0%), *rice* (93.8%), and *maize* (94.6%). Important misclassification errors are observed for *soybean* class across the inter-annual dataset.

4.4. Error Reduction Strategy

Since Lev1 map assessment showed some misclassification for summer crop classes (especially on 2014 dataset), we tested an expert-based ex-post pruning of the $CT_{min-max-ave}$ scheme, implemented by restructuring the set of rules for branches with higher class attribution cumulated error (>20%). The rationale is to test the performance of a classifier which reduces the overall classification error at the expenses of the detail of the thematic level. This way, we generate a crop map which is only partially at Lev1 detail (see Table 7 for Lev1 areal coverage percentage), by grouping summer crops, which are in the left branches of the schemes shown in Figure 6 together into a generic crop type label (generic *summer crop*, SCg) (Figure 9). These re-structured crop mapping schemes, implemented for both ERN and ERN+s are respectively named: $CT'_{min-max-ave}$(ERN), shown in Figure 9a, and $CT'_{min-max-ave}$(ERN+s) scheme, shown in Figure 9b.

Validation was carried out using the same validation sets used for assessing $CT_{min-max-ave}$ results, by excluding the areas labelled as SCg, which are classified now at Lev0. As a consequence, $CT'_{min-max-ave}$ early crop maps do not cover the whole study area at Lev1 and some of the cropland is mapped at Lev0. Table 8 summarizes accuracy metrics of the $CT'_{min-max-ave}$ scheme calculated for two validation datasets: case *t* and case *r*. The Lev1 classified coverage ranges from a minimum of 84% (obtained for the ERN 2013 map), to a maximum of 94% of the total study site cropland area (for the ERN 2014 map). As expected, the global accuracy scores achieved with the ex-post pruned schemes are higher than the ones derived with the original $CT_{min-max-ave}$ schemes: OA (κ) increases by 4.6–4.7% (0.047–0.062) for the year 2013 and by 3.7% (0.051–0.053) for the year 2014. As observed for the original scheme, the best performance for $CT'_{min-max-ave}$ is still achieved over the development set (2013), with OA~95% (κ~0.94). For the transferability set (2014), OA decreases to 70.6% using ERN input, but still a strong rebound of +19.7% (up to 90.3%)

is achieved by adding X-band $\sigma°$ (ERN+s set), with κ increasing from 0.625 to 0.877; the positive contribution of CSK based information for 2014 dataset is thus confirmed.

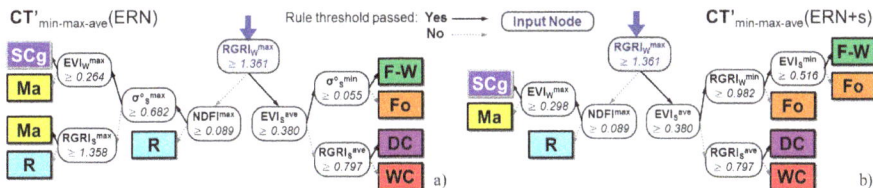

Figure 9. $CT'_{min-max-ave}$ classification tree schemes for: **(a)** ERN, and **(b)** ERN+s input dataset. Ma = maize; R = rice; SCg = summer crop (generic); WC = winter crop; DC = double crop; Fo = forages; F-W = forestry-woodland.

Table 8. Accuracy performance at Lev1 assessed using the $CT'_{min-max-ave}$ scheme (excluding the generic *summer crop* class), with either ERN or ERN+s input dataset, expressed in terms of OA and κ. The cropland area percentage classified at Lev1 is included. case r = validation set with class distribution proportional to real case crop acreage percent coverage; case t = validation set with class distribution same as training set; ERN = EVI+NDFI+RGRI; ERN = EVI+NDFI+RGRI+$\sigma°$ = ERN+s.

	Validation Set	Input Features			
		2013 Dataset		2014 Dataset	
		ERN	ERN+s	ERN	ERN+s
OA	case t	95.0%	95.5%	-	-
	case r	94.8%	95.4%	70.6%	90.3%
κ	case t	0.939	0.945	-	-
	case r	0.937	0.944	0.625	0.877
Lev1 coverage		88%	84%	94%	90%

Per-class error analysis (Figure 10) shows that the re-structured schemes provide an improvement only to CE of the summer crop classes, *maize* and *rice* (being *soybean* excluded as target here) with a significant reduction for *rice* CE (4.4%) over 2014 using ERN+s input set. In summary, the $CT'_{min-max-ave}$ showed slight but consistently better performance of the original scheme, and could be adopted as error reduction strategy when the proposed crop mapping approach is applied to different conditions; this would lead to a crop map with different thematic levels: Lev1, *i.e.*, distinguishing the majority of *rice* and *maize* fields , and Lev0.

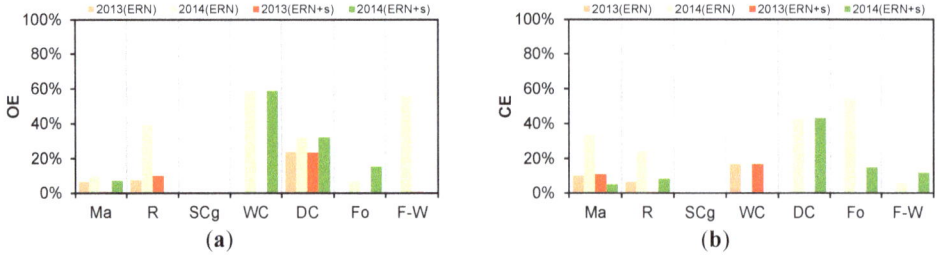

Figure 10. Class error at Lev1 for the CT'$_{min-max-ave}$ scheme over development (2013) and transferability (2014) sets, using different input features (ERN and ERN+s): (**a**) OE, (**b**) CE. Ma = maize; R = rice; SCg = summer crop (generic); WC = winter crop; DC = double crop; Fo = forages; F-W = forestry-woodland; ERN = EVI+NDFI+RGRI; ERN = EVI+NDFI+RGRI+σ° = ERN+s.

5. Conclusions

This study describes a classification tree approach for in-season crop mapping over Northern Italy during the early summer season (mid-July) from the integration of optical (Landsat 8 OLI) and X-band SAR (COSMO-SkyMed) synoptic seasonal features. A rule-based approach offers the advantage of being interpretable through rules/conditions applied to input features, which are representative of crop conditions and development. Results described could be applied to Northern Italy and, with minimal check and tuning by local experts, also to areas with similar environmental and agricultural characteristics (*i.e.*, European temperate to Mediterranean areas).

Key findings and conclusions relevant for crop mapping applications are listed below:

- The proposed approach produces early in-season (mid-July) crop type maps at two levels of thematic detail with the greatest accuracy obtained when both optical and SAR features (ERN+s set) are used as input: overall accuracy is 91.8% for the 2013 season and 86.6% for the 2014 season;

- Best performing input features for effectively distinguishing 7 crop types (*maize, rice, soybean, winter crop, double crop, forages, forestry-woodland*) can be extracted from synoptic seasonal features calculated for winter and summer crops and derived from two combinations of remote sensing proxies for vegetation and soil conditions: i) EVI, NDFI and RGRI from OLI data (ERN set), and ii) the integration of OLI-derived proxies with CSK backscattering (ERN+s set).

- The contribution of X-band σ° (HH polarization) is relevant for promoting the transferability of the approach over a season (2014) different from the one used for developing the classification rules, with an increment of 19.7% in OA compared to crop maps produced using only optical input features;

320

- The integration of CSK $\sigma°$ reduces class errors (omission and commission) especially for crop types characterized by more seasonally stable agronomic patterns (*i.e., maize, rice, forages, forestry-woodland*);
- Expert-based tuning and ex-post pruning are key assets when dealing with operational monitoring and can be used as error reduction strategy, delivering a modified early crop mapping scheme with hybrid thematic level output, and higher overall accuracy (90.3% using ERN+s input for 2014 season);
- In the framework of agriculture management, the achieved overall accuracy at mid-July is considered satisfactory given the fact that the information on crops is provided early during the growing season as management requires, *i.e.,* 2–3 months before end of season and harvesting;
- The proposed the approach is interpretable and flexible enough for being exploited for mapping crops at different levels of detail and possibly exploiting different input data with similar spectral bands (e.g., Sentinel-2 MSI).

Acknowledgments: The authors thank Mirco Boschetti (CNR-IREA) for the fruitful discussion during the preliminary phase of the work, and Alberto Crema (CNR-IREA) for the support in checking and preparing the crop type reference data. We are grateful to Francesco Holecz and team of Sarmap SA for having provided the software suite MAPscape-RICE used for processing CSK data and for their support in calculating interferometric coherence. Finally, we must thank the academic editor and four anonymous reviewers for their comments and suggestions, which have helped us in enhancing the manuscript. This work has been conducted in the frame of Space4Agri research project funded and supported by the AQ CNR-Regione Lombardia (CNR, Convenzione Operativa n. 18091/RCC, 05/08/2013). The work was carried out using CSK®Products, © of the Italian Space Agency (ASI), delivered under a license to use by ASI (project title: "Serie temporali di dati Cosmo SkyMed combinate con dati ottici per la mappatura delle varietà e dello stato colture in regione Lombardia").

Author Contributions: Paolo Villa, Daniela Stroppiana, Giacomo Fontanelli, and Pietro Alessandro Brivio contributed to the experimental design of the study. Paolo Villa, Giacomo Fontanelli, and Ramin Azar run the data preparation and processing. Paolo Villa, Daniela Stroppiana, Giacomo Fontanelli, and Pietro Alessandro Brivio carried out the analysis of results. Paolo Villa and Daniela Stroppiana prepared and wrote the manuscript, and all the authors revised it.

Conflicts of Interest: The authors declare no conflict of interest.

References

1. Potgieter, A.B.; Lawson, K.; Huete, A.R. Determining crop acreage estimates for specific winter crops using shape attributes from sequential MODIS imagery. *Int. J. Appl. Earth Obs.* **2013**, *23*, 254–263.
2. Hao, P.; Zhan, Y.; Wang, L.; Niu, Z.; Shakir, M. Feature selection of time series MODIS data for early crop classification using Random Forest: A case study in Kansas, USA. *Remote Sens.* **2015**, *7*, 5347–5369.

3. Becker-Reshef, I.; Vermote, E.; Lindeman, M.; Justice, C. A generalized regression-based model for forecasting winter wheat yields in Kansas and Ukraine using MODIS data. *Remote Sens. Environ.* **2010**, *114*, 1312–1323.

4. Rosegrant, M.W.; Cline, S.A. Global food security: challenges and policies. *Science* **2003**, *302*, 1917–1919.

5. Atzberger, C. Advances in remote sensing of agriculture: Context description, existing operational monitoring systems and major information needs. *Remote Sens.* **2013**, *5*, 949–981.

6. Mo, X.; Liu, S.; Lin, Z.; Xu, Y.; Xiang, Y.; McVicar, T.R. Prediction of crop yield, water consumption and water use efficiency with a SVAT-crop growth model using remotely sensed data on the North China Plain. *Ecol. Model.* **2005**, *183*, 301–322.

7. Reichstein, M.; Ciais, P.; Papale, D.; Valentini, R.; Running, S.; Viovy, N.; Cramer, W.; Granier, A.; Ogée, J.; Allard, V.; *et al.* Reduction of ecosystem productivity and respiration during the European summer 2003 climate anomaly: A joint flux tower, remote sensing and modelling analysis. *Glob. Change Biol.* **2007**, *13*, 634–651.

8. Ozdogan, M.; Yang, Y.; Allez, G.; Cervantes, C. Remote sensing of irrigated agriculture: Opportunities and challenges. *Remote Sens.* **2010**, *2*, 2274–2304.

9. McNairn, H.; Brisco, B. The application of C-band polarimetric SAR for agriculture: A review. *Can. J. Remote Sens.* **2004**, *30*, 525–542.

10. Carlson, R.E.; Aspiazu, C. Cropland acreage estimates from temporal, multispectral ERTS-1 data. *Remote Sens. Environ.* **1976**, *4*, 237–243.

11. Badhwar, G.B. Automatic corn-soybean classification using Landsat MSS data. II. Early season crop proportion estimation. *Remote Sens. Environ.* **1984**, *14*, 31–37.

12. Oetter, D.R.; Cohen, W.B.; Berterretche, M.; Maiersperger, T.K.; Kennedy, R.E. Land cover mapping in an agricultural setting using multiseasonal Thematic Mapper data. *Remote Sens. Environ.* **2001**, *76*, 139–155.

13. Price, K.; Egbert, S.; Lee, R.; Boyce, R.; Nellis, M.D. Mapping land cover in a high plains agroecosystem using a multi-date landsat thematic mapper modeling approach. *Trans. Kans. Acad. Sci.* **1997**, *100*, 21–33.

14. Martínez-Casasnovas, J.A.; Martín-Montero, A.; Casterad, M.A. Mapping multi-year cropping patterns in small irrigation districts from time-series analysis of Landsat TM images. *Eur. J. Agron.* **2005**, *23*, 159–169.

15. Vyas, S.P.; Oza, M.P.; Dadhwal, V.K. Multi-crop separability study of Rabi crops using multi-temporal satellite data. *J. Indian Soc. Remote Sens.* **2005**, *33*, 75–79.

16. Leite, P.B.C.; Feitosa, R.Q.; Formaggio, A.R.; da Costa, G.A.O.P.; Pakzadc, P.; Del'Arco Sanches, L. Hidden Markov Models for crop recognition in remote sensing image sequences. *Pattern Recogn. Lett.* **2011**, *32*, 19–26.

17. Wulder, M.A.; Masek, J.G.; Cohen, W.B.; Loveland, T.R.; Woodcock, C.E. Opening the archive: How free data has enabled the science and monitoring promise of Landsat. *Remote Sens. Environ.* **2012**, *122*, 2–10.

18. Rembold, F.; Maselli, F. Estimation of inter-annual crop area variation by the application of spectral angle mapping to low resolution multitemporal NDVI images. *Photogramm. Eng. Remote Sens.* **2006**, *72*, 55–62.

19. Chang, J.; Hansen, M.C.; Pittman, K.; Carroll, M.; DiMiceli, C. Corn and soybean mapping in the United States using MODIS time-series data sets. *Agron. J.* **2007**, *99*, 1654–1664.

20. An, Q.; Gao, W.; Yang, B.; Wu, Y.; Yu, L.; Liu, Z. Research on feature selection method oriented to crop identification using remote sensing image classification. In Proceedings of the Sixth International Conference on Fuzzy Systems and Knowledge Discovery, Tianjin, China, 14–16 August 2009; Volume 5, pp. 426–432.

21. Dutta, S.; Sharma, S.A.; Khera, A.P.; Yadav, M.; Hooda, R.S.; Mothikumar, K.E.; Manchanda, M.L. Accuracy assessment in cotton acreage estimation using Indian remote sensing satellite data. *ISPRS J. Photogramm.* **1994**, *49*, 21–26.

22. Dutta, S.; Patel, N.K.; Medhavy, T.T.; Srivastava, S.K.; Mishra, N.; Singh, K.R.P. Wheat crop classification using multidate IRS LISS-I data. *J. Indian Soc. Remote Sens.* **1998**, *26*, 7–14.

23. Aschbacher, J.; Pongsrihadulchai, A.; Karnchanasutham, S.; Rodprom, C.; Paudyal, D.R.; Le Toan, T. Assessment of ERS-1 SAR data for rice crop mapping and monitoring. In Proceedings of the IEEE International Geoscience and Remote Sensing Symposium, Firenze, Italy, 10–14 July 1995; Volume 3, pp. 2183–2185.

24. Le Toan, T.; Ribbes, F.; Wang, L.-F.; Floury, N.; Kung-Hau, D.; Kong, J.; Fujita, M.; Kurosu, T. Rice crop mapping and monitoring using ERS-1 data based on experiment and modeling results. *IEEE Trans. Geosci. Remote Sens.* **1997**, *35*, 41–56.

25. Dong, Y.; Sun, G.; Pang, Y. Monitoring of rice crop using ENVISAT ASAR data. *Sci. China* **2006**, *49*, 755–763.

26. Wang, X.; Shi, X.; Ling, F. Images difference of ASAR data for rice crop mapping in Fuzhou, China. *Geo. Spat. Inf. Sci.* **2010**, *13*, 123–129.

27. Satalino, G.; Mattia, F.; Le Toan, T.; Rinaldi, M. Wheat crop mapping by using ASAR AP data. *IEEE Trans. Geosci. Remote Sens.* **2009**, *47*, 527–530.

28. Stankiewicz, K.A. The efficiency of crop recognition on ENVISAT ASAR images in two growing seasons. *IEEE Trans. Geosci. Remote Sens.* **2006**, *44*, 806–814.

29. McNairn, H.; Ellis, J.; Van Der Sanden, J.J.; Hirose, T.; Brown, R.J. Providing crop information using RADARSAT-1 and satellite optical imagery. *Int. J. Remote Sens.* **2002**, *23*, 851–870.

30. Moran, M.S.; Vidal, A.; Troufleau, D.; Qi, J.; Clarke, T.R.; Pinter, P.J.; Mitchell, T.A.; Inoue, Y.; Neale, C.M.U. Combining multifrequency microwave and optical data for crop management. *Remote Sens. Environ.* **1997**, *61*, 96–109.

31. Liu, C.; Shang, J.; Vachon, P.W.; McNairn, H. Multiyear crop monitoring using polarimetric RADARSAT-2 data. *IEEE Trans. Geosci. Remote Sens.* **2013**, *51*, 2227–2240.

32. Moran, M.S.; Alonso, L.; Moreno, J.F.; Pilar Cendrero Mateo, M.; de la Cruz, D.F.; Montoro, A. A RADARSAT-2 quad-polarized time series for monitoring crop and soil conditions in Barrax, Spain. *IEEE Trans. Geosci. Remote Sens.* **2012**, *50*, 1057–1070.

33. McNairn, H.; Shang, J.; Jiao, X.; Champagne, C. The contribution of ALOS PALSAR multipolarization and polarimetric data to crop classification. *IEEE Trans. Geosci. Remote Sens.* **2009**, *47*, 3981–3992.

34. Larrañaga, A.; Alvarez-Mozos, J.; Albizua, L. Crop classification in rain-fed and irrigated agricultural areas using Landsat TM and ALOS/PALSAR data. *Can. J. Remote Sens.* **2011**, *37*, 157–170.

35. Bargiel, D.; Herrmann, S. Multi-temporal land-cover classification of agricultural areas in two European regions with high resolution spotlight TerraSAR-X data. *Remote Sens.* **2011**, *3*, 859–877.

36. Balenzano, A.; Satalino, G.; Belmonte, A.; D'Urso, G.; Capodici, F.; Iacobellis, V.; Gioia, A.; Rinaldi, M.; Ruggieri, S.; Mattia, F. On the use of multi-temporal series of COSMO-SkyMed data for land cover classification and surface parameter retrieval over agricultural sites. In Proceedings of the IEEE International Geoscience and Remote Sensing Symposium, Vancouver, Canada, 24–29 July 2011; pp. 142–145.

37. McNairn, H.; Kross, A.; Lapen, D.; Caves, R.; Shang, J. Early season monitoring of corn and soybeans with TerraSAR-X and RADARSAT-2. *Int. J. Appl. Earth Obs. Geoinf.* **2014**, *28*, 252–259.

38. Satalino, G.; Panciera, R.; Balenzano, A.; Mattia, F.; Walker, J. COSMO-SkyMed multi-temporal data for land cover classification and soil moisture retrieval over an agricultural site in Southern Australia. In Proceedings of the IEEE International Geoscience and Remote Sensing Symposium, Munich, Germany, 22–27 July 2012; pp. 5701–5704.

39. Sonobe, R.; Tani, H.; Wang, X.; Kobayashi, N.; Shimamura, H. Random forest classification of crop type using multi-temporal TerraSAR-X dual-polarimetric data. *Remote Sens. Lett.* **2014**, *5*, 157–164.

40. Gerstl, S.A.W. Physics concepts of optical and radar reflectance signatures A summary review. *Int. J. Remote Sens.* **1994**, *11*, 1109–1117.

41. Forkuor, G.; Conrad, C.; Thiel, M.; Ullmann, T.; Zoungrana, E. Integration of optical and Synthetic Aperture Radar imagery for improving crop mapping in Northwestern Benin, West Africa. *Remote Sens.* **2014**, *6*, 6472–6499.

42. Pohl, C.; Van Genderen, J.L. Review article multisensor image fusion in remote sensing: Concepts, methods and applications. *Int. J. Remote Sens.* **1998**, *19*, 823–854.

43. Stafford, J.V. Implementing precision agriculture in the 21st century. *J. Agric. Eng. Res.* **2000**, *76*, 267–275.

44. Michelson, D.B.; Liljeberg, B.M.; Pilesjo, P. Comparison of algorithms for classifying Swedish landcover using LANDSAT TM and ERS-1 SAR data. *Remote Sens. Environ.* **2000**, *71*, 1–15.

45. Blaes, X.; Vanhalle, V.; Defourny, P. Efficiency of crop identification based on optical and SAR image time series. *Remote Sens. Environ.* **2005**, *96*, 352–365.

46. McNairn, H.; Champagne, C.; Shang, J.; Holmstrom, D.; Reichert, G. Integration of optical and Synthetic Aperture Radar (SAR) imagery for delivering operational annual crop inventories. *ISPRS J. Photogramm.* **2009**, *64*, 434–449.

47. Skakun, S.; Kussul, N.; Shelestov, A.; Lavreniuk, M.; Kussul, O. Efficiency assessment of multitemporal C-band Radarsat-2 intensity and Landsat-8 surface reflectance satellite imagery for crop classification in Ukraine. *IEEE J. Sel. Topics Appl. Earth Obs. Remote Sens.* **2015**. in press.

48. De Oliveira Pereira, L.; da Costa Freitas, C.; Sant´Anna, S.J.S.; Lu, D.; Moran, E.F. Optical and radar data integration for land use and land cover mapping in the Brazilian Amazon. *GISci. Remote Sens.* **2013**, *50*, 301–321.

49. Zhu, Z.; Woodcock, C.E.; Rogan, J.; Kellndorfer, J. Assessment of spectral, polarimetric, temporal, and spatial dimensions for urban and peri-urban land cover classification using Landsat and SAR data. *Remote Sens. Environ.* **2012**, *117*, 72–82.

50. Stroppiana, D.; Azar, R.; Calò, F.; Pepe, A.; Imperatore, P.; Boschetti, M.; Silva, J.M.N.; Brivio, P.A.; Lanari, R. Integration of optical and SAR data for burned area mapping in Mediterranean Regions. *Remote Sens.* **2015**, *7*, 1320–1345.

51. Friedl, M.A.; Brodley, C.E. Decision tree classification of land cover from remotely sensed data. *Remote Sens. Environ.* **1997**, *61*, 399–409.

52. Lawrence, R.L.; Wright, A. Rule-based classification systems using classification and regression tree (CART) analysis. *Photogramm. Eng. Remote Sens.* **2001**, *67*, 1137–1142.

53. Vieira, M.A.; Formaggio, A.R.; Rennó, C.D.; Atzberger, C.; Aguiar, D.A.; Mello, M.P. Object Based Image Analysis and Data Mining applied to a remotely sensed Landsat time-series to map sugarcane over large areas. *Remote Sens. Environ.* **2012**, *123*, 553–562.

54. Azar, R.; Villa, P.; Stroppiana, D.; Crema, A.; Boschetti, M.; Brivio, P.A. Multi-temporal assessment of crop classification performance using Landsat 8 OLI data: A test case in Northern Italy. *Int. J. Remote Sens.* **2015**. submited.

55. Odenweller, J.B.; Johnson, K.I. Crop identification using Landsat temporal-spectral profiles. *Remote Sens. Environ.* **1984**, *14*, 39–54.

56. Murthy, C.S.; Raju, P.V.; Badrinath, K.V.S. Classification of wheat crop with multi-temporal images: Performance of maximum likelihood and artificial neural networks. *Int. J. Remote Sens.* **2003**, *24*, 4871–4890.

57. Wardlow, B.D.; Egbert, S.L. Large-area crop mapping using time-series MODIS 250 m NDVI data: An assessment for the US Central Great Plains. *Remote Sens. Environ.* **2008**, *112*, 1096–1116.

58. Foerster, S.; Kaden, K.; Foerster, M.; Itzerott, S. Crop type mapping using spectral–temporal profiles and phenological information. *Comput. Electron. Agric.* **2012**, *89*, 30–40.

59. Asilo, S.; de Bie, K.; Skidmore, A.; Nelson, A.; Barbieri, M.; Maunahan, A. Complementarity of two rice mapping approaches: Characterizing strata mapped by hypertemporal MODIS and rice paddy identification using multitemporal SAR. *Remote Sens.* **2014**, *6*, 12789–12814.

60. Nelson, A.; Setiyono, T.; Rala, A.B.; Quicho, E.D.; Raviz, J.V.; Abonete, P.J.; Maunahan, A.A.; Garcia, C.A.; Bhatti, H.Z.M.; Villano, L.S.; *et al.* Towards an operational SAR-based rice monitoring system in Asia: Examples from 13 demonstration sites across Asia in the RIICE project. *Remote Sens.* **2014**, *6*, 10773–10812.

61. Giuca, S.; Giannini, M.S.; Nebuloni, A.; Pretolani, R.; Pieri, R.; Cagliero, R.; Marras, F.; Gay, G. *Lombardy Agriculture in Figures—2013*; INEA: Milan, Italy, 2014; Available online: http://dspace.inea.it/bitstream/inea/846/1/Lombardy_agric_figures_2013.pdf (accessed on 15 May 2015).

62. ERSAF, Regione Lombardia. Carta dell'utilizzo agricolo annuale. Available online: http://www.ersaf.lombardia.it/servizi/Menu/dinamica.aspx?idArea=16914& idCat=17255&ID=22103 (accessed on 13 July 2015).

63. Barsi, J.A.; Lee, K.; Kvaran, G.; Markham, B.L.; Pedelty, J.A. The spectral response of the Landsat-8 operational land imager. *Remote Sens.* **2014**, *6*, 10232–10251.

64. Richter, R.; Schläpfer, D. *Atmospheric/Topographic Correction for Satellite Imagery*; DLR report DLR-IB 565-01/13; ReSe Applications: Wil, Switzerland, 2014; Available online: http://atcor.com/pdf/atcor3_manual.pdf (accessed on 18 May 2015).

65. Huete, A.R.; Liu, H.Q.; Batchily, K.; Van Leeuwen, W.J.D.A. A comparison of vegetation indices over a global set of TM images for EOS-MODIS. *Remote Sens. Environ.* **1997**, *59*, 440–451.

66. Huete, A.; Didan, K.; Miura, T.; Rodriguez, E.P.; Gao, X.; Ferreira, L.G. Overview of the radiometric and biophysical performance of the MODIS vegetation indices. *Remote Sens. Environ.* **2002**, *83*, 195–213.

67. Boschetti, M.; Nutini, F.; Manfron, G.; Brivio, P.A.; Nelson, A. Comparative analysis of normalised difference spectral indices derived from MODIS for detecting surface water in flooded rice cropping systems. *PLoS ONE* **2014**, *9*.

68. Gamon, J.A.; Surfus, J.S. Assessing leaf pigment content and activity with a reflectometer. *New Phytol.* **1999**, *143*, 105–117.

69. Fontanelli, G.; Crema, A.; Azar, R.; Stroppiana, D.; Villa, P.; Boschetti, M. Agricultural crop mapping using optical and SAR multi-temporal seasonal data: A case study in Lombardy region, Italy. In Proceedings of the IEEE International Geoscience and Remote Sensing Symposium, Quebec city, Canada, 13–18 July 2014; pp. 1489–1492.

70. Holecz, F.; Collivignarelli, F.; Barbieri, M. Estimation of cultivated area in small plot agriculture in Africa for food security purposes. In Proceedings of the ESA Living Planet Symposium, Edinburgh, UK, 9–13 September 2013.

71. Srivastava, H.S.; Patel, P.; Navalgund, R.R. Application potentials of synthetic aperture radar interferometry for land-cover mapping and crop-height estimation. *Curr. Sci. India* **2006**, *91*, 783–788.

72. De Grandi, G.F.; Leysen, M.; Lee, J.S.; Schuler, D. Radar reflectivity estimation using multiple SAR scenes of the same target: technique and applications. In Proceedings of the IEEE International Geoscience and Remote Sensing Symposium, Singapore, 3–8 August 1997; Volume 2, pp. 1047–1050.

73. Vijaya, V.; Niveditha, G.J. Classification of COSMO SkyMed SAR data based on coherence and backscattering coefficient. *Int. J. Comput. Sci. Inf.* **2012**, *1*, 60–63.

74. Witten, I.H.; Frank, E. *Data Mining: Practical Machine Learning Tools and Techniques*, 2nd ed.; Morgan Kaufmann: San Francisco, CA, USA, 2005; pp. 373–379.

75. Quinlan, J.R. Improved Use of Continuous Attributes in C4.5. *J. Artif. Intell. Res.* **1996**, *4*, 77–90.

76. Ruggieri, S. Efficient C4.5. *IEEE Trans. Knowl. Data Eng.* **2002**, *14*, 438–444.

77. Congalton, R.G. A review of assessing the accuracy of classifications of remotely sensed data. *Remote Sens. Environ.* **1991**, *37*, 35–46.

78. Breiman, L. Random Forests. *Mach. Learn.* **2001**, *45*, 5–32.

79. Chan, J.C.W.; Paelinckx, D. Evaluation of Random Forest and Adaboost tree-based ensemble classification and spectral band selection for ecotope mapping using airborne hyperspectral imagery. *Remote Sens. Environ.* **2008**, *112*, 2999–3011.

80. Gislason, P.O.; Benediktsson, J.A.; Sveinsson, J.R. Random Forests for land cover classification. *Pattern Recogn. Lett.* **2006**, *27*, 294–300.

81. Pal, M. Random Forest classifier for remote sensing classification. *Int. J. Remote Sens.* **2005**, *26*, 217–222.

82. Richards, J.A.; Jia, X. *Remote Sensing Digital Image Analysis: An Introduction*, 3rd ed.; Springer-Verlag: Berlin, Germany, 1999.

83. Swain, P.H.; King, R.C. Two Effective Feature Selection Criteria for Multispectral Remote Sensing. In Proceedings of the 1st International Joint Conference on Pattern Recognition, Washington, DC, USA, 30 October–1 November 1973; pp. 536–540.

Rapid Assessment of Crop Status: An Application of MODIS and SAR Data to Rice Areas in Leyte, Philippines Affected by Typhoon Haiyan

Mirco Boschetti, Andrew Nelson, Francesco Nutini, Giacinto Manfron,
Lorenzo Busetto, Massimo Barbieri, Alice Laborte, Jeny Raviz, Francesco Holecz
Mary Rose O. Mabalay, Alfie P. Bacong and Eduardo Jimmy P. Quilang

Abstract: Asian countries strongly depend on rice production for food security. The major rice-growing season (June to October) is highly exposed to the risk of tropical storm related damage. Unbiased and transparent approaches to assess the risk of rice crop damage are essential to support mitigation and disaster response strategies in the region. This study describes and demonstrates a method for rapid, pre-event crop status assessment. The *ex-post* test case is Typhoon Haiyan and its impact on the rice crop in Leyte Province in the Philippines. A synthetic aperture radar (SAR) derived rice area map was used to delineate the area at risk while crop status at the moment of typhoon landfall was estimated from specific time series analysis of Moderate Resolution Imaging Spectroradiometer (MODIS) data. A spatially explicit indicator of risk of standing crop loss was calculated as the time between estimated heading date and typhoon occurrence. Results of the analysis of pre- and post-event SAR images showed that 6500 ha were flooded in northeastern Leyte. This area was also the region most at risk to storm related crop damage due to late establishment of rice. Estimates highlight that about 700 ha of rice (71% of which was in northeastern Leyte) had not reached maturity at the time of the typhoon event and a further 8400 ha (84% of which was in northeastern Leyte) were likely to be not yet harvested. We demonstrated that the proposed approach can provide pre-event, in-season information on the status of rice and other field crops and the risk of damage posed by tropical storms.

Reprinted from *Remote Sens.* Cite as: Boschetti, M.; Nelson, A.; Nutini, F.; Manfron, G.; Busetto, L.; Barbieri, M.; Laborte, A.; Raviz, J.; Holecz, F.; Mabalay, M.R.O.; Bacong, A.P.; Quilang, E.J.P. Rapid Assessment of Crop Status: An Application of MODIS and SAR Data to Rice Areas in Leyte, Philippines Affected by Typhoon Haiyan. *Remote Sens.* **2015**, *7*, 6535–6557.

1. Introduction

1.1. Rice Crops in Asia and Their Exposure to Tropical Storms

Rice is the only staple crop suited to humid, high rainfall environments. Rice is predominantly grown in regions and seasons in Asia that are highly prone to extreme weather events such as tropical storms (called Typhoons in the northwestern Pacific ocean; Tropical Cyclones in the Indian, southwest Pacific and southern Atlantic oceans, and; Hurricanes in the northern Atlantic and northern Pacific oceans).

Figure 1 shows the approximate area occupied by standing rice crops in Asia for every month of the year, based on crop calendars [1] and national statistics [2]. About 70% of the region's rice is grown in the monsoon season from June to October, reaching a peak of almost 100 million hectares in August [1]. The same figure also shows the average monthly frequency of tropical storms in the Western Pacific between 1959 and 2011 [3]. There is a striking correlation suggesting that the main rice crop season in Asia is highly exposed to the risk of storm-related damage.

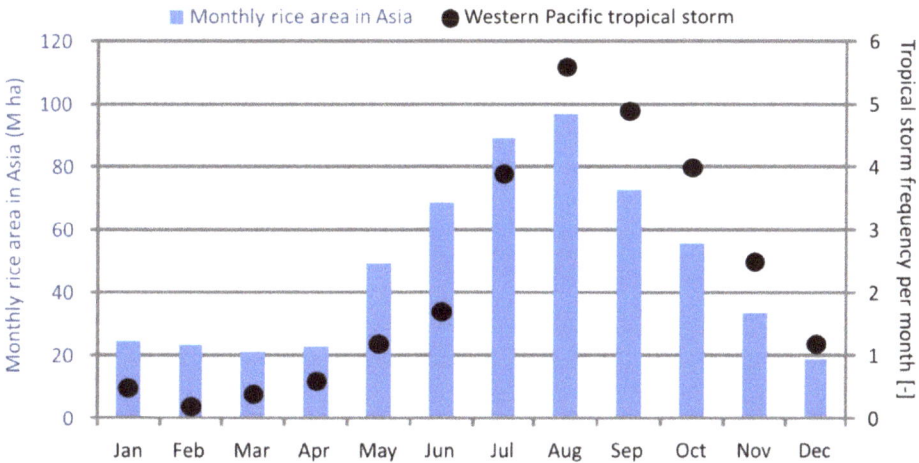

Figure 1. Monthly rice area (M ha) in Asia *vs.* average number of tropical storms per month in the western Pacific between 1959 and 2011.

Cyclones make landfall in eastern India, Bangladesh and Myanmar, while typhoons affect the Philippines, Vietnam, southern and eastern China, Taiwan and Japan. Specifically, the most vulnerable areas are the insular region of Southeast Asia and the coastal areas of mainland Asian countries (although severe storms can affect inland areas too) since significant amounts of rice are produced in floodplains, deltas and other low lying areas in coastal regions. Storm related crop losses in the main growing season can have significant negative impacts on rice production, imports, exports, and prices. Inevitably, these have a disproportionately high impact on the

329

more vulnerable sectors of society, which depend on rice for a substantial proportion of their income (producers) and calories (consumers).

1.2. Typhoon Haiyan and Its Impact in Philippines

The Philippines in particular is affected by an average of 20 tropical storms every year, some of which develop into devastating typhoons. Typhoon Haiyan (also called Typhoon Yolanda in the Philippines) developed into a category five 'super-typhoon' shortly before it made landfall on the eastern Visayas on the 8 November 2013. The eye of the typhoon passed directly over Leyte and its capital city of Tacloban as the typhoon moved from the Western Pacific across the Philippines and then into the South China Sea by the 9 November. The resulting devastation from sustained winds of up to 230 km/h, widespread flooding and the 5–6 m storm surge were well documented in the following weeks and months.

From official reports, casualties in the country reached over 6000 with over 1000 still missing two months after the typhoon [4]. More than 3.4 million families were affected, with nearly a million displaced in 44 provinces in the country. Of the total casualties in the country, 86% were from the province of Leyte. Total cost of damages was estimated at 39.8 Billion (B) Pesos (0.9B USD) and nearly one-fourth were damages to crops including rice and corn. In Leyte, cost of damages to agriculture and infrastructure was estimated at 6.8B Pesos (154M USD) with 1.4B Pesos (32M USD) of damage to rice and corn crops in the province.

1.3. Assessing the Risk of Crop Damage and Actual Crop Damage from Tropical Storms

Tropical storms can affect thousands of hectares of crop at any time during the main rice-growing season. Mitigation and disaster response strategies related to food security require pre- and post-event information on the likely impact and actual impact on crops in the path of a tropical storm.

Accurate information pre-event can assist decision-makers to take appropriate actions to safeguard recently harvested crops or to advance the harvest period so that crops are removed from the field in time. Pre-event assessments would need to be part of routine crop information collection procedures and may not be a high priority for local agricultural officers with many competing demands on their resources. Accurate information post-event can help to assess damages more accurately and determine appropriate compensation, as this is directly linked to inputs and investments such as seeds, fertilizers and labor which will vary depending on the crop stage at time of loss. Post-event assessments can be challenging if access to the area is limited or dangerous. Furthermore, there are often conflicting reports from different agencies and media outlets and there can be pressure for local offices to inflate damage reports to secure greater emergency assistance.

Thus, there is a need to provide unbiased and timely estimates of crop status with sufficient lead-time for agencies and governments to act upon this information. There are several examples of the use of remote sensing to provide such information, which we briefly review below. We focus on examples and applications in Asia related to flooding from extreme or prolonged rainfall, storm surges or tsunamis—all of which can result in submergence related damages to crops.

1.4. Remote Sensing as a Source of Unbiased and Timely Information on Crop and Vegetation Status

Affected areas can be identified and damages from floods or tropical storms can be evaluated using multi-temporal satellite imagery (both optical and radar), analysis of ancillary spatial data and ground based reporting [5]. Also, non-authoritative data, such as volunteered geographical data (sourced from online video, photos and social media streams *etc.*) and crowd sourced data (e.g., voluntary photo interpretation of aerial images), have been used to provide additional information that is integrated with authoritative data (from agencies mandated to collect such information), in order to perform flood impact assessments [6].

There are several examples of the use of remote sensing and spatial data to assess post-event impact [7–9]. Chau *et al.,* [9] assessed the potential impacts of extreme floods on agriculture in Vietnam by overlaying historical flood inundation maps (produced from flood depth markers recorded for each past flood event) and land use maps. This form of assessment results in risk maps that evaluate the potential impact on natural resources, which, in turn, aids planning activities. However, when an extreme event occurs, it is also important to have a rapid—and if possible a pre-event—estimation of actual agricultural area and actual crop growth stages that would complement such risk maps.

Geospatial and remote sensing based damage assessments have been used in post-tsunami impact assessments (e.g., [10,11]). On 26 December 2004, earthquakes in the Indian Ocean triggered massive waves that caused vast destruction of many coastal areas in the region [12]. This well documented and tragic event massively impacted coastal areas on both sides of the Indian Ocean, from Indonesia to Sri Lanka, and resulted in significant loss of lives, damaged coastal infrastructure and flooding. Some of these impacts were documented through geospatial information. Specific examples include the use of high-resolution imagery (IKONOS satellite datasets) to map changes in vegetation near Aceh, Indonesia, immediately after the event [13], mapping of coastal vegetation changes in Phang Nga province, Thailand [14,15] and assessments of the protection potential of mangrove vegetation cover along the west coast of Thailand [16]. ASTER/Landsat imagery has been used to estimate tsunami-damaged areas [12] and to map and assess vegetation changes—using vegetation indices such as the Normalized Difference Vegetation

Index (NDVI)—due to short-term [11] and long-term tsunami effects [17]. These studies are extremely useful for estimating the damages to natural resources. They can support rehabilitation, help to prioritize interventions and draw attention to the impact of these events in a visually compelling and quantitative manner. However, the direct and rapid evaluation of the crop losses incurred in the current season was not the aim of these studies.

In response to Typhoon Haiyan, the Food and Agriculture Organization (FAO) collected and published available information on the impact of the typhoon on the Philippines. A map with an assessment of the damages was provided in their report [18], but this did not include an assessment of the crop condition at the time of the event.

All of the above approaches are reactive and take place post-event. Figure 2 shows the general timeframe of post-event remote sensing assessments. Once an event occurs, pre-event information on land cover and archives of earth observation (EO) imagery are compared to EO images acquired post-event, any changes are detected and summarized through spatial analysis.

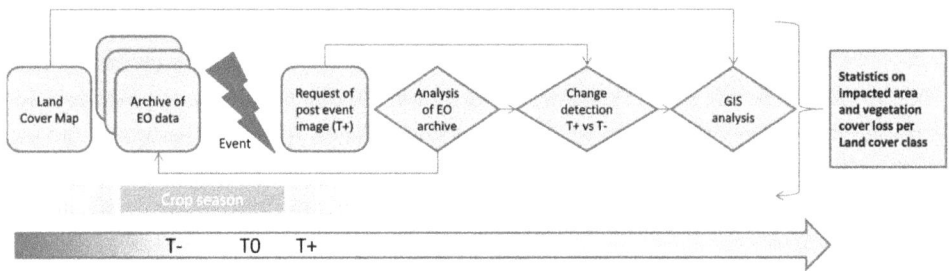

Figure 2. The timeline of post-event remote sensing assessments.

As outlined in Section 1.3 there is a need to complement this approach with a more proactive approach to crop information to assess the risk of damage before the event takes place. Our aim in this study is to develop and demonstrate a methodology that can provide near real time information on the likely impact of a specific event.

1.5. A Remote Sensing-based Proposal for a Proactive Approach on Rice Crop Status Assessment

There are two conditions that need to be met to deliver such near real time information [19]. First, there must be a high spatial resolution rice crop area map of the location and season at risk. Second, there must be a suitable source of timely, high temporal resolution remote sensing information that can be analyzed to provide accurate information on the current rice crop status within that mapped area [20,21]. Availability of these two complementary layers of remote sensing information can be exploited to derive spatially explicit estimates of crop status. This information can be related to the predicted or observed date of a tropical storm event, for immediate damage mitigation strategies or post-event damage estimation. Alternatively, it can be used to assess the frequency of exposure to storms at different crop growth stages and hence develop longer term mitigation strategies such as growing shorter duration varieties, choosing earlier/later planting windows, or adopting alternative crop rotations/land use.

Figure 3 shows the general timeframe of this pre-event (proactive) approach combined with a post-event (reactive) approach. As in Figure 2, a crop map (or land cover map), a pre-event image and a post event image are acquired and analyzed reactively to assess the actual damage after the event. The proactive information comes from the crop map and continual analysis of high temporal information on the crop status, analyzed before the event to assess the risk of damage to the standing crop. The two approaches are complementary.

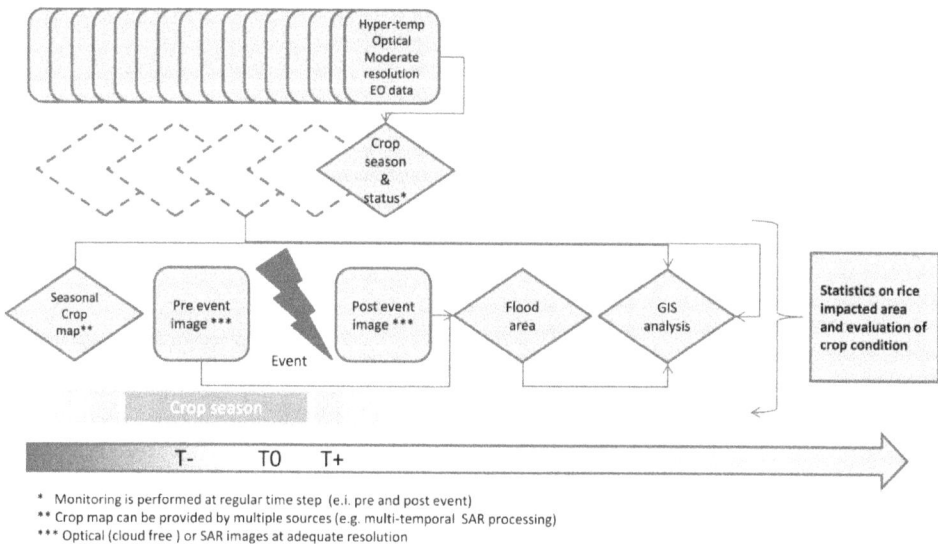

* Monitoring is performed at regular time step (e.i. pre and post event)
** Crop map can be provided by multiple sources (e.g. multi-temporal SAR processing)
*** Optical (cloud free) or SAR images at adequate resolution

Figure 3. A combined proactive and reactive remote sensing approach to evaluate the risk of crop losses and the actual crop losses due to flooding from tropical storms.

In this study, we propose a proactive crop status assessment method that can provide near real time information on the likely impact of a specific event. We apply *ex-post* in parallel with a reactive damage assessment for the case of Typhoon Haiyan and its impact on the rice growing areas of Leyte province, Philippines. We first describe the study site, our data (Section 2) and methodology (Section 3). The results section (Section 4) assesses the derived remotely sensed information on crop status/growth stage at the time of the event and compares these estimates to available field data. Finally, we discuss the significance of the results and the potentials of the proposed approach within the framework of planned operational satellite missions.

2. Study Site and Data

2.1. Leyte Province and Its Rice Production Systems

Leyte is a province in the eastern part of the Visayas island group of the Philippines. The study site is the northern part of Leyte, which was the most strongly affected by Typhoon Haiyan. Figure 4 shows the study site, the typhoon track and rice cultivated area during the 2013 wet season (July to December) (from [22]). The source and generation of the rice area map are described in Section 3.

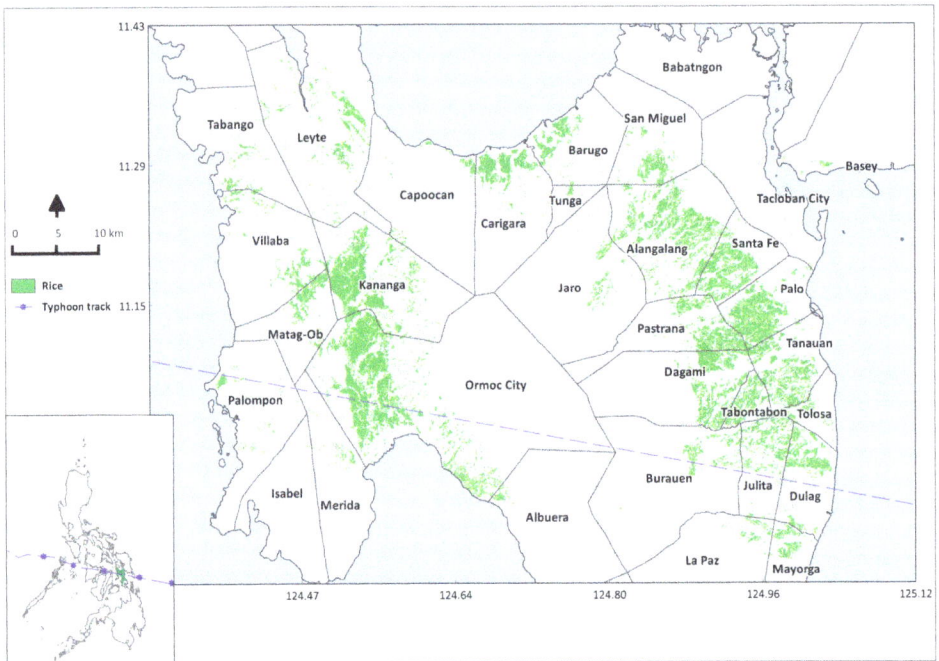

Figure 4. Northeastern part of Leyte Province. Wet season rice area in green, track of typhoon Haiyan in blue, municipal boundaries in black. Inset map shows the location of Leyte in the Philippines.

Leyte has a tropical climate with regular rainfall through the year. Poverty incidence is high (31% in 2012 [23]). Rice is cultivated on the east and west coasts of the island, while the hilly central area is occupied mainly by forests. The total rice area in the province in 2012 was 133000 hectares, 65% of which is irrigated [24]. Most farmers establish their crop by transplanting 21-day old seedlings. Popular varieties include NSIC Rc-222, NSIC Rc-238 and NSIC Rc-216, which are inbred varieties with durations ranging from 110 to 114 days.

Rice is grown in two distinct seasons per year and here we focus on the main or wet season. The transplanting window for the wet season in this region spans several months with farmers establishing their crop anytime between May and August, though the peak transplanting month is June. Such a large range in transplanting dates over a small area reinforces the need for a method that can provide spatial and temporal information on crop status. Since the dominant varieties mature in approximately four months, harvesting takes place between September and December, peaking in October and November.

2.2. Remote Sensing Data

Two sources of Earth Observation (EO) data were used to perform the analysis (Table 1). Very High Resolution (VHR, 3 m) synthetic aperture radar (SAR) data were used to map (i) rice cultivated areas in the 2013 wet season and (ii) post-typhoon flooded areas, while multi-temporal Moderate Resolution (MR, 250 m) optical data were used to assess crop seasonality and crop growth stage at the time of the typhoon event.

Table 1. EO data exploited in the analysis.

Location	Northwest Leyte	Northeast Leyte	Northeast Leyte		Leyte
Purpose	Rice area map	Rice area map	Flooded area detection		Rice crop status
Satellite or instrument	Cosmo-SkyMed (CSK 1, 3 and 4)	Cosmo-SkyMed (CSK 1 and 4)	Cosmo-SkyMed (CSK 1 and 2)		Terra and Aqua
Sensor mode	Stripmap	Stripmap	Stripmap		MODIS
Product	SLC	SLC	SLC		MOD13Q1 and MYD13Q1
Band	X (3.12 cm)	X (3.12 cm)	X (3.12 cm)		Red (620–670 nm), NIR (841–876 nm), Blue (459–479 nm)
Resolution (m)	3	3	3		250
Swath (km)	40 × 40	40 × 40	40 × 40	40 × 40	1200 × 1200
Scene center	11.18°N 124.56°E	11.11°N 124.89°E	11.11°N 124.89°E	11.08°N 124.93°E	14.9° N 129.41° E
Polarization	HH	HH	HH	HH	-
Look	Right	Right	Right	Right	-
Orbit	Descending	Descending	Descending	Descending	Descending
Incidence angle	48	46	46	54	-
Cycle (days)	16	16	-	-	16 (8)
Start day	12 May 2013	15 May 2013	20 September 2013	8 November 2013	6 December 2012
End day	24 September 2013	20 September 2013	11 November 2013	16 November 2013	11 December 2013
Images	9	10	2	2	70

Cultivated rice area of the 2013 wet season was produced by analyzing multi-temporal Cosmo-SkyMed (CSK) data acquired every 16 days from 15 May to 20 September 2013, as described in [22].

Flooded area was analyzed only for northeast Leyte because it was the area where the typhoon first made landfall and was expected to be the area most heavily affected by the typhoon. Where possible, flood assessments should rely on identical sources of pre- and post-event imagery, but a rapid post-event image depends on the orbit frequency of the satellite. Fortunately, most SAR platforms have the ability to change the viewing angle to allow nearby orbits to observe the required area. We acquired flood assessment post-event CSK image with the same acquisition geometry as the May–September data stack on 11 November 2013 and two CSK images with a different viewing angle on 8 and 16 November 2013. Footprints of the SAR data are provided in the supplementary materials (Supplementary Figure S1).

We used the MOD13Q1 (Terra satellite) and MYD13Q1 (Aqua satellite) 250 m resolution vegetation indices products available for Leyte from December 2012 to December 2013. These data are free to download from the USGS Land Processes Distributed Active Archive Centre (LP DAAC) [25]. For both sensors, these data are provided as 16-day composites with an 8 days nominal shift (Terra 16-day composites start at the day of the year (DOY) 001, while the Aqua compositing period starts at DOY 009). MOD/MYD13Q1 products are developed using the Constrained View Angle-Maximum value Composite (CV-MVC). Seventy composites (35 each for MOD13Q1 and MYD13Q1 products) were downloaded to analyze the entire 2013 crop year, and used to create a synthetic 8-day time series exploiting the nominal composite date to create a temporal series [26].

2.3. Field Data and Additional Spatial Information

Ground data were available as described in [22]. These data were specifically acquired to support the analysis of Very High Resolution SAR data. Field observations were performed throughout the 2013 wet season in 40 paddy fields in northwest and northeast Leyte (20 parcels within each CSK footprint) across eight municipalities. These fields were selected, with the farmers' consent, prior to the start of the rice growing season and the SAR image acquisition schedule. Observations were conducted on (or as close as possible to) the image acquisition date, using a standardized protocol. Observations included latitude and longitude from handheld GPS receivers, descriptions and photos of the status of the field, plant height, water depth, weather conditions and crop growth stage. At the end of the season, farmers were also interviewed to collect information on the rice variety, water source, crop management and establishment practices, and inputs such as pesticides and fertilizers. A total of 400 field observations were made through the season.

A further 184 validation points were collected in the same footprints for rice map accuracy assessment as described in [22].

Administrative municipal boundaries used in aggregating results were obtained from the Global Administrative Areas Database (GADM) [27].

3. Methods

We first describe the analysis and derived information from the proactive assessment and briefly describe the same for the reactive assessment although there are common inputs used in both.

3.1. Proactive Assessment of Rice Crop Status and Risk of Damage

The proactive assessment describes the pre-event status of the crop using processed MODIS time series data to estimate key rice crop stages within a predefined rice crop mask derived from SAR data (see Sections 3.1.1 and 3.1.4 for the crop mask). The SAR data and MODIS time series data were processed as follows.

3.1.1. Rice Crop Mask from SAR Data

A fully automated processing chain was developed to convert the multi-temporal space-borne SAR Single Look Complex (SLC) data into terrain-geocoded σ^0 values. Then, a multi-temporal σ^0 rule-based rice detection algorithm was applied to the time series using thresholds derived from 40 in-season monitoring locations within the two SAR footprints where monitoring took place. A confusion matrix was used to estimate the classification accuracy of the rice maps based on observed field data from 184 locations in rice and non-rice areas. The resulting maps show the detected rice area for the 2013 wet season and were estimated to have a classification accuracy of 87% for the northeastern footprint and 89% for the northwestern footprint. The two rice maps were mosaicked into one. Further details on the methodology and accuracy of the rice area maps can be found in [22]. The rice area was summarized per municipality.

3.1.2. MODIS Data Pre-Processing

Boschetti *et al.* [20] demonstrated that it is possible to identify the dates of crop establishment (transplanting or direct seeding) and heading (maximum plant development at the end of vegetative period) from time series analysis of moderate resolution EO data [20,21,28]. However, the time series data needs to be smoothed before this information can be extracted.

The synthetic 8-day time series of Terra and Aqua HDF files (See Section 0) were used to extract the necessary information to build a time series of vegetation indices and to evaluate the residual rate of noise (mainly due to residual cloud contamination) that still affected the observation after compositing. In particular, the Enhanced Vegetation Index (EVI; MOD/MYD13Q1 HDF layer 2; [29]), VI Quality information (HDF layer 3), the blue reflectance band (HDF layer 6) and the Pixel Reliability flags data (HDF layer 12) were extracted and analyzed.

EVI time series were smoothed following a two-step approach. The first step involved the detection and cleaning of outliers (*i.e.*, anomalous 'spikes') in the raw EVI profile following the approach proposed in the TIMESAT algorithm [30]. The checked EVI time series were then smoothed using a weighted Savitzky-Golay filter [31] with weights assigned on the basis of data quality. This algorithm allows data smoothing without forcing a given mathematical function (e.g., Gaussian or logistic curves) to fit the data time series thus reducing artifacts creation [31,32]. Data quality was derived from analysis of MODIS quality indicators (Pixel Reliability and VI Usefulness derived from HDF layers 12 and 3, respectively), complemented with blue reflectance data as proposed by other authors [33] (see Supplementary Table S2 for details). For each acquisition date (t), pixels (i) were classified as Clean, Contaminated or Cloudy. The Savitzky-Golay filter was then applied, using a symmetrical smoothing window of $\pm$ 3 periods, a 2nd order polynomial fitting function, and weighting EVI values of the different dates by associating them with different expected measurements errors (*Clean* $\varepsilon_t^i = 0.02$; *Contaminated* $\varepsilon_t^i = 0.11$; *Cloudy* $\varepsilon_t^i = 0.3$).

3.1.3. Phenological Metric Extraction from Smoothed MODIS Times Series Data

Following Boschetti *et al.* [20], a rule based method was implemented to identify the occurrence of the main phenological stages from the smoothed EVI time series. We first calculated and analyzed the derivative of the smoothed signal in order to identify all the points of local (relative) minima and local (relative) maxima. Following Manfron *et al.* [21], these minima and maxima were then evaluated with agronomically based criteria to identify which ones correspond to the transplanting and heading dates, respectively.

The transplanting dates were assumed to correspond to the local minima followed by a rapid and strong increase of the EVI smoothed signal (a sequence of at least three positive derivative points in a temporal window of five composites). The heading dates were assumed to correspond to the absolute maxima of the curve satisfying the following criteria:

i) located between 56 and 120 days after the estimated transplanting date, based on known durations of vegetative stages of rice crops grown in this area and season;
ii) showing an EVI value greater than 0.4; and
iii) followed by a rapid reduction of EVI (a decrease of 1/3 of max-EVI value within 40 days) following [21].

3.1.4. Spatial and Statistical Analysis of the Phenology Metrics

Several steps were taken to ensure that the crop status information extracted from the phenological metrics was both robust and representative.

Firstly, phenological analysis was performed only on MODIS pixels identified as rice by the SAR rice crop mask. The "detection rate" of the algorithm was analyzed for each municipality by comparing the total number of 250 m "rice pixels", derived resampling the high resolution SAR map, with the number of pixels for which the algorithm estimated the transplanting and heading dates. The phenological stages estimated from the MODIS time series were compared to crop stage information from the field. Due to the small dimension of the fields, which were selected for the analysis of VHR SAR data [22], it was not possible to perform a direct field to pixel comparison. Instead, transplanting and heading dates from MODIS were summarized at municipal level and their statistical distributions were compared against the distribution of field observed dates.

Secondly, we identified and removed unreliable estimates. We calculated the confidence intervals of the average heading occurrence in each municipality using a bootstrapping method for non-normally distributed data [34]. Municipalities where half the width of the confidence interval was greater than 8 days (*i.e.*, corresponding to the MODIS composite time span) exhibited phenological estimates with high variance due to a noisy and scattered sample and were discarded from the analysis. Additionally, any municipality with 10 or fewer pixels with heading estimates was also discarded from the analysis.

3.1.5. Indicator of Risk of Standing Rice Crop Loss from a Typhoon

The final step in the proactive assessment estimates the risk of standing rice crop loss from a typhoon with an indicator that identifies which areas of Leyte were more likely to have been impacted by the typhoon. The time span (Δ_{doy}, in days) between the estimated heading date and the typhoon occurrence was calculated for each pixel. Since harvesting can only start at the end of the ripening phase, which in the tropics occurs 30 to 40 days after heading [35], areas characterized by rice crop with a detected heading close to the typhoon event (Δ_{doy} of 30–50 days) were considered to be at risk of rice production loss due to the typhoon. The average number of days between heading and typhoon Haiyan was summarized per municipality.

Not all MODIS pixels in the rice area will meet the criteria in Section 3.1.2 for phenological metric extraction. Thus, to estimate the total rice area potentially subject to production loss, we multiplied the area of MODIS pixels with a given Δ_{doy} value by the MODIS detection rate, for each possible Δ_{doy} (Equation (1)).

$$S_{rice}\left(\Delta_{doy}\right) = S_M\left(\Delta_{doy}\right) \times DR = N_M\left(\Delta_{doy}\right) \times S_{pixel} \times DR \qquad (1)$$

where $S_{rice}\left(\Delta_{doy}\right)$ is the estimate of total rice area with a given Δ_{doy}, $S_M\left(\Delta_{doy}\right)$ is the area of MODIS pixels with a given Δ_{doy} (computed as the product of the number of pixels $N_M\left(\Delta_{doy}\right)$ by the surface of a MODIS pixel), and DR is the MODIS detection rate.

The rice area potentially subject to production loss was summarized per municipality.

3.2. Reactive Assessment of Rice Crop Damage

The reactive damage estimate is based on a simple spatial analysis of pixels classified as rice (Section 3.1.1) and as being flooded due to the typhoon. The flood detection relies on change detection applied to SAR images pre- and post-event. The relevant images for flood mapping are: 8 and 11 November (flooded images); and 16 November and 20 September (reference or non-flooded images). Because the scenes on 8 and 16 November were acquired with a different geometry, they were filtered using the single-date Gamma distribution entropy maximum *a-posteriori* method (a different method to the filtering used for the multi-date time series in [22] but all other processing to derive terrain geocoded and calibrated images were as in [22].

Flooded areas were detected with a rule-based classifier applied to the backscatter (or σ°) pixel values in the terrain geocoded and calibrated images.

1. σ^0 value for the flooded date (*i.e.*, 11 or 8 November) is less than -13 dB; this value is associated with surface water for X-band, HH data at this incidence angle as described in [22].
2. σ^0 ratio between the reference or non-flooded images and the corresponding flooded acquisitions, is larger than 2.0; this represents a strong backscatter decrease over a short period of time that is not consistent with normal rice crop practices in this region and season.

The flood damaged rice area was summarized per municipality.

4. Results and Discussion

We focus on the results of the proactive approach but place the results in context with the flooded rice area estimates from the reactive approach. We split the results into four sections. We first visually demonstrate the results of the phenological stage detection algorithm using two exemplar pixels in the study area. This is followed by an assessment of the representativeness of the detection across the entire study area. We then assess the accuracy of the detection by comparing the detected phenological stages against field observations. Finally, we estimate the standing rice crop area at risk of damage from typhoon related flooding and relate this to the flood affected area.

4.1. Example Rice Crop Phenological Stages from MODIS Time Series Data

Figure 5 shows an example for one pixel in Kananga municipality in Northwest Leyte (Figure 5a) and Pastrana municipality in Northeast Leyte (Figure 5b). The figure shows the 8-day raw EVI data (thin gray line), the effect of the smoothing process (green line), the date of landfall of the typhoon (blue diamond), the detection of transplanting date (red point), and the heading occurrence (dark-green point). The temporal profile extracted in Kananga (Figure 5a detected two seasons. In particular, the wet season, the one closer to the typhoon event, shows a rice crop establishment with transplanting at the beginning of June (DOY 153) and a peak EVI (*i.e.*, heading [20]) in late July (DOY 209). On the other hand, the time series analyzed in Pastrana. Figure 5b clearly shows a delayed crop establishment in the wet season with rice transplanting in late July (DOY 201) and a crop peak in late September (DOY 265). The preceding dry season (from January to May) was not detected as rice due to an anomalous and slow decreasing senescence period that does not match with the expected rice behavior and time series analysis criteria [21])

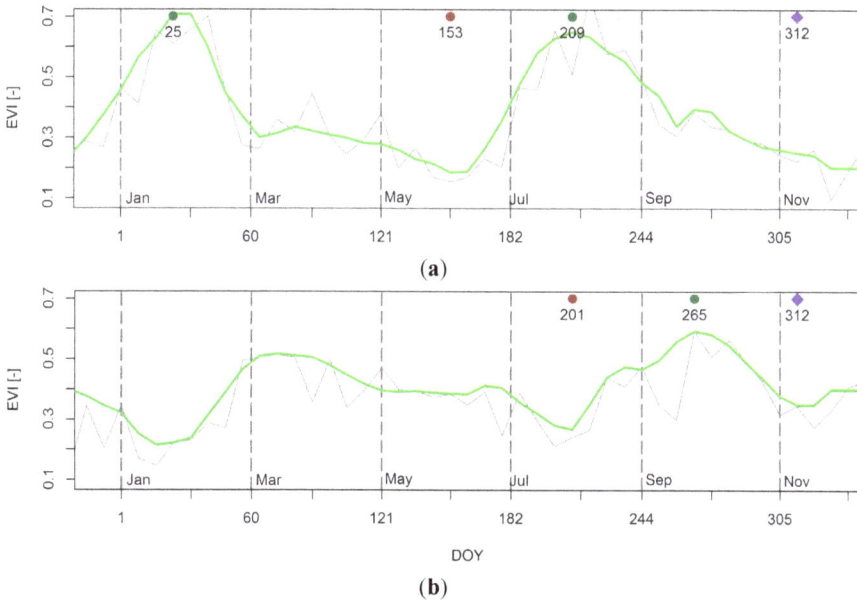

Figure 5. Time series analysis of EVI for a pixel located in the municipality of Kananga (a); 11°9.875′N 124°30.631′E) and Pastrana (b); 11°14.375′N 124°51.303′E) showing differences in planting dates during the wet season. The graphs show raw (thin black line) and smoothed EVI (green line); detected transplanting date (red point), heading occurrence (green point) and the typhoon Haiyan (blue diamond) together with their date of occurrence in DOY. Y-axis is EVI [–] and X-axis is time in DOY.

4.2. Analysis of Detection Rate

The phenological detection algorithm estimated the transplanting and heading dates for about 1700 MODIS pixels; more than 30% of the 250 m pixels in the SAR rice area. Across the study site the detection rate varied between 8% and 57%, and was found to be greater than 10% in the majority of the municipalities analyzed (21 out of 29). The detection rate is generally lower in areas characterized by low rice presence where the rice growing areas are more fragmented (low resolution bias) [36].

Figure 6 is a scatterplot of the number of rice pixels per municipality at 250 m resolution rescaled from the SAR derived rice map and the number of MODIS detected pixels per municipality. Figure 6 shows that the method provided automatic phenological estimation for a number of MODIS pixels that is proportional to the rice area in each analyzed administrative unit.

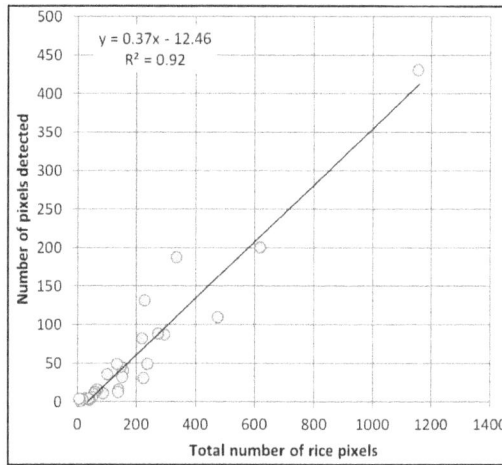

Figure 6. The relationship between the number of MODIS pixels where rice phenological dates were retrieved and the corresponding SAR derived rice area resampled at MODIS resolution in each municipality.

4.3. Analysis of Transplanting and Heading Date Accuracy

Eleven municipalities with small areas planted to rice and less than 10 MODIS pixel detections were excluded. The 18 retained municipalities account for 89% of the total rice cultivated area in the 2013 wet season and are uniformly distributed over the study area.

The box plots in Figure 7 show the estimated transplanting dates (a) and heading dates (b) derived from MODIS data for the main rice growing municipalities in Leyte. Box plots in the graph are colored in light and dark grey to highlight northwestern and northeastern municipalities, respectively, with the average longitude of the municipality increasing from left to right. The red dashed horizontal lines indicate the average transplanting and heading dates for the entire study area. Field data acquired in eight of the municipalities are also reported as colored points. Red points report the exact date recorded by ground observations. Blue points are cases where field observations did not explicitly provide the heading and the date was estimated based on crop growth stages observed across several field observations days through the season.

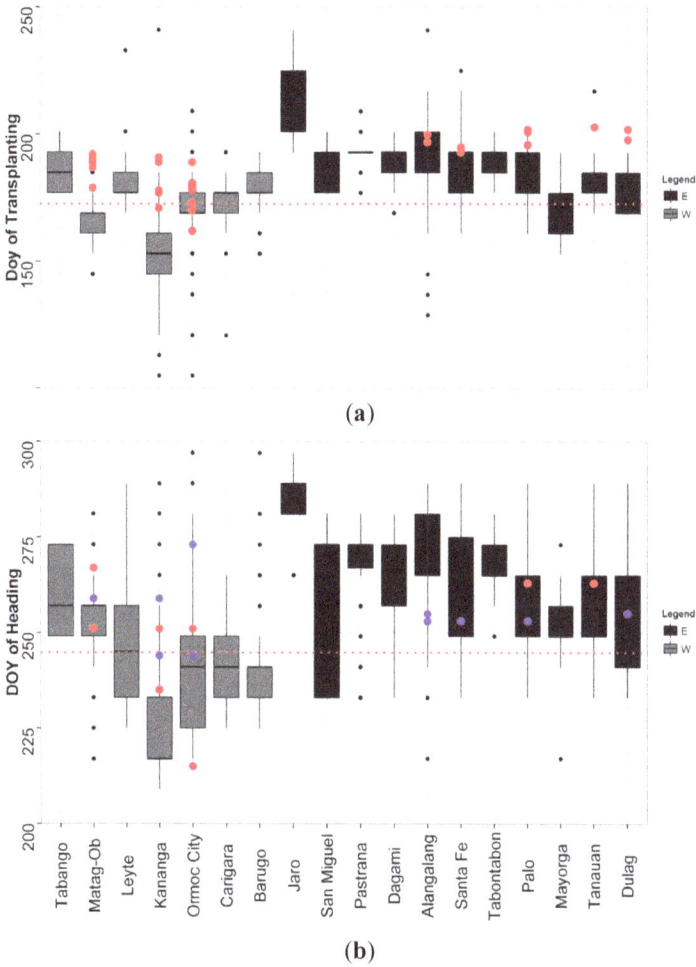

(a)

(b)

Figure 7. Box plots of the transplanting (**a**) and heading (**b**) dates in 18 municipalities for the 2013 wet season as derived from MODIS data. Red and blue points refer to observed and estimated field dates, respectively. Red dotted line represents the average value for the entire study area. Municipalities are reported from west (Tabango) to east (Dulag).

A *point in pixel* validation of MODIS estimated dates was not possible due to the small size of the fields with respect to MODIS spatial resolution. Instead, field observations are reported in this figure as a qualitative reference for the evaluation of the capability of the remote sensing method to identify crop establishment variability in time and space for a wide area.

Figure 7a shows how the method is able to identify variation in the timing of crop establishment within and among municipalities. Estimated dates were within reasonable agreement with field observations. Satellite estimates for transplanting fall in the expected range of dates for the study area: between May and August (DOY 120–240) with a mean value (red dotted line) in June (DOY 170). The estimates in the northwest municipalities (Kananga, Matag-Ob and Ormoc City) have earlier estimated transplanting dates (DOY 150–175) than those in the northeast (DOY 180–200), which is confirmed by field observations. In Figure 7b it is clear that satellite estimates of heading date show a better agreement with field observations confirming that the detection of peak of season date from time series analysis is more robust than that of crop establishment [20]. Figure 7b shows the same longitudinal trend identified in Figure 7a, with heading dates in northeastern municipalities occurring around 20 days later than in northwestern ones.

Not all monitored fields in northeast Leyte had reached the heading stage by the last in-season field visit (24 September 2013). This is also reflected in the farmer declarations of harvesting, whereby 10 out of 20 monitored fields in the northeastern footprint were not harvested before the typhoon event (See Supplementary Figure S1 and Table S1).

This qualitative assessment confirms that this approach can highlight spatial differences in agricultural practices (late and early cultivation) and plant development (variety crop cycle).

4.4. Standing Crop Area at Risk

Figure 8 provides a synthesis of the remote sensing and spatial analyses conducted in northern Leyte for the 2013 wet season.

Figure 8a,b show the high resolution rice map and flooding map from SAR, respectively, which are the basis of the accurate rice area estimation and typhoon damage estimation. Figure 8c,d show the same data aggregated to municipal area totals. Figure 8d shows that about 6500 ha were flooded and that Tanauan was the most affected municipality (1545 ha, or 24% of the total flooded area) followed by Alangalang (938 ha, or 14% of the total flooded area).

Figure 8e shows the MODIS derived heading date and Figure 8f shows the difference in days between heading date and the day typhoon Haiyan made landfall in Leyte (8 November). Figure 8g,h are the same data averaged to municipal level. Gray municipalities indicate areas with no rice according to the SAR rice area map or municipalities excluded from the analysis because the criteria for selection of robust phenological estimation were not satisfied (see Section 3.1.4). Figure 8e,g provide a spatial confirmation of the results in Figure 7; rice transplanting took place later in northeastern Leyte (DOY 161–225) than in northwestern Leyte (DOY 137–201) by an average of 24 days, a substantial duration relative to the maturity of the dominant rice varieties (110–114 days). The effect of this late transplanting in the northeast is highlighted in Figures 8f and 8h, where there are fewer days between heading and the typhoon than in the northwest. On average MODIS estimated heading dates in the northeastern municipalities occurred less than 50 days before the typhoon struck.

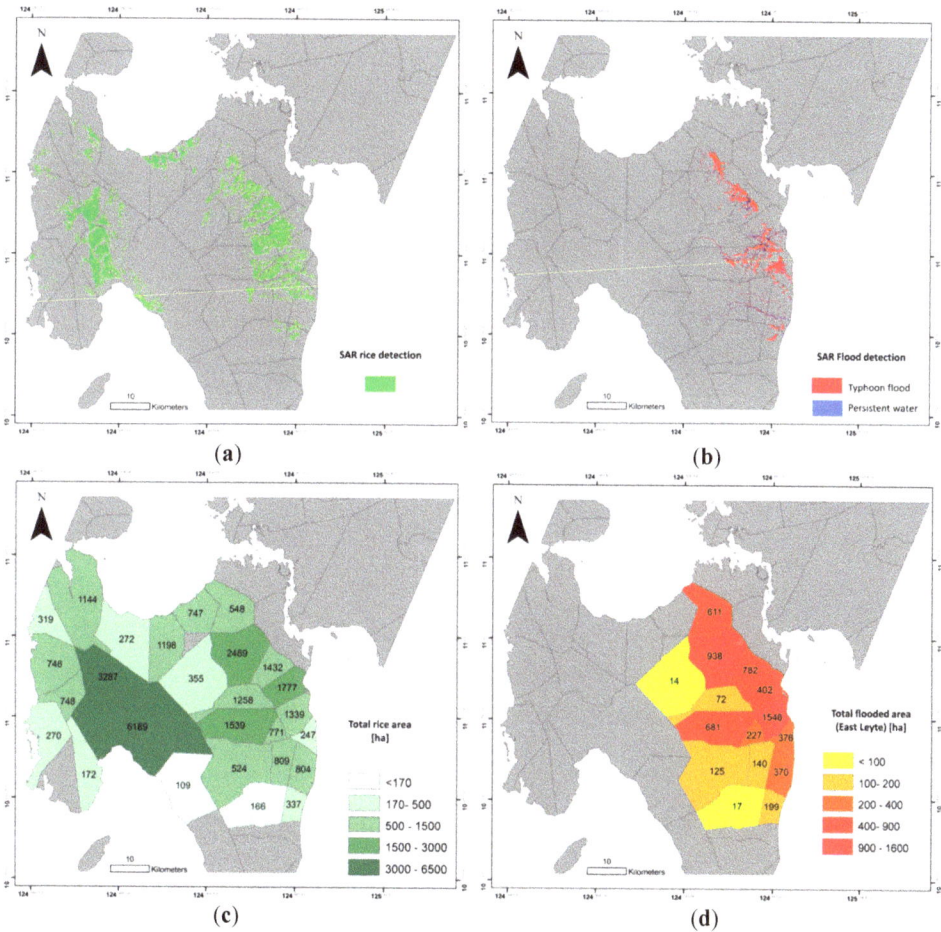

Figure 8. *Cont.*

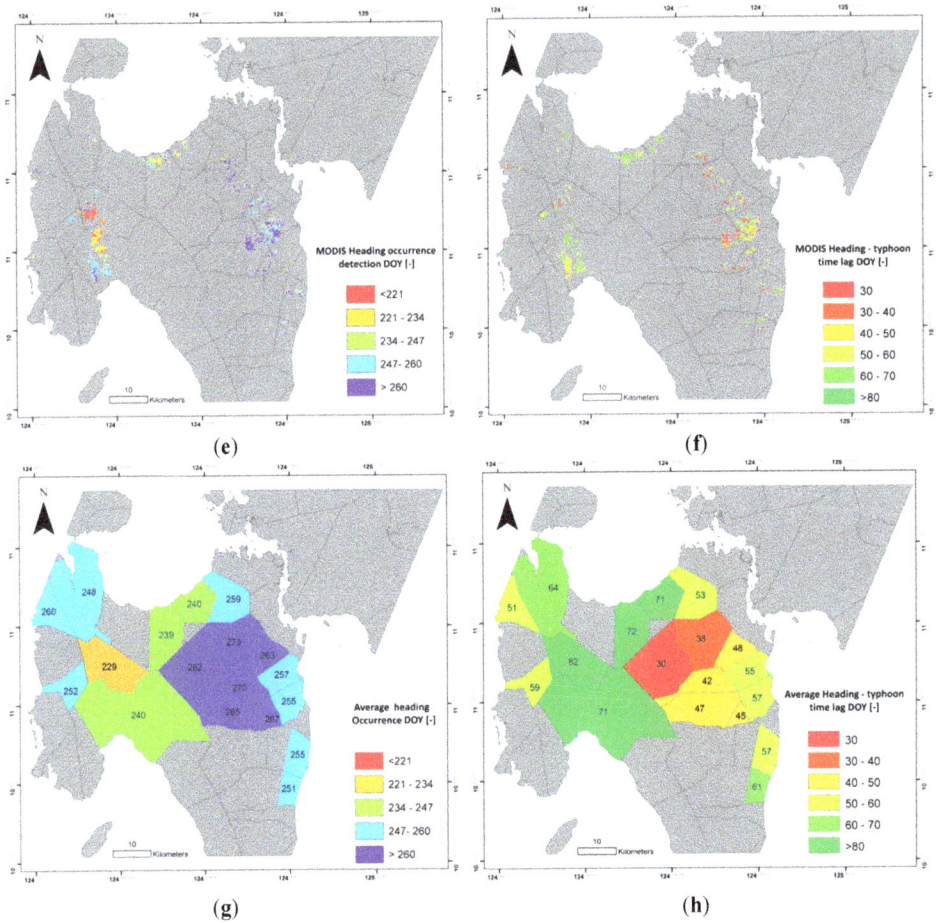

Figure 8. (**a**) Rice cultivated area from SAR; (**b**) flooded area from SAR; (**c**) rice area per municipality; (**d**) flooded area per municipality; (**e**) heading dates from MODIS; (**f**) days between heading and typhoon Haiyan per MODIS pixel; (**g**) averaged MODIS heading dates per municipality; and (**h**) average number of days between heading and typhoon Haiyan.

Overall, the panels in Figure 8 show that rice grown in northeastern Leyte was most affected by flooding due to typhoon Haiyan and at the same time those rice areas were less advanced in the rice season, thus exposing more of the standing crop to typhoon related damage.

Figure 9 shows the rice cultivated area grouped as a function of the time between the MODIS estimated heading date and the typhoon date (see Section 3.1.5). We identified two critical periods: from Typhoon (T) to T minus 30 days (T-30), where rice in the field was not yet mature, and from T minus 30 to T minus 50 days (T-50), where harvesting activities were probably not yet performed according to common practices in the area.

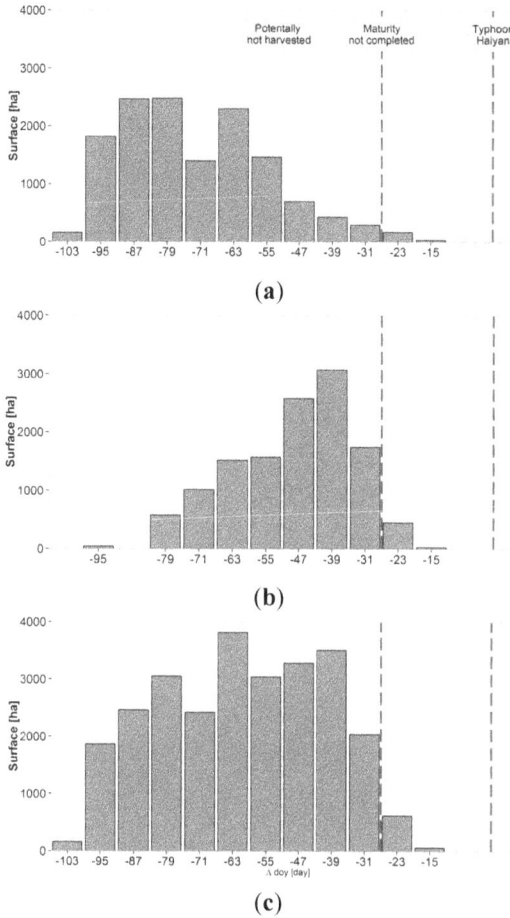

Figure 9. Distribution of rice area by time span Δ_{doy}— in days) between the MODIS estimated heading date and the typhoon occurrence in Leyte for the 2013 wet season. Northwestern and northeastern rice areas are reported in panels (**a**) and (**b**), respectively. Panel (**c**) provides the results for the entire study area. Vertical dashed lines are 30 (brown) and 50 (yellow) days before the typhoon date (red line).

Figure 9a,b highlight that the distribution of potentially affected areas is not homogeneous between northeast and northwest Leyte. For northwest Leyte, only 1% of the rice cultivated area (corresponding to about 200 ha) was estimated not to have reached maturity at the time of the typhoon event and only 10% (corresponding to about 1400 ha) were probably not yet harvested. Conversely in northeast Leyte, about 4% of the rice cultivated area (corresponding to about 500 ha) was estimated not to have reached maturity and about 59% (corresponding to about 7000 ha) was estimated not to have been harvested yet.

The analysis highlights a much higher exposure of standing rice crop in northeastern Leyte to typhoon Haiyan. Conversely, rice production in northwestern Leyte was much less affected. This kind of assessment can provide valuable information for mitigation and disaster response strategies. Moreover, it could be used as an important (although 'qualitative') basis in the analysis of crop damage reports coming from the different parts of the province. One caveat is that the assessment is a conservative evaluation of potential production loss, due to the exclusion of some administrative units for which no reliable MODIS estimation was available.

4.5. A Proposal for an Operational Monitoring System

Rice is the most important food security crop in Asia and the major rice-growing season is particularly prone to tropical storm related losses that are likely to increase in intensity and frequency in a changing climate. Unbiased and transparent approaches to crop damage assessments are essential to reduce moral hazard and to guide appropriate investments that mitigate risk and respond to disasters. The results of this *ex-post* analysis suggest that it is feasible to provide near real time information about crop status before a tropical storm event. The fundamental aspects of the proposed system are (i) the existence of a reliable seasonal crop map; and (ii) the availability of methods able to handle hyper-temporal optical data to derive periodic crop status information. A robust proactive operational system should be able to provide crop status information regularly at specific time steps by analyzing the continuous change of temporal signal as soon as a new image is provided (e.g., [37]). The new generation of operational SAR and optical satellite mission, such as the European Sentinel program, will contribute to the necessary EO data flow to support the proposed approach. At the time of writing, the first operational Sentinel-1 SAR images are already being provided to users, while the MODIS platform, and new PROBA-V and the foreseen Sentinel-3 satellites are the best option to perform near real time agricultural monitoring and derivation of crop status condition.

5. Conclusions

This research is a contribution to literature on crop damage assessments from natural disasters. Our innovation has been the development of a proactive assessment of the risk of damage that exploit the synergy of different satellite sensors: all weather capacity of very high resolution SAR sensor are fundamental to properly map rice crop in tropical area and hypertemporal information from moderate resolution optical sensor are the only way to perform an operation crop seasonal monitoring to retrieve reliable phenological and crop practices information on large areas.

We have demonstrated a method that:

i) Can provide automatic phenological estimation using vegetation index time series derived from MODIS for a representative sample of MODIS pixels.

ii) Is conservative yet captures rice crop status information on an area that is proportional to the rice area.

iii) Can highlight spatial differences in agricultural practices and plant development.

iv) Can be used to monitor the crop in-season and provide timely information on rice crop status.

Our test case confirmed that northeastern Leyte was identified as the region most at risk to storm related damage to the standing rice crop due to the confluence of a late established rice crop in the part of the province most affected by the typhoon.

This approach can provide unbiased and transparent pre-event information that is complementary to post-event damage assessments that characterize the majority of previous studies. Ongoing and new satellite platforms can provide the required information on seasonal rice crop area, cropping calendars, crop status and area affected by typhoon related damage.

Acknowledgments: This work has been undertaken within the framework of the Remote Sensing based Information and Insurance for Crops in Emerging economies (RIICE) project financially supported by the Swiss Agency for Development and Cooperation (SDC) and by the Philippine Rice Information System (PRiSM) project funded by the Philippine Department of Agriculture—Bureau of Agricultural Research (DA-BAR). It was also supported by the CGIAR Global Rice Science Partnership (GRISP) program and the European Union Seventh Framework Programme (FP7-SPACE/2007-2013) project ERMES (grant agreement n° 606983). The authors are grateful to Raffaele Argiento for help provided in the statistical analysis. Satellite data were provided by ASI/e-GEOS from COSMO-SkyMed satellites.

Author Contributions: Conceived and designed the experiments: M.B., A.N. Performed the experiments: M.B., F.N., G.M. Analyzed SAR data: M.Ba., F.H., J.V.R. Provided field data: M.R.O.M., A.P.B., E.J.P.Q. Analyzed and interpreted the results: M.B., A.N., F.N., L.B., A.L. Wrote the paper: M.B., A.N., F.N., L.B., F.H., A.L.

Conflicts of Interest: The authors declare no conflict of interest.

References

1. Global Rice Science Partnership (GRiSP). *Rice Almanac*, 4th ed.; International Rice Research Institute: Los Banos, Philippines, 2013; p. 283.

2. Statistics Division of the Food and Agriculture Organization of the United Nations (FAOSTAT). Avaliable online: http://faostat3.fao.org/faostat-gateway/go/to/home/E (accessed on 16 March 2015).

3. Angove, M.D.; Falvey, R.J. *Annual Tropical Cyclone Report 2011*; Joint Typhoon Warning Center: Pearl Harbor, HI, USA, 2011.

4. National Disaster Reduction and Risk Management Council (NDRRMC). NDRRMC Update: Site Rep No. 104 Effects of Typhoon "Yolanda" (HAIYAN). Avaliable online: www.ndrrmc.gov.ph (accessed on 16 March 2015).

5. Brivio, P.A.; Colombo, R.; Maggi, M.; Tomasoni, R. Integration of remote sensing data and GIS for accurate mapping of flooded areas. *Int. J. Remote Sens.* **2002**, *23*, 429–441.

6. Schnebele, E.; Cervone, G.; Waters, N. Road assessment after flood events using non-authoritative data. *Nat. Hazards Earth Syst. Sci.* **2014**, *14*, 1007–1015.

7. Lee, M.-F.; Lin, T.-C.; Vadeboncoeur, M.A.; Hwong, J.-L. Remote sensing assessment of forest damage in relation to the 1996 strong typhoon Herb at Lienhuachi Experimental Forest, Taiwan. *For. Ecol. Manag.* **2008**, *255*, 3297–3306.

8. Sanyal, J.; Lu, X.X. Application of Remote Sensing in Flood Management with Special Reference to Monsoon Asia: A Review. *Nat. Hazards* **2004**, *33*, 283–301.

9. Chau, V.N.; Holland, J.; Cassells, S.; Tuohy, M. Using GIS to map impacts upon agriculture from extreme floods in Vietnam. *Appl. Geogr.* **2013**, *41*, 65–74.

10. Dahdouh-Guebas, F.; Jayatissa, L.P.; di Nitto, D.; Bosire, J.O.; Lo Seen, D.; Koedam, N. How effective were mangroves as a defence against the recent tsunami? *Curr. Biol.* **2005**, *15*, R443–R447.

11. Kamthonkiat, D.; Rodfai, C.; Saiwanrungkul, A.; Koshimura, S.; Matsuoka, M. Geoinformatics in mangrove monitoring: damage and recovery after the 2004 Indian Ocean tsunami in Phang Nga, Thailand. *Nat. Hazards Earth Syst. Sci.* **2011**, *11*, 1851–1862.

12. Belward, A.S.; Stibig, H.J.; Eva, H.; Rembold, F.; Bucha, T.; Hartley, A.; Beuchle, R.; Khudhairy, D.; Michielon, M.; Mollicone, D. Mapping severe damage to land cover following the 2004 Indian Ocean tsunami using moderate spatial resolution satellite imagery. *Int. J. Remote Sens.* **2007**, *28*, 2977–2994.

13. Chen, P.; Liew, S.C.; Kwoh, L.K. Tsunami damage assessment using high resolution satellite imagery: a case study of Aceh, Indonesia. In Proceedings of the 2005 IEEE International Geoscience and Remote Sensing Symposium, Seoul, Korea, 25–29 July 2005; pp. 1405–1408.

14. Römer, H.; Kaiser, G.; Sterr, H.; Ludwig, R. Using remote sensing to assess tsunami-induced impacts on coastal forest ecosystems at the Andaman Sea coast of Thailand. *Nat. Hazards Earth Syst. Sci.* **2010**, *10*, 729–745.

15. Römer, H.; Jeewarongkakul, J.; Kaiser, G.; Ludwig, R.; Sterr, H. Monitoring post-tsunami vegetation recovery in Phang-Nga province, Thailand, based on IKONOS imagery and field investigations—A contribution to the analysis of tsunami vulnerability of coastal ecosystems. *Int. J. Remote Sens.* **2011**, *33*, 3090–30121.

16. Sirikulchayanon, P.; Sun, W.; Oyana, T.J. Assessing the impact of the 2004 tsunami on mangroves using remote sensing and GIS techniques. *Int. J. Remote Sens.* **2008**, *29*, 3553–3576.

17. Villa, P.; Boschetti, M.; Morse, J.L.; Politte, N. A multitemporal analysis of tsunami impact on coastal vegetation using remote sensing: a case study on Koh Phra Thong Island, Thailand. *Nat. Hazards* **2012**, *64*, 667–689.

18. FAO/Philippines Crop damages after Typhoon Haiyan in the Philippines. Available online: http://www.fao.org/emergencies/crisis/philippines-typhoon-haiyan/crop-damages-map/en/ (accessed on 15 March 2015).

19. Holecz, F.; Barbieri, M.; Collivignarelli; Francesco; Gatti, L.; Nelson, A.; Setiyono, T.D.; Boschetti, M.; Manfron, G.; Brivio, P.A.; *et al.* An operational remote sensing based service for rice production estimation at national scale. In Proceedings of ESA Living Planet Symposium, Edinburgh, UK, 9–11 September 2013.

20. Boschetti, M.; Stroppiana, D.; Brivio, P.A.; Bocchi, S. Multi-year monitoring of rice crop phenology through time series analysis of MODIS images. *Int. J. Remote Sens.* **2009**, *30*, 4643–4662.

21. Manfron, G.; Crema, A.; Boschetti, M.; Confalonieri, R. Testing automatic procedures to map rice area and detect phenological crop information exploiting time series analysis of remote sensed MODIS data. *Proc. SPIE* **2012**, *8531*, 85311E:1–85311E:11.

22. Nelson, A.; Setiyono, T.; Rala, A.; Quicho, E.; Raviz, J.; Abonete, P.; Maunahan, A.; Garcia, C.; Bhatti, H.; Villano, L.; *et al.* Towards an operational SAR-based rice monitoring system in Asia: Examples from 13 demonstration sites across Asia in the RIICE Project. *Remote Sens.* **2014**, *6*, 10773–10812.

23. National Statistical Coordination Board (NSCB). 2012 Full Year Official Poverty Statistics. Available online: www.nscb.gov.ph (accessed on 16 March 2015).

24. Philippine Statistical Authority CountrySTAT. Available online: http://countrystat.bas.gov.ph/ (accessed on 16 March 2015).

25. Land Processes Distributed Active Archive Centre (LP DAAC). Available online: http://e4ftl01.cr.usgs.gov/ (accessed on January 2015).

26. Von Grebmer, K.; Ringler, C.; Rosegrant, M.W.; Olofinbiyi, T.; Wiesmann, D.; Fritschel, H.; Badiane, O.; Torero, M.; Yohannes, Y.; Thompson, J.; *et al.* 2012 Global Hunger Index | International Food Policy Research Institute (IFPRI). Available online: http://www.ifpri.org/publication/2012-global-hunger-index (accessed on 16 March 2015).

27. Global Administrative Areas Database (GADM). Available online: http://www.gadm.org/ (accessed on 18 May 2015).

28. Boschetti, M.; Nelson, A.; Manfrom, G.; Brivio, P.A. An automatic approach for rice mapping in temperate region using time series of MODIS imagery: First results for Mediterranean environment. *EGU Geophys. Res. Abstr.* **2012**, *14*, EGU2012-14068-1.

29. Huete, A.; Didan, K.; Miura, T.; Rodriguez, E.P.; Gao, X.; Ferreira, L.G. Overview of the radiometric and biophysical performance of the MODIS vegetation indices. *Remote Sens. Environ.* **2002**, *83*, 195–213.

30. Jönsson, P.; Eklundh, L. TIMESAT—A program for analyzing time-series of satellite sensor data. *Comput. Geosci.* **2004**, *30*, 833–845.

31. Chen, J.; Jönsson, P.; Tamura, M.; Gu, Z.; Matsushita, B.; Eklundh, L. A simple method for reconstructing a high-quality NDVI time-series data set based on the Savitzky–Golay filter. *Remote Sens. Environ.* **2004**, *91*, 332–344.

32. White, M.A.; Nemani, R.R. Real-time monitoring and short-term forecasting of land surface phenology. *Remote Sens. Environ.* **2006**, *104*, 43–49.

33. Xiao, X.; Boles, S.; Liu, J.; Zhuang, D.; Frolking, S.; Li, C.; Salas, W.; Moore, B. Mapping paddy rice agriculture in southern China using multi-temporal MODIS images. *Remote Sens. Environ.* **2005**, *95*, 480–492.

34. Kundzewicz, Z.; Mondiale, O. *Detecting Trend and other Changes in Hydrological Data*; Zbigniew Kundzewicz, Z., Robson, A., Eds.; World Metereological Organization: Geneva, Swizerland, 2000; p. 158.

35. IRRI Growth Stages of the Rice Plant. Available online: http://www.knowledgebank.irri.org/ericeproduction/0.2._Growth_stages_of_the_rice_plant.htm (accessed on 16 March 2015).

36. Boschetti, L.; Flasse, S.P.; Brivio, P.A. Analysis of the conflict between omission and commission in low spatial resolution dichotomic thematic products: The Pareto Boundary. *Remote Sens. Environ.* **2004**, *91*, 280–292.

37. Combal, B.; Bartholomè, E. Retrieving Phenological Stages from Low Resolution Earth Observation Data. In *Remote Sensing Optical Observation of Vegetation Properties*; Maselli, F., Menenti, M., Brivio, P.A., Eds.; Research Signpost: Kerala, India, 2010; pp. 115–129.

Mapping of Daily Mean Air Temperature in Agricultural Regions Using Daytime and Nighttime Land Surface Temperatures Derived from TERRA and AQUA MODIS Data

Ran Huang, Chao Zhang, Jianxi Huang, Dehai Zhu, Limin Wang and Jia Liu

Abstract: Air temperature is one of the most important factors in crop growth monitoring and simulation. In the present study, we estimated and mapped daily mean air temperature using daytime and nighttime land surface temperatures (LSTs) derived from TERRA and AQUA MODIS data. Linear regression models were calibrated using LSTs from 2003 to 2011 and validated using LST data from 2012 to 2013, combined with meteorological station data. The results show that these models can provide a robust estimation of measured daily mean air temperature and that models that only accounted for meteorological data from rural regions performed best. Daily mean air temperature maps were generated from each of four MODIS LST products and merged using different strategies that combined the four MODIS products in different orders when data from one product was unavailable for a pixel. The annual average spatial coverage increased from 20.28% to 55.46% in 2012 and 28.31% to 44.92% in 2013.The root-mean-square and mean absolute errors (RMSE and MAE) for the optimal image merging strategy were 2.41 and 1.84, respectively. Compared with the least-effective strategy, the RMSE and MAE decreased by 17.2% and 17.8%, respectively. The interpolation algorithm uses the available pixels from images with consecutive dates in a sliding-window mode. The most appropriate window size was selected based on the absolute spatial bias in the study area. With an optimal window size of 33×33 pixels, this approach increased data coverage by up to 76.99% in 2012 and 89.67% in 2013.

Reprinted from *Remote Sens.* Cite as: Huang, R.; Zhang, C.; Huang, J.; Zhu, D.; Wang, L.; Liu, J. Mapping of Daily Mean Air Temperature in Agricultural Regions Using Daytime and Nighttime Land Surface Temperatures Derived from TERRA and AQUA MODIS Data. *Remote Sens.* **2015**, *7*, 8728–8756.

1. Introduction

Air temperature is an important parameter of the climate system and useful for a wide range of agriculture applications, including crop growth simulation [1,2], yield prediction [3,4], estimation of heat accumulation during the growing season [5], assessment of high-temperature damage [6], evaluation of crop freeze injury [7,8], and crop insect development prediction [9]. Currently, near-surface temperature data is collected by meteorological stations, and although such measurements offer the advantage of high accuracy and temporal resolution, their spatial resolution may be low and they may not adequately represent surface temperatures in areas with rugged or heterogeneous surfaces [10].These limitations can bias estimates of the spatial distribution of air temperature, even when researchers use advanced spatial interpolation methods [11].With the development of remote sensing technology, it has become possible to use thermal images from satellites to obtain land surface temperatures (LSTs) over wide areas, and this data can be used to instantaneously estimate spatially contiguous air temperatures [12–15]. By combining remote sensing data with meteorological station data, it becomes possible to upscale point data from meteorological stations to create meso-scale maps of the distribution of LSTs.

The advent of the Advanced Very High Resolution Radiometer (AVHRR) sensors on board the NOAA satellites series in the 1970s provided an opportunity to estimate air temperatures by means of remote sensing [16–21]. In 1999 and 2002, the Moderate Resolution Imaging Spectroradiometer (MODIS) sensor was launched as a payload on the TERRA and AQUA satellites. MODIS improved upon the performance of AVHRR by providing both higher spatial resolution and greater spectral resolution, and therefore represents an excellent sensor for monitoring the temporal and spatial variation of air temperatures over large areas [22,23]. Colombi *et al.* [24] explored the feasibility of the estimation of instantaneous air temperature measured at the corresponding time of satellite overpass using MODIS LST product (MOD11_L2), and used this data to estimate the daily mean air temperature in the Italian Alps. Vancutsem *et al.* [10] found that the MODIS nighttime products provided a good estimation of daily minimum air temperature over different ecosystems in Africa using the AQUA 8-day nighttime LST (MYD11A2), but that developing robust retrieval methods for daily maximum temperature using the TERRA 8-day daytime LST product (MOD11A2) will require further study. Zhang *et al.* [6] demonstrated that night-time LST was the optimal factor for estimating daily minimum, maximum and mean air temperatures in China. Benali *et al.* [25] noted that the integration of MODIS TERRA and AQUA data has great potential for air temperature estimation. Tomlinson *et al.* [12] compared the nighttime LST from MODIS with ground-measured air temperature across a conurbation and found that the measured air temperature was always greater than the MODIS-derived LST. Hachem *et al.* [26] found that the mean daily LST was more strongly correlated with near-surface air temperature in an area with continuous permafrost when the TERRA/AQUA MODIS data were combined than when these values were considered separately (TERRA or AQUA, daytime or nighttime). Zhu *et al.* [13] showed that daily maximum and minimum air temperatures could be retrieved effectively from MODIS LST products by using temperature-vegetation index method in the Xiangride River basin of the northern Tibetan Plateau. However, cloud contamination of satellite thermal images makes it challenging to apply estimation models to map spatially continuous daily mean air temperatures on a regional scale using LST datasets as predictors, which is an important goal for agricultural applications. Few studies have focused on merging daily mean air temperatures estimated by daytime and nighttime LST products derived from TERRA/AQUA MODIS to increase the spatial coverage. There have been even fewer studies of creating a map of daily mean air temperature with wide spatial coverage using advanced gap-filling techniques. Therefore, it is necessary to develop a model for estimation of daily mean air temperature using LST datasets. One promising option would be to merge four daily LST products (nighttime and daytime data from both TERRA and AQUA) and apply gap-filling techniques to produce spatially continuous maps of daily mean air temperature, especially in rural areas.

The main objective of the present paper was to develop a systematic method to create spatially continuous maps of daily mean air temperature by merging daytime and nighttime TERRA/AQUA MODIS LST products and using gap-filling techniques. Specifically, we first developed, calibrated and validated estimation models of daily mean air temperature (TA) using TERRA and AQUA MODIS LST data for China's Shaanxi province. Next, we tested the possible combinations of four MODIS datasets: daily mean air temperature estimated from the TERRA daytime LST (TA_{TD}), daily mean air temperature estimated from the TERRA nighttime LST (TA_{TN}), daily mean air temperature estimated from AQUA daytime LST (TA_{AD}), and daily mean air temperature estimated from AQUA nighttime LST (TA_{AN}). We used data from 2003 to 2011 in this analysis, then used 2012 and 2013 datasets to identify the optimal combination. Finally, we developed a merging strategy to fill spatial gaps created by cloud-contaminated pixels by using spatially and temporally adjacent data to create spatially continuous maps of daily mean air temperature.

2. Materials and Methods

2.1. Study Area and Ground Observation Data

The study area is located in central China's Shaanxi Province, and covered 205,800 km^2 (Figure 1). Shaanxi extends from 31°42′N to 39°35′N and from 105°29′E to 111°15′E. This represents a distance of 880 km from north to south and 160 to 490 km from west to east. The area includes three distinct natural regions: the mountainous southern region (Qinling), the Wei River valley (the Guanzhong plains), and the northern upland Loess plateau.

Shaanxi has a continental monsoon climate, but the climate varies widely due to its large span in latitude and altitude. The northern parts, including the Loess

Plateau, have either a cold arid or cold semi-arid climate. The middle area in the Guanzhong plains is mostly warm and semi-arid and the southern portion lies in the humid subtropical zone. Due to the influence of the monsoon climate, Shaanxi has a hot summer and cold winter. The annual mean air temperature ranges between 8 °C and 16 °C, with January mean air temperatures ranging from −11 °C to 3.5 °C and July mean air temperatures ranging from 21 °C to 28 °C. The annual precipitation range between 500 and 1000 mm in the southern mountain area, between 500 and 640 mm in the Wei River valley, and is only about 250 mm on the Loess Plateau. The daily mean air temperature (*TA*) data from 23 meteorological stations belonging to the Shaanxi Provincial Meteorological Bureau were downloaded from the China Meteorological Data Sharing Service System [27]. Based on the MODIS Land Cover Type product (MCD12Q1) in 2012, the land cover in Shaanxi includes mixed forest (38.65%), cropland (28.23%), grassland (26.71%), deciduous broadleaf forest (4.72%), urban and built-up area (0.85%), and other land use types (water, evergreen needle-leaf forest, evergreen broadleaf forest, deciduous needle-leaf forest, closed shrub-lands, open shrub-lands, savannas, permanent wetland, and barren or sparsely vegetated areas). There is no snow and ice in the study region.

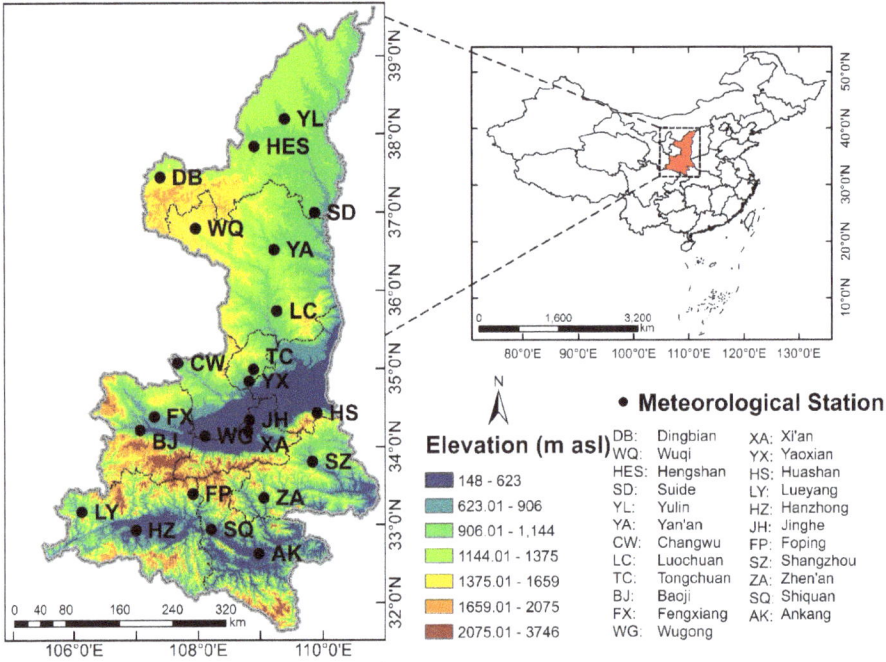

Figure 1. Location of the 23 meteorological stations in Shaanxi province and range of elevations in the study area.

2.2. Daytime and Nighttime MODIS LST Data

To provide more comprehensive support for studies of the Earth, the U.S. National Aeronautics and Space Administration (NASA) developed its earth observation system (EOS) program in the 1990s. The TERRA (EOS-AM) and AQUA (EOS-PM) satellites were specifically designed to support this program. The TERRA satellite was launched in December 1999 and the AQUA satellite was launched in May 2002. TERRA descends past the equator at about 10:30 AM and ascends at about 10:30 PM; in contrast, AQUA passes in the opposite directions over the equator at around 1:30 AM and 1:30 PM, respectively. The MODIS sensors on board the TERRA and AQUA satellites have 36 spectral channels that cover the electromagnetic spectrum from 0.4 μm to 14 μm with a viewing swath width of 2330 km [28], and their orbital parameters provide global coverage for 1 to 2 days.

We used two MODIS LST products (Collection 5) in this study: (i) the MOD11A1 daily Land Surface Temperature & Emissivity product derived from MODIS on board the TERRA satellite and its corresponding information from quality control (QC); and (ii) the MYD11A1 daily Land Surface Temperature & Emissivity product derived from MODIS onboard the AQUA satellite and its corresponding information from QC. The MODIS LST is generated using a split-window algorithm [29,30] with two thermal infrared bands: band 31 (10.78 to 11.28 μm) and band 32 (11.77 to 12.27 μm). The MOD11A1 from TERRA and the MYD11A1 from AQUA are created in tiles that contain 1200 rows by 1200 columns for each tile at approximately 1-km resolution. The MODIS Cloud Mask algorithm, which is based on a series of visible and infrared threshold tests, is used to determine the confidence of the satellite's view of the Earth's surface, because clouds often obscure parts or even the entirety of the satellite images. The LST data will not be available for a location if clouds are present [31]. These data can be downloaded from the Land Processes Distributed Active Archive Center [32].

2.3. Preprocessing of the MODIS LST Data

The MODIS LST products are created in tiles with 1-km resolution and their accuracy has been assessed and found to be satisfactory using several ground reference and validation efforts [29,30]. In the present study, the MODIS LST products were preprocessed to build valid LST maps of Shaanxi Province. First, we used the MODIS Reprojection Tool to extract the corresponding bands (LST_Day_1km, QC_Day, LST_Night_1km, QC_Night) from MOD11A1 and MYD11A1. Next, we created a mosaic of two tiles of LST products (h26v05 and h27v05) that covered the study area and reprojected the geographic coordinates to use the Albert Conic Equal Area projection (SD1 = 25, SD2 = 47, CM = 105). Using an Interactive Data Language (IDL) program, we clipped the re-projected MODIS LST datasets using the boundary polygon that defined the study area. The valid LST values were stored for subsequent processing only when the QC values equaled zero. In the final step, we converted the LST values in the satellite products from Kelvin to Celsius values using the following formula:

$$C = 0.02\,T - 273.15 \tag{1}$$

where C is Celsius temperature ($°C$), T is the absolute temperature (in Kelvins), and 0.02 is a scale factor that converts the scientific data sets values to real LST values in Kelvin degrees [30]. Table 1 summarizes the key terms used in this study and their descriptions.

Table 1. Descriptions of the key terminology used in this study.

Terms	Description
LST ($°C$)	land surface temperature derived from the remotely sensed data
LSTTD ($°C$)	Daytime LST derived from the TERRA MODIS data
LSTTN ($°C$)	Night-time LST derived from the TERRA MODIS data
LSTAD ($°C$)	Daytime LST derived from the AQUA MODIS data
LSTAN ($°C$)	Night-time LST derived from the AQUA MODIS data
TA ($°C$)	Daily mean air temperature observed at the meteorological stations
TATD ($°C$)	Daily mean air temperature estimated using LSTTD
TATN ($°C$)	Daily mean air temperature estimated using LSTTN
TAAD ($°C$)	Daily mean air temperature estimated using LSTAD
TAAN ($°C$)	Daily mean air temperature estimated using LSTAN

2.4. Calibration and Validation of the Estimation Models for Daily Mean Air Temperature

In previous studies, linear regression has been the most common method used to infer daily mean air temperature (*TA*) directly from satellite thermal infrared data [33–37]. Therefore, we used linear regression to estimate the daily mean air temperature from the MODIS LST data:

$$TA = a\,LST + b \tag{2}$$

where a and b are regression coefficients estimated by means of ordinary least-squares regression. *TA* for specific date t is calculated as:

$$TA_t = \frac{TA_{t-1,20} + TA_{t,2} + TA_{t,8} + TA_{t,14}}{4} \tag{3}$$

where TA_t is the daily mean air temperature on date t. $TA_{t-1,20}$ is the air temperature at 8 PM on date t-1. $TA_{t,2}$, $TA_{t,8}$, and $TA_{t,14}$ are the air temperatures at 2 AM, 8 AM, and 2PM on date t respectively. Therefore, the LST derived from MODIS data at 8 PM on date t-1 ($LST_{t-1,20}$) was used to estimate the daily mean air temperature on date t (TA_t).

The daytime and nighttime LST derived from the TERRA MODIS data (LST_{TD} and LST_{TN}) and from the AQUA MODIS data (LST_{AD} and LST_{AN}) can be used as independent variables. This offers the possibility of four types of estimation model for daily mean air temperature on a clear day. This will greatly increase the potential data coverage and estimation accuracy.

The first type of model was constructed using year-round daily mean air temperatures from the 23 meteorological stations. Because previous studies have shown that the relationships between daily mean air temperature and LST may change seasonally [24,38,39], we created a second type of model based on the use of separate temperature data for spring (March, April, and May), summer (June, July, and August), fall (September, October, and November), and winter (December, January, and February). Because air temperatures in urban areas are higher than those in rural and agricultural areas by an average of 2 to 5 °C and because the urban structure influences air temperatures and the relationship between LST and vegetation cover [40], we built a third type of model that reduces the effect of this phenomenon by using only seasonal data from meteorological stations in agricultural region. This reduced the total number of stations that provided data from 23 to 14 for our study area including 14 meteorological stations in rural areas (Changwu, Dingbian, Fengxiang, Foping, Wuqi, Hengshan, Jinghe, Lueyang, Luochuan, Shiquan, Suide, Wugong, Yaoxian and Zhen'an).

To avoid problems with autocorrelation, we divided the available data from 2003 to 2013 into two parts: we used data from 2003 to 2011 to calibrate the estimation models, and then used data from 2012 and 2013 to validate the calibrated models. We evaluated the models' performance using the coefficient of determination (R^2), the root-mean-square error (RMSE), the mean absolute error (MAE), and the bias [25]. These parameters were calculated as follows:

$$R^2 = \frac{\sum_{t=1}^{n}\left(TA_{est,t} - \overline{TA_{ob}}\right)^2}{\sum_{t=1}^{n}\left(TA_{ob,t} - \overline{TA_{ob}}\right)^2} \tag{4}$$

$$RMSE = \sqrt{\frac{1}{n}\sum_{t=1}^{n}\left(TA_{ob,t} - TA_{est,t}\right)^2} \tag{5}$$

$$MAE = \frac{1}{n}\sum_{t=1}^{n}\left|TA_{ob,t} - TA_{est,t}\right| \tag{6}$$

$$bias = \frac{1}{n}\sum_{t=1}^{n}\left(TA_{ob,t} - TA_{est,t}\right) \tag{7}$$

where $TA_{ob,t}$ is the daily mean air temperature observed at a meteorological station on date t, $TA_{est,t}$ is the estimated daily mean air temperature from the MODIS LSTs at that station on date t, n is the number of observations at that station, and $\overline{TA_{ob}}$ is the mean of the observed daily mean air temperatures at that station $\left(\overline{TA_{ob}} = \frac{1}{n}\sum_{t=1}^{n} TA_{ob,t}\right)$.

3. Results and Discussion

We generated the estimation models for daily mean air temperature based on the year-round LST data (Model I), the seasonal LST data (Model II), and the seasonal LST data for rural areas (Model III).

3.1. Calibration and Validation of the Estimation Models Using MODIS LSTs

3.1.1. Model I: The Estimation Models of Daily Mean Air Temperature Using Year-Round LST Data

Figure 2 shows the relationships between daily mean air temperatures (TA) based on data from all meteorological stations and the LSTs derived from the TERRA and AQUA MODIS products. The results show that TA was lower than daytime LST but higher than nighttime LST derived from both satellites. This is because the land surface is the source of heat for air near the surface, and absorbs solar radiation during the daytime and releases that heat at night through long-wave radiation. Differences between TA and LST would be accentuated by heat absorption during the insolation period. Therefore, the differences between TA and LST derived from AQUA MODIS (with an overpass at 1:50 PM local time) are greater than those derived from TERRA MODIS (with an overpass at 10:50 AM local time) during the daytime.

Figure 2 also shows a clear linear relationship between TA and the MODIS LSTs. Therefore, we used linear regression to establish the estimation models for TA as a function of the MODIS LST. The model fit was good ($R^2 > 0.77$, $p < 0.001$), with low errors (RMSE < 4.2 and biases < 0.001) for all models. When nighttime LSTs were used as estimators instead of daytime LSTs, the R^2 improved to values greater than 0.86. This means that LST_{TD} or LST_{AD} can explain at least 77% of the variance in TA, whereas LST_{TN} or LST_{AN} can explain at least 86% of the variance. In addition, the RMSEs for the regression equations based on nighttime LSTs as estimators were at least 20% smaller than those using daytime LSTs as estimators. This means that the night-time LSTs are better estimators of TA than the daytime LSTs. These results are consistent with other studies conducted in various regions [6,10,41].

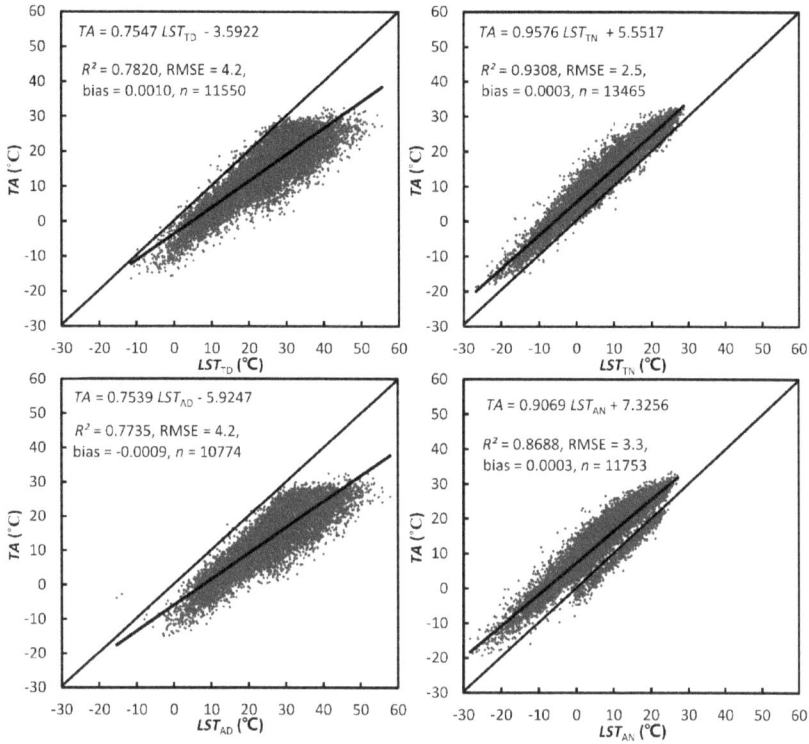

Figure 2. The relationship between observed TA at the 23 meteorological stations and the daytime and nighttime MODIS LSTs (2003 to 2011). The diagonal line from the origin represents the relationship $TA = LST$; the shorter lines represent the linear regression lines.

There were differences between the AQUA and TERRA night point clouds. These point clouds are from measured and estimated data at the Huashan station. Huashan, located about 120 km east of Xi'an, is one of China's Five Great Mountains. The highest point is the South Peak (Huashan station) at 2154.9 m. The complicated interlinks between ambient TA and LST can be explained by the balance between incoming shortwave radiation, incoming and outgoing long wave radiation, the surface albedo, the ground heat flux, and the sensible and latent heat fluxes. As a general rule, the land surface cools rapidly during the night to yield a negative $LST-TA$ difference, and the longer after sunset, the bigger the difference. Therefore, the intercept of the equation between TA and LST_{AN} (7.3256) is greater than the intercept of the equation for the relationship between TA and LST_{TN} (5.5517) because the average overpass time for AQUA at night (1:50 AM) is later than that of TERRA at night (10:50 PM) for our study area. This is similar to the results of Shamir and Georgakakos' [8]. However, TA is lower at the Huashan station because atmospheric

361

temperature decreases with increasing altitude. Thus, the scatter plot between TA and LST_{AN} at Huashan station differs from that at the other stations and is closer to the line for $TA = LST_{AN}$. Therefore, LST_{TN} is a better predictor of TA than LST_{AN} in this context.

3.1.2. Model II: Estimation of Daily Mean Air Temperature Using Seasonal LST Data

Table 2 presents the linear regression equations for the relationship between TA and LST based on separate data for the spring, summer, fall, and winter. As in the case of Model I, all models were statistically significant ($p < 0.001$), but the nighttime LSTs were better estimators of TA than the daytime LSTs. The biases for all models were less than 0.04. We found that the RMSE values for the model estimates during the summer were equivalent to or slightly better than those in the other seasons. RMSE $\leqslant$ 2.7 for the summer estimation models, *versus* RMSEs $\leqslant$ 4.4, 3.2, and 3.1 for models in the spring, fall, and winter, respectively. The lowest R^2 and RMSE occurred during the summer. However, the diurnal temperature variation differs between the land surface and the air above it.

Compared with the RMSEs of estimation models based on year-round data (*i.e.*, Model I), using seasonal LST_{TD}, LST_{TN}, LST_{AD}, *and* LST_{AN} as the independent variables (Figure 2) improved the model's performance greatly in the summer, fall, and winter (Table 2). The RMSE of the estimation model for TA was 4.2 when year-round LST_{TD} was used as estimator, *versus* 2.6, 3.2, and 3.0, respectively, for the summer, fall, and winter models. The corresponding RMSEs for the LST_{TN} model were 1.9 (summer), 1.9 (fall), and 2.2 (winter), and were less than the RMSE for the year-round data (2.5) The RMSE for the model based on year-round LST_{AD} was 4.2 and was greater than those of the models for summer (2.6), fall (3.2), and winter (3.1). With the LST_{AN} models, the RMSEs for models in the summer (2.7), fall (2.8), and winter (2.7) were less than the RMSE (3.3) with the year-round data. The models for spring had estimation power similar to that of the models using the year-round data.

3.1.3. Model III: Estimation of Daily Mean Air Temperature for Agricultural Regions Using Seasonal LST Data

In the past few decades, with the accelerating rate of urbanization in China, the increasing amount of buildings, public squares, and roads in Shaanxi Province has decreased the amount of green space, water, and other natural surfaces. The surface thermodynamic properties differ greatly between urban and rural land. Buildings and artificial pavement represent the dominant urban surface, and because their materials have a high thermal conductivity, they absorb more of the incident solar radiation. This can cause greater atmospheric warming than would occur over natural surfaces such as vegetation, creating what is known as the "urban heat island" effect. As a result of urbanization, the urban temperature has therefore risen steadily [39]. In summer, when the surface temperature of grass lawns is 32 °C, cement roads may have a surface temperature of up to 57 °C, and asphalt surface temperatures may rise as high as 63 °C [42]. These high-temperature surfaces become a huge heat source for the surrounding atmosphere. High concentrations of air pollutants and increasing levels of aerosol particles also retain this heat by acting to some extent as insulators that trap outgoing radiation [43]. Kawashima *et al.* [44] pointed out that LSTs generally determine the variations of the surrounding air temperature. Therefore, the relationship between *TA* and LST derived from satellite data for urban environments will be quite different from that in rural regions. To enhance the estimation accuracy of *TA* for rural areas, we repeated our analysis after excluding data from urban stations and mountainous area.

Table 2. Estimation models for daily mean air temperature (*TA*) in the spring, summer, fall, and winter using the MODIS-derived land surface temperatures (LSTs) as the independent variable.

Independent Variable	Season	Model II	R^2	RMSE	bias	N
LST_{TD}	Spring	$TA_{TD} = 0.529LST_{TD} + 1.185$	0.4390	4.4	0.0002	3698
	Summer	$TA_{TD} = 0.232LST_{TD} + 15.73$	0.2280	2.6	0.0237	3007
	Fall	$TA_{TD} = 0.8149LST_{TD} - 4.5194$	0.7487	3.2	0.0168	3256
	Winter	$TA_{TD} = 0.8542LST_{TD} - 6.0169$	0.6620	3.0	0.0007	1589
LST_{TN}	Spring	$TA_{TN} = 0.8626LST_{TN} + 7.7517$	0.8241	2.5	0.0051	4213
	Summer	$TA_{TN} = 0.7113LST_{TN} + 10.503$	0.6355	1.9	0.0082	3525
	Fall	$TA_{TN} = 0.9039LST_{TN} + 4.7326$	0.9015	1.9	0.0071	3654
	Winter	$TA_{TN} = 0.8714LST_{TN} + 4.0055$	0.879	2.2	−0.0011	2073
LST_{AD}	Spring	$TA_{AD} = 0.512LST_{AD} - 0.044$	0.4070	4.2	0.0169	3473
	Summer	$TA_{AD} = 0.239LST_{AD} + 14.86$	0.2530	2.6	0.0338	2545
	Fall	$TA_{AD} = 0.784LST_{AD} - 6.012$	0.7370	3.2	0.0180	3186
	Winter	$TA_{AD} = 0.716LST_{AD} - 7.752$	0.6400	3.1	0.0048	1570
LST_{AN}	Spring	$TA_{AN} = 0.801LST_{AN} + 9.198$	0.6710	3.3	0.0014	4095
	Summer	$TA_{AN} = 0.567LST_{AN} + 13.57$	0.3941	2.7	0.0130	2971
	Fall	$TA_{AN} = 0.817LST_{AN} + 6.532$	0.7620	2.8	0.0041	3134
	Winter	$TA_{AN} = 0.795LST_{AN} + 4.924$	0.8250	2.7	−0.0057	1553

Table 3 presents the *TA* estimation models based on the data from meteorological stations in rural regions in each season and the corresponding LSTs. There was generally a strong and statistically significant positive linear relationship between *TA* and the LSTs derived from the TERRA or AQUA MODIS products during both the daytime and nighttime. Estimates of *TA* improved (lower RMSE; Model III in Table 3) compared to those that included data from urban areas (Model II in Table 2) when using LST_{AN} or LST_{AD} as the independent variable. The RMSEs of Model III with LST_{AN} as the estimator were 2.3 (spring), 1.9 (summer), 1.9 (fall), and 2.3 (winter), which were less than the corresponding values of 3.3 (spring), 2.7 (summer), 2.8 (fall), and 2.7 (winter) for Model II. The RMSEs of Model III using

LST_{AD} as the estimator were 4.0 (spring), 2.4 (summer), 3.0 (fall), and 3.0 (winter), *versus* RMSEs of 4.2 (spring), 2.6 (summer), 3.2 (fall), and 3.1 (winter) with Model II. With LST_{TD} or LST_{TN} as the estimator, Model III produced results similar to Model II. The RMSEs of Model III using LST_{TN} as the estimator were 2.5 (spring), 1.9 (summer), 1.7 (fall), and 2.2 (winter), *versus* 2.5 (spring), 1.9 (summer), 1.9 (fall), and 2.2 (winter) using Model II. The RMSEs of Model III using LST_{TD} as the estimator were 4.2 (spring), 2.7 (summer), 3.4 (fall), and 2.9 (winter), *versus* RMSEs of 4.4 (spring), 2.6 (summer), 3.2 (fall), and 3.0 (winter) using Model II.

Table 3. The seasonal estimation models for daily mean air temperature (*TA*) based on the corresponding land surface temperature (LST) data from meteorological stations in rural regions (*i.e.*, Model III).

Independent Variable	Season	Model III	R^2	RMSE	bias	N
LST_{TD}	Spring	$TA_{TD} = 0.602LST_{TD} - 0.223$	0.4860	4.2	0.0184	2460
	Summer	$TA_{TD} = 0.238LST_{TD} + 15.52$	0.2370	2.7	0.0260	2426
	Fall	$TA_{TD} = 0.801LST_{TD} - 4.224$	0.7250	3.4	0.0010	2657
	Winter	$TA_{TD} = 0.844LST_{TD} - 5.819$	0.6770	2.9	0.0047	1360
LST_{TN}	Spring	$TA_{TN} = 0.864LST_{TN} + 7.772$	0.8249	2.5	0.0002	3300
	Summer	$TA_{TN} = 0.7215LST_{TN} + 10.279$	0.6317	1.9	0.0195	2811
	Fall	$TA_{TN} = 0.9223LST_{TN} + 4.5128$	0.9140	1.7	0.0032	2869
	Winter	$TA_{TN} = 0.8868LST_{TN} + 4.1513$	0.8744	2.2	−0.0032	1575
LST_{AD}	Spring	$TA_{AD} = 0.519LST_{AD} - 0.134$	0.4560	4.0	0.0099	2253
	Summer	$TA_{AD} = 0.249LST_{AD} + 14.50$	0.2330	2.4	0.0307	1656
	Fall	$TA_{AD} = 0.848LST_{AD} - 6.987$	0.7650	3.0	0.0057	2107
	Winter	$TA_{AD} = 0.737LST_{AD} - 7.824$	0.6430	3.0	0.0011	1304
LST_{AN}	Spring	$TA_{AN} = 0.886LST_{AN} + 9.105$	0.8120	2.3	0.0024	2412
	Summer	$TA_{AN} = 0.666LST_{AN} + 12.29$	0.6520	1.9	0.0086	1926
	Fall	$TA_{AN} = 0.918LST_{AN} + 6.228$	0.8770	1.9	0.0019	1908
	Winter	$TA_{AN} = 0.865LST_{AN} + 5.610$	0.8760	2.3	−0.0049	810

3.1.4. Validation of the Estimation Models for Daily Mean Air Temperature

We validated the estimation models for *TA* using data from 2012 and 2013 at all available meteorological stations for models I and II, and data only from rural meteorological stations for model III. Figure 3 shows the relationship between the *TA* estimated using Model I, Model II, and Model III with the MODIS daytime and nighttime LSTs as independent variables and the *TA* measured at the meteorological stations. Most of the points were distributed around the line *TA* = *LST*. This indicates a good agreement between the estimated and measured *TA*. Model III tended to be more accurate than models I and II (Table 4).

Table 4 shows that Model III performed best. Model III generally had the highest coefficient of determination and lowest RMSE and MAE, or values comparable to those in the other models. Pairwise tests showed that the *TA* values estimated by Model III differed significantly ($p < 0.01$) from those estimated using Model I when LST_{TD}, LST_{AD}, or LST_{AN} was used as the estimator, but not for the model with LST_{TN}

as the estimator (Table 5). The *TA* estimated by Model III differed significantly from that of Model II when LST_{AD} and LST_{AN} were used as the estimator ($p < 0.05$); for the TERRA datasets, the difference was not significant. In addition, Model II differed significantly from Model I ($p < 0.01$) only when the daytime LST data were used.

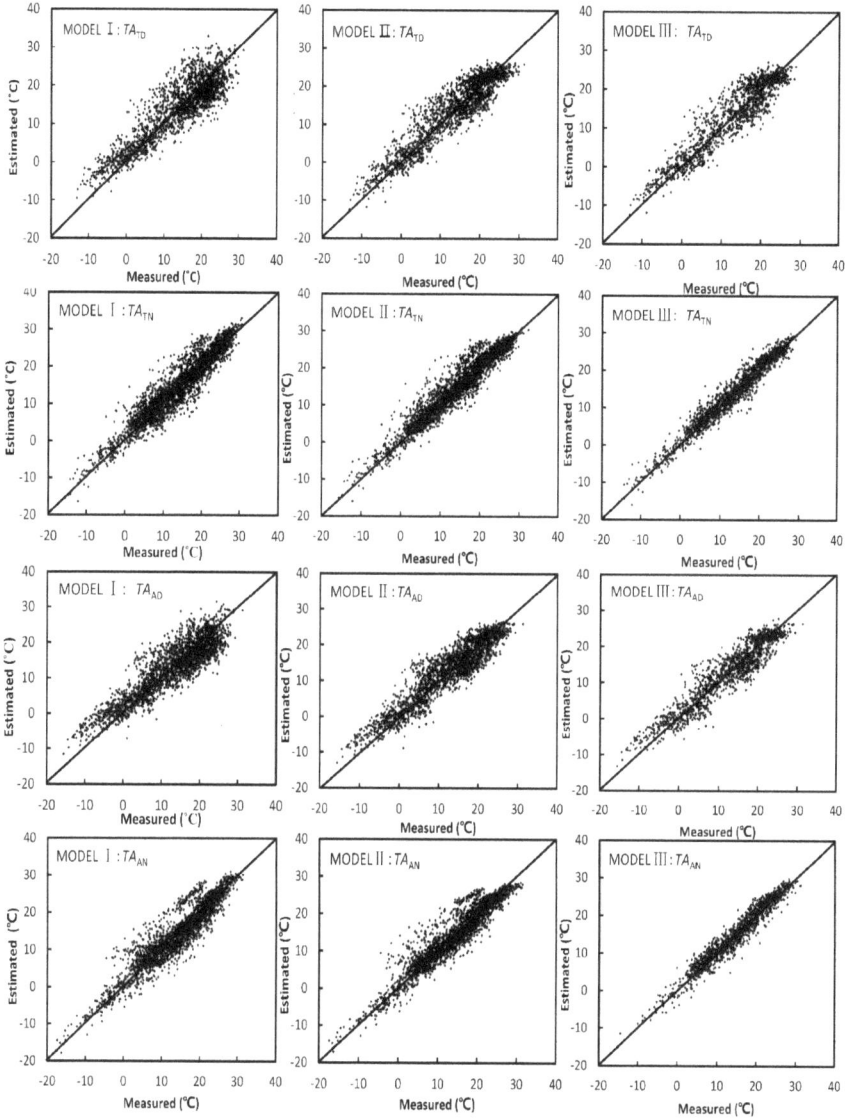

Figure 3. Relationships between the estimated and measured daily mean air temperature (*TA*) based on daytime (D) and nighttime (N) data for the three models (I = all data, II = seasonal data, III = seasonal data from rural stations). Table 4 provides statistical data on the results of each linear regression.

Table 4. Validation of the estimated daily mean air temperature (TA) predicted using land surface temperature (LST) data from meteorological stations for the three models in 2012 and 2013.

Independent Variable	MODEL I				MODEL II				MODEL III			
	R^2	RMSE	MAE	Bias	R^2	RMSE	MAE	Bias	R^2	RMSE	MAE	Bias
LST_{TD}	0.7791	4.3	3.38	0.2345	0.8622	3.4	2.66	−0.3462	0.867	3.4	2.7	0.4096
LST_{TN}	0.8941	2.9	2.20	0.7119	0.9176	2.6	1.87	−0.7445	0.9516	2.0	1.56	−0.6007
LST_{AD}	0.795	4.1	3.27	−0.1203	0.8649	3.3	2.62	0.2830	0.887	3.2	2.52	0.3737
LST_{AN}	0.8633	3.2	2.32	0.2886	0.8919	2.8	2.01	0.2572	0.94	2.0	1.54	0.3467

Table 5. Significant test between the different estimation values of daily mean air temperature using the three model types with the MODIS LSTs as the predictor.

	Pairwise Test Result (p level)		
	Model III/Model I	Model III/Model II	Model II/Model I
LST_{TD}	0.0000	0.0933	0.0022
LST_{TN}	0.1795	0.0513	0.2021
LST_{AD}	0.0000	0.0106	0.0001
LST_{AN}	0.0093	0.0109	0.4708

3.2. Optimal Strategies for Merging Images of Daily Mean Air Temperature Estimated from LSTs

We created maps of TA values for each day from 2003 to 2013. We obtained a maximum of four TA images on each clear day (*i.e.,* the daytime and nighttime LSTs from the TERRA and AQUA MODIS products). However, because of cloud cover, one or more of these datasets was often unavailable for certain parts of the study area, and the resulting map of estimated TA suffered from gaps. Figures 4 and 5 present the proportion of the data available for each pixel in the study area based on the daytime and nighttime TERRA and AQUA MODIS LSTs in 2012 and 2013, respectively. Most of the pixels in the study area had availability values of less than 50%. For example, Figure 6 shows that on 21 June 2012, the available data amounted to only 18.40, 40.81, 51.04, and 24.29% of the pixels for TA_{TN}, TA_{AN}, TA_{TD}, and TA_{AD}, respectively. Fortunately, the available data from the different MODIS products both overlap and complement each other, which makes it possible to merge the data to increase data availability for the study area. Figure 7 provides examples of merged TA data for 21 June 2012. The available data coverage in the merged image totaled 75.58%, which represents an increase of 57.18, 34.77, 24.54, and 51.29 percentage points compared with the coverage based only on TA_{TN}, TA_{AN}, TA_{TD}, and TA_{AD}, respectively.

Table 4 shows that the RMSEs using Model III were 3.4, 2.0, 3.2, and 2.0 when using LST_{TD}, LST_{TN}, LST_{AD}, and LST_{AN} as the estimator, respectively, and the corresponding MAEs were 2.70, 1.56, 2.52, and 1.54. This means that combining the different TA images calculated using the different LST products will result in a different accuracy. Table 6 shows the possible strategies for combining TA images.

In strategy 1, for instance, the TA_{TN} image was the initial basis for the merged data. When the TA_{TN} value was missing, it was replaced by the TA_{AN} value. If both TA_{TN} and TA_{AN} were missing, they were replaced by the TA_{AD} value. When all three were missing, they were replaced by the TA_{TD} value. The values in Table 6 were calculated independently for each grid cell throughout the study area.

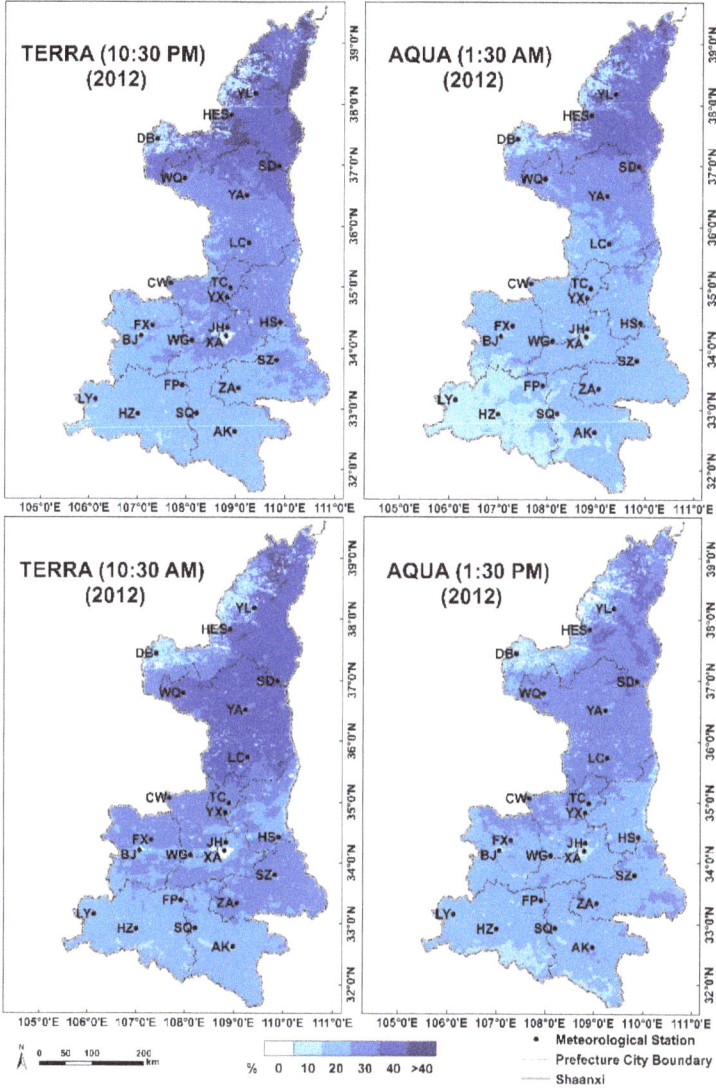

Figure 4. Percentages of available data for daily mean air temperature (*TA*) in 2012 based on daytime and nighttime TERRA and AQUA MODIS land surface temperatures (LSTs).

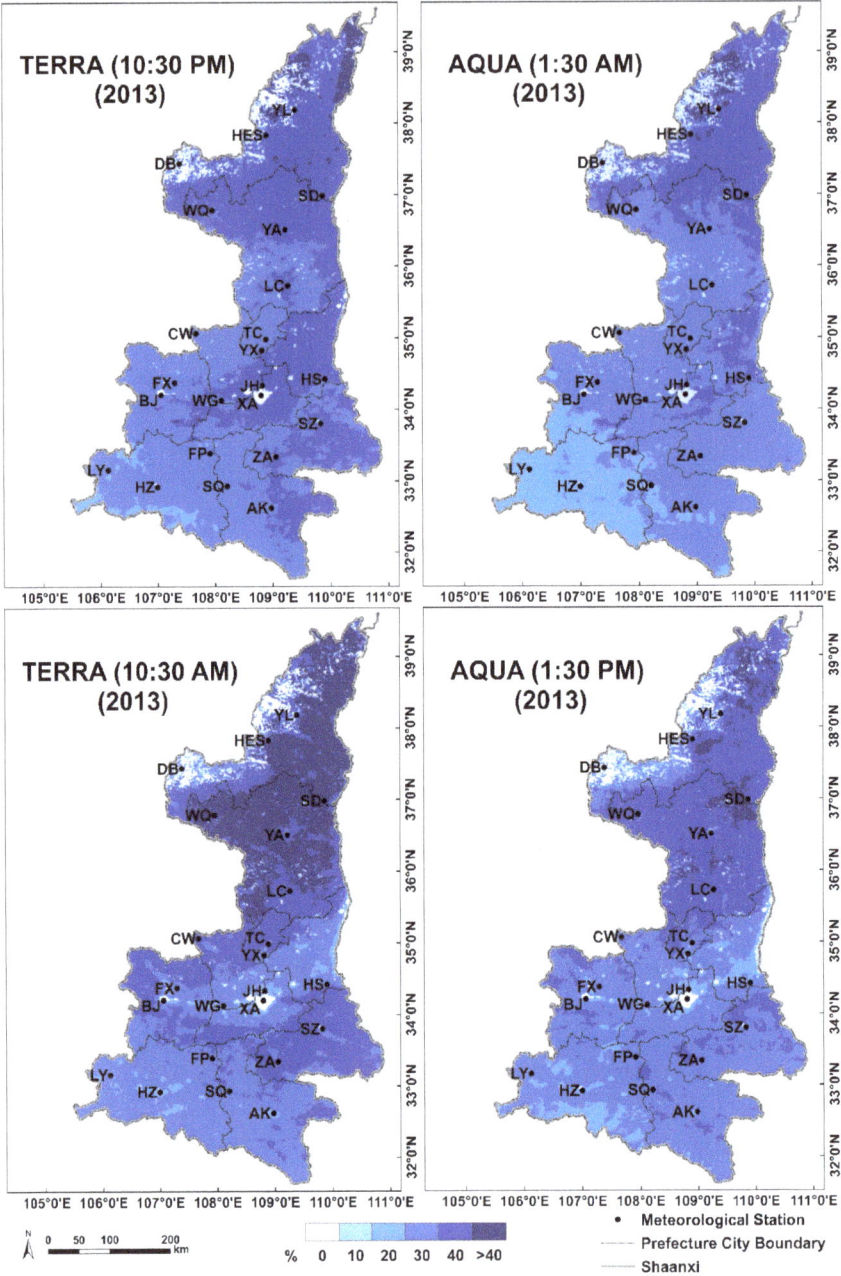

Figure 5. Percentages of available data for daily mean air temperature (*TA*) in 2013 based on daytime and nighttime TERRA and AQUA MODIS land surface temperatures (LSTs).

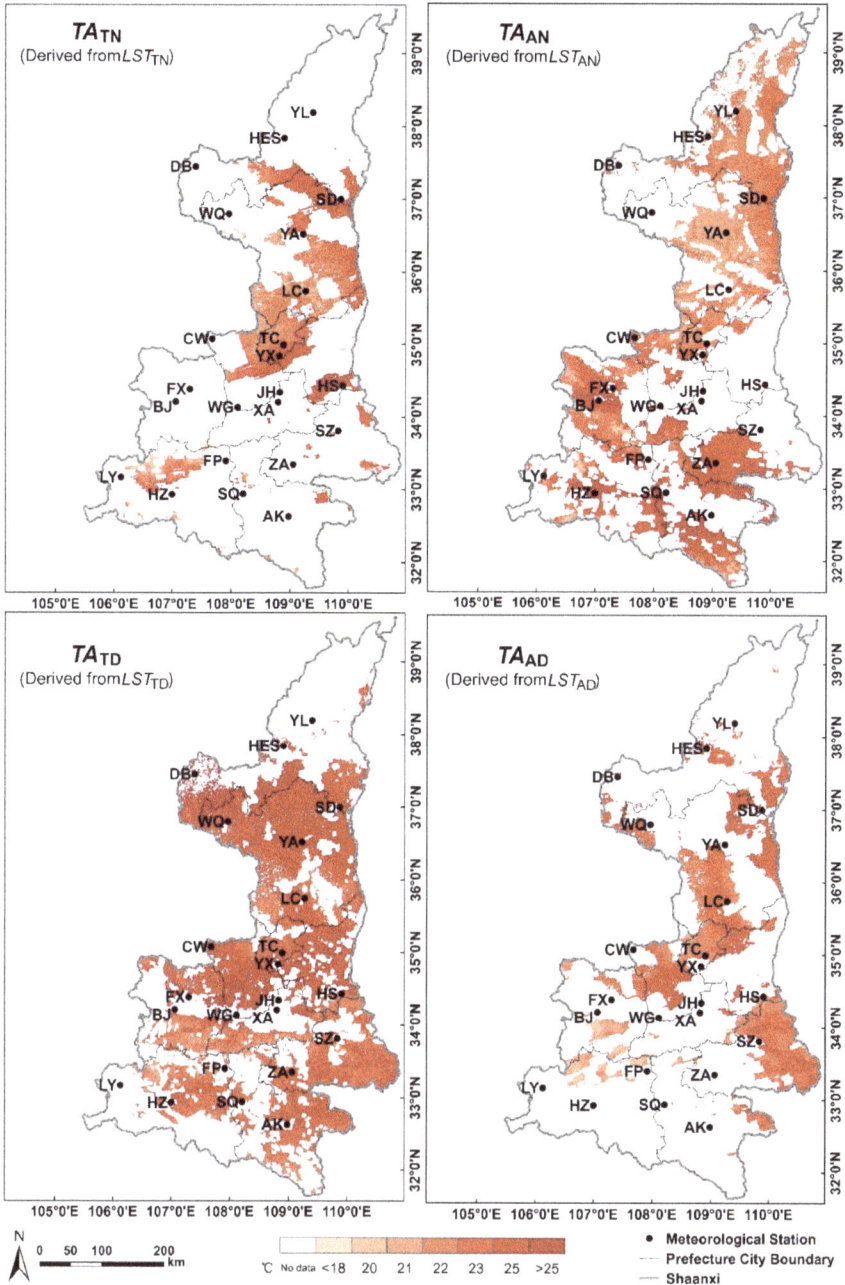

Figure 6. Daily mean air temperature (*TA*) estimated using the daytime and nighttime land surface temperatures (LSTs) from the TERRA and AQUA MODIS products for 21 June 2012.

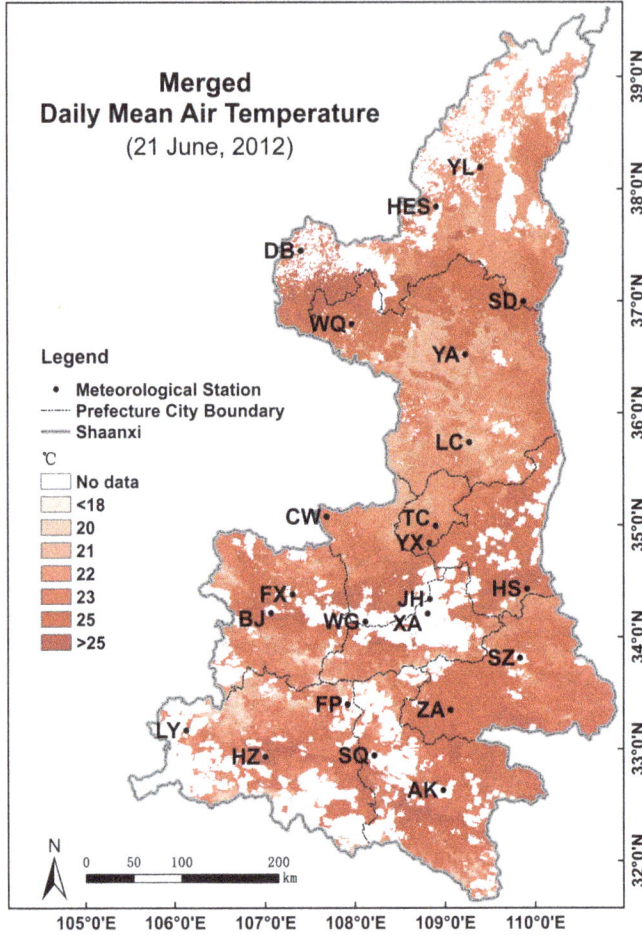

Figure 7. Merged image using the daily mean air temperatures (*TAs*) estimated using the daytime and nighttime land surface temperatures (LSTs) from the TERRA and AQUA MODIS products for 21 June 2012.

Based on results in Table 6, we found that R^2 varies between 0.9073 and 0.9383, the RMSEs range from 2.41 to 2.91, MAE change from 1.84 to 2.24, and bias varies between −0.5049 and −0.3421. The merged results had a higher R^2 and a lower RMSE and MAE if TA_{TN} and TA_{AN} were used as the first two merged images (strategies 1, 2, 7, and 8). The prediction ability was high in each case, with $R^2 > 0.93$, RMSE $\leqslant 2.43$, and MAE $\leqslant 1.87$ in all cases. In contrast, if we used TA_{TD} and TA_{AD} as the first two merged images (strategies 17, 18, 23, and 24), R^2 decreased (to values <0.92) and RMSE and MAE increased (to values $\geqslant 2.76$ and 2.11, respectively). This is because the R^2 between *TA* and nighttime LSTs was stronger than those between *TA* and daytime LSTs (Section 3.1). The estimation models based on nighttime LSTs had

370

lower RMSEs and MAEs. The optimal combination of TA images was provided by Strategy 2; although Strategy 18 produced similar results, the R^2 value was higher for Strategy 2. Therefore, we chose Strategy 2 as the optimal combination of the four TA images.

Table 6. Possible image merging strategies to create maps of daily mean air temperature (TA) using the TERRA and AQUA MODIS land surface temperatures (LSTs) and their accuracy using the validation datasets from 2012 to 2013.

Strategy	Base Image	Second Image	Third Image	Fourth Image	R^2	RMSE	MAE	bias
1	TA_{TN}	TA_{AN}	TA_{AD}	TA_{TD}	0.9381	2.42	1.84	-0.4549
2	TA_{TN}	TA_{AN}	TA_{TD}	TA_{AD}	0.9383	2.41	1.84	-0.4190
3	TA_{TN}	TA_{AD}	TA_{AN}	TA_{TD}	0.9352	2.48	1.90	-0.5033
4	TA_{TN}	TA_{AD}	TA_{TD}	TA_{AN}	0.9338	2.51	1.92	-0.5049
5	TA_{TN}	TA_{TD}	TA_{AN}	TA_{AD}	0.9361	2.45	1.87	-0.4215
6	TA_{TN}	TA_{TD}	TA_{AD}	TA_{AN}	0.9341	2.49	1.91	-0.4541
7	TA_{AN}	TA_{TN}	TA_{AD}	TA_{TD}	0.9367	2.43	1.87	-0.4501
8	TA_{AN}	TA_{TN}	TA_{TD}	TA_{AD}	0.9369	2.42	1.87	-0.4141
9	TA_{AN}	TA_{AD}	TA_{TN}	TA_{TD}	0.9299	2.54	1.94	-0.3851
10	TA_{AN}	TA_{AD}	TA_{TD}	TA_{TN}	0.9253	2.63	2.00	-0.4348
11	TA_{AN}	TA_{TD}	TA_{TN}	TA_{AD}	0.9281	2.57	1.96	-0.4073
12	TA_{AN}	TA_{TD}	TA_{AD}	TA_{TN}	0.9250	2.62	2.00	-0.3756
13	TA_{AD}	TA_{AN}	TA_{TN}	TA_{TD}	0.9170	2.76	2.13	-0.4245
14	TA_{AD}	TA_{AN}	TA_{TD}	TA_{TN}	0.9124	2.84	2.19	-0.4743
15	TA_{AD}	TA_{TN}	TA_{AN}	TA_{TD}	0.9190	2.74	2.10	-0.4443
16	TA_{AD}	TA_{TN}	TA_{TD}	TA_{AN}	0.9124	2.84	2.19	-0.4743
17	TA_{AD}	TA_{TD}	TA_{TN}	TA_{AN}	0.9097	2.89	2.22	-0.4794
18	TA_{AD}	TA_{TD}	TA_{AN}	TA_{TN}	0.9079	2.91	2.24	-0.4647
19	TA_{TD}	TA_{AN}	TA_{TN}	TA_{AD}	0.9167	2.76	2.11	-0.3739
20	TA_{TD}	TA_{AN}	TA_{AD}	TA_{TN}	0.9136	2.81	2.15	-0.3421
21	TA_{TD}	TA_{TN}	TA_{AN}	TA_{AD}	0.9183	2.74	2.09	-0.3769
22	TA_{TD}	TA_{TN}	TA_{AD}	TA_{AN}	0.9162	2.78	2.13	-0.4095
23	TA_{TD}	TA_{AD}	TA_{AN}	TA_{TN}	0.9073	2.91	2.24	-0.3562
24	TA_{TD}	TA_{AD}	TA_{TN}	TA_{AN}	0.9091	2.89	2.22	-0.3709

The LSTs from MODIS TERRA and AQUA are retrieved only under clear-sky conditions. LSTs under cloudy conditions would differ from those obtained under a clear sky. In contrast, TA is available irrespective of cloud conditions. It is therefore important to analyze the influence of clouds on the estimation of TA. Table 7 summarizes the difference between the measured and estimated TA under four different cases of cloud conditions. In Case 1, pixels from all four datasets are cloud-free. In Case 2, pixels from three of the four datasets are cloud-free. In Case 3, pixels from two of the four datasets are cloud-free. In Case 4, pixels from only one of the four datasets are cloud-free. Case 5 means that pixels are contaminated with clouds in all four LST products, so LST is instead estimated by means of interpolation between adjacent pixels (we will describe the interpolation method in Section 3.3). In Case 2, Case 3 and Case 4, missing pixels due to cloudiness in different datasets were filled with corresponding cloud-free pixels in another dataset which means that TA could have been overestimated (*i.e.*, because LST would be lower as a result of the shade created by clouds). In Case 5, missing pixels were interpolated from surrounding cloud-free pixels, leading to a positive bias. The bias also varied seasonally. In winter, as the vegetation coverage is lower, bare soil and built structures receive more solar radiation, which causes a slightly higher bias when daytime *LST* is merged into the map. In summer, vegetation and water reduce the temperature difference during the daytime and nighttime. The results show no obvious relationship among the four cases because the LSTs from the MODIS Terra and Aqua have already been filtered to eliminate cloud cover using the QC data. Under these circumstances, the data merging method proposed in this study has no consistent bias that must be adjusted to account for cloudiness.

Table 7. Bias between measured and estimated daily mean air temperature (*TA*) under different cloud conditions for the merged data from four MODIS datasets using data from 2012 and 2013. Cases: 1 = pixels from all four datasets are cloud-free; 2 = pixels from three of the four datasets are cloud-free; 3 = pixels from two of the four datasets are cloud-free; 4 = pixels from one of the four datasets are cloud-free; 5 = pixels are contaminated with clouds in all four datasets and must be estimated by means of interpolation of data from adjacent pixels.

Case	Whole Year	Spring	Summer	Autumn	Winter
1	−0.4837	−0.8475	−0.3088	−0.3207	−0.2039
2	−0.6341	−0.3766	−1.0319	−0.4779	−0.7821
3	−0.3775	−0.2540	−0.6404	−0.0127	−1.0149
4	−0.3979	−0.0953	−0.4046	−0.4798	−0.7197
5	0.0646	−0.1483	−0.1535	0.4003	0.3944

3.3. Temporal and Spatial Fusion of Daily Mean Air Temperature Using Time Series Images

Figure 8 demonstrates the spatial distribution of data availability for each pixel using data for the whole year in 2012 and 2013. The merged image greatly improved the spatial coverage by the available data. Table 8 presents the annual data availability percentages for the *TA* images throughout the study area for the merged images and for images derived from daytime and nighttime TERRA and AQUA MODIS LSTs in 2012 and 2013. The results show that the merged image greatly improved the spatial coverage by the available data, reaching values of 55.46% in 2012 and 44.92% in 2013. But the percentages of available data were 22.08%, 17.94%, 22.48%, and 18.65% for TA_{TN}, TA_{AN}, TA_{TD}, and TA_{AD} images in 2012. They were 28.85%, 25.23%, 31.30%, and 27.84%, respectively, in 2013. The available data in the merged images therefore increased by 33.38, 37.52, 32.98, and 36.81 percentage points in 2012 and by 16.07, 19.69, 13.62, and 17.08 percentage points in 2013 compared with the corresponding TA_{TN}, TA_{AN}, TA_{TD}, and TA_{AD} images. However, given the fact that data for an average of half of the pixels were unavailable even after merging the images, it is clearly necessary to obtain fuller spatial coverage to improve the accuracy of estimation of *TA*.

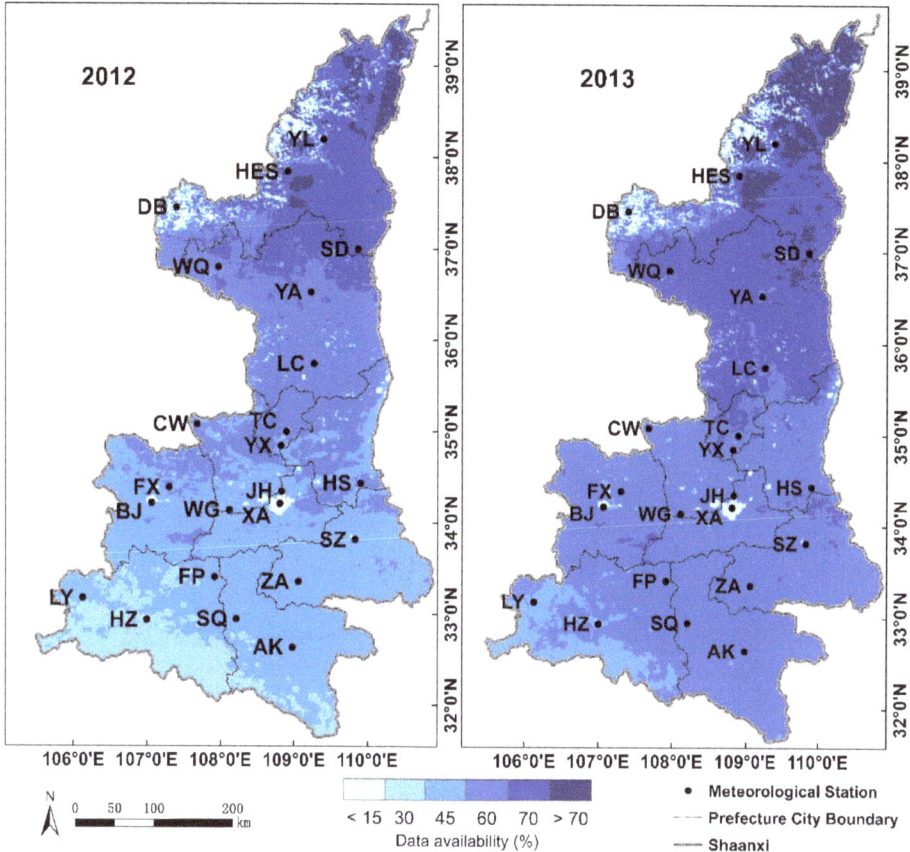

Figure 8. Percentages of data availability for daily mean air temperature (*TA*) in 2012 and 2013 for merged images based on the validation dataset.

Table 8. Percentages of the daily mean air temperature (*TA*) images for the merged dataset (all four MODIS LST products combined using strategy 2 in Table 6) and for images derived from daytime and nighttime TERRA and AQUA MODIS LSTs in 2012 and 2013.

Year	Merged coverage	Data Availability (%for Coverage, Percentage Points for Increase)							
		TA_{TN}		TA_{AN}		TA_{TD}		TA_{AD}	
		coverage	increase	coverage	increase	coverage	increase	coverage	increase
2012	55.46	22.08	33.38	17.94	37.52	22.48	32.98	18.65	36.81
2013	44.92	28.85	16.07	25.23	19.69	31.30	13.62	27.84	17.08

373

Ideally, the *TA* for crop growth monitoring and model simulation should take advantage of complete datasets. In reality, noise and missing pixels create gaps in the data that must be filled somehow. Aiming to achieve improved accuracy of *TA* will require efforts to reduce the loss of data and fill gaps, thereby providing better coverage of the whole study region. The daily mean temperature images before the date of the estimation are obtained if their data are available. These images carry important information for the *TA* estimation for the present day. We therefore proposed a method to fill gaps in the data based on the assumption that atmospheric conditions would be uniform within a relatively small window surrounding a pixel for which data is missing. This means that the *TA* difference of a target pixel between data *t* and data *t-1* is equal to the mean *TA* difference of surrounding pixels between data *t* and data *t-1*. To minimize the uncertainty in the error introduced by cloudiness and gaps between swaths, we exploited a possible strategy based on finding the optimal window size by extending the process into a larger geographic area. The optimum window size was obtained from statistical analysis of the difference between *TA* estimated from the MODIS LST and *TA* measured by the meteorological stations. Figure 9 displays the change in the quality of the estimate as a function of the window size used for the spatial filling. Figure 10 illustrates this estimation procedure (using a window size of 9 $\times$ 9 pixels as an example). The R^2 increased with increasing window size. In the contrast, RMSE, MAE, and bias decreased with increasing window size. The magnitude of the bias reached its minimum at a window size of 33 $\times$ 33 pixels. We therefore used a grid of 33 $\times$ 33 pixels centered on the pixel with missing data in our subsequent analysis. The mean difference in *TA* among these pixels is calculated as follows:

$$\overline{TA} = \frac{1}{N} \sum_{j=i-4}^{j+4} \sum_{i=i-4}^{i+4} \left(TA_{i,j}^t - TA_{i,j}^{t-1} \right) \tag{8}$$

where $\overline{TA}$ is the mean *TA* difference among the pixels with available data both at the date of estimation (*i.e.*, at time *t*) and the date before the estimation (*i.e.*, at time *t*–1). *N* is the number of pixels with available *TA* values both at the date of estimation and the date before estimation. $TA_{i,j}^t$ and $TA_{i,j}^{t-1}$ are the *TA* values for the pixel in line *i* and column *j* of the image at times *t* and *t*–1.

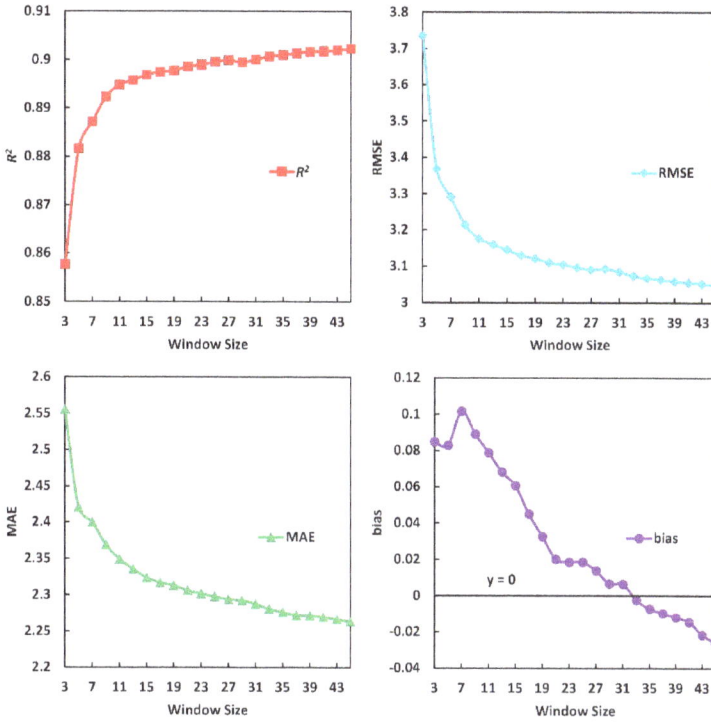

Figure 9. Analysis the optimal size (pixels) of the window used for the spatial filling method illustrated in Figure 10.

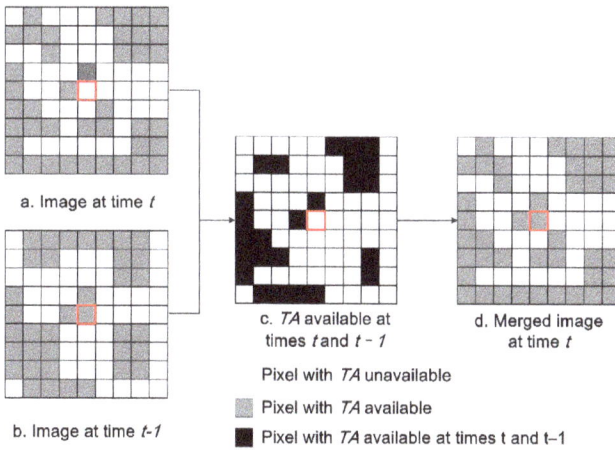

a. Image at time *t*

b. Image at time *t-1*

c. *TA* available at times *t* and *t - 1*

d. Merged image at time *t*

Pixel with *TA* unavailable

Pixel with *TA* available

Pixel with *TA* available at times t and t–1

Figure 10. Flowchart for calculation of missing pixel values caused by cloud cover and other problems using the merged images from the day before the estimation date The red square represents the pixel for which *TA* will be calculated based on data from the previous day.

$TA_{i,j}^t$ can be estimated as follows:

$$TA_{i,j}^t = TA_{i,j}^{t-1} + \overline{TA} \tag{9}$$

We used this approach to generate a time series of filled pixels using data for the whole year derived from the 14 rural meteorological stations in both 2012 and 2013. Figure 11 shows the resulting relationship between the estimated and measured TA. Overall, the approach produced good results: a strong and statistically significant relationship ($R^2 = 0.8971$) with a low MAE (2.35) for data from all seasons. Using the filled dataset increased data coverage to 78.23% and 86.02% in 2012 and 2013, respectively. The model showed a seasonal pattern of error. In summer, the R^2 was relatively poor and the distribution of the data was more concentrated than in other seasons which led to a lower RMSE. In turn, the higher LST increased turbulence in the atmospheric boundary layer, thereby affecting heat transfer from the land surface into the ambient air and subsequently to the upper atmosphere. In contrast, the land surface receives less solar radiation in winter, thereby weakening turbulence. The retrieval of TA is simpler at night because solar radiation does not affect the thermal infrared signal. Figure 12 demonstrates the day-to-day variation of the measured and estimated daily mean air temperature in 2012 and 2013 at the 14 rural meteorological stations. Judging from the similarity of the measured and estimated air temperatures (*i.e.*, the difference was small and centered on 0 °C), the estimation model generally showed good agreement with the measured TA and was able to reflect the annual pattern of TA fluctuation. We found no systematically positive or negative biases between the estimated and measured values (Figures 11 and 12).

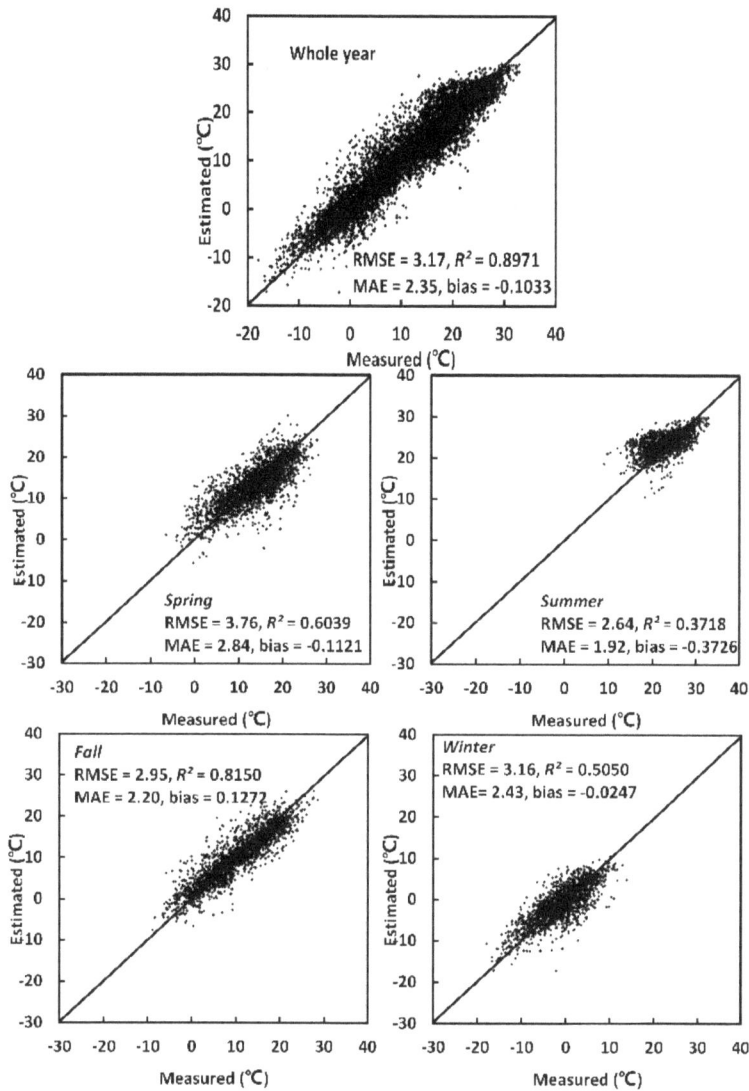

Figure 11. Relationships between the estimated and measured daily mean air temperature (*TA*) for whole-year data and for data from the spring, summer, fall, and winter.

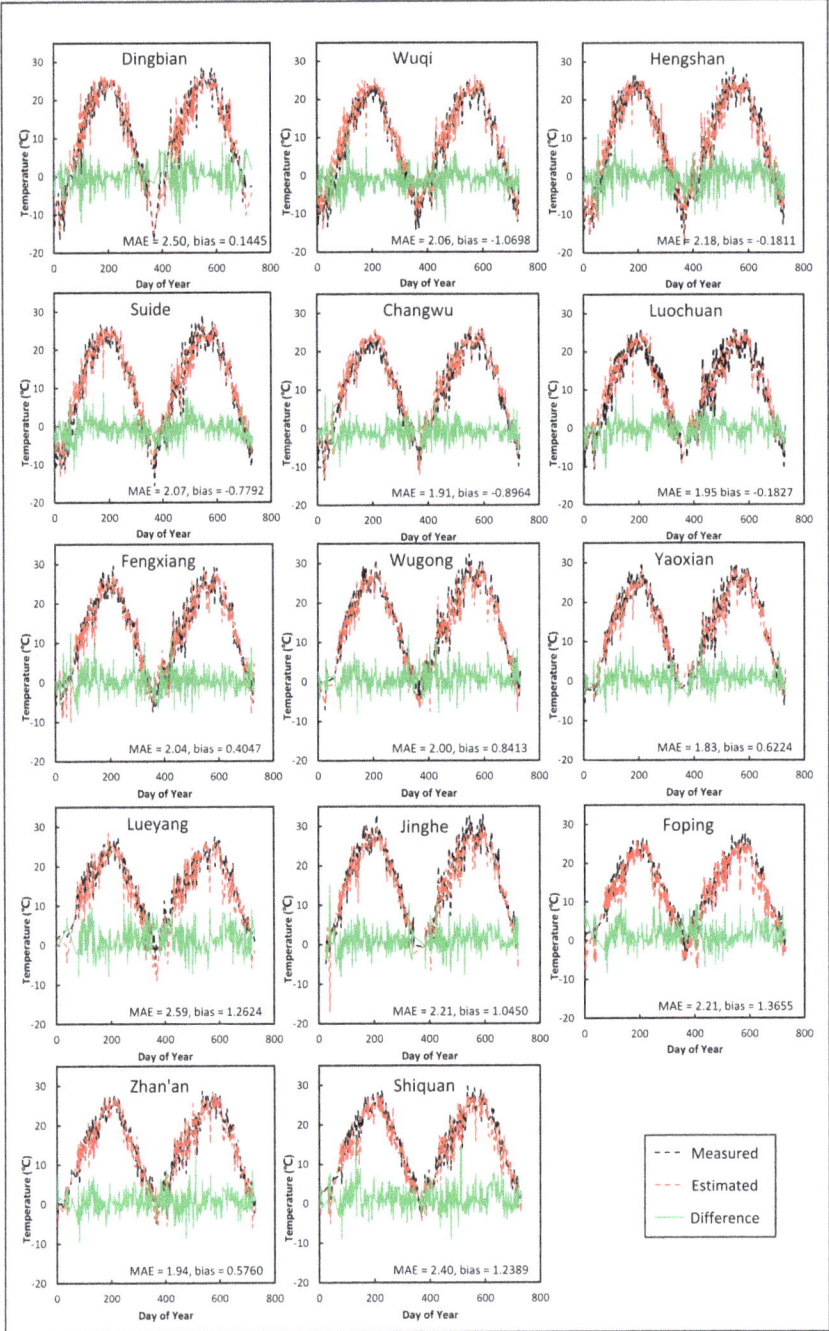

Figure 12. Annual variation in the measured and estimated daily mean air temperature (*TA*) in 2012 and 2013 based on the data from the 14 rural meteorological stations.

4. Conclusions

If clouds are present at a given location, then LST data will not be available. In this paper, we describe a systematic method for filling such gaps in the data based on spatial and temporal data fusion techniques. The first step in this method was to find the optimal strategy for merging images of daily mean air temperature (TA) estimated from daytime and nighttime TERRA and AQUA LST data. The second step was to select the optimal spatial window size to use in interpolation and gap-filling based on estimated TA from the previous day. This process generates high spatial and temporal coverage for the study area. The calibration results showed that the annual average spatial coverage could be improved significantly. Using this approach, the proportion of the pixels with available data increases from 20.28% to 76.99% in 2012 and 28.31% to 89.67% in 2013.

The relationship between TA from the meteorological stations and LSTs derived from the daytime and nighttime TERRA and AQUA MODIS LST data was strong and significant. The nighttime LSTs from TERRA and AQUA MODIS provided a better TA estimator than the daytime LSTs. By comparing different strategies for merging the four TA images calculated using the daytime and nighttime TERRA and AQUA MODIS LST data in different orders, we found an optimal merging strategy (Table 6, Strategy 2). That is, TA_{TN} was used as the initial image, followed by the TA_{AN} value if the TA_{TN} value was missing; if both were missing, the TA_{TD} value was used, and if all three were missing, the TA_{AD} value was used. This strategy greatly increased the spatial coverage, and achieved the highest R^2 and lowest RMSE and MAE among the 24 possible merging strategies. Since this method depends on the availability of TERRA and AQUA daytime and nighttime data, it is of the greatest value under conditions of partial or short-lived cloud cover.

The validation results demonstrate that the data availability was only 55.46% in 2012 and 44.92% in 2013 after the first processing step. Therefore, more effort should be made to increase the spatial coverage of the available data. The relative proximity of air temperature difference between the estimated date and the previous days near the estimated pixels provides an opportunity to predict the missing data, most of which resulted from cloud contamination. The second step was to determine the difference in TA between the estimation date and previous days for every pixel within a selected window size. Adding the mean differences for these pixels to the value for the center pixel from the previous day replaces the missing data. Our analysis found an optimal window size of 33 × 33 pixels. The spatial coverage increased to 76.99% in 2012 and 89.67% in 2013.

The TA values obtained using this method can be employed as input data in crop growth simulation models to monitor the crop growth, predict the timing of crop development stages and forecast the crop yield at the regional scale [45]. Combined with indicators of a potential agricultural disaster such as extreme temperature, these data can improve the ability to predict the development and spatial distribution of damage caused by cold [5,46], freezing [7] or high temperatures [6]. In addition, the growth degree days (GDD), an important indicator for the cropping system in a region, can be calculated with high spatial-temporal daily mean air temperature data.

Additional research should be carried out to test other strategies for filling in data gaps, such as accounting for the effects of solar declination and vegetation indices on *TA*, and to validate the mapping method. In our future research, we hope to improve *TA* estimates based on LSTs by considering the influence of local variables such as land use or cover types, soil moisture, snow cover, frozen ground, regional microclimatic conditions, terrain characteristics, and local landscape features on the relationship between *TA* and LST [8,16,47,48]. We believe that the estimation accuracy of *TA* will be improved by improving both the mapping strategy and the estimation model.

Acknowledgments: We thank the China Meteorological Data Sharing Service System of the China Meteorological Administration for providing the *TA* data and the Land Processes Distributed Active Archive Center for providing the MODIS LST data. We also thank the journal's anonymous reviewers for their valuable comments, which greatly improved our paper. This work was financially supported by the National Key Technology R&D Program of China (Grant No. 2012BAHB04/2012BAHB02), the National Natural Science Foundation of China (41171276, 41371326), the China Special Fund for Meteorological Research in the Public Interest (GYHY201306036), and the Agricultural Project of Scientific and Technological Research of Shanghai (2011-2-11), the National high technology research and development 863 program in China (2013AA10230103), and the Fundamental research funds for the Chinese central universities (No. 2015XD004). We also thank Huang Bo, and Zhang Hankui, Zhang Liwen, and Huang Weijiao for their help with the data analysis and experimental design.

Author Contributions: Ran Huang proposed the original idea for the study and wrote the original manuscript. Chao Zhang supervised the research and manuscript revision processes. Jianxi Huang offered valuable advice on an early draft of the manuscript and was responsible for responding to the reviewers and final revision of the manuscript. Limin Wang and Liu Jia provided some of the source code used for the image analysis. Dehai Zhu contributed to data analysis and manuscript revision. All authors read and approved the final manuscript.

Conflicts of Interest: The authors declare no conflict of interest.

References

1. De Wit, A.J.W.; van Diepen, C.A. Crop growth modelling and crop yield forecasting using satellite-derived meteorological inputs. *Int. J. Appl. Earth Obs.* **2008**, *10*, 441–425.
2. Ma, H.; Huang, J.; Zhu, D.; Liu, J.; Su, W.; Zhang, C.; Fan, J. Estimating regional winter wheat yield by assimilation of time series of HJ-1 CCD NDVI into WOFOST-ACRM model with Ensemble Kalman Filter. *Math. Comput. Model.* **2013**, *58*, 759–770.
3. Huang, J.; Wang, X.; Li, X.; Tian, H.; Pan, Z. Remotely sensed rice yield prediction using multi-temporal NDVI data derived from NOAA's-AVHRR. *PLoS ONE* **2013**, *8*, e70816.

4.   Huang, J.; Tian, L.; Liang, S.; Ma, H.; Becker-Reshef, I.; Huang, Y.; Su, W.; Zhang, X.; Zhu, D.; Wu, W. Improving winter wheat yield estimation by assimilation of the leaf area index from Landsat TM and MODIS data into the WOFOST model. *Agric. Forest. Meteorol.* **2015**, *204*, 106–121.

5.   Zhang, L.W.; Huang, J.F.; Guo, R.F.; Li, X.X.; Sun, W.B.; Wang, X.Z. Spatio-temporal reconstruction of air temperature maps and their application to estimate rice growing season heat accumulation using multi-temporal MODIS data. *J. Zhejiang Univ. Sci. B* **2013**, *14*, 144–161.

6.   Zhang, J.; Yao, F.; Li, B.; Yan, H.; Hou, Y.; Cheng, G.; Boken, V. Progress in monitoring high-temperature damage to rice through satellite and ground-based optical remote sensing. *Sci. China Earth Sci.* **2011**, *54*, 1801–1811.

7.   She, B.; Huang, J.; Guo, R.; Wang, H.; Wang, J. Assessing winter oilseed rape freeze injury based on Chinese HJ remote sensing data. *J. Zhejiang Univ. Sci. B* **2015**, *16*, 131–144.

8.   Shamir, E.; Georgakakos, K.P. MODIS land surface temperature as an index of surface air temperature for operational snowpack estimation. *Remote Sens. Environ.* **2014**, *152*, 83–98.

9.   Pasotti, L.; Maroli, M.; Giannetto, S.; Brianti, E. Agrometeorology and models for the parasite cycle forecast. *Parassitologia* **2006**, *48*, 81–83.

10. Vancutsem, C.; Ceccato, P.; Dinku, T.; Connor, S.J. Evaluation of MODIS land surface temperature data to estimate air temperature in different ecosystems over Africa. *Remote Sens. Environ.* **2010**, *114*, 449–465.

11. Cresswell, M.P.; Morse, A.P.; Thomson, M.C. Estimating surface air temperature from Meteosat land surface temperature using an empirical solar zenith angle model. *Int. J. Remote Sens.* **1999**, *20*, 125–132.

12. Tomlinson, C.J.; Chapman, L.; Thornes, J.E.; Baker, C.J.; Prieto-Lopez, T. Comparing night-time satellite land surface temperature from MODIS and ground measured air temperature across a conurbation. *Remote Sens. Lett.* **2012**, *3*, 657–666.

13. Zhu, W.B.; Lu, A.F.; Jia, S.F. Estimation of daily maximum and minimum air temperature using MODIS land surface temperature products. *Remote Sens. Environ.* **2013**, *130*, 62–73.

14. Wan, Z.M.; Dozier, J. A generalized split-window algorithm for retrieving land-surface temperature from space. *IEEE Trans. Geosci. Remote* **1996**, *34*, 892–905.

15. Wan, Z.M.; Li, Z.L. A physics-based algorithm for retrieving land-surface emissivity and temperature from EOS/MODIS data. *IEEE Trans. Geosci. Remote* **1997**, *35*, 980–996.

16. Prihodko, L.; Goward, S.N. Estimation of air temperature from remotely sensed surface observations. *Remote Sens. Environ.* **1997**, *60*, 335–346.

17. Lakshmi, V.; Susskind, J.; Choudhury, B.J. Determination of land surface skin temperatures and surface air temperature and humidity from TOVS HIRS2/MSU data. *Adv. Space Res.* **1998**, *22*, 629–636.

18. Jang, J.D.; Viau, A.A.; Anctil, F. Neural network estimation of air temperatures from AVHRR data. *Int. J. Remote Sens.* **2004**, *25*, 4541–4554.

19. Florio, E.N.; Lele, S.R.; Chang, Y.C.; Sterner, R.; Glass, G.E. Integrating AVHRR satellite data and NOAA ground observations to predict surface air temperature: A statistical approach. *Int. J. Remote Sens.* **2004**, *25*, 2979–2994.

20. Chokmani, K.; Viau, A.A. Estimation of the air temperature and the vapour quantity in atmospheric water with the help of the AVHRR data of the NOAA. *Can. J. Remote Sens.* **2006**, *32*, 1–14.

21. Riddering, J.P.; Queen, L.P. Estimating near-surface air temperature with NOAA AVHRR. *Can. J. Remote Sens.* **2006**, *32*, 33–43.

22. Justice, C.O.; Townshend, J.R.G.; Vermote, E.F.; Masuoka, E.; Wolfe, R.E.; Saleous, N.; Roy, D.P.; Morisette, J.T. An overview of MODIS land data processing and product status. *Remote Sens. Environ.* **2002**, *83*, 3–15.

23. Tatem, A.J.; Goetz, S.J.; Hay, S.I. Terra and Aqua: New data for epidemiology and public health. *Int. J. Appl. Earth Obs. Geoinf.* **2004**, *6*, 33–46.

24. Colombi, A.; de Michele, C.; Pepe, M.; Rampini, A. Estimation of daily mean air temperature from MODIS LST in Alpine areas. *EARSeL eProc.* **2007**, *6*, 38–46.

25. Benali, A.; Carvalho, A.C.; Nunes, J.P.; Carvalhais, N.; Santos, A. Estimating air surface temperature in Portugal using MODIS LST data. *Remote Sens. Environ.* **2012**, *124*, 108–121.

26. Hachem, S.; Duguay, C.R.; Allard, M. Comparison of MODIS-derived land surface temperatures with ground surface and air temperature measurements in continuous permafrost terrain. *Cryosphere* **2012**, *6*, 51–69.

27. China Meteorological Administration. Available online: http://cdc.nmic.cn/ (accessed on 24 January 2015).

28. MODIS Web. Available online: http://modis.gsfc.nasa.gov/ (accessed on 24 January 2015).

29. Wan, Z.M.; Zhang, Y.L.; Zhang, Q.C.; Li, Z.L. Validation of the land-surface temperature products retrieved from Terra Moderate Resolution Imaging Spectroradiometer data. *Remote Sens. Environ.* **2002**, *83*, 163–180.

30. Wan, Z. Collection-5 MODIS Land-Surface Temperature Products Users' Guide. Available online: http://www.icess.ucsb.edu/modis/LstUsrGuide/MODIS_LST_products_Users_guide_C5.pdf (accessed on 24 January 2015).

31. MODIS Atmosphere. Available online: http://modis-atmos.gsfc.nasa.gov/ (accessed on 24 January 2015).

32. Land Processes Distributed Active Archive Center. Available online: https://lpdaac.usgs.gov/ (accessed on 24 January 2015).

33. Jin, M.; Dickinson, R.E. A generalized algorithm for retrieving cloud sky skin temperature from satellite thermal infrared radiances. *J. Geophys. Res.* **2000**, *105*, 27037–27047.

34. Mostovoy, G.V.; King, R.L.; Reddy, K.R.; Kakani, V.G.; Filippova, M.G. Statistical estimation of daily maximum and minimum air temperatures from MODIS LST data over the state of Mississippi. *Gisci. Remote Sens.* **2006**, *43*, 78–110.

35. Qi, S.; Luo, C.; Wang, C.; Niu, Z. Pre-study on reverse air temperature from remote sensing relationship between vegetation index, land surface temperature and air temperature. *Remote Sens. Technol. Appl.* **2006**, *21*, 130–136.

36. Cristóbal, J.; Ninyerola, M.; Pons, X. Modeling air temperature through a combination of remote sensing and GIS data. *J. Geophys. Res.* **2008**, *113*.

37. Kloog, I.; Nordio, F.; Coull, B.A.; Schwartz, J. Predicting spatiotemporal mean air temperature using MODIS satellite surface temperature measurements across the Northeastern USA. *Remote Sens. Environ.* **2014**, *150*, 132–139.

38. Sun, D.; Kafatos, M. Note on the NDVI-LST relationship and the use of temperature-related drought indices over North America. *Geophys. Res. Lett.* **2007**, *34*.

39. Crosson, W.L.; al-Hamdan, M.Z.; Hemmings, S.N.J.; Wade, G.M. A daily merged MODIS Aqua-Terra land surface temperature data set for the conterminous United States. *Remote Sens. Environ.* **2012**, *119*, 315–324.

40. Adams, M.P.; Smith, P.L. A systematic approach to model the influence of the type and density of vegetation cover on urban heat using remote sensing. *Landsc. Urban Plan* **2014**, *132*, 47–54.

41. Zhang, W.; Huang, Y.; Yu, Y.Q.; Sun, W.J. Empirical models for estimating daily maximum, minimum and mean air temperatures with MODIS land surface temperatures. *Int. J. Remote Sens.* **2011**, *32*, 9415–9440.

42. Yang, W.J.; Gu, H.R.; Shan, Y.T. Influence of pavement in urban heat island. *J. Highw. Transp. Res. Dev.* **2008**, *25*, 147–152.

43. Bowler, D.; Buyung-Ali, L.; Knight, T.M.; Pullin, A.S. Urban greening to cool towns and cities: A systematic review of the empiri-cal evidence. *Landsc. Urban Plan* **2010**, *97*, 147–155.

44. Kawashima, S.; Ishida, T.; Minomura, M.; Miwa, T. Relations between surface temperature and air temperature on a local scale during winter nights. *J. Appl. Meteorol.* **2000**, *39*, 1570–1579.

45. Franch, B.; Vermote, E.F.; Becker-Reshef, I.; Claverie, M.; Huang, J.; Zhang, J.; Justice, C.; Sobrino, J.A. Improving the timeliness of winter wheat production forecast in the United States of America, Ukraine and China using MODIS data and NCAR growing degree day information. *Remote Sens. Environ.* **2015**, *161*, 131–148.

46. Cheng, Y.; Huang, J.; Han, Z.; Guo, J.; Zhao, Y.; Wang, X.; Guo, R. Cold damage risk assessment of double cropping rice in Hunan, China. *J. Integr. Agric.* **2013**, *12*, 352–363.

47. Mildrexler, D.J.; Zhao, M.S.; Running, S.W. A global comparison between station air temperatures and MODIS land surface temperatures reveals the cooling role of forests. *J. Geophys. Res. Biogeo* **2011**, *116*.

48. Jin, M.L.; Dickinson, R.E. Land surface skin temperature climatology: Benefitting from the strengths of satellite observations. *Environ. Res. Lett.* **2010**, *5*.